Methods in Biotechnology

Methods in Biotechnology

Seung-Beom Hong
M. Bazlur Rashid
Lory Z. Santiago-Vázquez

WILEY Blackwell

Published by John Wiley & Sons, Inc., Hoboken, New Jersey
Published simultaneously in Canada

Library of Congress Cataloging-in-Publication Data applied for.

Cover image: Courtesy of freedigitalphotos.net

10 9 8 7 6 5 4 3 2 1

Contents

Preface

Biotechnology is one of the fastest growing fields of science and industry and is proving to be very promising in the twenty-first century, going along with new advances in automated instrumentation and the emerging synergy among genomics, transcriptomics, proteomics, metabolomics, and bioinformatics. New products based on discoveries through research and development continue to transform the way we live. Because biotechnology makes use of living organisms or their products for the benefit of humans and the environment, it has several subdisciplines depending on the type of research organism or system being studied. Whatever the source organism or system, it remains an experimental science where most basic and applied researches are performed in the laboratory until they are translated into commercial or publishable products approved by regulatory agencies. It is therefore vital that students of future biotech scientists become familiar with the up-to-date experimental methods, experimental designs, and data analyses. Employers, both in industry and academia, have a preference for a qualified workforce with good hands-on laboratory experience and who can work together as part of a team with good critical thinking complemented by good communication skills. This course is designed to address such transferable knowledge, skills, and ability by adopting an inquiry-based self-directed active learning approach to the laboratory-intensive course.

This manual is based on a 15-week semester schedule with a four-hour laboratory session per week. However, it also can be used for a 10-week quarter schedule by selecting appropriate lab exercises at the discretion of instructors. The lab exercises were developed for the core graduate curriculum of the Biotechnology Program at the University of Houston Clear Lake. Students take three biotechnology core laboratory courses, named Methods in Biotechnology (MB) and Advanced Methods in Biotechnology (AMB) 1 and 2. The MB course is a prerequisite to two other AMB 1 and 2 courses. Each lab or more than one lab exercise can be finished within four hours. We offered this laboratory course in the afternoon or evening so that the instructor can begin incubation in the morning to prepare cultures and other laboratory materials. The course content can be tailored to suit the need of the individual instructor. Throughout the experiments, time-saving procedures were implemented in order to complete them within the time period, though students often need to observe results and collect data after overnight or several days of incubation to check their plates. Although we have successfully used protocols in laboratory exercises, we do not claim that they are flawless. They may need modifications in response to new technological innovations and some other changing needs.

Study assignments (pre-lab and post-lab questions) are listed in each lab protocol to help students prepare for each laboratory session and apply knowledge to solve its related problems. This self-directed study will encourage students to extend their knowledge from simply performing experiments to a higher level of critical thinking and troubleshooting. The purpose of each experiment is also intentionally not given to promote the student thinking. It is important that students prepare in advance so that they can clearly understand the protocols and concepts while they are doing the experiments. Lack of preparation will increase the time spent performing experiments as well as the chances of making careless mistakes. We have experienced the difference between prepared and unprepared student groups in terms of the time they spend completing the exercises successfully. Data tables are provided to aid in data collection and step-by-step procedures for data analysis are described in detail.

Throughout the lab exercises we have used the two most widely used model organisms, *Escherichia coli* and *Saccharomyces cerevisiae*, because not only are they much easier and safer to handle compared to other organisms but they are also applicable to diverse cellular and molecular biology experiments. The concept and principle underlying each experiment are explained in the introduction part, and sometimes additional background information on specific details is provided at the end of a protocol. In order to understand how an experiment works, it is important to know what materials and equipment are used. The reagents and microbiological strains required for each experiment are listed at the beginning of each experiment. We often mentioned specific brand names of reagents because we have had satisfactory experiences with those products; however, comparable products from competitor companies can also be used, allowing the users to work within their budget or personal preferences.

To facilitate preparedness and instruction of this course, we have provided detailed notes for instructors in Appendix 1 sections. We also have included laboratory schedules outlining the individual experiments we have performed in each lab session. This requires very labor-intensive teaching and supervision. It is a great advantage to both the prospective employers and students if the students learn as much as possible before entering the workforce. The schedules will be helpful, especially when instructors want to choose certain experiments tailored to their own needs at their discretion. In Appendix 2 sections, we have included a lab math practice problem set for the students to master laboratory mathematics skills because these skills are crucial for success in experiments conducted by technicians and bench scientists.

Acknowledgements

The authors express their gratitude to the reviewers: Dr. Om V. Singh at the University of Pittsburgh, Dr. Stephen C. Kempf at Auburn University, and Dr. Kathleen M. Susman at Vassar College. We also express our gratitude to the editors of John Wiley & Sons, Dr. Gregor Chicchetti, Senior Commissioning Editor; Mindy Okura-Marszycki, Senior Acquisitions Editor; and Stephanie Dollan, Senior Editorial Assistant for reviewing process. We greatly appreciate Dorathy Steve who managed to coordinate all the schedules in time throughout the publishing process and Patricia Bateson, who read every word of the text, for editing process.

Although we do not endorse the particular products of any company over others, we are grateful to the Agilent Technologies, Thermo Fisher Scientific, GE Healthcare Bio-Sciences, Takara-Clontech Laboratories, and New England BioLabs companies that allowed us to use their photo images and illustrations related to the materials used for laboratory exercises.

The development of this course would not have been possible without the undivided support from Dr. Larry H. Rode, Division Chair for Natural Sciences at the University of Houston Clear Lake. We especially thank Dr. Stephens Brian for editing equipment operation protocols and teaching assistants and staffs at UHCL Department of Biotechnology who helped make this course work run smoothly in a timely manner.

Introduction to biotechnology lab

To work in biotechnology requires you to perform experiments using a variety of biochemicals, tools, and instruments and record your result data. The quality of work is more important than the quantity of work to generate a reliable and acceptable product of new findings as a result of scientific research experiments. Regardless of the work environment, attention to detail is particularly important to work in a professional manner for carrying out procedures using necessary equipment and instruments correctly. During this process, you must also keep to safety rules that deal with potential risks and hazards associated with output and input materials of each experiment. Lab safety influences lab productivity because quality work is highly dependent on both safe performance and health status of workers. Unsafe performance or operational activity is a leading cause of the loss of or damage to laboratory resources. Perhaps one of the most important elements of Good Laboratory Practice (GLP) is the accurate documentation of written records. However, injuries or illness caused by neglecting precautionary safety measures will in turn have a negative impact on the investigator's mental and physical activities for written records. Safety should be a priority always regardless of the experiment. Because science is a very human activity, workers should maintain the integrity in conducting experiments and reporting scientific findings. Accordingly, you should familiarize yourself with safety issues and take necessary measures to prevent work-related injuries and incidents. In addition, you should also practice good work habits along with lab etiquette to promote a cheerful and productive working atmosphere. The compiled lists of lab safety, good work habits, lab etiquette, and record keeping guidelines, which are routinely practiced by skilled workers, are described in the following pages.

Lab safety

The instructor should distribute a copy of the safety issues and have students read and sign up to them so that they can understand the risk of accidents associated with lab exercises and accept personal responsibility for risk prevention. The instructor should introduce a biotechnology laboratory facility showing safety-related materials and equipment and explaining how to implement safety measures during the first class meeting.

1. Place all clothing, unnecessary books, handbags, and backpacks away from your work bench area. Your work space must be kept free of articles not actually in use.

2. Eating, drinking, smoking, chewing gum, and cosmetic application are forbidden in the lab at all times.

3. Wear a laboratory coat and safety goggles to protect clothing and eyes from UV exposure, vaporized mist, and accidental splashing or rubbing of chemical solutions.

4. Wear closed-toed shoes. Sandals are prohibited.

5. Do not pipette by mouth. Keep your hands away from your mouth, eyes, or face.

6. Do not dispose of broken glass or contaminated items in a regular garbage can. Place all broken glass into a labeled glass waste box only, all used plastic pipettes into a labeled cardboard box, needles and razor blades into a sharps container, all contaminated biological items into an autoclave biohazard bag, and all hazardous chemicals in labeled waste containers at a designated area.

7. Wear gloves whenever instructed in the lab procedures.

8. Remove gloves before touching shared surfaces, such as light switches, door knobs and handles, phones, and keyboards.

9. Clean your workbench area with a disinfectant before and after each lab period.

10. Clean up any spilled chemicals, solutions, media, and cultures (balance, pH meter, centrifuge, incubator, microwave oven, etc.) promptly.

11. Tie long hair back to minimize fire hazard and contamination of experiments and cultures.

12. Do not wear loose or flowing clothing or dangling jewelry in the laboratory.

13. Be aware of Bunsen burner flame and turn off immediately when not in use. Never leave the open flame unattended. Shut off the gas outlet valve to turn off the burner.

14. Place a danger sign of a hot surface after you use heat to dissolve in a hot plate magnetic stirrer.

15. Never heat (autoclave, hot plate stir, or microwave) a solution in an airtight container. Pressure buildup may cause the container to explode.

16. Do not tighten the bottle caps soon after autoclaving as this will stress the vessel, resulting in either deforming plastic ware or breaking glassware.

17. Do not put hot glassware on cold surfaces, or cold glassware on hot surfaces. It will crack after a sudden temperature change. Cool or warm all glassware slowly to avoid breakage.

18. Always balance a centrifuge rotor by placing the tubes containing samples of nearly equal weight symmetrically inside the rotor, and cap the tubes tightly before centrifugation.

19. Always open the bottles containing hazardous (flammable, explosive, corrosive, and toxic) materials under the ventilated fume hood, keeping its sash down as far as possible.

20. Locate fire extinguisher, fire blanket, eyewash station, safety shower, fume hood, Material Safety Data Sheet (MSDS) reference folder, hazardous chemical waste containers, chemical spill kit, first aid kit, emergency exit, and emergency telephone.

21. Comply with all listed safety and waste disposal procedures with regards to biological and biochemical hazardous materials imposed by the regulatory agency of your institution. Ask the instructor before you discard if you are not sure.

22. Never uncover the lid of Petri plates contaminated with fungi and molds to prevent the air-borne spread of spores that not only contaminate lab space but also cause negative effects on human health.

23. Wash your hands thoroughly to remove any residual chemicals and biological organisms before leaving the laboratory.

24. Report all accidents immediately to the instructor or teaching assistant for assistance.

I have read and agreed to abide by the above safety precautions agreement.

Student Name and ID:

Sign:

Date:

Good work habits

Students with good work habits tend to be more successful in their duties than poorly organized individuals. Below are recommended examples of good work habits for students to manage time and effort, as well as organize materials and work space for productivity and quality of their lab activity.

1. Keep your assigned work area clear of unnecessary items (backpack, computer, etc.).

2. Keep everything you need within reach, gather all materials, label as needed, and set up disposal areas before you begin. This requires planning ahead with protocols.

3. Sign in a logbook of commonly used equipment before you start working if it is needed.

4. Never do protocols from memory; always read every step every time you follow a procedure and then check it off as it is completed.

5. Always replace the lid immediately after dispensing stock solutions and chemicals. Otherwise evaporation of solutions, hydration of chemicals, and contamination will occur.

6. Keep test tubes and microcentrifuge tubes in racks at all times.

7. Not all compounds are soluble in water; check whether it is the correct solvent before use.

8. Make sure to use clean, dry vessels including bottle, beaker, flask, and graduated cylinder.

9. Rinse with tap water and wash the vessels after use as quickly as possible; invisible chemical residue film or agar solution film dried does not easily come off. The longer it is left unwashed, the harder it will be to clean.

10. Use an appropriate container for storage (e.g., a dark one for light-sensitive compounds).

11. Store stock solutions and media at an appropriate temperature (25 °C room, 4 °C refrigerator, – 20 °C or – 80 °C freezer).

12. Make small aliquots if necessary to avoid contamination and repeated freezing–thawing that can destabilize and cause damage to nearly all biological components.

13. Eliminate or minimize air bubbles that interfere with nearly all biological experiments (enzyme reactions, agar plate, running gel, membrane blotting, chromatography, etc.).

14. Label the reagent name, concentration, pH if it is important, date of preparation, and the name of the person who prepared it.

15. Always label the bottom edge of the Petri plate since the lid may become separated from the actual plate containing the growth medium.

16. Save the lab materials, including disposable items; calculate how many and how much you will need economically and take only necessary numbers and amounts of supplies.

17. Do not return the leftover reagents to the original stock bottles after use. This is not the way to save. Keep them separately if you want to use them again.

18. Return all reagents, cultures, vessels, and pipettes to their proper storage places before leaving the lab.

Lab etiquette

The guidelines below are intended to foster an amicable and trustful environment in the laboratory where many lab colleagues share common facilities and thus conflict of interest may often arise. Below is the list of ethical issues made based on the experiences often encountered in the laboratory. Professional courtesy includes but is not limited to the following:

1. Sign the logbook (user name, date, start and end time) before you plan to use the lab for an extended time with changes in operating conditions of commonly used instruments such as autoclave, centrifuge, shake incubator, and thermal cycler.

2. Check whether any culture or reaction tubes are being used inside a shaking or water bath incubator before you turn it off.

3. Do not take culture or reaction tubes other than yours away from the instruments being used. If you do so accidentally, you must honestly tell your fellow lab mates.

4. Do not leave your culture tubes, flasks, or agar plates in the incubator and sample tubes in the centrifuge for an unnecessary prolonged time; take them out in a timely manner.

5. Clean up any spills or your mess in common areas after use (balance, pH meter, laminar flow hood, centrifuge, incubator, refrigerator, microscopy, gel photography, autoclave, etc.).

6. Prepare stock solutions (e.g., electrophoresis buffer) or culture media (e.g., LB, YPD) that are commonly used by all lab colleagues when you have used up these solutions.

7. Make sure to securely close doors after you take out samples from a refrigerator (4 °C) or freezer (– 20 °C, – 80 °C).

8. Do not put near-empty reagent bottles on a storage shelf if the content runs out after use but put a note and notify the lab manager.

9. If a piece of equipment breaks or needs more consumables while you are using them, notify the lab manager and leave a note or e-mail so that it can be repaired or replenished promptly.

10. Do not take private stocks of reagents, media, or plates from your colleagues or anything out of their freezers or benchtops without asking in person.

11. Turn off your smartphone and refrain from texting while working in the lab. However, the smartphone may be left on vibration mode for emergency notification purposes. You may use a smartphone or other cameras to photograph projects in progress.

12. You are expected to arrive on time for the laboratory class and stay for your scheduled work time unless you are under certain extenuating circumstances.

Lab record

Each student should keep a separate notebook. Good recordkeeping is an essential part of Good Laboratory Practice (GLP) and is imperative for reproducibility of experiments when they are conducted by others. Honesty and accuracy are critical for documentation. You should record failures as well as successes. Your notebook also should be legible and complete so that anyone can easily tell why, what, and when you did something and follow in your footsteps without your help. You may also need to refer to the notebook for small details that may be critically important at a later date. According to the old Chinese proverb, "The faintest ink is more powerful than the strongest memory."

Advanced information technology has allowed us to easily capture images of drawings, barcodes, and documents or photographs or digitize notes written with digital pen and upload electronic data files. Nowadays laboratories become more computerized. Clarity and organization of data recording and inventory management can be facilitated by electronic lab notebook (ELN) software with traceability of the editing of data files. ELN comes in many forms suited for specific science disciplines, with cost ranging from free to very expensive. Some specific web-based ELNs are able to collect data directly from instruments such as the spectrophotometer, pH meter, and microscope, to generate graphs and tables, as well as to execute a query search for all study data. Other commercial biology ELNs have integrated DNA and protein analysis informatics. The direct collection and processing of data through the connected devices could be a standard feature of ELN in the near future as the technology of Internet of Things (IoT) advances. A free cloud version of Microsoft OneNote or Evernote can be used as a general ELN. The ELN system is currently becoming more and more adopted by research laboratories. Although storing data in the cloud offers many benefits with respect to sharing, collaborating in real time, accessing data, and being editable from anywhere on any mobile electronic device, there may be security and privacy concerns of data storage by cyber hacking. If you choose to use ELN, print out a hard copy once in a while, compile in chronological order, and store in a secure place.

The following key elements are listed for a paper notebook from the perspective of an academic laboratory setting.

1. Use a bound notebook with pre-numbered pages.

2. Write using a black ink ball-point only (no felt-tip pens or pencils).

3. Do not rip the page out or white-out your mistakes. Instead, cross out errors with your initials and date, and write legible corrections nearby. For mistakes found at a later date with not enough space for writing, you may mark a cross-line (×) through the error and write "See page xx for explanation."

4. Any large blank areas or unused pages between the entries must have a cross-line marking.

5. Record any deviations from the protocol given by the instructor or taken by you.

6. Do not staple, but glue or tape photographs, graphs, and other printouts to the notebook.

7. Do not record your data on a separate sheet of paper; record all data and observations directly into your notebook as soon as they become available.

8. If notebook page space is limited for the data that are obtained later, attach to the unused separate page and indicate the page number in the Results section. If you have a large electronic data file that cannot be accommodated in the notebook, record the name and the production date of an electronic file in the notebook.

9. Make a table of contents with date and title of experiment and page numbers on the first page so that it can be periodically updated with each experiment.

10. The write-up format of each experiment should include:

A. Date

B. Title

C. Purpose. Be specific and concise, focusing on why you did something. Do not repeat the title.

D. Materials and methods. Write in a narrative form what you will be doing. Do not make a copy of all the procedures but write in your own words what and how you plan to accomplish something without going into too much detail. Show your calculations if required. If you follow procedures as essentially described in the lab protocol, you may simply write a summary of steps. However, you have to write down any minor changes that differed from the lab protocol.

E. Results. All data including pictures, tables, and graphs must be clearly labeled. Make sure to specify units on the abscissa and the ordinate of graphs. All graphs and tables should have their own titles. A gel photograph should include the following information:

(a) Gel composition, voltage, running time, buffer used.

(b) Table displaying contents of each lane with volume and mass loaded.

(c) Label each sample lane and bands of size marker.

(d) Briefly describe the collected data and explain observations made in the experiment instead of showing just the result data.

Before you come to the lab, you should prepare a handwritten pre-lab writing of A, B, C, and D sections. You should also prepare blank data tables of the E section in your pre-lab writing when they appear in the lab protocol. At the beginning of each lab session, the instructor will briefly explain the objective, concepts, and principles behind each experiment, and an overview of a step-by-step procedure with a flowchart outlining the lab activity, while the teaching assistant will check each student's pre-lab writing notebook. If necessary, the instructor will go over calculations and data analysis. You will then conduct experiments as instructed and record result data (E) or observations directly in your notebook during the lab work. For multitasking procedures, members of a group can do their part of each task to get things done. This collaborative work approach will increase the efficiency of the experiment to finish within the allocated time. In this case, you should record the names and roles of individuals who conducted a part of the experiment. After you finish the experiment, you must have a witness signature in your notebook from the teaching assistant.

F. Discussion and conclusion. Interpret your result and draw a conclusion.

(a) Write in a paragraph form.

(b) Evaluate results in comparison to controls. If there is any incorrect or negative result, you should first check with other groups' results and then discuss it. Because several different groups use identical protocols with the same supplies in the classroom, you do not need to repeat experiments confirming your results but instead you may borrow their results. Never simply write "Our experiment failed and thus needs to be repeated." Even if all other groups failed, explore the most likely reasons. Do not be afraid of incorrect reasoning. To find possible reasons for negative or unexpected results is a key part of the critical thinking process, as well as becoming proficient in a technique.

(c) Compare and contrast your results with the previous ones if they are available from relevant references.

(d) Describe future directions for experiments and how an experiment can be improved.

G. References. Cite the relevant literature or website you consulted for the source of information as appropriate. Use quotation marks to reproduce words, phrases or paragraphs.

Lab report and assignment

During this course you have to hand in three typewritten materials as well as a pre-lab notebook, pre-lab and post-lab assignments, and post-lab report. Assignments are the homework questions given in each lab protocol that encourage students to think and study so that they can better understand the process and principle of experiment as well as to apply knowledge to solve the related problems as an independent investigator. For the lab report, you present your work in a general scientific writing format on the basis of your lab notebook record. All written materials are submitted in both a printout and an electronic form on the designated due date. A pre-lab assignment must be turned in before the lab begins. A post-lab assignment is submitted along with a post-lab report, but both should be filed separately with a cover page indicating your name, student ID, and group number. The format of the post-lab report is the same as that of the lab notebook record, but do not include photographs when submitting an electronic post-lab report. It is strongly recommended that you proofread your assignments and report using plagiarism and grammar/spell checkers before submission. Plagiarism detection programs such as Turnitin and SafeAssign are made available to students in some academic schools. If they are not available, you may use the following free online checkers:

https://academichelp.net/check-paper-for-plagiarism/
(Plagiarism check)
https://www.languagetool.org/
(Language Tool Style and Grammar Check)
http://www.grammarcheck.net/editor/
(Grammar Check)

Further reading

Marcrina, F.L. (1995). Scientific integrity: an introductory text with cases, Chapter 3. In *Scientific Record Keeping*, pp. 41–63. American Society for Microbiology (ASM) Press. ISBN 1-55581-069-1.

Rubacha, M., Rattan, A.K., and Hosselet, S.C. (2011). A review of electronic laboratory notebooks available in the market today. *Journal of Laboratory Automation*, **16** (1): 90–98.

Seidman, L.A. and Moore, C.J. (2000). Unit I: Introduction to the Biotechnology Workplace, pp. 1–33; Unit II: Product Quality and Biotechnology, pp 45–73. Unit VII: Safety in the Laboratory, pp. 594-662. In *Basic Laboratory Methods for Biotechnology*. Prentice Hall, Inc. ISBN 0-13-795535-9.

Internet Resources for MSDS: http://www.ilpi.com/msds/index.html

Care and Safe Handling of Laboratory Glassware: https://www.sigmaaldrich.com/content/dam/sigma-aldrich/docs/Sigma-Aldrich/Usage/1/glass_care_safe_handling.pdf

Safe Handling of Equipment: http://mdk12.org/instruction/curriculum/science/safety/handling.html

Safety and Health Topics: https://www.osha.gov/SLTC/molds/

Notebook Software Sources: OneNote ELN: http://www.onenote.com/

Evernote ELN: https://evernote.com/

Laboratory Schedule (Methods in Biotechnology (MB))

MB Laboratory Schedule (Spring Semester)				
Week	**Date**	**Lab Experiment (4 hours)**	**Lab Report**	**Exam/Quiz**
1	1/13	Course Introduction Lab Notebook Guideline Lab Safety, Lab Math		
2	1/20	**Holiday (Martin Luther King): No Lab**		
3	1/27	Exp. 1: Lab Measurements		
4	2/03	Exp. 2: Use of the Spectrometer and Beer's Law		Quiz #1: Lab Math
5	2/10	Exp. 3: Making Solutions and Buffer Efficacy		
6	2/17	Exp. 4: Acid–Base Titration	Exp. 1–4	Quiz #2: Exp. 1–4
7	2/24	Exp. 5: Protein Denaturation and Precipitation		
8	3/03	Exp. 6: Bacterial Transformation		
9	3/10	**Spring Break: No Lab**		
10	3/17	Exp. 7: GFP Purification	Exp. 5 and 6	Quiz #3: Exp. 5–7
10	3/21	**Midterm Exam Friday**		**Midterm (Exp. 1–7)**
11	3/24	Exp. 8: SDS-PAGE Analysis		
13	3/31	Exp. 9: DNA Isolation		
13	4/07	Exp. 10: PCR-Based *Alu*-Human DNA Typing Exp. 11: Restriction Enzyme Digestion Exp. 12: Agarose Gel Electrophoresis		Quiz #4: Exp. 8 and 9
14	4/14	Exp. 13: Ouchterlony and ELISA Immunoassays	Exp. 8–12	
15	4/21	Exp. 14: Testing Plant Substances for Antimicrobial Activity Exp. 15: Peroxidase Enzyme Activity Assay		
16	4/28	Review Session	Exp. 13–15	Quiz #5: Exp. 10–15
17	5/02	**Final Exam: Friday**	Lab Notebook	**Final (Exp. 8–15)**

MB Laboratory Schedule (Fall Semester)				
Week	**Date**	**Lab Experiment (4 hours)**	**Lab Report**	**Exam/Quiz**
1	8/26	Course Introduction Lab Notebook Guideline Lab Safety, Lab Math		
2	9/02	**Labor Day: No Lab**		
3	9/09	Exp. 1: Lab Measurements		
4	9/16	Exp. 2: Use of the Spectrometer and Beer's Law		Quiz #1 (Lab Math)
5	9/23	Exp. 3: Making Solutions andBuffer Efficacy		
6	9/30	Exp. 4: Acid–Base Titration	Exp. 1–4	Quiz #2 (Exp. 1–4)
7	10/07	Exp. 5: Protein Denaturation and Precipitation		
8	10/14	Exp. 6: Bacterial Transformation		
9	10/21	Exp. 7: GFP Purification	Exp. 5 and 6	Quiz #3 (Exp. 5 and 6)
9	10/25	**Midterm Exam Friday**		**Midterm (Exp. 1–7)**
10	10/28	Exp. 8: SDS-PAGE Analysis		
11	11/04	Exp. 9: Plasmid and Genomic DNA Isolation		
12	11/11	Exp. 10: PCR-Based *Alu*-Human DNA Typing Exp. 11: Restriction Enzyme Digestion Exp. 12: Agarose Gel Electrophoresis		Quiz #4 (Exp. 8 and 9)
13	11/18	Exp. 13: Ouchterlony and ELISA Immunoassays	Exp. 8–12	
14	11/25	Exp. 14: Testing Plant Substances for Antimicrobial Activity Exp. 15: Peroxidase Enzyme ActivityAssay		
15	12/02	Review Session	Exp. 13–15	Quiz #5: (Exp. 10–15)
16	12/09	**Final Exam: Monday**	Lab Notebook	**Final (Exp. 8–15)**

Laboratory Schedule (Advanced Methods in Biotechnology (AMB) 1)

AMB 1 Laboratory Schedule (Spring Semester)				
Week	**Date**	**Lab Experiment (4 hours)**	**Lab Report**	**Exam/Quiz**
1	1/14	Course Introduction Lab Safety, Lab Math		
2	1/21	Exp. 16: Aseptic Technique and Culture Handling Exp. 17: Yeast Culture Media Preparation		Quiz #1: Lab Math
3	1/28	Exp. 18: Growth Curve Exp. 19: Mini Plasmid Prep		
4	2/04	Exp. 20: Restriction Digestion, Purification, Concentration, and Quantitation of DNA Exp. 21: Polymerase Chain Reaction (PCR)		Quiz #2: Exp. 16–19
5	2/11	Exp. 22: TA, Blunt End, SLIC, and CPEC	Exp. 16–20	
6	2/18	Exp. 23: One-Step Multifragment Assembly Cloning		
7	2/25	Exp. 24: Restriction Enzyme Digestion and Fast Agarose Gel Electrophoresis Exp. 25: Southern Blot Transfer Exp. 26: Probe Labeling and Purification		
8	3/04	Exp. 27: Dot Blot Assay Exp. 28: Pre-hybridization, Hybridization, and Detection	Exp. 21–23	Quiz #3: Exp. 20–26
9	3/11	**Spring Break: No Lab**		
10	3/18	Exp. 28: Southern Band Detection Exp. 30: Isolation of Overlapping DNA Fragments for *in vivo* Recombination Cloning (Part I)		
10	3/21	**Midterm Exam Friday**		**Midterm (Exp. 16–27)**
11	3/25	Exp. 29: Total Yeast RNA Isolation and RT-PCR	Exp. 24 and 25	
12	4/01	Exp.30: Yeast Transformation (Part II) Exp. 31: Plasmid Isolation from Yeast Exp. 32: *E. coli* Transformation with Yeast Plasmid DNA		Quiz #4: Exp. 28–30
13	4/08	Exp. 33: X-Gal Filter Lift Assay Exp. 34: Protein Quantitation Assays		
14	4/15	Exp. 35: Quantitative β-Galactosidase Assays in Yeast	Exp. 26–29	
15	4/22	Exp. 36: Gel Filtration Chromatography Exp. 37: Ion Exchange Chromatography		
16	4/29	Review Session		Quiz #5: Exp. 31–37
17	5/02	**Final Exam: Friday**	Exp. 30–37 and Lab Notebook	**Final (Exp. 28–37)**

AMB I Laboratory Schedule (Fall Semester)				
Week	**Date**	**Lab Experiment (4 hours)**	**Lab Report**	**Exam/Quiz**
1	8/26	Course Introduction, Lab Safety, Lab Math		
2	9/02	**Labor Day**		
3	9/09	Exp. 16: Aseptic Technique and Culture Handling Exp. 17: Yeast Culture Media Preparation		Quiz #1 Lab Math
4	9/16	Exp. 18: Growth Curve Exp. 19: Mini Plasmid Prep		
5	9/23	Exp. 20: Restriction Digestion, Purification, Concentration, Quantitation of DNA Exp. 21: PCR		Quiz #2: Exp. 16–19
6	9/30	Exp. 22: TA, Blunt End, SLIC, CPEC	Exp. 16–20	
7	10/07	Exp. 23: One-Step Multifragment Assembly Cloning		
8	10/14	Exp. 24: Restriction Enzyme Digestion and Fast Agarose Gel Electrophoresis Exp. 25: Southern Blot Transfer Exp. 26: Probe Labeling and Purification		
9	10/21	Exp. 27: Dot Blot Assay Exp. 28: Pre-hybridization, Hybridization, and Detection	Exp. 21–23	Quiz #3: Exp. 20–26
9	10/25	**Midterm Exam**		**Midterm (Exp. 16–27)**
10	10/28	Exp. 28: Southern Band Detection Exp. 30: Isolation of Overlapping DNA Fragments for *in vivo* Recombination Cloning (Part I)		
11	11/04	Exp. 29: Total Yeast RNA Isolation and RT-PCR	Exp. 24 and 25	
12	11/11	Exp. 30: Yeast Transformation (Part II) Exp. 31: Plasmid DNA Isolation from Yeast Exp. 32: *E. coli* Transformation with Yeast Plasmid DNA		Quiz #4: Exp. 28–30
13	11/18	Exp. 33: X-Gal Filter Lift Assay Exp. 34: Protein Quantification Assays		
14	11/25	Exp. 35: Quantitative β-Galactosidase Assays in Yeast	Exp. 26–29	
15	12/02	Exp. 36: Gel Filtration Chromatography Exp. 37: Ion Exchange Chromatography		
16	12/09	**Final Exam: Friday**	Exp. 30–37 and Lab Notebook	**Final (Exp. 28–37)**

Laboratory Schedule (AMB 2)

AMB 2 Laboratory Schedule (Spring Semester)				
Week	**Date**	**Lab Experiment (4 hours)**	**Lab Report**	**Exam/Quiz**
1	1/16	Course Introduction Lab Safety, Lab Record, Lab Math		
2	1/23	Exp. 38: *E. coli* Culture Media Preparation		Quiz #1: Lab Math
3	1/30	Exp. 39: Site-Directed Mutagenesis		
4	2/06	Exp. 40: Protein Expression in *E. coli*		
5	2/13	Exp. 41: Protein Purification by Affinity Column Chromatography	Exp. 38–40	Quiz #2: Exp. 38--40
6	2/20	Exp. 42: SDS-PAGE Analysis of Affinity Column Fractions Exp. 43: Western Blot Transfer of Affinity Column Fractions		
7	2/27	Exp. 43: Western Blot Detection Exp. 44: Yeast Culture Media Preparation and Phenotypic Analysis of Yeast Strains		
8	3/06	Exp. 45: Yeast Transformation for Y2H Assay and Genome Editing by CRISPR-Cas System	Exp. 41–43	Quiz #3: Exp. 41–43
9	3/13	**Spring Break: No Lab**		
10	3/20	Exp. 46: Yeast Mating-Mediated Y2H Assay and Genomic PCR		
10	3/21	**Midterm Exam**		**Midterm (Exp. 38–45)**
11	3/28	Exp. 47: Yeast Colony PCR and Cycle DNA Sequencing		
12	4/03	Exp. 48: DNA Sequencing Electrophoresis Exp. 49: RNA Interference	Exp. 44–46	
13	4/10	Exp. 50: Protein Preparation for 2D Gel Electrophoresis		
14	4/17	Exp. 51: Rehydration of IPG Strip and IEF	Exp. 47–48	Quiz #4: Exp. 46--48
15	4/24	Exp. 51: Two-Dimensional Gel Electrophoresis		Quiz #5: Exp. 49–51
16	5/02	**Final Exam**	Exp. 49–51 and Lab Notebook	**Final (Exp. 46–51)**

AMB 2 Laboratory Schedule (Fall Semester)				
Week	**Date**	**Lab Experiment (4 hours)**	**Lab Report**	**Exam/Quiz**
1	8/29	Course Introduction Lab Safety, Lab Record, Lab Math		
2	9/05	Exp. 38: *E. coli* Culture Media Preparation		Quiz #1: Lab Math
3	9/12	Exp. 39: Site-Directed Mutagenesis		
4	9/19	Exp. 40: Protein Expression in *E. coli*		
5	9/26	Exp. 41: Protein Purification by Affinity Column Chromatography	Exp. 38–40	Quiz #2: Exp. 38–40
6	10/03	Exp. 42: SDS-PAGE Analysis of Affinity Column Fractions Exp. 43: Western Blot Transfer of Affinity Column Fractions		
7	10/10	Exp. 43: Western Blot Detection Exp. 44: Yeast Media Preparation and Phenotypic Analysis of Yeast Strains		
8	10/17	Exp. 45: Yeast Transformation for Y2H Assay and Genome Editing by CRISPR-Cas System	Exp. 41–43	Quiz #3: Exp. 41–43
9	10/24	Exp. 46: Yeast Mating-Mediated Y2H Assay and Genomic PCR		
9	10/25	**Midterm Exam**		**Midterm (Exp. 38–45)**
10	10/31	Exp.47: Yeast Colony PCR and Cycle DNA Sequencing		
11	11/07	Exp. 48: DNA Sequencing Electrophoresis Exp. 49: RNA Interference	Exp. 45–47	
12	11/14	Exp. 50: Protein Preparation for 2D Gel Electrophoresis		
13	11/21	Exp. 51: Rehydration of IPG Strip and IEF		Quiz #4: Exp. 46–48
14	11/28	**Thanksgiving Holiday**		
15	12/05	Exp. 51: Two-Dimensional Electrophoresis		Quiz #5: Exp. 49–51
16	12/09	**Final Exam**	Exp. 48–51 and Lab Notebook	**Final (Exp. 46–51)**

About the companion website

This book is accompanied by a companion website:

www.wiley.com\go\hong\Methodsinbiotechnology

The website includes:

- Instructor's solutions manual
- Student's site with figures and tables
- PowerPoint lecture discussion files related to laboratory experiments

Methods in Biotechnology (MB)
Laboratory Exercises

1 MB experiment 1: Lab measurements

Purpose: This is up to you to write down.

Introduction

Experiments frequently require measurement of volume and mass. Accurate and precise measurements are necessary to produce the reliable and reproducible data. In a biotechnology lab, volume is often measured in microliters (μL) and mass is calculated in terms of microgram (μg), nanogram (ng), or picogram (pg) in order to set up reactions in microcentrifuge tubes. You will be using micropipettors throughout the experiments in order to draw up small volumes ranging from 0.1 to 1000 μL. Depending on the volume to be withdrawn, you have to choose the correct model of micropipettor (P2, P10, P20, P200, or P1000). You must familiarize yourself with setting the volume and the effective range of volume that can be taken by each micropipettor. The maximum microliter volume that can be withdrawn is indicated by each pipettor model number, and the minimum volume is typically one-tenth of the model number. Most of the digital volume indicator consists of three numbers reading from top to bottom, as shown below.

In P1000, the top, middle and bottom numbers are for 1000s, 100s, and 10s of μL, respectively. In P200, the top, middle, and bottom numbers denote 100s, 10s, and 1s of μL, respectively. In P20 and P10, the top, middle, and bottom numbers indicate 10s, 1s, and 0.1s of μL, respectively. In P2, the top, middle, and bottom numbers refer to 1s, 0.1s, and 0.01s of μL, respectively.

There are five subdivision line marks between the last digits in all the micropipettors. For example, five subdivisional increments between 125 and 126 or between 124 and 125 exist for P2, P20, and P200. The same subdivisional increments between 055 and 056 or between 054 and 055 exist for P10 and P1000. Accordingly, each subdivision mark equals 2 μL for P1000, 0.2 μL for P200, 0.02 μL for P20 and P10, and 0.002 μL for the P2 micropipettor.

The volume is set by either turning the plunger or thumbwheel to read the numerical settings displayed. To use the plunger button makes it easier and faster to set volumes, especially when you wear gloves. However, you must not rotate the plunger or thumbwheel beyond the upper or lower range limit of the pipettor, though the 0.1 μL setting for P2 and 0.5 μL setting for P10 can be used when a sample is taken carefully according to the manufacturer's instruction. If you want to draw up the volume that goes beyond the upper limit of a pipettor, you may first set the lower volume within the range to draw a sample and then set the additive volume to draw the sample using the same pipettor and tip instead of changing to a larger volume pipettor. Of course, you need to use a lower volume pipettor when the minimum volume of a pipettor is higher than the volume you want to take. Most micropipettors have two stop points on the plunger to draw and expel a liquid, as shown below.

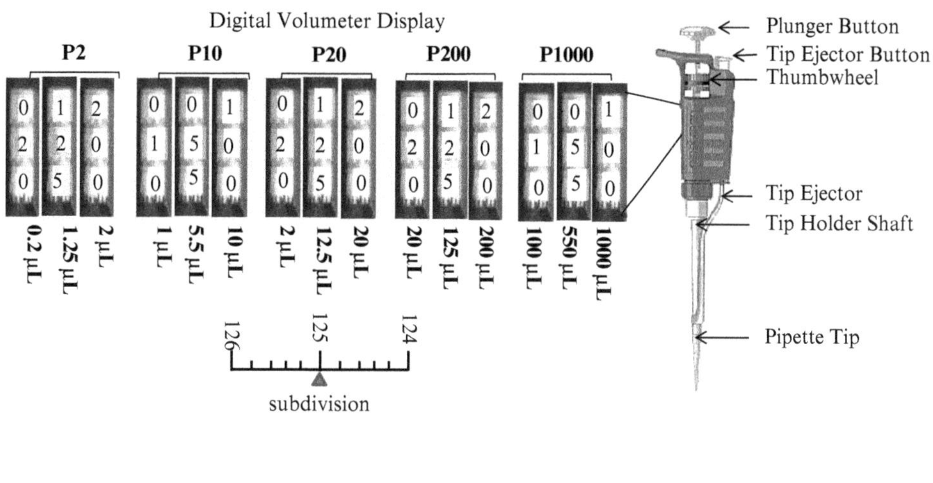

Methods in Biotechnology, First Edition. Seung-Beom Hong, M. Bazlur Rashid and Lory Z Santiago-Vázquez.

Companion Website: www.wiley.com\go\hong\Methodsinbiotechnology

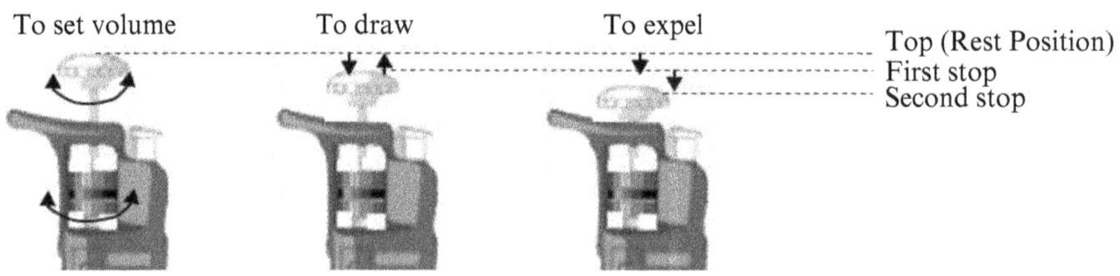

First, fit the tip to the end of the shaft and press down to ensure an airtight seal. Make sure that the pipet is securely placed on the micropipettor. If the tip is loose on the pipettor, the volume will not be correct and the tip may fall down. However, do not impose an excessive pressure pushing down on the pipette tip as this may cause breakage of the shaft or pipette tip box. Second, press the plunger to the first stop point, immerse the pipet tip about 2 mm below the surface of the solution (not too deeply into the solution!) and slowly release the plunger to fill the tip. If you release the plunger too quickly, the liquid may splash up into the micropipettor and contaminate it. If you draw up viscous (thick) liquids, such as dense glycerol, sucrose, and detergent solutions, too quickly, the liquid will not enter the tip fast enough and your measurement will be inaccurate. Sometimes this happens with thin liquids as well, so you should pipette slowly and wait a few seconds to ensure that the tip is filled with the full volume. Lastly, insert and touch the tip end to the bottom of tube, and press the plunger down to the second final stop point to expel the fluid. Neither drop the liquid nor shake the tip to expel the last drop of tiny liquid adhering to the tip point.

If you are working with a power-assisted pipette-aid to deal with larger volumes (>5 mL), the upper button fills the pipet and the lower button expels the fluid. Do not draw up the liquid beyond the upper marking of the pipet because this will wet the cotton in the neck of the pipet and the pipet will no longer work.

When working under sterile condition, the following rules apply:

- Anything that your hand (even glove-worn hand) touches is no longer sterile. If your hand passes over an open sterile plate or bottle, it is likely that you have contaminated these contained items.
- If the pipet tip touches anything other than a sterile surface, it is no longer sterile.
- Keep your workspace clean and in good order. A messy workspace could cause contamination of your experiment.
- When transferring sterile liquids from a stock bottle, wipe out the micropipettor plastic shaft with 70% ethanol paper towel, remove the cap with one hand, and insert the pipet at a slight angle into the bottle without touching the neck of the bottle with the micropipettor shaft. While transferring the fluid to the plate or tube, replace the lid on the bottle. If multiple transfers are needed from one bottle to several plates or tubes, the lid is placed on the bottle loosely and replaced after all transfers. Do not pass your hand over the open bottle.

It is recommended to watch the following YouTube videos as to how to use a micropipettor and Spectronic 20 prior to coming to the lab:

- https://www.youtube.com/watch?v=tL0acTneiNY (Auto pipet technique)
- https://www.youtube.com/watch?v=y-OHnnhWCdo (How to pipette: lab survival skills)
- http://www.youtube.com/watch?v=jmZomizSPxw&NR=1 (Using the Spec-20)

Pre-lab assignment

(Typing and submission must be completed before lab work begins.)

1. You have P2, P10, P20, P200, and P1000 micropipettors, as shown below. Fill out the table to indicate the μL volume for each of the five pipettors according to the volume setting, as shown below. Indicate "NO" if the volume setting cannot be used for the pipettor chosen.

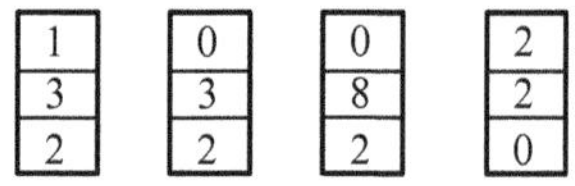

Pipettor / Volume setting	P2	P10	P20	P200	P1000
132 (top to bottom)					
032 (top to bottom)					
082 (top to bottom)					
220 (top to bottom)					

2. How would you draw 205 μL using a P200 pipettor? How would you draw 1010 μL using a P1000 pipettor?
3. What micropipettor would it be best to take all of the individual volumes in each of Part I, C1 and C2 exercises?
4. How would you prepare 1.5 mL of 1 mg/mL blue dextran solution using a 10 mg/mL blue dextran stock solution (at Part I, step D8)? Show your work.
5. How much 3 M NaCl is needed to conduct the Part II experiment? Show your calculations.

Materials and equipment

- Four different dye solutions, dH_2O
- 50% glycerol
- Solutions A, B, C, D (different color solutions)
- Blue dextran (10 mg/mL in dH_2O)
- 3 M NaCl, unknown NaCl solutions I and II
- Parafilm
- P2, P10, P20, P200, P1000 micropipettors, micropipette tips
- 0.5-mL and 1.5-mL microcentrifuge tubes (non-sterile)
- Borosilicate glass test tubes (13 mm × 100 mm)
- Test tube racks and microcentrifuge tube racks
- 70% ethanol wash bottle
- Spectronic 20, cuvette
- Analytical balance

Procedure

Part I. Pipetting skill

A. Volume setting practice

Pipetting error is a major contributor to poor laboratory results. Two important sources of pipetting error are the use of an uncalibrated micropipettor causing systematic error and the practice of inaccurate pipetting introducing random error.

1. Practice setting the volume on the micropipettors; each person in your group should set at least one and have it checked by other group members and/or your instructor. Look at the plunger button top of the micropipettor to identify its measuring range. Remember that the value listed on the top is the largest volume you can measure on that pipettor and that the lowest volume is 1/10 of the largest volume. On a P1000 micropipettor, the largest and lowest measurable volumes are 1000 and 100 µL, respectively; 200 and 20 µL on a P200 micropipettor; 20 and 2 µL on a P20 micropipettor. Set the P1000 pipettor to 0.45 mL (450 µL), the P200 micropipettor to 0.15 mL (150 µL), and the P20 micropipettor to 0.015 mL (15 µL). You should practice doing that kind of conversion in your head.

2. Insert a tip into the end of the pipettor by pushing the end of the micropipettor firmly into the tip in the box. Do not touch the tip with your hands. The smaller tips fit both the P20 and the P200 micropipettors. They are often yellow or clear. The larger tips are for the P1000 micropipettor and are often blue.

3. Take up the water volume (microliter) you want, empty into a waste beaker, and discard the tip in a waste beaker by pressing the eject button. You may want to practice this technique a few times as it is a very important skill to master.

B. Checking pipettor calibration

- 1 mL H_2O = 1.0 (0.995–0.998) g at 25 °C
- Check the mass of each volume using an electronic analytical balance:

– P1000: 500 µL = 0.5 mL = 0.5 g
– P200: 100 µL = 0.1 mL = 0.1 g
– P20: 10 µL = 0.01 mL = 0.01 g

1. Place a strip of parafilm on the balance pan.
2. Set the balance to zero.
3. Dispense the water on to the balance and read the reading.

C. Pipetting practice (accuracy and precision)

Accuracy is how close a measurement value is to the true value or accepted value, referring to a mean value of measurements. Precision is the consistency of a series of measurements or tests, referring to standard deviation. Familiarize yourself with the amount of liquid filled in a tip. Try to develop a sense for the volume of liquid you are pipetting by looking at the amount of liquid in the tip. It is easy to grab the wrong pipettor or to set it incorrectly; however, you should be able to tell the difference between 1 µL in a tip and 10 µL in a tip just by looking at the amount of liquid in the tip.

1. Add the following solutions to two separate 0.5-mL microcentrifuge tubes:

Solution A	6.5 µL
50% glycerol	9.0 µL
Solution B	1.5 µL
Solution C	1.0 µL
Solution D	2.0 µL

The total volume should be 20 µL. Check the accuracy of your measurements by setting a micropipettor to the total volume in the tube and slowly withdrawing all of the solution. Your pipetting is accurate if you leave no solution behind and have no air bubble trapped in the liquid solution in your tip.

2. Add the following solutions to two separate 1.5-mL microcentrifuge tubes:

Solution A	75 µL
50% glycerol	50 µL
Solution B	125 µL
Solution C	150 µL
Solution D	250 µL

The total volume should be 650 µL. Check the accuracy. Is the tip just filled? If so, is any solution left in the tube? Is the tip underfilled with air space?

D. Spectrophotometric test for pipetting

1. Turn on the Spectronic 20 instrument to warm up.

2. Set the wavelength control knob to 630 nm. Make sure that the filter lever on the bottom left is positioned at 600 to 950 nm.

3. Using the zero control knob on the left side, set the absorbance to read 0% transmittance (% T) on the top of the meter. Nothing should be in the sample cuvette holder.

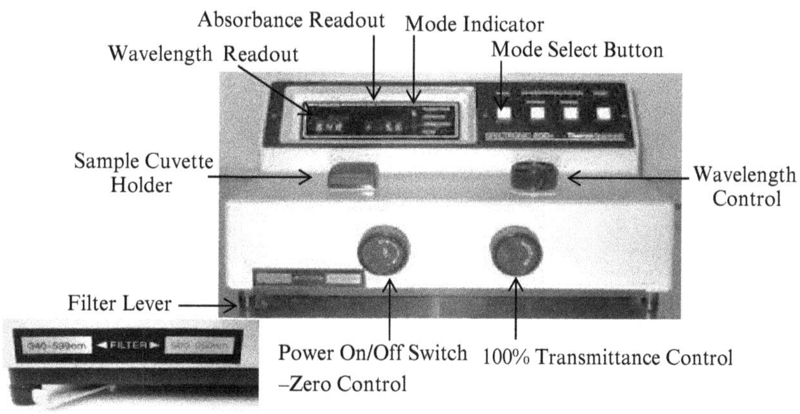

4. Prepare the following triplicate solutions (A1, A2, A3) in test tubes # 1 to 5.

Tube #	Volume of blue dextran (10 mg/mL)*	Volume of water needed† (μL)	Total volume (μL)	μg of blue dextran
0	0 μL (A1, A2, A3)	3000	3000	
1	2 μL (A1, A2, A3)		3000	
2	4 μL (A1, A2, A3)		3000	
3	6 μL (A1, A2, A3)		3000	
4	10 μL (A1, A2, A3)		3000	
5	20 μL (A1, A2, A3)		3000	

*Remove each dextran volume of water, add each volume of dextran starting from tube #1 to tube #5, and rinse pipette tip by pipetting up and down. Do not change the pipette tip.
†Add 3 mL of water first to all tubes using a 5-mL pipette.

5. Fill one cuvette with H_2O as a blank and insert it in the sample cuvette holder with the line of the cuvette facing the front. Close the top.

6. Adjust % transmittance to read 100 using the transmittance control knob, and push the "mode" button to get 0.000 absorbance. Do not touch the wavelength control, zero control, and transmittance control knobs once the instrument is calibrated.

**If you accidently touch them during sample measurements, re-set following steps 5 and 6.*

7. Take out the cuvette, pour off the blank water, and add each sample to the empty cuvette, and read the absorbance of each sample at 630 nm.

Tube #	μg of blue dextran	A_{630} sample A1	A_{630} sample A2	A_{630} sample A3	Mean value	Standard deviation
0						
1						
2						
3						
4						
5						

8. Prepare 1.5-mL of a 1-mg/mL blue dextran from a 10-mg/mL stock solution, and then make the following triplicate solutions (B1, B2, B3) in glass test tubes # 1 to 5.

Tube #	Volume of blue dextran (1 mg/mL)	Volume of water needed (μL)	Total volume (μL)	μg of blue dextran
0	0 μL (B1, B2, B3)	3000	3000	
1	20 μL (B1, B2, B3)		3000	
2	40 μL (B1, B2, B3)		3000	
3	60 μL (B1, B2, B3)		3000	
4	100 μL (B1, B2, B3)		3000	
5	200 μL (B1, B2, B3)		3000	

9. Repeat steps 5 to 7 to read the absorbance of each sample at 630 nm.

Tube #	μg of blue dextran	A_{630} sample B1	A_{630} sample B2	A_{630} sample B3	Mean value	Standard deviation
0						
1						
2						
3						
4						
5						

10. Using the table data obtained from step 7 and 9, plot a graph indicating μg of blue dextran on the *x* axis and A_{630} on the *y* axis. Use the mean A_{630} values of three samples to display data points. Connect the data points using different colors or symbols to distinguish between the two best-fit lines on the same graph. Put a standard deviation bar (I) on to each data point. Compare the two graphs.

Part II. Density and specific gravity

Density (ρ) is defined as mass (m) per unit volume (v): $\rho = m/v$ with units of kg/m^3, g/cm^3, or g/mL. The specific gravity for a liquid has the same numerical value as the density of that liquid if the unit of the density is g/mL. The specific gravity is used to find out the purity of a drug compound since each chemical has a distinct specific gravity, as well as to calculate volume of a fluid using the specific gravity of the fluid. Conversely, the weight can be calculated if the volume is known.

In this experiment, you will determine the densities of different concentrations of NaCl solutions by measuring the weight of a known volume of liquid. You will then determine the concentration of unknown NaCl solutions from the standard curve.

Procedure

1. Label each duplicate tube 1 and 2 for six NaCl molarity solutions (see the table below showing the effect of NaCl concentration on solution density), measure the individual weights of 8 empty 1.5-mL microcentrifuge tubes, and record the weights.

NaCl molarity	Mass of tube		Volume of solution		Mass of tube plus solution		Mass of solution		Mean value of solution mass	Density of solution
	1	2	1	2	1	2	1	2		
3.0										
1.5										
0.75										
0.375										
Unknown 1										
Unknown 2										

2. Prepare 1 mL of 3.0 M, 1.5 M, 0.75 M, and 0.375 M NaCl solutions in the above-measured microcentrifuge tubes. (*Show calculations of how to prepare them using 3 M stock solution in the pre-lab notebook.*) Add unknown NaCl solutions 1 and 2 to the measured microcentrifuge tubes.
3. Measure the individual weights of the microcentrifuge tubes containing solutions and determine the weight of the solution by subtracting the weight of the microcentrifuge tube from the weight of the microcentrifuge tubes containing solution.
4. Calculate the density of the solution.
5. Plot the density and solution concentration (M NaCl) data on the graph with the density on the y axis and the M NaCl on the x axis.

**Adjust the axis scales so that the data points are spread out and occupy as much of the graph space as possible with a 45 degree slope and the major and minor tick marks on both axes.*

6. When the data points have been plotted, draw a best-fit line through them. Do not connect the dots. Use this graph to determine the concentrations of NaCl in unknown samples. You may use a Microsoft Excel program to plot a calibration curve. Regarding how to use, visit the online website (http://www.youtube.com/watch?v=NJYAMNlBGb4).

Discussion

(Do not copy the number and discussion point. Write a paragraph in your own words.)

1. If your micropipetting technique was good, the graph you produced should be a nearly straight line from start to end. The more points that deviate from the line, the more inaccurate is the pipetting skill.
2. The standard deviation for the three values for triplicate solutions provides a measure of the reproducibility of your pipetting skills. The smaller the standard deviation, the more precise is your pipetting skill.
3. Compare the two best-fit lines on the same graph. Do the two lines overlap? If not, explain why.
4. Suggest something you could do to ensure that your micropipettor is measuring correctly.
5. Describe the relationship between the concentration and the density of solution. How do they differ from each other?

Post-lab assignment

1. Why are there no units for specific gravity?
2. How do you determine the density of water-insoluble solids like gold and the density of water-soluble solids like sugars and salts?
3. Convert the molar NaCl solutions used in Part II into g/mL and compare with their respective densities of NaCl solution measured in Part II. How does solution density differ from solution concentration in g/mL?
4. If the density of propylene glycol (antifreeze) is 1.04 g/mL, what is the volume in mL of 4.92 lb of antifreeze (1 lb = 454 grams)?
5. An aqueous vinegar sample contains 4.4% acetic acid (v/v) and has a density of 1.006 g/mL. Calculate the amount of acetic acid in grams in 750 mL of this vinegar.
6. What instrument other than analytical balance is used to determine the concentrations of salt solutions and purity of water?
7. Draw the position lines of 10, 20, 50, and 100 μL volumes when you pipetted a solution into 0.5-mL and 1.5-mL microcentrifuge tubes. Sketch the lines as accurately as possible on the basis of *actual size* of the microcentrifuge tubes. (Hand drawings to the actual scale of both the volumes and centrifuge tubes are required.)
8. Draw the position lines of 2, 5, and 10 μL volumes in P20 pipette tip. Sketch the lines as accurately as possible on the basis of *actual size* of a P20 pipette tip. (Hand drawings to the actual scale of both the volumes and P20 pipette tip are required.)

Further reading

Miller, J.S., Sass, M.E., Wong, S.J., and Nienhuis, J. (2004). Micropipetting: an important laboratory skill for molecular biology. *The American Biology Teacher*, **66** (4): 291–296.

Seiman, L.A. and Moore, C.J. (2000). The measurement of volume. In *Basic Laboratory Methods for Biotechnology*, pp. 301–339. Prentice Hall, Inc. ISBN 0-13-795535-9.

2 MB experiment 2: Use of the spectrophotometer and Beer's law

Purpose: This is up to you to write down.

Introduction

Spectroscopy or spectrophotometry is one of the most widely used methods for performing both qualitative and quantitative analyses of biochemical substances. Each chemical species has a unique spectral fingerprint of absorbance spectrum, the extent to which different wavelengths of light are absorbed. Biochemists commonly use absorbance spectroscopy to obtain both qualitative (identity) and quantitative (amount) information. The identification is done by scanning a sample for absorbance at all wavelengths of visible light. The quantitative measurement is achieved because each photon of light absorbed corresponds to the excitation of a single electron.

Consider a solution containing red dye molecules. This mixture appears red because the dye molecules mainly absorb wavelengths from 400 to 580 nm and the wavelengths corresponding to red color at 580 to 660 nm are not absorbed. As a result, red color wavelengths are transmitted or reflected back to the observer. A spectrophotometer is capable of producing a single wavelength light termed as monochromatic light and then accurately measure the amount of that light absorbed by a given sample. It is recommended to watch the following videos as to how a spectrophotometer works:

- https://www.youtube.com/watch?v=pxC6F7bK8CU
- http://www.jove.com/science-education/5038/introduction-to-the-spectrophotometer

In the laboratory, analyses are performed on very large numbers of atoms or molecules. Therefore, a relationship must be established to obtain quantitative information. Initial spectrophotometric readings measure transmittance, which is defined as the fraction of light that passes through the sample:

$$T = I/I_0, \quad \%T = I/I_0 \times 100, \quad I/I_0 = \%T/100$$

where I_0 is the intensity of light passing through the blank solvent and I is the intensity of light that passes through the sample solution. A more useful quantity in performing quantitative analyses is the absorbance:

$$A = -\log_{10} T = \log_{10}(1/T) = \log(I_0/I) = \log(100/\%T)$$
$$= 2 - \log_{10}(\%T)$$

A linear relationship exists between absorbance and concentration. This relationship is expressed as Beer's Law equation: $A = \varepsilon \times c \times l$, where A is absorbance (also called optical density (OD)), ε is the molar extinction coefficient (also called molar absorption or absorptivity coefficient) of the solute expressed in terms of M^{-1} cm^{-1} (i.e., liters/mole cm^{-1}), c is solute concentration, and l is the length of the light path; ε is a measurement of how strongly a chemical species absorbs light at a given wavelength and is an intrinsic property of the species that varies with wavelength and solvent. It also can be expressed in other units, especially when the molecular weight of a substance is not known. For example, $\%^{-1}$ cm^{-1} or $(mg/mL)^{-1}$ cm^{-1} is numerically equivalent to the absorbance of a 1% (w/v) or 1 mg/mL solution in a 1-cm path length cuvette.

A calibration curve is prepared using standard solutions of known concentrations versus absorbance. A straight line relationship will result, from which you can determine the slope of the line. The slope of the line equals εl in the Beer's Law equation $A = \varepsilon cl$. However, if the ε value of a chemical species in a given solution at a specific wavelength is already known, concentration can be determined simply by measuring its absorbance. There is one potential pitfall in determining concentration from absorbance. Beer's Law breaks down at high concentrations, and such readings are inaccurate. If absorbance values are greater than 1.0, a second dilution is recommended.

Before using Beer's Law as an analytical tool, it is necessary to select a suitable wavelength and determine whether Beer's Law is valid (linear) at the wavelength selected. The wavelength at maximum absorbance (called λ_{max}) allows detection of lower concentrations of sample. The cuvette for measuring samples should be of the proper optical material; for example, a quartz cuvette must be used for ultraviolet (UV light. A glass cuvette is used for the colorimetric detection and measurement of many groups of organic molecules such as carbohydrates, amino acids, lipids, proteins, and nucleic acids. In addition to determination of sample concentration, absorption spectroscopy is helpful to detect the conformational changes caused by a ligation binding such as oxidized or reduced forms of hemoglobin, cytochrome C, and $NAD(P)^+/NAD(P)H$.

In this lab exercise, you will prepare a series of diluted phenolphthalein solutions and measure the absorption spectrum of each solution. Based on this spectrum profile, you will determine λ_{max} and read the absorbance of a set of standard solutions using the determined λ_{max} to obtain a Beer's Law plot. The concentration of phenolphthalein in an unknown sample can be

Methods in Biotechnology, First Edition. Seung-Beom Hong, M. Bazlur Rashid and Lory Z Santiago-Vázquez.

Companion Website: www.wiley.com\go\hong\Methodsinbiotechnology

determined using this calibration plot. Note that the assumption is made that the unknown sample contains no other substances that absorb at this wavelength. You will also perform a colorimetric assay for different concentrations of a protein solution using Biuret's reagent.

It is recommended to watch the following YouTube videos as to how to use Spectronic 20 prior to coming to the lab:

- http://www.youtube.com/watch?v=jmZomizSPxw&NR=1 (Use of Spectronic 20)
- http://www.youtube.com/watch?v=Il9hlMT0aUQ (Beer–Lambert Law)

Microvolume samples can be directly quantified without the use of cuvettes using a NanoDrop 2000c spectrophotometer by reducing a conventional 1 cm pathlength to a microvolume pathlength of 1 mm to 0.05 mm:

http://www.jove.com/video/1610/microvolume-protein-concentration-determination-using-nanodrop-2000c/

Pre-lab assignment

(Typing and submission must be completed before lab work begins.)

1. How much do you have to weigh out in order to prepare 250 mL of a 3 × 10^{-3} M solution of phenolphthalein (MW 318.33 g/mole)? What solvent do you have to use to dissolve it?
2. Each group needs 3 mL of 10 mg/mL BSA solution in 0.9% NaCl. You need to prepare the solution for a total of eight groups. How would you prepare this solution from BSA and NaCl?
3. Draw a diagram or table to show how the dilution in the Part D experiment is accomplished with BSA and 0.9% NaCl.
4. How much 3.0 × 10^{-3} M phenolphthalein solution is needed to conduct today's experiment? Show your calculation at each part and total the exact amount.
5. How much 1 M NaOH solution is needed to conduct today's experiment? Show your calculation at each part and the exact total amount.

Materials and equipment

- 3.0 × 10^{-3} M stock solution of phenolphthalein (MW = 318.33) in 50% ethanol
- Unknown phenolphthalein solutions #1, #2
- 1 M NaOH, 1M HCl
- Biuret reagent
- BSA (high quality: 10 mg/mL in 0.9% NaCl)
- Spectronic 20, cuvette, ruler, test tubes, pH paper

Procedure

Part A. Obtaining absorbance spectrum to determine λ_{max}

1. Prepare 6 mL of solution that is 1 × 10^{-5} M phenolphthalein and 0.01 M NaOH using 3.0 × 10^{-3} M phenolphthalein and 1 M NaOH stock solutions (use $M_1V_1 = M_2V_2$). *Calculation is needed in the pre-lab notebook.* Caution: 1.0 M NaOH solution is caustic.

2. Let this solution react at room temperature for ~10 min.

3. Prepare 6 mL blank solution that is 0.01 M NaOH. *Calculation is needed in the pre-lab notebook.*

4. Turn on the spectrophotometer (Spectronic 20) and warm up for ~10 minutes.

5. Set the wavelength control knob to 400 nm. Make sure that the filter lever on the bottom left is positioned at 340 to 599 nm.

6. Using the zero control knob on the left side, set the absorbance to read 0% transmittance (% T) on the top of the meter. Nothing should be in the sample cuvette holder.

7. Fill one cuvette with a blank solution and insert it in the sample cuvette holder with the line of the cuvette facing the front. Close the top.

8. Select the absorbance mode using the mode select button and use the 100% transmission adjustment knob on the right-hand side to set the reading to 0.000 absorbance. Remove the cuvette and save this blank solution.

9. Fill the cuvette with solution that is 1 × 10^{-5} M phenolphthalein and 0.01 M NaOH. Insert the cuvette into the instrument and close the cover. Read the absorbance from the bottom scale on the meter. Record the wavelength and absorbance readings in the notebook.

10. Remove the cuvette, close the top, and increase the wavelength by 25 nm. (To save time, initially make measurements at increments of 25 nm rather than 10 nm.) Reset the 0% transmission if it has changed when the sample compartment is empty.

11. Recalibrate at each wavelength. (Insert the cuvette of the blank solution and reset the 0 Absorbance.)

12. Replace the blank solution cuvette with your sample-containing cuvette and read the absorbance again, recording your results on your data sheet.

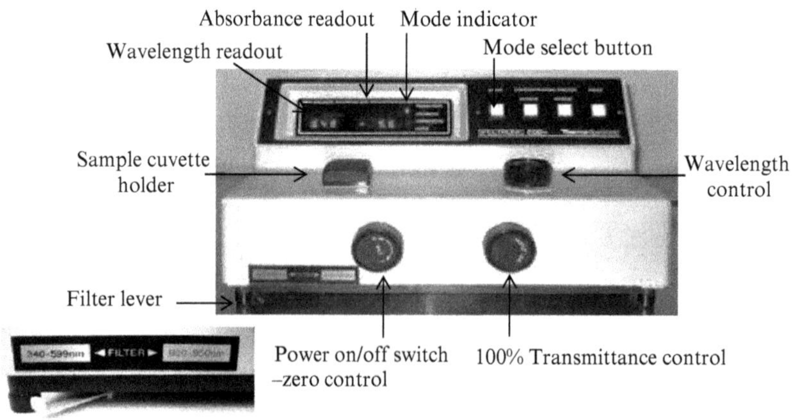

13. Repeat steps 8 through 10 until you have recorded the absorbance of all the wavelengths to 595 nm.
14. Once the reading values from 400 to 595 nm have been measured, determine the maximum wavelength (λ_{max}) by graphing absorbance (y axis) versus wavelength (x axis).

Part B. Preparation of Beer's Law plot (calibration curve) and determining the concentration of unknown solutions

15. Make 3 mL 0.01 M NaOH solutions that are 4.8×10^{-5} M, 2.4×10^{-5} M, 1.2×10^{-5} M, 0.6×10^{-5} M, 0.4×10^{-5} M, and 0.2×10^{-5} M in phenolphthalein using 3.0×10^{-3} M phenolphthalein and 1 M NaOH stock solutions (use $M_1V_1 = M_2V_2$). *Calculation is needed in the pre-lab notebook.*
16. Make 3 mL of two unknown solutions 1 and 2 by adding 0.03 mL of 1 M NaOH to 2.97 mL of unknown solutions 1 and 2.
17. Let all the solutions react at room temperature for ~10 min.
18. Set the spectrometer to the selected λ_{max} and read and record the absorbance for each solution starting from the lowest concentration to the highest concentration and then unknown solutions.
19. Plot the data for all known concentrations on the graph. Draw the best-fit line throughout the four known data points on the Absorbance (y axis) versus Concentration (x axis) graph. The data point in which both absorbance and concentration are a 0.00 reading also should be included for the best-fit line.
20. Determine the concentration of your unknown solutions from the best-fit line graph. To determine the concentration of an unknown, use the absorbance of the unknown solution. Find that value on the absorbance scale, and then move across the graph parallel to the concentration axis until you meet the best-fit line. At that point, move directly down to the concentration axis and read the concentration of the unknown.

Part C. Effect of pH and reaction time on absorbance

21. Calibrate Spectronic 20 at the selected λ_{max}, using a blank solution of 0.01M NaOH.
22. One person in a group prepares 3.0 mL of 1.0×10^{-5} M phenolphthalein and 0.05 M NaOH directly in a cuvette using 3.0×10^{-3} M phenolphthalein and 1 M NaOH stocks. *Calculation is needed in the pre-lab notebook.*

**Add phenolphthalein at last to start the reaction.*

23. Immediately place it into the Spectronic 20 and record the Absorbance at 0 time point.

**Since the time at which phenolphthalein and NaOH are mixed is time zero for the reaction, note the time (h: min: s) when phenolphthalein is added to mix with NaOH.*

24. Read the absorbance every 1 min without taking the cuvette from the sample holder for a total of 10 min. Check the final pH using pH paper after 10 min.
25. Another person in a group prepares 3.0 mL of 1.0×10^{-5} M phenolphthalein and 0.01 M NaOH directly in a cuvette using 3.0×10^{-3} M phenolphthalein and 1 M NaOH stocks. *Calculation is needed in the pre-lab notebook.*
26. Proceed as in steps 23 and 24. Share all the data.

Part D. Biuret test of protein sample

27. Label 12 test tubes (#1 to #12).
28. To each tube, add the following volumes of BSA stock (10 mg/mL): 0, 0.0025, 0.005, 0.01, 0.025, 0.05, 0.1, 0.2, 0.4, 0.6, 0.8, and 1.0 mL.
29. Bring a final total volume in each tube to 1.0 mL by adding an appropriate volume of 0.9% NaCl.
30. • Group 1 and 2. Add 4.0 mL Biuret reagent to each tube, mix gently, and let stand at room temperature for 10 min.
 • Group 3 and 4. Add 2.0 mL Biuret reagent to each tube, mix gently, and let stand at room temperature for 10 min.
31. Calibrate Spectronic 20 at 540 nm wavelength using blank solution: 1.0 mL of 0.9% NaCl + 4.0 mL Biuret reagent for Groups 1 and 2; 1.0 mL of 0.9% NaCl + 2.0 mL Biuret reagent for Group 3 and 4.
32. Record the absorption data for each known concentration of BSA–Biuret mixture samples.
33. Share all the data between the group data of 2 and 4 mL Biuret reagent volumes.
34. Plot a graph of the concentration (x axis) and absorbance (y axis) of each known sample using Microsoft Excel. Draw the best-fit line through the data points.

You Should Turn In:

- A plot of absorbance versus wavelength using all the values obtained in Part A (use Microsoft Excel to create a plot). When you make the graph, adjust the axis scales so that the data points are spread out and occupy as much of the graph space as possible with the major and minor tick marks on both axes.
- Your Beer's Law plot calibration curve from Part B.
- Compare the rate of changes in absorbance value (OD versus time) between the two experiments in Part C. Draw the two independent data on the same graph so that the two best-fit lines can be distinguished using the appropriate scale of tick marks.
- The extinction coefficient value determined from the plot and calculation. (For this, you must determine the path length to the nearest 0.01 cm by measuring the inside diameter of your cuvette.)
- The data sheets and post-lab assignment.

Result

Show all work when performing calculations.

Part A. Determination of λ_{max}

Wavelength (nm)	Absorbance[a]
400	
425	
450	
475	
500	
525	
550	
575	
595	

[a] 1×10^{-5} M phenolphthalein in 0.01 M NaOH.

Wavelength selected for Beer's Law plot ________________.

Part B. Data for beer's law plot

Phenolphthalein concentration (moles/L)	Absorbance[†] at λ_{max}
4.8×10^{-5}	
2.4×10^{-5}	
1.2×10^{-5}	
0.6×10^{-5}	
0.4×10^{-5}	
0.2×10^{-5}	

[b] Dilute 1/10 in water if reading value is higher than 1.9.

Part B. Analysis of unknowns

Unknown #	Absorbance	Concentration (moles/L)
1		
2		

Part C. Effect of pH and reaction time

Time (min)	Absorbance	
	0.05 M NaOH	0.01 M NaOH
0		
1		
2		
3		
4		
5		
6		
7		
8		
9		
10		
pH		

Part D. Biuret reagent testing of protein

Test tube	BSA concentration (mg/mL)	Biuret reagent (mL)	Absorbance at 540 nm	Color type	Protein detected (Y or N)
1		2			
		4			
2		2			
		4			
3		2			
		4			
4		2			
		4			
5		2			
		4			
6		2			
		4			
7		2			
		4			
8		2			
		4			
9		2			
		4			
10		2			
		4			
11		2			
		4			
12		2			
		4			

Discussion

(Do not copy the number and discussion point. Write a paragraph in your own words.)

1. Establish the absorption calibration curve at your determined λ_{max}.
2. Determine unknown concentrations of phenolphthalein.
3. What is the extinction coefficient value (ε) of phenolphthalein at your determined λ_{max}?
Calculate using the plotted graph of absorbance versus molar concentration.

$$\text{Slope of line} = \in \times l = \frac{\Delta\text{Absorbance (no unit in } y \text{ coordinate)}}{\Delta\text{ Concentration (molarity* in } x \text{ coordinate)}}$$

**Pay attention to the exponent on 10 in molarity.*

4. Make a graph of absorbance (y axis) versus reaction time (x axis) for the two different solutions. On the basis of the Part C experiment result, is phenolphthalein an acid or a base? Explain your choice. What lesson can you take in the Part C experiment?
5. On the basis of the Part D experiment result, what is the detection limit of concentration in Biuret testing of a protein sample? What is the effect of the Biuret reagent volume on the detection sensitivity and linearity between absorbance and concentration?
6. What are possible sources of experimental errors in your results?

Post-lab assignment

1. Does the extinction coefficient vary depending on pH and salt concentration? Explain why.
2. Why does phenolphthalein change color in alkaline pH?
3. What is the relationship between the color of a solution and the wavelengths the solution absorbs?
4. A substance's molar extinction coefficient at 260 nm is $34.7 \times 10^3\ M^{-1}\ cm^{-1}$. You have a 0.01 mM solution of the substance. What would be the absorbance at 260 nm measured in a spectrophotometer with a 1-cm light path?

5. What is the advantage and disadvantage of Biuret testing of protein samples?

6. How can the concentrations of DNA and RNA be determined using a spectrophotometer?

7. You were given a 500 mL of a solution at pH 7.0 containing an unknown concentration of adenosine-5′-phosphate (ATP; $\varepsilon = 14\,300\ M^{-1}\ cm^{-1}$ at 260 nm and pH 7.0). You placed 4 mL of this solution in a 1 cm wide, 1 cm long, and 4 cm high quartz cuvette, and determine the A_{260} of this solution to be 0.143.

(a) What is the molarity of the ATP solution?

(b) How many moles of ATP are in the 500 mL of the solution?

(c) How many μmoles of ATP are in your cuvette?

8. Six samples of caffeine are prepared at the same concentration and analyzed spectroscopically. The absorbances at 270 nm were found to be:

0.2625, 0.2635, 0.2590, 0.2700, 0.2610, 0.2715

(a) Calculate the mean absorbance.

(b) Calculate the standard deviation of the measurements.

(c) A 7th sample is prepared and its absorbance is found to be 0.2439. Use the *Q*-test to determine if this sample can be rejected or accepted using the 95% confidence limit. (*Hint:* http://en.wikipedia.org/wiki/Dixon's_Q_test.)

9. Two students have obtained the following reading values for the same sample using a well-established method.

Student 1: 0.745, 0.798, 0.854, 0.826, 0.72.4

Student 2: 0.882, 0.869, 0.753, 0.875, 0.894, 0.795

Use the *t*-test to determine if there is a statistically important difference between these two means at the 95% confidence level.

- Visit online GraphPad QuickCalcs *t*-test calculator:
    - http://graphpad.com/quickcalcs/ttest1.cfm.
- Choose data entry format.
- Enter data.
- Choose a test – Paired *t*-test.
- Click "Calculate now" to view the results.

10. A certain spectrophotometer utilizes a photodiode array detector (PDA) system. What is PDA and how is it useful for assaying a compound (what is the advantage)?

Further reading

Desjardins, P., Hansen, J.B., and Allen, M. (2009). Microvolume protein concentration determination using the NanoDrop 2000c Spectrophotometer. *Journal of Visual Experiments*, (**33**), e1610. DOI: 10.3791/1610.

Seidman, L.A. and Moore, C.J. (2000). Measurements involving light. In *Basic Laboratory Methods for Biotechnology*, pp. 371–439. Prentice Hall, Inc. ISBN 0-13-795535-9.

3 MB experiment 3: Making solutions and buffer efficacy

Purpose: This is up to you to write down.

Introduction

All living systems contain buffer solutions to sustain the structure and activity of biological components such as DNA, RNA, proteins, carbohydrates, and lipids. Buffer solutions are remarkably resistant to pH changes and generally consist of a weak acid and its conjugate base or a weak base and its conjugate acid. In the laboratory, artificially made buffers are often used to help maintain a biological system at the proper pH.

A laboratory buffer should be inert in the system being studied. For example, Tris buffer is unsuitable for some protein assays such as Biuret and Lowry because it reacts with the assay reagent components. Phosphate buffers contribute phosphate ions to a solution, which inhibit some types of enzyme reactions such as alkaline phosphatase. Tris-borate-EDTA (TBE) buffer and Tris-acetate-EDTA (TAE) buffers are most commonly used for DNA gel electrophoresis. However, because the borate reacts with the hydroxyl group of carbohydrates in the gel, TAE buffer is the buffer of choice when DNA is to be extracted from gels after electrophoresis.

Buffer solutions must always be prepared with high-quality water. They are often sterilized by autoclaving or filtration to prevent microbial growth from degrading the buffer compounds. When a buffer is prepared, correct amounts of the buffer components are weighed and dissolved in high-quality water, and the solution must not be brought to its final volume to get the desired concentration before it is adjusted to the proper pH with a strong acid or base. If you accidentally overshoot the pH, it is better to remake the entire solution rather than offsetting by adding some extra acid or base because additional acid or base adds extra ions, changes the solution's composition and causes an inconsistency. If you overshoot the volume, it is better to remake to avoid changing the concentration and introducing inconsistency. When the buffer pH is adjusted, an acid or base component of solution must be used. For example, sodium acetate buffer with pH 5.2 must be prepared with glacial acetic acid but not with HCl.

The pH often affects the solubility of a compound; ionic charged (either protonated or deprotonated) form of molecules are more soluble than nonionized form due to the strong polar property of water. When pH is higher than pK_a, deprotonation occurs, and vice versa. However, depending on the species, either the protonated or deprotonated form is its neutral uncharged form. Ethylene-diamine-tetraacetic acid (EDTA) is essentially insoluble in water and will only dissolve when pH is at least 8.0. Phenolphthalein and bromophenol blue are not soluble in water but soluble in basic solutions or in ethanol.

It is often convenient to prepare buffers as concentrated stock solutions, which will be diluted before use. The equation $C_1V_1 = C_2V_2$ ($\text{Concentration}_{\text{-stock}} \times \text{Volume}_{\text{-stock}} = \text{Concentration}_{\text{-final}} \times \text{Volume}_{\text{-final}}$) is useful to determine how much of the concentrated stock solution is needed for making a dilute solution. It is important to plug in the identical units of concentration and volume of both stock and final solutions into the equation in order to get a correct answer. Some buffers change pH when diluted. If the pH change caused by dilution is not acceptable, then use of stock solution should be avoided. Some buffers, especially Tris, change pH when the temperature changes. It is therefore important to adjust the pH at the same temperature as the buffer that will be used. Solubility of buffer also should be taken into consideration; sodium phosphates are quite insoluble at low temperature unlike potassium phosphate. However, potassium phosphate buffer is useless if sodium dodecyl sulfate (SDS) is added to the buffer because potassium ions react with SDS molecules to make an insoluble complex.

Most buffer solutions consist of a weak acid (HA) and its conjugate base (A^-) from salt of the weak acid. Examples are [acetic acid ↔ acetate], [phosphoric acid ↔ phosphate], and [Tris-HCl ↔ Tris], where HA reacts with small amounts of added base and A^- reacts with small amounts of added acid to keep the pH constant. The maximum buffer capacity occurs when $pH = pK_a$. Most buffering compounds are not at their pK_a when they are dissolved and thus must be adjusted to the desired pH. For more detailed background information on buffers, study the material in the following website:

- http://www.wiley.com/college/pratt/0471393878/student/review/acid_base/4_strong_and_weak.html (Review exercises of acids, bases, and pH)

It is also highly recommended to watch YouTube videos regarding buffers and pH meter:

Methods in Biotechnology, First Edition. Seung-Beom Hong, M. Bazlur Rashid and Lory Z Santiago-Vázquez.

Companion Website: www.wiley.com\go\hong\Methodsinbiotechnology

- http://www.youtube.com/watch?v=HZFIdpThd-s (Buffers and pH meters)
- http://www.youtube.com/watch?v=cckAwavEKA0 (Making up a standard solution)

The selection and proper use of cylinders and beakers are important for measuring liquids. The volumes marked in Erlenmeyer flasks and beakers are approximate and are not accurate enough to measure volumes for solution preparation. The required volume must be measured in a graduated cylinder and then poured into a beaker or flask for mixing. Due to the bad maintenance and some low-quality pH meter electrodes, buffers prepared in the lab often result in inconsistent pH (variations in different batches). If the pH meter readings differ greatly from the calculated pH, the electrode should be either cleaned and reconditioned or replaced with a new one. The pH calculation-based buffer preparation provides the independent confirmation of pH. Once the recipe is established, it makes you more easily and consistently prepare. The recipe tables for the commonly used Tris, phosphate, and acetate buffers are provided in MB Appendix 4. The pH of buffers may vary appreciably depending on the temperature and concentration. You may also use several online buffer calculators that provide recipes for different buffers over a continuous range of concentrations, pH values, volume and temperature:

- http://www.currentprotocols.com/WileyCDA/CurPro3Tool/toolId-3.html
- http://www.liv.ac.uk/buffers/buffercalc.html

Pre-lab assignment

(Typing and submission must be completed before lab work begins.)

1. What is a buffer solution? Why do buffer solutions resist changes in pH?
2. What causes the inaccurate readings by pH meter?
3. Determine the pH for each of the following aqueous solutions and indicate if the solution is neutral, acidic, or basic.
 (a) 2.0×10^{-5} M NaOH
 (b) 500 mL of 6.0×10^{-6} M HCl
4. Calculate gram quantity of Tris and the volume of HCl to prepare 100 mL of 1 M Tris buffer at pH 7.5. The MW and pK_a of Tris are 121.14 and 8.06, respectively. The concentrated HCl is always 12 M. Show your work using the Henderson–Hasselbalch equation pH = pK_a + log ([conjugated base A^-]/[acid HA]) where [ ] indicates molarity.

Materials and equipment

- 1 M HCl, 1 M NaOH, 3 M KCl
- Glacial acetic acid, concentrated hydrochloric acid (11.6 M)
- Sodium acetate
- Tris or Trizma base [Tris (hydroxymethyl) methylamine]
- Sodium phosphate: monobasic NaH_2PO_4 and dibasic Na_2HPO_4
- Borosillicate glass test tube (16 × 150 mm)
- Weighing balances, beakers, magnetic hot plate stirrer, magnetic bars
- pH meters, pH standard solutions (pH 4, 7, and 10), wash bottle, Kimwipe

Procedure

Part I. pH meter calibration

1. Obtain about 10 to 15 mL each of pH 7 and pH 4 standard solutions in separate clean 30-mL beakers.
2. Remove the electrode from the electrode storage bottle and rinse it well in a stream of dH_2O using a wash bottle. Use one large beaker as a waste beaker for all electrode washes. Gently blot off the excess water with Kimwipe.
3. Insert the electrode into a standard pH 7 solution. The glass bulb of the electrode must be completely immersed into the solution. Stir gently to mix the solution. Be careful not to damage the electrode.

**A pH meter is usually in a "standby" mode. If not, turn on and warm up before use.*

4. Press "CAL" key on the pH meter key panel. The decimal point flashes during the calibration measurement.
5. Remove the electrode, rinse it with dH_2O, and blot with Kimwipe (do not rub or wipe the electrode bulb). Repeat 3 and 4 using a standard pH 4 solution.
6. Switch the pH meter to "Autoread" mode and check the pH 10 standard or other known pH solutions.
7. Remove the electrode from the standard solution, rinse it with dH_2O and put back into the storage bottle.

**Never let the electrode probe dry.*

Part II. Making solutions (to share four groups)

Calculations are needed in the pre-lab.

Group 1: 25 mL of 1 M Tris (MW = 121.14)
100 mL of 1 M Tris-HCl (pH 7.5, adjust pH with concentrated HCl)
Group 2: 100 mL of 1 M sodium acetate (CH_3COONa, MW = 82.03)
100 mL of 1 M acetic acid (MW = 60.05, density = 1.049 g/mL)
Group 3: 50 mL of 2 M monobasic NaH_2PO_4 (MW = 119.98)
Group 4: 100 mL of 2 M dibasic Na_2HPO_4 (MW = 141.96)
Groups 3 and 4: Prepare 200 mL of 1 M sodium phosphate (pH 7.5) by mixing 16 mL of 2 M monobasic NaH_2PO_4, 84 mL of 2 M dibasic Na_2HPO_4, and 100 mL of dH_2O

**Note: keep 2M dibasic Na_2HPO_4 solution warm to prevent re-crystallization. When making solutions, dissolve chemicals in 70–80% final volume of water and bring up to the final volume once all chemicals are dissolved.*

Part III. Buffering capacity (addition of strong acid and strong base to buffered and unbuffered solutions)

1. Use a 50-mL beaker for each of the following solutions shown in the table below. Mix well. Measure and record the pH.

Unbuffered solution	Measured pH*	Calculated pH
25 mL dH_2O		
25 mL dH_2O + 10 μL 3 M KCl		
25 mL dH_2O + 1 mL 1.0 M HCl		
25 mL dH_2O + 1 mL 1.0 M NaOH		

*A solution of very low ionic strength takes much longer to stabilize the pH reading.

2. Use a 100-mL beaker to make a 50 mL of the sodium acetate (NaA)/acetate (HA) buffer solution, as shown in the table. Measure and record the pH. Pipette 25 mL of this buffer solution and place it in a new 50-mL beaker. Add 1 mL of 1 M HCl or 1 M NaOH to each separate beaker and mix well. Measure and record pH.

Buffered solution (acetate)	Measured pH	Calculated pH*
25 mL 1.0 M NaAc + 25 mL 1.0 M HA		
25 mL above mix buffer + 1 mL 1.0 M HCl		
25 mL above mix buffer + 1 mL 1.0 M NaOH		

*The pK_a of acetic acid is 4.75.

3. Use a 100-mL beaker to make 50 mL of 0.1 M Tris-Cl buffer by diluting 1 M Tris buffer stock in dH_2O (*A calculation is needed in the pre-lab*) and adding 12 M HCl as shown in the table. Measure and record the pH. Pipette 25 mL of this buffer solution and place it in a new 50-mL beaker. Add 1 mL of 1 M HCl or 1 M NaOH to each separate beaker and mix well. Measure and record pH.

Buffered solution (tris)	Measured pH	Calculated pH*
50 mL 0.1 M tris + 0.22 mL 12 M HCl		
25 mL above mix buffer + 1 mL 1.0 M HCl		
25 mL above mix buffer + 1 mL 1.0 M NaOH		

*The pK_a of tris is 8.06.

4. Use a 100-mL beaker to make 50 mL of sodium phosphate solution by mixing 0.8 mL of 2 M monobasic NaH_2PO_4, 4.2 mL of 2 M dibasic Na_2HPO_4, and 45 mL of H_2O as shown in the table. Measure and record the pH. Pipette 25 mL of this buffer solution and place it in a new 50-mL beaker. Add 1 mL of 1 M HCl or 1 M NaOH to each separate beaker and mix well. Measure and record the pH.

Buffered solution (phosphate)	Measured pH	Calculated pH*
0.8 mL of 2 M monobasic NaH_2PO_4 + 4.2 mL of 2 M dibasic Na_2HPO_4 + 45 mL H_2O		
25 mL above mix buffer + 1 mL 1.0 M HCl		
25 mL above mix buffer + 1 mL 1.0 M NaOH		

*The pK_{a2} and pK_{a3} of phosphate groups are 6.8 and 12.3, respectively.

5. Rinse the pH electrode with dH_2O and place it back into a storage bottle.

Part IV. Dilution effect on buffer pH and buffer capacity

1. Using 1 M Tris-Cl (pH 7.5) stock solution, make 5 mL each of 0.5 M, 0.25 M, 0.1 M, 0.05 M, 0.01 M, and 0.001 M in the test tubes.
2. Using 1 M sodium phosphate (pH 7.5) stock solution, make 5 mL each of 0.5 M, 0.25 M, 0.1 M, 0.05 M, 0.01 M, and 0.001 M in the test tubes.

Calculations are needed in the pre-lab.

3. Measure and record the pH of all the diluted solutions.
4. Add 0.1 mL of 1 M HCl to each of the above tubes and measure the pH.

Discussion

(Do not copy the number and discussion point. Write a paragraph in your own words.)

1. Show all your work for the calculated pH. Calculate using the Henderson–Hasselbalch equation: $pH = pK_a + \log([\text{conjugated base } A^-]/[\text{acid HA}])$.
2. Is the measured pH of pure water the same as the calculated pH? If not, is it more acidic or basic? What is attributed to the pH difference? Why is it difficult (slow, non-reproducible, noisy) to read the pH of pure water? Assume that the vessel containing water is perfectly clean.
3. Plot all data points on the same graph of pH (y axis) versus molarity (x axis) in the Part IV experiment. What is the slope of the line on your graph? What is the significance of the slope? Determine the percent change in pH of the buffers after changes in concentration, [absolute value of (initial pH – final pH)/initial pH] × 100, where initial pH is the pH of 1 M solution and final pH is the individual pH of the diluted solutions.

4. Which buffer is more sensitive to the dilution? Which buffer increases or decreases its pH as the buffer becomes diluted? Based on the results, what ionic strength (molarity) of the buffer remains within ± 0.1 pH unit with respect to the initial pH of undiluted buffer?

5. Why do you think that the buffer's pH changes after further dilutions? Discuss the effect of dilution on the pH and buffer capacity.

Post-lab assignment

1. The amino group of glycine is often used as the main ingredient of a buffer in biochemical experiments. The amino group of glycine with a pK_a of 9.6 can exist either in the protonated form ($-NH_2^+$) or in the free base ($-NH_2$), because of the reversible equilibrium: $R\text{-}NH_3^+ \leftrightarrow R\text{-}NH_2 + H^+$. Use the Henderson–Hasselbalch equation $pH = pK_a + \log$ ([conjugated base A^-]/[acid HA]) to answer (b), (c), and (d) questions.

- **(a)** In what pH range can glycine be used as an effective buffer due to its amino group? Explain your choice.
- **(b)** When 99% of the glycine is in its $-NH_3^+$ form or $-NH_2$ form, what is the numerical relation between the pH of the solution and the pK_a of the amino group?
- **(c)** In a 0.1 M solution of glycine at pH 9.0, what molar fraction of glycine has its amino group in the $-NH_3^+$ and NH_2 form?
- **(d)** How much 10 M KOH must be added to 1.0 liter of 0.1 M glycine at pH 9.0 to bring its pH to exactly 10.0?

2. **(a)** Why does the pH not change when a buffer is diluted in theory?

- **(b)** How is buffer capacity defined? What determines the buffer capacity? Does the dilution change the buffer capacity? Explain.

3. A 5 mL solution of 0.65 M formic acid was diluted with dH_2O to bring to a total volume of 100 mL. Calculate the pH of the resulting solution. The K_a of formic acid is 1.77×10^{-4}. Show your work.

4. A 300 mL solution of 1.8 M sodium formate (HCOONa) was added to 700 mL of 0.65 M formic acid (HCOOH). Calculate the pH of the resulting solution. The K_a of formic acid is 1.77×10^{-4}. Show your work.

5. A laboratory technician has 1 M acetic acid and 1 M sodium acetate. Calculate the volumes of acetic acid and sodium acetate solutions that he needs to use in order to prepare 1 L of 0.1 M buffer solution at pH 5.5. The pK_a of acetic acid is 4.75. Show you work.

6. Calculate the pH of a buffer solution prepared by dissolving 242.2 mg of Tris (hydroxymethyl) aminomethane in 10 mL of 0.17 M HCl and diluting it to 100 mL with water. The MW of Tris = 121.1 and Tris has a $pK_a = 8.08$. (*Hint: Tris is equivalent to the ionized A^- and Tris-HCl is equivalent to the acid HA.*) Show your work.

7. How would you prepare 200 mL of 1 M phosphate solution at pH 7.5 using 2 M potassium phosphate dibasic (K_2HPO_4) and 2 M potassium phosphate monobasic (KH_2PO_4)? The pK_{a2} of phosphate is 6.8. Show your work.

8. What volume of acetic acid (17.4 M) and what weight of sodium acetate (MW 82) would be required to make 100 mL of 3 M buffer, pH 5.2. The pK_a of acetic acid is 4.75. Show you work.

9. Aspirin is a weak acid with a pK_a of 3.5:

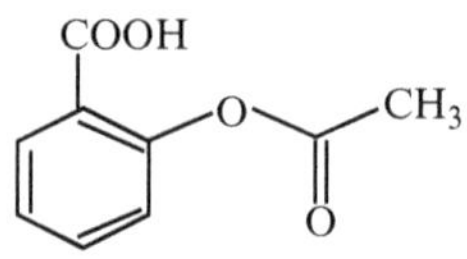

$C_9H_8O_4$

It is absorbed into the bloodstream through the cells. The rate of its passing through the hydrophobic plasma membrane varies depending on the polarity of the molecule; the more ionic the molecule, the more polar it becomes and the more slowly it moves through. The pHs of the stomach and small intestine are ~1.5 and ~ 6, respectively. Is more aspirin absorbed into the blood from the stomach or from the small intestine? Explain your choice.

10. The pH of a 0.1 M solution of a pure weak acid (no other solutes) was measured at 3.68. What is the pK_a of this acid?

Further reading

Montegomery, R. and Swenson, C.A. (1976). *Quantitative Problems in the Biochemical Sciences*, 2nd edition). W.H. Freeman and Company. ISBN 0-7167-0178-2.

Seidman, L.A. and Moore, C.J. (2000). Preparation of laboratory solutions. In *Basic Laboratory Methods for Biotechnology*, pp. 449–482. Prentice Hall, Inc. ISBN 0-13-795535-9.

Timberlake, K.C. (2006). *Laboratory Manual for General, Organic, and Biological Chemistry*. Benjamin-Cummings Publishing Company. ISBN 0-8053-4904-9.

4 MB experiment 4: Acid–base titration

Purpose: This is up to you to write down.

Introduction

Acid–base reactions belong to a very important class of biochemical reactions. For example, in living systems the control of pH exerted by acid–base equilibrium is crucial for survival. A bicarbonate buffer system functions to maintain pH 7.4 in the bloodstream and extracellular fluids, whereas a phosphate buffer system functions to maintain the same pH in the urine and intracellular compartment in mammals. A 0.2 pH unit shift results in serious changes in blood chemistry, causing severe illness. Acid–base titration is one of the most common analytical procedures by which the concentration and dissociation constant of a wide variety of chemical substances can be determined. Titration involves the addition of a standard solution of known molarity (called the titrant) to a solution whose concentration is unknown (called the analyte). By knowing how much of the titrant is required to react completely with a known volume of the analyte solution we can calculate the analyte concentration.

The point at which stoichiometrically equal amounts of acid and base have been reacted is called the equivalence point. At the equivalence point, moles of H^+ in the analyte are equal to moles of OH^- in the titrant used when a strong acid (H^+) is titrated with a strong base (OH^-); moles of HA in the analyte = moles of OH^- in the titrant used when a weak acid (HA) is titrated with a strong base (OH^-); moles of B in the analyte = moles of H^+ in the titrant used when a weak base (B) is titrated with a strong acid (H^+). An appropriate indicator that exhibits a sharp color change as close to the equivalence point as possible is typically added to the solution in order to mark the point at which the two quantities reach equivalence. For example, phenolphthalein is a colorimetric pH indicator that is pink in a basic solution (at $pH > 8.2$) and colorless in an acidic solution. Phenol red is yellow at a $pH < 6.8$ and red at a pH of > 8.4. Indicators usually exhibit intermediate colors at pH values inside the transition range. The actual point at which the indicator changes color is termed the end point. However, the end point is not the same as the equivalence point in the strict sense. The difference between the two is the titration error. Obviously, for an accurate titration, care must be taken to select an indicator for which the difference between the equivalence point and the end point is very small.

In this lab exercise, you will add a few drops of indicator to a volume of hydrochloric acid or acetic acid solution, and then add sodium hydroxide (NaOH) base to it slowly until the acid is neutralized, as indicated by a sudden color change to achieve the end point titration. To calculate the concentration of the unknown acid we must begin with a balanced equation. The reaction between the strong acid HCl and the strong base NaOH is: $HCl + NaOH \rightarrow NaCl + H_2O$. The reaction between the weak acid CH_3COOH and the strong base NaOH is: $CH_3COOH + NaOH \rightarrow CH_3COONa + H_2O$. Stoichiometrically one mole of the acid reacts with one mole of the base. Because of this one-to-one relationship we can use the following formula to calculate the unknown concentration: $M_{acid} \times V_{acid} = M_{base} \times V_{base}$. Rearrange the equation to solve for the unknown concentration of the base:

$$M_{acid} = \frac{M_{base} \times V_{base}}{V_{acid}}$$

where V_{acid} (analyte) and M_{base} (titrant) are already known and V_{base} is determined by measuring the volume at the end point (or equivalence point).

Unlike strong acids, when you add a weak acid to pure water, weak acids dissociate (ionize) only slightly in aqueous solution:

$$\underset{\text{acid}}{HA} + \underset{\text{base}}{H_2O} \rightleftharpoons \underset{\text{conjugate acid}}{H^+ \text{ (more correctly } H_3O^+)} + \underset{\text{conjugate base}}{A^-}$$

The dissociation (ionization or equilibrium) constant of HA in aqueous solution is written as follows:

$$K_a = \frac{[H^+][A^-]}{[HA]}$$

$\log_{10} K_a = \log_{10} [H^+] + \log_{10} [A^-] - \log_{10} [HA]$
$-\log_{10} [H^+] = -\log_{10} K_a + \log_{10} [A^-] - \log_{10} [HA]$
$-\log_{10} [H^+] = -\log_{10} K_a + \log_{10} ([A^-]/[HA])$
$pH = pK_a + \log_{10} ([A^-]/[HA])$
or
$pH = pK_a + \log_{10} ([\text{conjugate base}]/[\text{acid}])$

The last mathematic expression is called the Henderson–Hasselbalch equation.

Methods in Biotechnology, First Edition. Seung-Beom Hong, M. Bazlur Rashid and Lory Z Santiago-Vázquez.

Companion Website: www.wiley.com\go\hong\Methodsinbiotechnology

When a weak acid HA is titrated with a strong base, the acid HA reacts with OH^- according to the following balanced equation: $HA + OH^- \rightarrow H_2O + A^-$. At the equivalence point in this titration, moles of HA in analyte = moles of OH^- of titrant, that is, all acid molecules HA are converted to conjugated base A^-. Hence, by titrating a weak acid sample with a known concentration of NaOH, the concentration of an acid can be determined. The titration of a weak base with a strong acid is very similar to that of the titration of a weak acid with a strong base: $B + H^+ \rightarrow BH^+$. At the equivalence point in this titration, moles of B in analyte = moles of H^+ of titrant used, that is, all bases (B) are converted to conjugate acid (BH^+).

At the half equivalence point in a titration of weak acid with strong base, half of HA is consumed to produce the conjugate base A^- and $[HA] = [A^-]$. This is because regardless of the strength of the acid to begin with, at the half reaction 50% of the acid is still there and 50% of it is neutralized to form its conjugate base. Thus, the ratio $[A^-]/[HA] = 1$, $\log(1) = 0$, and $pH = pK_a$ according to the Henderson–Hasselbalch equation.

A typical curve for titration of a weak acid with a strong base is shown below.

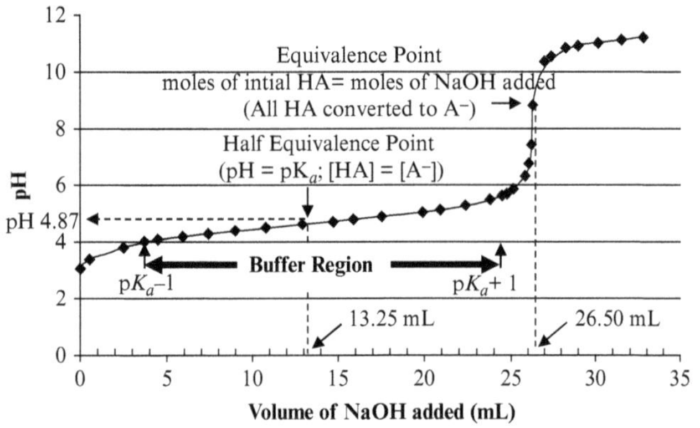

The equivalence point is where all ionizable H^+ of pK_a ionizable groups reacted with OH^- and occurs in the region where a steep increase in pH is observed. The corresponding volume of NaOH at this point is 26.50 mL. Thus, the volume of NaOH at the half equivalence point is 26.5/2 = 13.25 mL, which corresponds to a pH of 4.87. Because $pH = pK_a$ at the half equivalence point, the pK_a for this acid is 4.87. The point corresponding to the pK_a is an inflection point for the titration curve that occurs at midway between the beginning of the titration and the equivalence point. Note that just because the amounts of acid and conjugate base are the same at the pK_a, the pH is not 7 (neutral) at this point. Remember that pH is a function of the molar concentration of H_3O^+, not the acid. The titration curve also shows that buffer pH is most effective within $pK_a \pm 1$. By definition $pK_a = -\log K_a$. Thus, $K_a = 10^{-pKa} = 10^{-4.87} = 1.35 \times 10^{-5}$. This corresponds to propionic acid. Since different weak acids have different unique K_a values, identification of an acid is made possible.

Note that for a weak acid titrated with a strong base, the pH is >7.0 at the equivalence point. This is because all weak acid (HA) is converted to the conjugate base A^-, which has a strong affinity with H^+ and reacts with water to produce a basic solution ($A^- + H_2O \rightarrow HA + OH^-$). It is recommended to watch the following YouTube video as to how to titrate:

- http://www.youtube.com/watch?v=9DkB82xLvNE

Pre-lab assignment

(Typing and submission must be completed before lab work begins.)

$(M_{acid} \times \#H) \times V_{acid} = (M_{base} \times \#OH) \times V_{base}$, where M is molarity. This equation allows us to determine what combination of acid and base results in neutralization. Essentially, it employs the idea that neutralization occurs when the number of moles of H^+ is equal to the number of moles of OH^-.

1. What volume of 1 M HCl would be required to neutralize 40 mL of 0.5 M NaOH?

2. What volume of 1 M H_2SO_4 would be required to neutralize 40 mL of 0.5 M NaOH?

3. What volume of 1 M $Al(OH)_3$ would be required to neutralize 30 mL of 2 M H_2SO_4?

4. A 60 mL of 0.5 M $Mg(OH)_2$ is required to titrate a 50 mL of acetic acid (CH_3COOH) in order to reach the end point.

 (a) Write a balanced neutralization equation for the reaction.

 (b) Using the balanced equation, calculate the concentration of the acid.

5. What volume of 0.5 N HCl would be required to neutralize 20 mL of 0.5 N $Ba(OH)_2$?

Materials and equipment

- pH meters
- pH standard solutions (pH 4, 7, and 10)
- Squeeze wash bottle with deionized water, Kimwipe, pipettes, two beakers (100 mL), burette (50 mL), ring stand, burette clamp, funnel
- Phenolphthalein (0.5% in ethanol)
- Phenol red (0.05% in dH_2O)
- HCl solution with unknown concentration
- Acetic acid (vinegar) solution with unknown concentration
- 1.0 M NaOH

Note: use fresh deionized water throughout the experiment. Deionized water should have most of the carbon dioxide removed from it during its preparation. Solid NaOH is highly hygroscopic and its solutions absorb carbon dioxide from the air to form carbonate. The equilibrium level of dissolved CO_2 in water open to atmosphere is about 1.5×10^{-5} M. This causes a severe deviation in solution concentration and reduces the titer of the base. Fine powder that does not dissolve quickly when you prepare the NaOH solution is probably sodium carbonate, and the white particles should be removed by filtering prior to use. In this lab, we do not standardize NaOH solution to determine its true concentration.

Procedure

Part I. Two-point calibration of pH Meter

Calibrate the pH meter using pH 4 and 10 standard solutions.

Part II. Titration of acetic acid with NaOH

1. Clean a 50-mL burette with soap and water thoroughly. Rinse well with tap water and do a final rinse with dH_2O. Check for leaks.
2. Set up the ring stand, burette, and burette clamp. Make sure that the burette remains vertical at all times.
3. Prepare 150 mL of 0.1 M NaOH solution from 1.0 M NaOH stock. *Show calculations in your pre-lab notebook.*
4. Rinse the burette with 0.1 M NaOH and fill the burette with 0.1 M NaOH. Drain some of the NaOH solution through the stopcock into the waste beaker. Make sure that the burette tip is filled with the solution (no air bubbles in the tip). Fill the burette so that the bottom of the meniscus of the NaOH is sitting on the 0.00-mL line mark with the tip filled. Make sure no drop remains hanging on the burette.
5. Using the pipette, transfer 25 mL CH_3COOH solution into a clean dry 100-mL beaker.
6. Add 3 drops of phenolphthalein to the acid solution.
7. Place the beaker under the burette. Place a sheet of white paper under the beaker so that a color change will be more readily observed.
8. Remove the calibrated pH electrode from the storage bottle, rinse it with dH_2O, blot gently with Kimwipe, and immerse it into the liquid just low enough to completely cover the glass bulb of the electrode. Make sure there is enough space at the bottom of the beaker.
9. Measure and record the pH of the acetic acid solution. Remember to wait for the pH reading to stabilize.
10. You are now ready to begin titrating. Carefully add 1 mL of NaOH from the burette to the beaker and swirl to mix. Record the pH and volume in the burette. Continue adding 1-mL increments until you begin to see a faint pink color and record pH readings. The pink color should appear briefly and then quickly disappear.
11. Once you have seen the pink color, begin adding the NaOH solution slowly in smaller amounts (0.1 mL). Record the pH readings and volumes in the burette to the nearest 0.01 mL. Swirl the beaker after each drop is added.

Note: as you near the end point, a faint pink color should stay longer but will still disappear when the beaker is swirled. The longer the pink color stays, the smaller your next increment of NaOH should be. You want to eventually add one small drop of NaOH at a time until the palest pink color appears and stays.

Good endpoint

Bad endpoint

12. When the pale pink color appears and stays permanent, you have reached the end point. Record the pH reading and the volume in the burette for the end point. Now you can add larger volumes between readings as the pH starts to change less dramatically.
13. Continue the titration, and record the pH readings and the volumes in the burette until you reach a pH value of 12. (*Note:* as you continue titration, the pink color gets darker and darker.) Indicate the end point volume and pH in your data table.
14. Clean out the beaker and dispose of any remaining acid or base by flushing down the drain.
15. Repeat steps 4 to 12 without the pH meter and pH value recordings, and determine the end point volume.

**This step can be skipped if the first titration experiment results in a good end point detection.*

**If you have to repeat due to the bad end point, you may add the total volume that reached the bad end point minus 1 mL and then start to titrate by dropwise addition of titrant to analyte solution in order to save the titration time.*

Part III. Titration of HCl with NaOH

Rinse out your beaker and carry out titration of HCl as described in Part II (steps 4 to 14) with 0.1 M NaOH except for step 6 in which phenolphthalein is replaced with phenol red. When the pale orange color appears and stays permanent, you have reached the end point (step 12). After the experiment is completed, rinse each burette with dH_2O, invert, and hold it to the burette clamp, leaving the stopcock open.

Part IV. Data integration

Use Microsoft Excel to graph pH values along the *y* axis versus the cumulative volume of NaOH added along the *x* axis. Plot all the data points on one graph. Draw smooth curves using different colors or symbols to distinguish the titration of HCl from the titration of acetic acid. Denote both minor and major tick marks of x and y axes in the graph below.

Data from titration of HCl		Data from titration of CH_3COOH	
NaOH mL	pH	NaOH mL	pH

Discussion

(Do not copy the number and discussion point. Write a paragraph in your own words.)

1. From the acetic acid titration graph, determine the following:

(a) Volume of NaOH at the equivalence point
(b) Volume of NaOH at the half equivalence point
(c) pH at the half equivalence point
(d) pK_a at the half equivalence point
(e) Equilibration (ionization or dissociation) constant K_a

2. Check any difference between the end point determined by titration experiment and the equivalence point determined by graph in terms of the pH and titrant volume.
3. Calculate the molarity of HCl and acetic acid.
4. Discuss the major differences between strong acid and weak acid titration curves.
5. What are the possible errors in this experiment?

Post-lab assignment

1. Why does the pH of an end point differ from the pH of an equivalence point?
2. In doing a titration, the concentrations of the acid and base need to be considered. In this experiment the NaOH concentration is about 0.1 M. Why do you think this concentration is a better choice than 1.0 M NaOH or 0.01 M NaOH? What problems would be associated with using solutions of these concentrations?
3. Why does it not matter much how much water you add when dissolving the acid or base when carrying out the titration?
4. Amino acid glycine has two ionizable groups, the α-COOH group (pK_{a1} 2.35) and α-NH_3^+ group (pK_{a2} 9.78). Sketch the expected titration curve when 10 mL of 0.1 M glycine hydrochloride (Glycine-HCl) solution is titrated with 0.1 M NaOH. Indicate the titration volumes of NaOH to reach pK_{a1}, pK_{a2}, and the equivalence point in the curve.
5. Calculate the pH of the solution formed when 45 mL of 0.1 M NaOH is added to 50 mL of 0.1 M CH_3COOH ($K_a = 1.8 \times 10^{-5}$).
6. A 2.06 gram sample of ascorbic acid (vitamin C) was crushed and dissolved in water. It was titrated with 0.2 M NaOH (the ascorbic acid reacts with sodium hydroxide in a 1:1 ratio). The sample required 46 mL to reach an end point using phenolphthalein as the indicator.

(a) Calculate the number of moles of ascorbic acid in the sample, which is equal to the number of moles of NaOH.
(b) Calculate the molar mass of ascorbic acid in the sample.
(c) Calculate the % purity of the sample: (molar mass of ascorbic acid ÷ molar mass of sample) × 100. The molar mass of ascorbic acid is 176.12 g/mole.

7. What happens to the pH of blood in a diabetic with a high blood sugar count? Explain and make sure to cite a reference.
8. **(a)** Write a balanced equation for the reaction of phosphoric acid (H_3PO_4) with sodium hydroxide.
(b) If 20.14 mL of 0.1 M NaOH solution is required to completely neutralize 7.15 mL of the phosphoric acid solution, what is the molarity of the phosphoric acid solution? Show your work.
(c) You have 10 mL of 0.1 M phosphoric acid (H_3PO_4), a triprotic acid, which has 3 pK_a values of 2.14, 6.86, and 12.4. How many mmoles of NaOH are needed to get to pH 6.86 and 12.4? Show your work.

Further reading

Montegomery, R. and Swenson, C.A. (1976). *Quantitative Problems in the Biochemical Sciences*, 2nd edition. W.H. Freeman and Company. ISBN 0-7167-0178-2.

Timberlake, K.C. 2006. *Laboratory Manual for General, Organic, and Biological Chemistry*. Benjamin-Cummings Publishing Company. ISBN 0-8053-4904-9.

5 MB experiment 5: Protein denaturation and precipitation

Purpose: This is up to you to write down.

Introduction

Proteins are polymers of amino acids whose side chains contain polar, non-polar, neutral, acidic, and basic groups. The primary structure of a protein is determined by the sequence of amino acids, but the secondary and tertiary structures of proteins define their native conformation for a functional activity. Proteins are held in their native conformation by combined forces of hydrogen bonds, ionic bonds (salt bridges), disulfide bridges, and hydrophobic interactions. These forces can be disrupted by heat, pH changes, organic solvents (alcohol, acetone), heavy metal salts, thiol-reducing reagents (β-mercaptoethanol, dithiothreitol), chaotropic salts (urea, guanidine hydrochloride), and non-ionic or ionic detergents.

Once proteins are denatured by misfolding, they tend to aggregate. Protein aggregation can occur inside a cell due to the misfolding, which is caused by mutations and a wide variety of stresses (heat, redox, chemical, irradiation, etc.). Because it is toxic, it has been implicated in various diseases such as Alzheimer, Parkinson, Huntington, Mad Cow (Prion), and Cystic Fibrosis. Some misfolded proteins are passed from mother cells to their daughters, influencing biological phenotype with no change in its DNA (https://www.youtube.com/watch?v=_Q0oOcZminY).

Since every protein differs in the composition and sequence of amino acids, it has unique biochemical and physical characteristics in terms of size, solubility (hydrophobicity), isoelectric point, and stability. Most soluble proteins are precipitated out of the solution if they are denatured. When a protein solution is heated, the proteins may be denatured, lose their solubility, and precipitate. Since all proteins will not precipitate at the same temperature, heat treatment can be useful in the separation of proteins. Some proteins such as RNase and prion are highly resistant to heat, even at boiling or autoclaving. Unlike heat, pH changes, organic solvents, and heavy metal ions, both chaotropic salts and detergents not only denature proteins but also cause them to be soluble in aqueous solutions.

Protein stability and precipitation effectiveness by salts are affected by different salt species. Chaotropic salts denature and solubilize proteins by disrupting hydrogen bonding of water molecules at the surface of protein and disrupting intramolecular hydrogen bonds and hydrophobic interactions in protein. Antichaotropic salts affect the electrostatic ionic interactions between the side chains of amino acids. At low ionic strengths, ions may have little effect on these interactions. As ionic strengths increase, proteins tend to be stabilized due to the shielding of the unpaired charged groups on protein surfaces. However, at very high ionic strengths, many proteins are denatured since normal sites of electrostatic interactions between the charged groups on amino acids are neutralized by the salt ions. As a result, many proteins become more soluble in dilute salt solutions (salting-in effect) but become less soluble in concentrated salt solutions (salting-out effect). Unlike many other salts, ammonium sulfate causes proteins to precipitate without irreversibly denaturing them. Different proteins are soluble to varying degrees in ammonium sulfate. Because of this property, ammonium sulfate is often used to concentrate proteins as well as to fractionate proteins.

Many organic solvents such as alcohol, acetone, and trichloroacetic acid are used to precipitate proteins, but they often denature proteins because they are capable of forming intermolecular hydrogen bonds with protein molecules. Oxidant compounds such as potassium permangate, peroxide, chlorine, and iodine are used for disinfection since they can oxidize amino acids to denature proteins.

Proteins differ in solubility at a given pH. When the pH of a solution is brought to the isoelectric point (pI) of a protein, the protein has minimum solubility and precipitates out of the solution because the presence of an equal number of positive and negative charges are attracted to each other, thereby making its net charge become zero. This phenomenon is referred to as isoelectric precipitation. Once protein is isoelectrically precipitated, it is irreversibly denatured. At a pH above and below the pI, proteins have a net negative and positive charge, respectively, and repel one another. Therefore the proteins are very soluble at either acidic or basic pH. However, proteins are no longer precipitated if their peptide bonds are hydrolyzed by very acidic or basic pH. It is recommended to watch the following videos and animation:

- http://www.youtube.com/watch?v=HN4jD2MCKfg (isoelectric precipitation)
- http://highered.mcgraw-hill.com/sites/0072943696/student_view0/chapter2/animation__protein_denaturation.html
- http://www.youtube.com/watch?v=b39698t750c (Chaperone)
- http://www.youtube.com/watch?v=_Q0oOcZminY (protein folding I)
- http://www.youtube.com/watch?v=NFsCqqHUsoE&feature=relmfu (protein folding II)

Methods in Biotechnology, First Edition. Seung-Beom Hong, M. Bazlur Rashid and Lory Z Santiago-Vázquez.

Companion Website: www.wiley.com\go\hong\Methodsinbiotechnology

In this lab exercise, four purified proteins are used to study the precipitation. Albumin is a major component of serum (BSA: MW 66.5 kDa; pI 4.7) and hen egg white (MW 76 kDa; pI 6.0, 6.3, 6.6). Casein (MW ~25 kDa; pI 4.6) is a major component of milk proteins and is a heterogeneous mixture of phosphoproteins. Gelatin (MW ~60 kDa) is a protein produced by partial hydrolysis of an insoluble collagen protein. Bovine γ-globulin (150 to 200 kDa) is a blood plasma protein containing immunoglobulins. You will test these proteins for the solubility after treatment of heat, acid and base, and chemical substances. In addition, you will spectrophotometrically monitor the state of hemoglobin as you vary the concentration of its denaturant.

Pre-lab assignment

(Typing and submission must be completed before lab work begins.)

1. How many glass test tubes do you need to perform Part A I, II, III, and IV experiments today? Show your calculations that explain the number at each part.
2. How many transparent microcentrifuge tubes do you need to perform Part B experiments?
3. Draw and label all the test tubes for Part A II, III, and IV experiments.
4. How much (mL) albumin, casein, gelatin, γ-globulin, and hemoglobin solutions are needed to complete today's experiment? Show your calculations.
5. How many grams of albumin, casein, gelatin, γ-globulin, and hemoglobin are needed to complete today's experiment? Show your calculations.

Materials and equipment

- 1% Bovine serum albumin (BSA), casein, gelatin solutions
- 1% γ-Globulin (Sigma-Aldrich G7516-1G) in 1× PBS (phosphate buffered saline)
- 1% Casein (Sigma-Aldrich C4765-10ML: 5%) in 1× PBS
- 1 mg/mL hemoglobin in 1× PBS (phosphate buffered saline)
- 0.1 M HCl, 0.1 M NaOH, 2% $CuSO_4$, 2% $AgNO_3$, isopropanol, acetone, ammonium sulfate $(NH_4)_2SO_4$. 100% TCA (trichloroacetic acid), 6 M Guanidine-HCl
- Bio-Rad Bradford dye reagent (1/5 diluted in water) and Biuret reagent
- pH paper, microcentrifuge tubes, glass test tubes (13 × 100 mm) in tube racks
- Spectronic 20, water bath (60 °C)

Procedure

Part A. Solubility and protein denaturation

I. Effect of heat

1. Place 1 mL each of albumin, casein, gelatin, and γ-globulin solution into the four labeled glass test tubes.
2. Incubate all tubes at 37 °C first → transfer to 60 °C → transfer to 90 °C water baths for 10 min each.
3. Record the first signs of the precipitated protein appearing as milky white, with a hazy appearance, and final results.

II. Effect of inorganic and organic additives

4. Place 1 mL each of albumin, casein, gelatin, and γ-globulin solution into the four labeled glass test tubes.
5. Add 2% $CuSO_4$ drop by drop and mix until a change is observed. Record the volumes used and observations (one drop of Pasteur pipette = ~17 μL). **Do not add more than 10 drops.*
6. Repeat the step 1 and 2 using 2% $AgNO_3$ until a change is observed. Record the volume used and observations. **Do not add more than 10 drops.*
7. Add one volume (1 mL) of isopropanol to 1 mL of each protein solution (step 4) and mix. Record the observations. If no precipitation occurs, add another 1 mL isopropanol.
8. Add 4 volumes (4 mL) of cold acetone (–20 °C) to 1 mL of each protein solution (step 4) and mix. Record the observations.

III. Effect of chaotropic salt

9. Label 6 test tubes as Blank, 0 M, 0.5 M, 1.0 M. 1.5 M, and 3.0 M Gu-HCl.
10. Add the appropriate volumes of each stock solution of hemoglobin (1 mg/mL) and 6 M guanidine hydrochloric acid to a final hemoglobin concentration of 0.1 mg/mL with guanidine concentrations of 0 M, 0.5 M, 1.0 M, 1.5 M, and 3 M in a final total volume of 3 mL adjusted with 1× PBS. Prepare each blank solution as shown in the table. *Calculations are needed in the pre-lab notebook.*

Addition order*		Blanks					Test samples				
	Gu-HCl (M)	0	0.5	1.0	1.5	3.0	0	0.5	1.0	1.5	3.0
1	1× PBS (mL)										
2	6 M Gu-HCl (mL)										
3	1 mg/mL Hemoglobin (mL)	0	0	0	0	0					
	Total volume (mL)	3	3	3	3	3	3	3	3	3	3

*The order of addition matters. Mix well gently.

11. Incubate all the tubes at room temperature for 20 min.
12. Set the wavelength to 410 and calibrate the spectrometer using the blank solution.
13. Record absorbance for the tubes, starting from 0 M guanidine. After you are done, put all the tubes in a 60 °C water bath and incubate for 20 min.
14. Record absorbance for the tubes, starting from 0 M guanidine.
15. Plot the absorbance versus guanidine concentration for each temperature.

Part B. Solubility and "salting out" with ammonium sulfate

1. Weigh duplicate amounts of each 0.472 g (70% saturation), 0.314 g (50% saturation), and 0.176 g (30% saturation) of ammonium sulfate crystals and place them into two separate

1.5 mL transparent microcentrifuge tubes labeled for albumin and γ-globulin with percentage (TA will do this step).

2. Add 1 mL of albumin protein solution to each tube and gently shake until all crystals are dissolved. Do the same for γ-globulin protein solution. Do not vortex.

**Make sure that all salt crystals disappear by shining light on the tube and carefully looking at the tube bottom.*

3. Incubate the tube on ice for 5 min.
4. Spin at 13 000 rpm, 4 °C for 20 min.

**Be certain to always centrifuge microcentrifuge tubes with the cap hinge facing the outside of the centrifuge.*

5. Aspirate off the supernatant as much as possible from each tube. Record the size of pellet: +++ (big), ++ (medium), + (small), – (not detectable).

**You will see no pellet after centrifugation if protein is not precipitated.*

6. Add 1 mL dH_2O to each tube and gently suspend each pellet by pipetting up and down.

**Do not make foams during suspension.*

7. Transfer 0.5 mL each of $(NH_4)_2SO_4$ albumin suspension to two 1.5 mL clear microcentrifuge tubes labeled with each percentage of $(NH_4)_2SO_4$ (%A/Biuret and %A/Bradford).
8. Transfer 0.5 mL each of $(NH_4)_2SO_4$ γ-globulin suspension to two clear microcentrifuge tubes labeled with each percentage of $(NH_4)_2SO_4$ (%C/Biuret and %C/Bradford).
9. Transfer 0.5 mL each of albumin solution to two clear microcentrifuge tubes as positive controls (labeled +A/Biuret and +A/Bradford).
10. Add 0.5 mL each of γ-globulin solution to clear microcentrifuge tubes as positive controls (+C/Biuret and +C/Bradford).
11. Weigh 0.132 g (23% saturation = 1 M), 0.066 g (12% saturation), 0.033 g (6.0% saturation) of ammonium sulfate crystals, place in clear microcentrifuge tubes and dissolve in 1 mL of dH_2O. Transfer 0.5 mL of each solution to the labeled tube as negative controls (prepared by TA and shared by all groups):

- N-23%/Biuret, N-23% Bradford
- N-12%/Biuret, N-12% Bradford
- N-6%/Biuret, N-6% Bradford
- N-0%/Biuret, N-0%/Bradford

12. Add 1 mL of Biuret reagent to all the microcentrifuge tubes labeled Biuret. Mix well.
13. Add 1 mL of Bradford dye to all microcentrifuge tubes labeled Bradford. Mix well.
14. Arrange all the experimental, positive and negative control tubes on the white paper, compare, and record any changes in color (Tables IV, V, VI, and VII). Take a photograph.

Data table I. Effect of heat

Test tube	1	2	3	4
temperature	Albumin	Casein	Gelatin	γ-Globulin
37 °C				
60 °C				
90 °C				

Data table II. Effect of metal salts and organic solvents

Test tube reagent	1		2		3		4	
	Albumin		Casein		Gelatin		γ-Globulin	
	# D	A*	# D	A*	# D	A*	#D	A*
$CuSO_4$								
$AgNO_3$								
Isopropanol								
Acetone								

#Indicate the number of drop or volume.
*Record observations.

Data table III. Effect of guanidine-HCl on hemoglobin

Gu-HCl conc.	A_{410} at 25 °C	A_{410} at 60 °C
0 M		
0.5 M		
1.0 M		
1.5 M		
3.0 M		

Data table IV. Effect of ammonium sulfate on albumin precipitation

Ammonium sulfate	Pellet size*	Color of pellet suspension†	
		Biuret	Bradford
+ Control	N/A		
70% saturation			
50% saturation			
30% saturation			

*Indicate the size of pellet as +++ (large), ++ (medium), + (small), – (none) in microfuge tubes.
†Indicate the color after addition of Biuret and Bradford reagent dyes.

Data table V. Effect of ammonium sulfate on γ-globulin precipitation

Ammonium sulfate	Pellet size*	Color of pellet suspension†	
		Biuret	Bradford
+ Control	N/A		
70% saturation			
50% saturation			
30% saturation			

*Indicate the size of pellet as +++ (large), ++ (medium), + (small), – (none) in microfuge tubes.
†Indicate the color after addition of Biuret and Bradford reagent dyes.

Data table VI. Effect of ammonium sulfate on biuret color reaction (negative control)

Ammonium sulfate	Color observations*	
	Biuret	Bradford
0% saturation		
6% saturation		
12% saturation		
23% saturation		

*Indicate any changes in color.

Discussion

(Do not copy the number and discussion point. Write a paragraph in your own words.)

1. Which protein is most and least sensitive to heat?
2. Which metal salt ($CuSO_4$ and $AgNO_3$) precipitates albumin more effectively? Explain why.
3. What is the effect of guanidine concentration and temperature on hemoglobin denaturation? Are the observed absorbance values expected or not? If so, explain why.
4. How do Biuret and Bradford assays work? How does ammonium sulfate affect the Biuret reagent (copper sulfate in a strong base NaOH) and Bradford dye (Coomassie Blue G-250 in phosphoric acid) reaction for protein samples?
5. Compare the ammonium sulfate effects on the size of albumin and γ-globulin pellet. Can ammonium sulfate fractionation be used to separate albumin from γ-globulin in a protein mixture?

Post-lab assignment

1. What is the "salting-out" effect of ammonium sulfate on the protein?
2. The reaction of $CuSO_4$ with proteins in the Biuret assay and the reaction of Coomassie blue G-250 dye with proteins in Bradford assay are the basis of each protein assay. What do the results of Part B for the reaction of Biuret and Bradford reagents with albumin, γ-globulin, ammonium sulfate, and the protein pellet suspension tell you about the effectiveness of the Biuret and Bradford reagent with a protein solution containing ammonium sulfate?
3. Why is 70% alcohol used for disinfection rather than 95 to 100% alcohol?
4. What is the molecular basis of protein denaturation or precipitation by (a) heat, (b) pH changes, (c) alcohols, (d) heavy metal ions?
5. What happens when hemoglobin denatures by chaotropic agent?
6. Serum protein (blood protein) consists of 60% albumin, 35% globulins, 4% fibronectin, and 1% other proteins. When serum is simply diluted with water or dialyzed against water, globulins are precipitated out of solution, while albumin remains in solution. What property makes two proteins differ in their solubility?
7. People who have accidentally swallowed heavy metal salts (such as mercury, silver, or lead) are given egg white as an emergency treatment. Explain why egg whites would be used.
8. What are the commercial and technical uses of the casein and gelatin? Briefly summarize in a table format.
9. Hydrogen bonds are plentiful in proteins and are stronger than hydrophobic interactions. Nonetheless, hydrophobic interactions are more important than hydrogen bonding in driving the process of protein folding. Explain why.

Further reading

Bollag, D.M., Rozycki, M.D., and Edelstein, S.J. (1996). *Protein Methods*, 2nd edition. John Wiley & Sons, Inc., New York. ISBN 978-0-471-11837-4.

Loverien, R.V. and Matulis D. (2001). *Unit 4.5. Selective Precipitation of Proteins. Current Protocols in Protein Science*, pp. 4.5.1–4.5.36. John Wiley & Sons, Inc. ISSN 1934-3655.

6 MB experiment 6: Bacterial transformation

Purpose: This is up to you to write down.

Introduction

A majority of engineered plasmids in use typically contain a clustered region of multiple unique restriction enzyme sites, referred to as a multiple cloning site or polylinker site, so that a foreign DNA cut with such unique enzymes can be ligated to the linearized plasmid vector to generate recombinant DNA molecules. They are then introduced into bacteria by the process called bacterial transformation. This entire procedure is called cloning because many identical copies of the original DNA fragment can be produced as a result of replication of a small circular plasmid into many copies independently of chromosome replication. Thus, cloning experiments are almost always accompanied by *Escherichia coli* transformation for further *in vivo* propagation of the recombinant plasmid as well as for protein expression of a cloned gene. *E. coli* cells can be easily transformed in the laboratory using well-established chemical or electric pulse procedures. Electroporation is amenable to the uptake of large plasmid DNA and results in much higher transformation efficiency than $CaCl_2$-mediated transformation. Moreover, the electroporation method can be applied to a wide variety of bacterial cells, whereas the $CaCl_2$ method is limited to *E. coli* cells. However, unlike the chemical method, the electroporation system requires that samples be free of salts, rendering the direct use of a salt-rich ligation mixture difficult.

The plasmid vectors used to transform bacterial cells should have a selectable marker gene, which permits only transformed cells to grow on selection medium. The most commonly used selectable markers are the genes encoding antibiotic resistance. Often cloning plasmid carries an additional screening marker gene such as β-galactosidase (*lacZ*) and green fluorescent (*gfp*). Under the right conditions, transformed bacteria can express the foreign DNA unless its encoded protein is toxic to the bacterial cells. It is recommended to watch the following video regarding plasmid cloning:

- http://www.youtube.com/watch?v=acKWdNj936o

Scientists cloned a gene coding for green fluorescent protein (GFP) isolated from a bioluminescent jellyfish (*Aequorea victoria*), changed its codon usage, and shuffled DNA fragments in order to improve a whole cell fluorescence signal. Unlike wild-type GFP, which is expressed in inactive and insoluble inclusion bodies, the mutant GFP_{uv} is soluble and active in *E. coli*. Plasmid pGLO has both a negative and a positive gene regulation system that controls GFP_{uv} gene expression. In this expression system, the GFP_{uv} gene is placed behind the promoter sequence called P_{BAD}, which is recognized by a transcription activator (AraC)–coactivator (arabinose) complex. Plasmid pGLO carries the *araC* gene, which encodes a bifunctional regulatory protein that acts as either an activator or a repressor depending on the availability of catabolic substrate for arabinose operon. In the presence of L-arabinose, AraC acts as a transcription activator that helps RNA polymerase bind to the P_{BAD} promoter. In the absence of L-arabinose, AraC represses the activity of P_{BAD}. When sugar arabinose is present in the nutrient medium, bacteria will take it up. Once inside the cell, the arabinose interacts directly with AraC protein, and this interaction results in a conformational change in AraC protein, assisting RNA polymerase to stably bind the P_{BAD} promoter and actively transcribe the GFP_{uv} gene. Therefore, transformed cells will emit green fluorescence under UV light when arabinose is added to the nutrient agar but will remain white on plates lacking arabinose.

In this lab exercise, you will practice both electroporation and chemical transformation methods using the plasmid pGLO (5371 bp), as depicted below.

Methods in Biotechnology, First Edition. Seung-Beom Hong, M. Bazlur Rashid and Lory Z Santiago-Vázquez.

Companion Website: www.wiley.com\go\hong\Methodsinbiotechnology

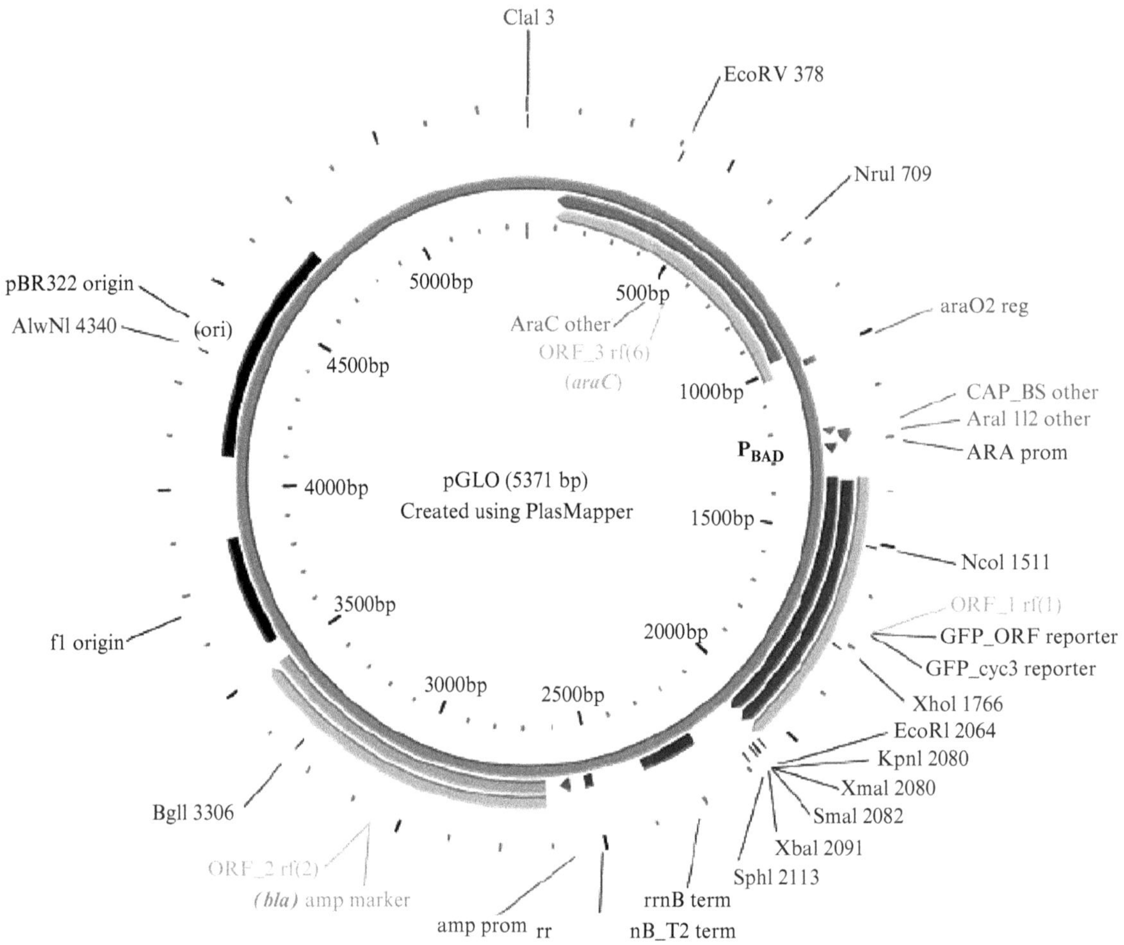

bla. The gene that encodes the enzyme β-lactamase, which breaks down the antibiotic ampicillin. Bacteria containing the *bla* gene can be selected by placing ampicillin in the growth medium.

ori. The origin of pGLO plasmid DNA replication.

araC. The gene that encodes the regulatory protein that binds to the P_{BAD} promoter. Only when arabinose binds to the AraC protein is the expression of GFP_{uv} turned on.

P_{BAD} promoter. The nucleotide sequence that binds AraC-arabinose and promotes RNA polymerase binding and transcription of the GFP_{uv} gene.

Multiple cloning sites. A group of known unique restriction (*SphI, XbaI, SmaI, KpnI, and EcoRI*) sites that permit insertion of the gene of interest.

For transformation, you will use *E. coli* strain that is able to transport but not to metabolize L-arabinose so that the concentration of L-arabinose remains constant for maximum GFP expression. In the chemical transformation method, the mixture of cells and plasmid incubated on ice is heat-shocked to 42 °C, followed by incubation at 37 °C before plating on to the selection medium. In the electroporation method, the mixture of cells and plasmid is briefly shocked in an electric field of 10 to 20 kV/cm to make holes in the cell membrane through which the plasmid DNA enters. After the electric shock, the holes are rapidly closed by the cell's repair mechanisms.

The transformation experiment also requires sterile agar plates containing an antibiotic as a selection marker. To weigh media components, you need to "TARE" electronic balance to the

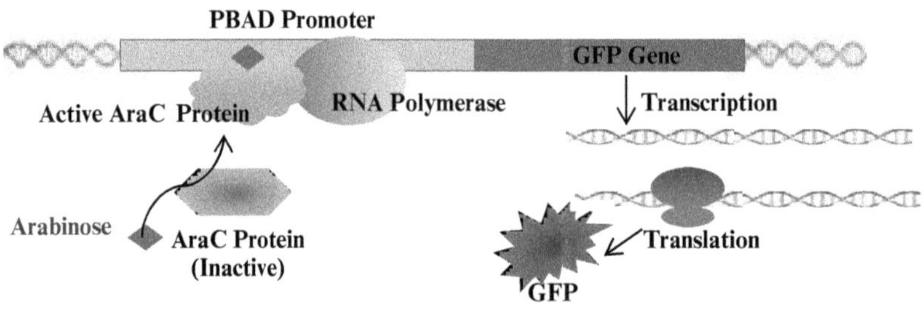

balance at "0". After the instrument is balanced at "0", place a weighing boat on the pan and press "TARE" again. When the LED reads "0.000" it is time to weigh the agar and other reagents for the respective recipes. Add the solid mixture to distilled water and bring the final volume to the mark on the flask. It is recommended to watch the following video regarding how to make an agar plate:

http://www.youtube.com/watch?v=OljTdYH_Wtg

Autoclave is a standard equipment to sterilize media and typically has a "slow exhaust" and a "fast exhaust" mode. These terms refer to how rapidly the steam exhaust valve releases the steam pressure after the sterilization cycle is completed. Under fast exhaust, the steam pressure is dumped all at once without cool-down time as soon as the sterilization cycle is finished. Fast exhaust is used only when you autoclave empty glassware or dry materials. Under slow exhaust, the steam pressure is released slowly in order to prevent media from boiling over. If a screw-cap bottle is used to autoclave medium, it is important not to tighten the cap in order to provide a release of pressure built up inside the bottle. Autoclaving is usually done at 121 °C, 15 to 20 psi. The time required for autoclaving varies depending on the volume. For small volumes (1 liter or less), the time required for autoclaving is 15 to 20 min, but for larger quantities (2 to 4 liters), 30 to 40 min is necessary to complete the cycle. Autoclaved media should not be left inside the autoclave chamber for a prolonged time after autoclave is completed; this may lead to the decomposition of carbohydrates and other components of a medium, as indicated by darkening of medium. When the autoclave is done, open the door carefully while wearing autoclave gloves, remove your vessels, and place them in the water bath (set at about 50 to 60 °C). This will allow the agar to cool, but not to go less than 45 °C, at which it will rapidly solidify. Let it cool to ~55 °C before you add antibiotic solution and pouring into Petri dishes.

Pre-lab assignment

(Typing and submission must be completed before lab work begins.)

1. If you harvested 250 mL of *E. coli* culture cells, what is the final volume of ice-cold 10% glycerol solution to prepare electrocompetent cells?
2. For *E. coli*, it has been calculated that 1 unit at A_{600} corresponds to 10^9 cells per mL. If a 250-mL *E. coli* culture that has 0.4 unit at A_{600} was used in today's experiment, what would be the final cell density of the electrocompetent ready cells?
3. Why is arabinose added to LB agar-ampicillin medium?
4. What is a negative control in the transformation experiment? Why is it important?
5. As compared to untransformed cells, transformed *E. coli* cells will express two new phenotypes in today's experiment. What are they?

Materials and equipment

- *E. coli* HB101 K-12 (*F- mcrB mrr hsdS20(rB-mB-) recA13 leuB6 ara-14 proA2 lacY1 galK2 xyl-5 mtl-1 rpsL20(SmR) glnV44 λ-*): Bio-Rad Cat. No. 166-0408EDU
- Chemically competent HB101
- Pre-chilled sterile 1.5-mL sterile microcentrifuge tubes
- Ice-cold sterile distilled water
- Ice-cold sterile 10% glycerol
- Sterile TE (10 mM Ttris, pH 8.0, 1 mM EDTA) buffer
- pGLO plasmid DNA (Bio-Rad Cat. No. 166-04505EDU)
- LB broth powder (BD Biosciences Cat. No. 244620; tryptone, yeast extract, NaCl)
- Bacto agar (Difco)
- Ampicillin (100 mg/mL; filter-sterilized), L-arabinose (20%: filter-sterilized), glucose (20%: filter-sterilized)
- 70% ethanol wash bottle
- Petri plates (100 mm × 15 mm)
- 250-mL and 100-mL Erlenmeyer flasks
- Bio-Rad Gene Pulser Xcell™, electro cuvette, cuvette holder, tabletop centrifuge (set up at 4 °C), microcentrifuge (set up at 4 °C), ice box, vortexer
- Autoclave, laminar flow hood, UV transilluminator

Procedure (day 1)

Part A. Preparation of LB agar medium for *E. coli* transformation

1. Each group prepares 150 mL of LB-agar medium in a 250-mL Erlenmeyer flask (label LB-Amp/Ara), 100 mL of LB-agar medium in a separate 250-mL flask (label LB-Amp), and 50 mL of LB-agar in a 100-mL flask (label LB-Amp/Ara/Glc). To prepare LB agar medium, suspend LB broth powder in distilled water (25 g/L) and add agar (15 g/L).
2. Autoclave for 15 min at 121 °C, 15–20 psi.
3. Cool the medium to ~60 °C in the water bath.
4. Spray and thoroughly clean the laminar flow hood with 70% ethanol and paper towel.
5. Using a Sharpie pen, label LB-Amp (4 plates), LB/Amp/Ara (6 plates), and LB-Amp/Ara + Glc (2 plates) on the bottom edge of each Petri plate, and place them in the laminar flow hood.
6. Add ampicillin (100 mg/mL stock) to one flask (LB-Amp/Ara) containing 150 mL of agar medium to a final 100 μg/mL and L-arabinose (20% stock) to a final 0.25% and mix well by gently swirling the flask. *Show your calculation for ampicillin and arabinose in the pre-lab notebook.*
7. Wrap a clean paper towel around the neck of the flask, and hold it to pour 20 to 25 mL into each of 6 Petri plates labeled LB/Amp/Ara.

**The paper towel around the flask neck is to catch dribbles that flow down the outside of the flask.*

**Remove the bubbles on the agar surface or push them to the edge using a sterile pipette tip.*

**Immediately rinse out the inside of the empty flask with tap water to remove agar film after pouring.*

8. To the second flask (LB-Amp) containing 100 mL of agar medium, add ampicillin (100 mg/mL stock) to a final 100 μg/mL, mix well, and pour into 4 Petri plates labeled LB-Amp.
9. To the third flask (LB-Amp/Ara/Glc) containing 50 mL agar medium, add ampicillin (100 mg/mL stock) to a final 100 μg/mL and L-arabinose (20% stock) to a final 0.25%, and glucose

(20% stock) to a final 0.25%, mix well, and pour into 2 Petri plates labeled LB-Amp/Ara/Glc. *Show your calculation in the pre-lab notebook.*

10. Leave the lid open in the laminar flow hood until the agar medium is solidified.

Part B. Growth of *E. coli* (TA will do this part)

11. Inoculate *E. coli* HB101 cells in a 2-mL LB media and grow overnight at 37 °C in a shaking incubator at 200 rpm.

12. Next morning, dilute the overnight culture to ~1:100 (0.2 mL overnight culture + 20 mL media).

13. Grow a cell at 37 °C shaker (250 rpm) to A_{600} units of 0.4 to 0.8 (i.e., early to mid-exponential phase; typically this takes 3 to 4 hours in LB medium).

Part C. Preparation of electrocompetent cells

14. Obtain a 10-mL culture and transfer the growing cell culture to a pre-chilled sterile 15-mL conical centrifuge tube.

**From now on, it is important to keep cells and electrocuvette in ice.*

15. Centrifuge at 2500 rpm, 4 °C for 10 min in a clinical tabletop centrifuge.

16. Decant the supernatant immediately and place the tube having a loose cell pellet on ice.

17. Add 1 mL of sterile ice-cold dH_2O and pipette up and down to suspend the cell pellet using a P1000 micropipettor.

18. Transfer the suspension to a pre-chilled sterile 1.5-mL microcentrifuge tube.

19. Spin at 13 000 rpm for 1 min in 4 °C microcentrifuge.

20. Wash the cell pellet with 1 mL of sterile ice-cold dH_2O (i.e., remove the supernatant, resuspend in 1 mL of sterile ice-cold dH_2O by pipetting up and down using a P1000 micropipettor until there are no clumped cells, and spin as in step 19).

21. Repeat the above washing step 20 three more times.

22. Resuspend in 100 μL of ice-cold sterile 10% glycerol.

**Use 1/100 of the initially harvested culture volume if a different culture volume is harvested.*

23. Use 20 μL and 40 μL cell suspensions for electroporation in 0.1 cm and 0.2 cm cuvettes, respectively. Transfer the suspension to a pre-chilled tube and proceed to step 27 of Part D.

24. If not used immediately, aliquot the electrocompetent cells into sterile microtubes, freeze by placing in a dry ice/ethanol bath or in liquid N_2, and store at –70 °C until use.

Part D. Electroporation

25. Locate the Bio-Rad Gene Pulser and cuvette holder. Turn on the Bio Rad Pulser Xcell™.

- For HB101 *E. coli* cells, use 25 μFD, 200 Ω, and 1.8 kV. The time constant (tau value) should be 3–4 ms.
- Bio-Rad Gene Pulser Xcell™ conditions:
  C = 25 μFD; PC = 200 Ω; V = 1.8 kV (in 0.1 cm cuvette)
  C = 25 μFD; PC = 200 Ω; V = 2.5 kV (in 0.2 cm cuvette)
- Select "Pre-set protocol screen"
  - Select "Bacterial Protocol"
    - Select "1 mm, 1.8 kV" when using 0.1 cm cuvette or "2 mm, 2.5 kV" when using 0.2 cm cuvette.

26. Thaw the frozen electrocompetent cells in ice. Place electrocuvettes in ice.

27. Add 1 μL of pGLO plasmid DNA to 20 μL of cells and transfer the cells/DNA mixture to a pre-chilled 0.1 cm cuvette, or add the plasmid to 40 μL of cells and transfer to a pre-chilled 0.2 cm cuvette. For negative control, add 1 to 2 μL of TE buffer to 20 or 40 μL of cells.

28. Flick the cuvette to settle the mixture all the way down into the bottom of the cuvette. Keep on ice for ~5 min to ensure its coldness.

29. Wipe out the moisture outside the cold cuvette using Kimwipe. Place the cuvette in the holder and slide it into position.

30. Press and hold the "pulse" button once until you hear a beep sound.

**If you see or hear sparking coming from your cuvette, then the cells are dead! Things that can cause sparking are a warm cuvette, excess moisture on the cuvette outside, too high a salt content in the DNA sample (try diluting the DNA tenfold), or poorly washed cells.*

31. Immediately add 1 mL of LB to the electropulsed cells in the cuvette.

**The period between applying the pulse shock and adding growth medium to the electropulsed cells is crucial for recovering transformants.*

32. Using a P1000 pipettor, transfer cells from the cuvette into a sterile 17 × 100 mm tube.

33. Incubate the tube at 37 °C, 200 rpm for 1 h.

34. Save the cuvette. Wash out the cuvette with 70% ethanol, rinse with dH_2O, and drain off the water on to a paper towel.

**The cuvette can be used several times after autoclave or UV irradiation for 5 min.*

35. Plate 100 μL of the incubated cells on to each plate:

- LB-Amp agar (Experiment 1)
- LB-Amp/Ara agar (Experiment 2) × 2 plates

**Use the same sterile (alcohol-flamed and cool down) glass L-shaped spreader for plating the transformed mixture on to all the above plates.*

36. Plate 100 μL of a negative control cell on to a LB-Amp (– Control) agar plate.

**Use a fresh sterile glass L-shaped spreader for the negative control plate.*

37. Stack up your plates, tape them together, and put your group name on the tape.

38. Turn the plates upside down and incubate at 37 °C for 14 to 16 h.

**Waste disposal. When finished, dispose of any materials that had contact with the bacteria in the biohazard container. This includes tips and microcentrifuge tubes.*

**E. coli transformants should show up after overnight incubation; if not, discard the plates.*

Part E. Chemical transformation

39. Obtain a frozen 100 μL chemically competent cell in a microcentrifuge tube and add 1 μL of pGLO plasmid DNA as soon as the cells are thawed.
40. Incubate the cells/DNA mixture on ice for 30 min without disturbing the tube.
41. Transfer the tube to a 42 °C water bath, incubate for 1 min, and return to the ice box.
42. Add 0.5 mL of LB medium and incubate at 37 °C, 200 rpm for 1 h.
43. Plate 100 μL of the incubated cells on to each plate:
- LB-Amp agar (Experiment 3)
- LB-Amp/Ara agar (Experiment 4) × 2 plates

44. Plate 100 μL of negative control cell on to an LB-Amp (− Control) agar plate.
45. Incubate the plates upside down at 37 °C for 14 to 16 h.

Procedure (day 2)

1. Count the number of colonies on each plate obtained from Parts D and E. Observe your plates under regular visible light and record the colony appearance. Then observe your plates under UV light in a dark area, and note the color and the number of the colonies. Record the results.

**Wear goggles to protect your eyes from UV irradiation.*

2. Using sterile toothpicks, lightly touch the colony surface and transfer 10 well-isolated colonies from LB-Amp and LB-Amp/Ara plates to fresh plates of LB-Amp/Ara and LB-Amp/Ara+Glc prepared in Part A, as shown below. The same number of streaks come from the same colony.
3. Incubate all the plates upside down at 37 °C for 14 to 18 h. Do not overincubate.

Procedure (day 3)

1. Observe your plates under UV light and note the color of the colonies. Record and explain the results.
2. Store the streaked LB/Amp/Ara plate at 4 °C. The fluorescent colonies will be used in GFP purification (Experiment 7).
3. Incubate the plate of LB/Amp/Ara+ Glc at 37 °C for 28 to 36 h and do the same as in step 1.

Discussion

(Do not copy the discussion point. Write a paragraph in your own words.)

1. Was your transformation successful? How do you know? If not successful, what mistakes do you think you made?
2. Determine the transformation efficiency using the formula:

$$\frac{\#\,\text{Transformants}}{\mu\text{g Plasmid DNA}} = \frac{\#\,\text{Transformants}}{\text{Cell plate volume (mL)}} \times \frac{\text{Total cell volume (mL)}}{\text{Total DNA added (}\mu\text{g)}}$$

where

- Cell plate volume = fraction of the incubated cells/DNA mixture spread
- Total cell volume = total cells/DNA mixture volume after addition of LB medium
- Total DNA added = total DNA amount added to cells

Calculate the average transformation efficiency for the two plates above.
3. Were there any colonies on the LB/amp (− pGLO) plate? If there were any colonies, how would you interpret the results obtained from the + pGLO plates?
4. Does the presence of glucose affect the green fluorescent intensity of the colonies on the LB/Amp/Ara plate? If so, explain why this happens.
5. Is there any difference in the intensity of fluorescence after colonies are transferred from the initial LB-Amp/Ara and LB-Amp plates to the new plates? If so, explain.

Post-lab assignment

1. Why are the harvested cells washed with cold water several times instead of once or twice in order to prepare electrocompetent cells?
2. Why does the LB medium immediately added to the electropulsed or heat-shocked cells not contain ampicillin?
3. Is the transformation efficiency dependent on or independent of arabinose in the nutrient LB agar? Explain your choice.
4. Compare and contrast the constitutive and inducible gene expression. Show an example in each case. Explain why living

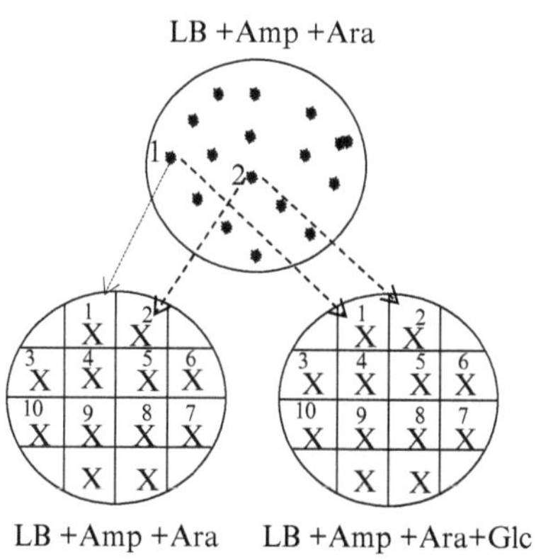

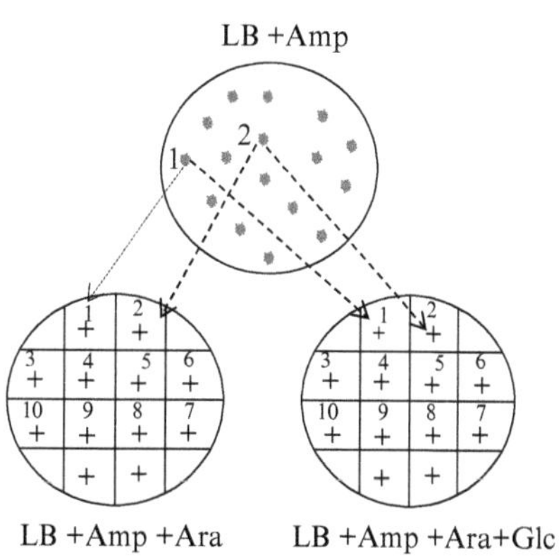

organisms have both expression systems in the regulatory gene network?

5. What is catabolite repression? Does this happen in your lab exercise?

6. Distinguish between the two following terms: (a) selection marker and screening marker; (b) vector and plasmid. Show an example in each case.

Further reading

Biotechnology Explorer™ pGLO Bacterial Transformation Kit. Bio-Rad Catalog No. 166-003EDU.

Crameri, A., Whitehorn, E.A., Tate, E., and Stemmer, W.P.C. (1996). Improved green fluorescent protein by molecular evolution using DNA shuffling. *Nature Biotechnology*, **14**: 315–319.

Zhang, X. and Schleif, R. (1998). Catabolite gene activator protein mutations affecting activity of the araBAD promoter. *Journal of Bacteriology*, **180**: 195–200.

7 MB experiment 7: GFP purification

Purpose: This is up to you to write down.

Introduction

GFP is a 26.9 kDa protein of 238 amino acids with a pI of 6.2. It is a single polypeptide chain that is tightly packed into a "β barrel" tertiary structure. It is fluorescent because it contains a fluorophore, a central portion of the molecule that fluoresces. GFP fluorescence is an autocatalytic process that requires no cofactor or enzymatic reaction. The bulk of the protein surrounds the fluorophore to protect it from water. Its ability to fluoresce comes from a remarkable reaction between three adjacent amino acids, serine, tyrosine, and glycine, at locations 65, 66, and 67 that occur spontaneously when they are folded into the right conformation. Wild-type GFP is not water soluble due to a high surface hydrophobicity. Several mutations causing a reduction in surface hydrophobicity made GFP variants become much more water soluble, along with various changes in fluorescent behavior that increases fluorescence without rapid quenching and shifts a major excitation peak from 395 nm to 488–490 nm with the emission peak staying at 507–509 nm (green). Some representative GFP variants ranging from the shortest to the longest emission spectra are: blue (BFP), cyan (CFP), green (EGFP, GFP_{uv}), and yellow (YFP). GFP_{uv} has the same excitation (395 nm) and emission (509 nm) maxima as wild-type GFP but is 16 to 18 times brighter than wild-type GFP.

Bacteria contain thousands of endogenous proteins from which GFP must be separated. Chromatography is the most popular method for separating proteins and other molecules in complex mixtures. In chromatography, a column is densely filled with microscopic spherical beads called a gel or matrix through which proteins must pass before being collected. Gels are available with a variety of different properties in size, charge, affinity, and hydrophobicity. Since overall surface hydrophobicity of GFP is higher than in most other proteins, it will bind more tightly to a hydrophobic resin than the less hydrophobic proteins in the host cell. The binding and elution of GFP protein are controlled by varying salt concentration in the column; the hydrophobic interaction force increases with high salt concentration and decreases with low salt concentration.

In this experiment, you will prepare cell lysate from the transformed *E. coli* cells and purify soluble GFP_{uv} in the supernatant using two different methods; one is hydrophobic interaction chromatography (HIC) and the other is ethanol extraction. In the course of the HIC fractionation, it is crucial to the success of purification that the column bed should not be disturbed as much as possible during the loading, binding, washing, and elution steps, while keeping the column wet or filled with buffer above it. The ethanol extraction method is very simple, yielding good purity of GFP without expensive reagents and equipment.

The unique fluorescent nature of GFP allows you to monitor all phases of the purification with a simple hand-held UV light, so you can better understand each process. You will test GFP for the sensitivity to heat, acid, and base treatment after HIC elution or organic extraction. It is recommended to watch the following YouTube videos regarding GFP:

- http://www.youtube.com/watch?v=90wpvSp4l_0

(Nobel Prize in Chemistry of GFP)

- http://www.youtube.com/watch?v=giYQAdExKdE

(GFP Purification)

Pre-lab assignment

(Typing and submission must be completed before lab work begins.)

1. Draw the flowchart diagram of the Part A experiment.

2. Draw the flowchart diagram of Parts B → D → E → F experiments.

3. Draw the flowchart diagram of the Part C experiment.

4. If GFP does not fluoresce due to the denaturation, what critical feature is lost from GFP?

Materials and equipment

- pGLO transformed bacterial colonies (Experiment 6)
- Ampicillin (100 mg/mL: filter-sterilized), L-arabinose (1.5 M: filter-sterilized)
- Lysozyme (50 mg/mL, freshly prepared in TE buffer, pH 7.5–8.0 just before use)
- Halt™ Protease Inhibitor Cocktail (Life Technologies, Cat. No. 87786; 100×)
- Hydrophobicity column [Hydrophobic Bead Resin: Macroprep Methyl HIC Bio-Rad #1580080]
- Equilibration buffer (2 M $(NH_4)_2SO_4$ in TE, pH 7.5–8.0: autoclaved)

Methods in Biotechnology, First Edition. Seung-Beom Hong, M. Bazlur Rashid and Lory Z Santiago-Vázquez.

Companion Website: www.wiley.com\go\hong\Methodsinbiotechnology

- Binding buffer (4 M $(NH_4)_2SO_4$ in TE, pH 7.5–8.0: autoclaved)
- Wash buffer (1.3 M $(NH_4)_2SO_4$ in TE, pH 7.5–8.0: autoclaved)
- Elution buffer (TE: 10 mM Tris, 1 mM EDTA, pH 7.5–8.0: autoclaved)
- Four sterile test tubes in a rack for collection of eluents
- Saturated $(NH_4)_2SO_4$ solution
- 96% ethanol
- 5 M NaCl, 1 M HCl, and 1 M NaOH
- pH paper
- n-Butanol
- Hand-held UV light lamps, UV light transilluminator
- Liquid nitrogen

Procedure

Part A. Bacterial cell growth (TA will perform this part)

1. Examine your LB/Amp and LB/Amp/Ara plates under UV light from transformation Experiment 6. Identify well-isolated green-glowing colonies by circling on the back of the LB/Amp/Ara plate. Also identify well-isolated colonies from the LB/Amp plate.
2. Obtain two sterile culture flasks: one (labeled "+ Ara") contains 20 mL of LB broth + ampicillin (100 μg/mL) + L-arabinose (1.5×10^{-3} M) and the other (labeled "– Ara") contains 10 mL of LB broth + ampicillin (100 μg/mL) only.
3. Inoculate the pGLO transformant into each culture tube using a sterile inoculation loop. To do this, touch the loop end to a circled single green colony and immerse the loop in the "+" labeled tube. Spin the loop between your index fingers and thumb to disperse the entire colony. Using a new sterile loop, repeat for a single white colony from the LB/amp plate and immerse it in the "–" labeled tube.
4. Grow overnight (12 to 16 h) at 37 °C, 250 rpm.

Part B. Bacterial cell lysis

5. Place 5 mL each of induced (+) and uninduced (–) overnight cultures in three 2.0-mL microcentrifuge tubes labeled +Ara (HIC), +Ara (EtOH), and –Ara, and centrifuge at 13 000 rpm for 1 min.
6. Discard the supernatant and observe the final pellets under an UV light transilluminator. Note any color differences that you observe.
7. Suspend the pellets completely in 0.5 mL TE buffer (pH 8.0) by pipetting up and down with the pipette.
8. Add 200 μL lysozyme. Gently mix the contents by inverting the tube several times and incubate at room temperature for 5 min.
9. Place the tube in liquid nitrogen for ~ 1 min so that the solution completely freezes (ice crystal formation helps disrupt cell walls, membranes, and intracellular vesicles), and thaw in a warm water bath.
10. As soon as the cells are thawed, add 7 μL of protease inhibitor cocktail (100×).
11. Sonicate cell suspension with a microtip at an intermediate setting for two cycles of 30-s sonication and 30-s pause on ice.

**Keep the sample in ice during sonication.*

12. Centrifuge at 13 000 rpm for 10 min at 4 °C to precipitate insoluble cellular debris. During centrifugation, prepare the chromatography column as depicted below.

Part C. Column chromatography (adapted from GFP purification kit, biotechnology explorer, instruction manual, Bio-Rad laboratories, inc.)

13. Shake one HIC column to resuspend the beads and tap the column to help settle the beads at the bottom. Attach the column to a ring stand high enough to allow for a test tube rack to place underneath the column. The column needs to be perpendicular to the workbench.
14. Remove the top cap and snap off the tab at the bottom of the hydrophobicity column. Allow the entire buffer to drain from the column (this will take ~3 to 5 min).

**Do not let the resin bed become dry at any time.*

15. Add two 1-mL aliquots of equilibration buffer to the top of the column, 1 mL at a time using your pipette. Drain the buffer from the column until it reaches just above the top of the white column bed. Close the bottom of the column until you are ready to do step 18. The column should be about 1.2 cm tall (~0.7 mL position in the column) and as level as possible.

**The top column bed should have a relatively flat upper surface. Subsequent steps of loading, washing, and eluting should minimize disrupting the column bed so that beads do not float into the buffer.*

16. After 10 min of the centrifugation of step 12, transfer ~350 μL of cell lysate supernatant to three new microcentrifuge tubes labeled "+Ara-T (HIC)", "+Ara-T (EtOH)," and "–Ara-T." Place on a UV light transilluminator and note any differences.

**Make a 100-μL aliquot of "+Ara-T" and "–Ara-T" and keep on ice. This will be used for Part E experiment.*

17. Transfer 150 μL of cell lysate supernatant of +Ara-T (HIC) tube to a new microfuge tube and add 150 μL of binding buffer to it (total volume = 300 μL). Mix gently and thoroughly by pipetting up and down to avoid foaming.
18. Remove the cap from the top and bottom of the column and let it drain completely into a liquid waste container. When the buffer meniscus just touches the top of the column matrix, place collection tube 1 (FT) under the column, and then add 300 μL of lysate supernatant in the binding buffer (step 17) on to the top of the column by resting the pipette tip against the side of the column *near the top bed surface* and letting the entire supernatant slowly drip down the side of the column wall. Do this by circling the inside of the column tube *using a Pasteur pipette*, slowly and carefully layering the sample on to the top bed surface.

**Do not drop the sample directly on to the column bed. Do not use a micropipettor for sample application because the pipette tip often falls down during this process.*

19. Let it drain into collection tube 1. Keep the FT fraction in ice.

**When the sample has completely entered the column, a green ring of fluorescence should be visible at the top of the bed when viewed with the hand-held UV light.*

20. Place collection tube 2 (W) under the column and add 750 μL (375 μL × 2) of wash buffer to the column using a Pasteur pipette. Collect the liquid and keep the W fraction in ice.

**The wash buffer will aide in the removal of weakly bound proteins.*

21. Place collection tube 3 (EB-I) under the column, add 750 μL (375 μL × 2) of TE buffer (elution buffer) slowly using a Pasteur pipette. Collect the liquid and keep the EB-I fraction in ice.
22. Monitor the green glow band using a hand-held UV light. When the green-glow band is about to reach the bottom of the column, immediately place collection tube 4 (EB-II) under the column, and keep the EB-II fraction in ice.
23. Place all four collection tubes on the UV light transilluminator and examine any differences in color between the tubes. Seal them with Parafilm, and store them at –20 °C if not used immediately.
24. Add about 3 column volumes (~4 mL) of wash buffer and drain. Add 5 column volumes (~7 mL) of sterile dH_2O and drain. Add 10 column volumes (~10 mL) of equilibration buffer and drain, add a small volume of the buffer on the column bed, and close the bottom of the column. Store the regenerated column at 4 °C.

**TA will do this regeneration step.*

Part D. GFP purification by organic extraction

25. Transfer 250 μL of cell lysate supernatant from the +Ara-T (EtOH) tube to a new microcentrifuge tube.
26. Add 0.3 volume (75 μL) of 5 M NaCl and 2.33 volume (583 μL) of saturated $(NH_4)_2SO_4$ (pH 7.8) and 1.2 volume (300 μL) of 96% ethanol.
27. Vortex for 30 s and spin at 3000 × g (5600 rpm in microcentrifuge) for 7 min.
28. Examine the tube under UV light and transfer the upper phase (~300 μL) to a new microcentrifuge tube.
29. Add 0.25 volume of n-butanol, vortex for 30 s, and spin as above in step 27.

**Do not use iso-butanol but n-butanol.*

**During the centrifugation, proceed to do Part E and F experiments.*

30. Examine the tube under UV light, remove the upper phase as much as possible, and transfer the lower phase (~25 μL) to a new microcentrifuge tube labeled GFP/EtOH and Group #.
31. Store the tubes at –20 °C until use (Experiment 8, Part B).

Part E. Heat denaturation

32. Place ~50 μL of +Ara-T (HIC) each in two separate 0.5-mL tubes, transfer one tube to a 70–75 °C water bath and the other tube to a boiling water beaker. Leave them for 5 min and cool on ice for 1 min.

**Save the heat-treated +Ara-T sample for Experiment 8, Part B.*

33. Examine under UV light and note any differences before and after heat denaturation.

Part F. Acid/base denaturation

34. Dispense ~50 μL of EB-II GFP fraction (step 22) into two 0.5-mL microcentrifuge tubes.
35. Add 2 μL drops of 1 M HCl to one tube until the pH is ~3, while checking the pH by spotting 2 μL on to pH paper. Examine under UV light, along with the untreated GFP fraction.
36. Add 2 μL drops of 1 M NaOH to one tube until the pH is ~11, while checking the pH by spotting 2 μL on to pH paper. Examine under UV light, along with the untreated GFP fraction.

Discussion

(Do not copy the discussion point. Write a paragraph in your own words.)

1. Discuss the principle of HIC to purify GFP.
2. Do you think denaturation by heat, acid, or base is helpful for GFP purification? Explain your choice. How would you use the GFP characteristics to purify this protein from a complex protein mixture of cell lysate?

Further reading

Biotechnology Explorer™ Protein Electrophoresis of GFP: A pGLO™ Bacterial Transformation Kit Extension, Instruction Manual, Bio-Rad Laboratories, Inc.

Living Colors® User Manual (Protocol No. PT2040-1, Version No. PR1Y691). Clontech Laboratories, Inc. Published on 26 November 2001.

Piston, D.W., Patterson, G.H., Lippincott-Schwartz, J., Claxton, N.S. and Davidson, M.W. (2013). *Introduction to Fluorescent Proteins.* Nikon Microscopy (https://www.microscopyu.com/articles/livecellimaging/fpintro.html).

Samarkina, O.N., Popova, A.G., Gvozdik, E.U., Chkalina, A.V., Zvyagin, I.V., Rylova, Y.V., Rudenko, N.V., Lusta, K.A., Kelmanson, I.V., Gorokhovatsky, A.U., Vinokurov, L.M. (2009). Universal and rapid method for purification of GFP-like proteins by the ethanol extraction. *Protein Expression and Purification*, **65**: 108–113.

8 MB experiment 8: SDS-PAGE analysis

Purpose: This is up to you to write down.

Introduction

Sodium dodecyl sulfate polyacrylamide gel electrophoresis (SDS-PAGE) is a standard technique for separating proteins. This gel can be made of single percentage (continuous) or two different percentages (discontinuous) of polyacrylamide or gradient percentage of polyacrylamide. In all of the gel system, SDS (detergent) and ß-mercaptoethanol (thiol reducing agent) are included for denaturing gel. SDS is a negatively charged hydrophobic molecule that binds roughly to two amino acid residues per molecule, disrupting secondary and tertiary protein structures and preventing protein shape and aggregation from influencing gel run. Hence, proteins associated with SDS have a relatively uniform charge-to-mass ratio as well as a uniform shape, which makes them migrate in an electric field toward anode (positive electrode) at speed inversely proportional to size.

The discontinuous SDS-PAGE system employs two different layers of buffer pHs and polyacrylamide percentages between the upper stacking and the lower resolving gel to generate a voltage gradient and a discontinuous pH. A lower percentage of stacking gel (pH 6.8) is poured on top of a higher percentage of resolving gel (pH 8.8). The stacking gel with a large pore size serves to concentrate all the proteins so that the large ones can catch up with the small ones, compressing into a thin starting band on top of the resolving gel. After entering the resolving gel of a smaller pore size, the individual proteins are separated according to the relative molecular size. The gradient gel has two advantages over the continuous gel; increasing the sieving effect leads to higher resolution of both lower and higher molecular weight proteins in a single gel. The percentage of polyacrylamide used in both gradient and non-gradient gels can be adjusted to maximize the resolution of target proteins. Whatever gel system is used, however, certain glycoproteins, highly acidic or highly basic proteins, and hydrophobic membrane proteins behave anomalously on SDS-PAGE gels due to the changes in the charge-to-mass ratio.

Protein samples are denatured by boiling for 5 minutes in the presence of SDS and ß-mercaptoethanol. Heat breaks ionic and hydrogen bonds; SDS disrupts hydrophobic interactions; and ß-mercaptoethanol breaks disulfide bonds. The sample buffer also contains glycerol to increase the density so that when the sample is loaded it sinks to the bottom of the well. Bromophenol blue dye is present in the sample buffer to monitor the progress of electrophoresis. For most analytical applications, mini-size gels are routinely used owing to the reduced amounts of time and materials without affecting much resolution.

In this lab exercise, you will separate protein samples of chromatography fractions and crude protein extracts obtained from Experiment 7 using vertical electrophoretic apparatus. Denatured protein samples are loaded and electrophoresed until bromophenol blue dye migrates to the bottom of the gel. The gel is stained in Coomassie blue dye that binds to protein bands, and then de-stained to remove the unbound dyes from the areas where no protein exists. Each distinct visualized band contains many protein molecules of the same molecular weight. It is recommended to watch the following video regarding SDS-PAGE:

- http://www.dnatube.com/video/4334/SDS-PAGE

Pre-lab assignment

(Typing and submission must be completed before lab work begins.)

1. Salts contaminants in a sample increase the local ionic strength in the electrophoresis. Does this increase or decrease the electric field (V/cm) force that is experienced by and needed to move sample proteins? What happens to the movement of tracking dyes and protein in the loading sample containing high amounts of salts? Explain your choice.

2. Glycine has a carboxyl group with pK_{a1} = 2.35 and an amino group with pK_{a2} = 9.78. Calculate the percentage of a glycine form with a protonated amino group (NH_4^+) at pH 6.8 and pH 8.8 buffer solutions, respectively. Show your work.

3. How do you know that a band on a gel contains the protein of your interest?

4. What do you think are possible causes when you have virtually no SDS-PAGE bands from samples despite the presence of a clear band of control protein BSA marker? What would you do to avoid this problem?

Materials and equipment

- Uninduced cell (–Ara-T) lysate, induced cell (+Ara-T) lysate, FT fraction, W fraction, EB I fraction, EB II fraction, GFP/EtOH sample (Experiment 7)
- 10× SDS running buffer
- 4× SDS-sample loading dye

Methods in Biotechnology, First Edition. Seung-Beom Hong, M. Bazlur Rashid and Lory Z Santiago-Vázquez.

Companion Website: www.wiley.com\go\hong\Methodsinbiotechnology

• PageRuler plus prestained protein ladder (Life Technologies, Cat. No. 26619)
• GelCode™ blue staining solution (Life Technologies, Cat No. 24592)
• Purified BSA (66.5 kDa; 1 mg/mL)
• Sephadex G-25 Medium (particle size 85 to 260 μm) swollen in TE buffer (pH 7.5–8.0)
• Bradford Dye Reagent Concentrate (Bio-Rad Cat. No. 500-0006 EDU)
• Spin columns (volume 1000 μL)
• TE buffer (pH 8.0)
• Safety cap locks for microcentrifuge tubes
• Mini-Protean 3 system (Bio-Rad): see Mini-PROTEAN® 3 Cell Instruction Manual (http://www.bio-rad.com/webroot/web/pdf/lsr/literature/4006157B.pdf)
• Power supply (capacity 200 V, 500 mA)
• 4-20% Mini-PROTEAN® TGX™ Precast Gel (Bio-Rad Cat. No. 456-1093SEDU)
• NanoDrop 2000 spectrophotometer
• Boiling water bath

Procedure

A. Desalting FT and W fraction

1. Mark a 0.6 mL position on the hinge side of the spin column and add a 700-μL slurry of Sephadex G-25 to each column.
2. Spin at 800 × g for 1 min and remove the flow-through in the collection tube.

**Place the column with the hinge side facing outwards in all subsequent centrifugation steps.*

**The packed bed volume should be ~0.6 mL. If not, add more slurry of Sephadex G-25.*

3. Add 400 μL of TE buffer (pH 8.0), spin as above, and remove the flow-through.
4. Repeat the above steps 2 and 3 four more times (total 5 times for column equilibration).
5. Replace the used collection tube with a new labeled 1.5-mL microtube (FT or W).
6. Apply 150 μL of each sample carefully to the middle of the packed bed.
7. Centrifuge at 800 × g for 2 min and keep the eluents in ice.

B. Protein quantitation using NanoDrop 2000 spectrophotometer (optional)

8. Mix 4 μL each of Ara-T, –Ara-T, FT, W, BSA (1 mg/mL) and BSA (0.5 mg/mL) proteins with 1800 μL of 1/50 diluted Bradford reagent, and spot 2 μL onto the lower measurement pedestal of NanoDrop 2000 to read absorbance. The blank solution is H_2O and the zero reference solution is 4 μL TE + 1800 μL of 1/50 diluted Bradford dye.

**Due to the time limitation, only one group's samples are measured, and the results are applied to all other groups' samples.*

9. Based on the concentrations of samples, dilute +Ara-T and –Ara-T protein samples only so that both samples are of the same concentration.

C. Assemble electrophoresis unit

10. Peel off the tape from the bottom of the Bio-Rad precast gel and attach the gel to the electrode assembly so that the shorter plate faces inward. Put the electrode assembly with gel into the clamping frame and secure by pushing the clamping frame cam inward. Lower the whole gel sandwich assembly into the electrophoresis chamber. For the description of each component, refer to the Mini-PROTEAN® 3 Cell Instruction Manual.

**You may use a home-made 12% discontinuous gel prepared beforehand.*

**Bio-Rad precast gels do not contain SDS.*

11. Pour 1× running buffer into the lower chamber (= outer chamber: Mini Tank in Bio-Rad Mini-PROTEAN 3 system: ~200 mL) until the level is above the bottom of the glass plate.
12. Fill the upper chamber (= inner chamber: gel sandwich + electrode assembly + clamping frame in Bio-Rad Mini-PROTEAN 3 system: ~125 mL) to the top with 1× running buffer so that sample wells are filled with buffer.
13. Thoroughly flush the sample wells with a stream of running buffer using a Pasteur pipette to clear off any polymerized and unpolymerized acrylamide gel. Check for the leakage of buffer into the lower chamber.

**If the buffer leaks, add more buffer to the upper and lower chambers.*

D. Prepare loading samples and electrophoresis

14. Transfer 12 μL each of your 7 samples to the safety-cap microfuge tubes labeled +Ara-T, –Ara-T, FT, W, EB I, EB II, and GFP/EtOH.
15. Add 3 μL of 4× SDS-sample loading dye to each tube.

**SDS in a sample loading dye will precipitate on ice. Thaw it at room temperature and vortex to mix before use.*

**Keep the protein sample on ice. Add the SDS sample dye directly to the 0 °C sample.*

**Do not leave the sample even in the SDS sample dye at room temperature before heating to 100 °C.*

16. Cap all seven tubes tightly with safety cap locks and boil for 5 min.
17. Load each sample (15 μL) into separate wells of a 4–15% pre-cast SDS-PAGE gel (or home-made 12% discontinuous gel), along with 5 μL of a protein molecular weight marker as follows:

Lane 1: protein marker (5 μL)
Lane 2: –Ara-T (12 μL) + 3 μL of 4 × SDS loading dye
Lane 3: +Ara-T (12 μL) + 3 μL of 4 × SDS loading dye
Lane 4: FT (12 μL) + 3 μL of 4 × SDS loading dye
Lane 5: W (12 μL) + 3 μL of 4 × SDS loading dye
Lane 6: EB I (12 μL) + 3 μL of 4 × SDS loading dye
Lane 7: EB II (12 μL) + 3 μL of 4 × SDS loading dye
Lane 8: GFP/EtOH (12 μL) + 3 μL of 4 × SDS loading dye
Lane 9: purified BSA (1.0μL) + 11.5 μL of TE + 3 μL of 4 × SDS loading dye

**The minimum protein loading per well (single protein band) is ~0.1 μg and the maximum protein loading per well (protein mixtures) is 20–40 μg for Coomassie staining.*

18. Connect electric lead wires, turn on the power supply, and run through the gel at 200 V until the BPB dye has moved to the bottom of the gel.

**Make sure that bubbles are produced, which indicates a good electric connection.*

E. Stain proteins in gel

19. Disconnect electric leads, pour off the running buffer, and remove the gel from the gel sandwich cassette by prying open the plates using a razor blade.
20. Wash the gel three times with dH_2O each for 10 min.
21. Discard the water and cover the gel with GelCode™ blue staining solution and shake gently at room temperature for 1 h to overnight.

**Cover the staining box with plastic wrap.*

22. Pour off the staining solution into a bottle and de-stain the gel with dH_2O. Shake gently until a clear background is obtained.
23. Take a photograph. Label each lane on the photograph, along with the MW marker.

**Plot a scatter diagram with the migration distance (mm) of MW marker bands (x axis) and $log_{10}MW$ (y axis) using Microsoft Excel. Use a linear regression to determine the best-fit line. The absolute value of correlation coefficient (r) should be greater than 0.95 for a reliable relationship. You can use the linear regression equation to determine $log_{10}MW$ of your interest bands.*

Discussion

(Do not copy the number and discussion point. Write a paragraph in your own words.)

1. Compare the separated protein patterns of SDS-PAGE between +Ara-T and −Ara-T and find what approximate size of a band is absent in −Ara-T and is present in +Ara-T.
2. Evaluate the overall success of the purification by looking at the number and intensity of protein bands in the individual lanes.
3. What is the approximate molecular weight of GFP based on the result? Does your experimentally determined molecular weight from SDS-PAGE analysis agree with the published MW (26.9 kDa)? If not, what do you think has attributed to this discrepancy?
4. Estimate the GFP protein amount by comparing the band thickness of GFP with that of BSA of known amount. Show your work.
5. Which method (HIC versus organic extraction) do you think gives a higher purity of GFP based on the SDS-PAGE result? Explain why it is.
6. What would you do to further improve the experiment to obtain a single GFP band?

Post-lab assignment

1. Several scientists who first discovered GFP and developed its variants were awarded the 2008 Nobel Prize. Today, GFP has found its broad use in almost all organisms. List at least three general applications of GFP in biological sciences.
2. List at least three other methods to independently confirm the protein molecular weight determined by SDS-PAGE. Briefly describe the principle of each method. Which method would give the most accurate information on molecular weight?
3. What other useful information than the size of proteins can be obtained from the results of SDS-PAGE? List at least three things.
4. After you have purified the target protein, you run the protein sample treated with β-mercaptoethanol along with the untreated protein sample on denaturing SDS-PAGE gel (A). In addition, you run the same untreated protein sample on non-denaturing native polyacrylamide gel (B). Based on the band patterns shown in the figures below, deduce the schematic quaternary structure including the disulfide linkage arrangement.

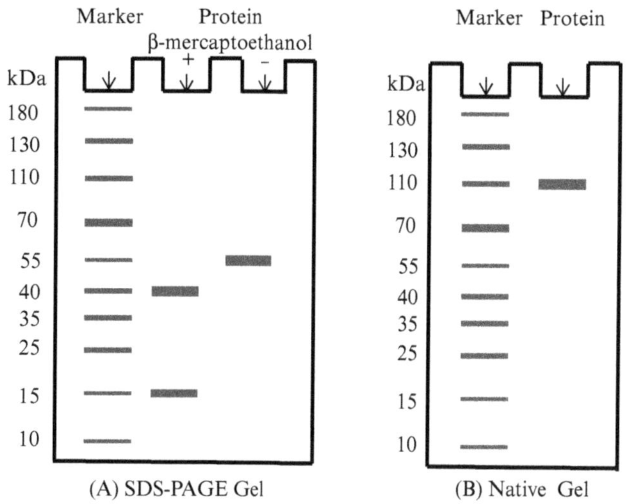

Further reading

A Guide to Polyacrylamide Gel Electrophoresis and Detection. Bio-Rad Bulletin Tech Note 6040A (http://www.bio-rad.com/webroot/web/pdf/lsr/literature/Bulletin_6040.pdf).

Bollag, D.M., Rozycki, M.D., and Edelstein, S.J. (1996). *Protein Methods*, 2nd edition. John Wiley & Sons, Inc. ISBN 0-471-118370-0.

Gallagher, S.R. (2000). One-dimensional SDS gel electrophoresis of proteins. In *Current Protocols in Molecular Biology*, pp. 10.2A.1–10A.34. John Wiley & Sons, Inc. ISBN 0-4715-0338-X.

Laemmli, U.K. (1970). Cleavage of structural proteins during the assembly of the head of bacteriophage T4. *Nature*, **227**: 680–685.

Molecular Weight Determination by SDS-PAGE. Bio-Rad Bulletin Technical Note 3133 (http://www.bio-rad.com/LifeScience/pdf/Bulletin_3133.pdf).

9 MB experiment 9: DNA isolation

Purpose: This is up to you to write down.

Part I. Plasmid DNA isolation

Introduction

Plasmid is the most popular vector DNA molecule for propagating a target gene fragment, constructing genomic or cDNA library, protein expression, PCR cloning, site-specific mutagenesis, and DNA sequencing. Therefore, plasmid isolation is a prerequisite to further manipulation and characterization of its cloned insert for the subsequent analysis and application in biotechnology. Isolation of plasmid DNA in a large and small scale is called maxi-prep and mini-prep, respectively. High purity of maxi-prep can be obtained using the cesium chloride density gradient method. However, this method is time consuming and often unnecessary for most experimental purposes such as restriction enzyme analysis, DNA ligation, transformation, PCR, and DNA sequencing because they can be routinely carried out using mini-prep DNA. The most widely used mini-prep method adopts the alkaline lysis procedure. In this protocol, bacterial cells are lysed by treatment with alkaline SDS solution. During this treatment, detergent SDS molecules disrupt the phospholipid bilayer of cell membrane to break open the cells. The released cellular proteins are denatured by SDS, while chromosomal and plasmid DNA are distorted by alkaline pH that disturbs hydrogen bonds between the bases. The denatured mixture is then neutralized with potassium acetate, rendering small plasmid DNAs to re-anneal rapidly. However, chromosomal DNA is too large to re-anneal easily unless it is broken down to smaller pieces. During the neutralization step, most chromosomal DNA, bacterial proteins, and membrane lipids precipitate and SDS forms an insoluble complex with potassium ions. Centrifugation of the neutralized lysate separates the precipitates from the supernatant that contains the re-annealed plasmid DNA. Plasmid DNA is then precipitated by addition of alcohol to the supernatant followed by centrifugation of the mixture. Despite the alkaline condition during the lysis step, a bulk of bacterial RNA molecules are present in the clarified supernatant. Alkaline incubation should not be extended because long incubation also denatures small plasmid DNA permanently. Because this protocol does not separate DNA from RNA, ribonuclease addition is incorporated either in the first cell lysis step or in the last suspension step of a nucleic acid pellet. However, this standard procedure does not eliminate or reduce polysaccharide contaminants. The level of purification required depends on the downstream applications for which DNA will be used.

In a commercial plasmid mini-prep kit, the clarified supernatant of the alkaline SDS lysates is loaded on to a mini-spin column containing a silica gel membrane that binds to dsDNA molecules under a high ionic strength condition, and the membrane-bound DNA is washed and eluted. During the washing and eluting step, salts, proteins, lipids, and polysaccharides are removed. Although the commercial kit generally yields a pure plasmid DNA from *E. coli*, it is still expensive when many DNA samples need to be prepared for a screening purpose. The yield of plasmid DNA depends on the copy number, the size of plasmid DNA, and bacterial strains. Some bacterial strains produce more polysaccharides than others. Plasmids larger than 25 kb in length are generally difficult to isolate and poor in the yield.

In this lab exercise, you will practice the alkaline lysis mini-prep procedure that works well in *E. coli*. The plasmid DNA in the clarified supernatant is concentrated by ethanol precipitation. The DNA pellet is then briefly rinsed with 70% ethanol, air-dried, and dissolved in TE buffer. Such a mini-prep DNA is often contaminated with lots of RNAs but is good enough for the screening purpose. Each group will prepare two plasmid DNA samples, one of which is treated with RNase and the other is not. It is recommended to watch the following YouTube video about plasmid DNA isolation:

- http://www.youtube.com/watch?v=R5EraRP3V-o

Pre-lab assignment

(Typing and submission must be completed before lab work begins.)

1. You have the following stock solutions: 1 M glucose, 1 M Tris-HCl (pH 8.0), 0.5 M EDTA (pH 8.0). How would you prepare 100 mL of GTE cell suspension solution using the above stocks?

2. How would you prepare 2 mL of 20 μg/mL RNase A in TE (pH 8.0) buffer from 10 mg/mL of stock RNase A solution?

3. What is size of the plasmid and chromosomal DNA to be isolated from *E. coli* in this lab exercise?

4. Step 5 of the Part I experiment below calls for "DO NOT VORTEX." What happens if you vortex?

5. Draw the flowchart diagram of the Part I experiment.

6. Draw the flowchart diagram of the Part II experiment.

Methods in Biotechnology, First Edition. Seung-Beom Hong, M. Bazlur Rashid and Lory Z Santiago-Vázquez.

Companion Website: www.wiley.com\go\hong\Methodsinbiotechnology

Materials and equipment

- Overnight culture of transformant HB101(pGLO)
- GTE cell suspension solution (50 mM Glucose, 25 mM Tris-HCl, pH 8.0, 10 mM EDTA; autoclave and store at 4 °C)
- SDS/NaOH senaturing solution (0.2 M NaOH, 1% (w/v) SDS; prepare just before use)
- Potassium acetate (~pH 4.8) neutralizing solution (11.5 mLof glacial acetic acid + 60.0 mL of 5M potassium acetate + 28.5 mL dH_2O; do not autoclave; store at 4 °C)
- TE buffer (pH 8.0)
- TE buffer (pH 8.0) containing 20 µg/mL of DNase-free RNase A
- 95% and 70% ethanol (molecular biology grade)
- 2.0-mL microcentrifuge tubes (sterile)
- Microcentrifuge

Procedure

Part I. Plasmid DNA isolation

1. Inoculate 4 mL sterile LB medium (+100 µg/mL of ampicillin) with a single transformant colony and shake-cultivate overnight at 37 °C, 200 rpm (TA will do this step).
2. Fill up two 2.0-mL microfuge tubes with overnight culture and spin at 13 000 rpm for 30 s.
3. Aspirate off the supernatant and suspend each cell pellet completely in 100 µL GTE solution by vortexing.
4. Add 200 µL of a freshly made SDS/NaOH solution to each tube, mix by tapping the tube with a finger, and incubate at room temperature for 3 to 5 min. Each group needs to make 1 mL of SDS/NaOH denaturing solution from 1 M NaOH and 10% SDS. *Show your calculations in the pre-lab notebook.*

**Do not extend the incubation. If you have many samples, you do not need to wait for 3–5 min.*

5. Add 150 µL of ice-cold potassium acetate solution to each tube, close the tube tightly, and mix by inverting 6 to 8 times (DO NOT VORTEX). Disperse white flocculent lysate materials by gentle flicking or shaking. Let it sit at room temperature for 2 min.

**If you have many samples, you do not need to wait.*

6. Spin at 13 000 rpm at 4 °C for 10 min.
7. Transfer the supernatant (~400 µL) to a fresh microcentrifuge tube.
8. Add 0.8 mL of 95% ethanol, mix by inversions, and let stand at room temperature for 2 min.
9. Spin at 13 000 rpm at room temperature for 10 min.

**Be certain to always position microcentrifuge tubes with the cap hinge facing the outside of the centrifuge.*

10. Decant the supernatant into a waste beaker and drain off the residual fluid by turning the tube upside down and tapping on a paper towel. Immediately go to the next step 11.
11. Rinse with 0.2 mL of 70% ethanol (i.e., gently layer the ethanol into the tube, immediately decant, and drain off the residual fluid by turning the tube upside down and putting on to a paper towel).

**Perform steps 10 and 11 consecutively for the same sample. Watch the pellet. It may become loose and flow down, especially for the samples processed later when you have many mini-prep samples.*

12. Air-dry the pellet in the ventilated hood.
13. Resuspend one pellet (DNA + RNA) in 50 µL of TE buffer. Resuspend the other pellet (DNA + RNA) in 50 µL of TE buffer containing 20 µg/mL of DNase-free RNase A.

**The TE buffer suspension volume depends on both the initial culture volume and the plasmid copy number. This protocol is for the microcentrifuge tube volume of E. coli cells harboring a high copy number plasmid.*

14. To facilitate suspension, incubate the tube at 37 °C and gently tap the tube frequently until the pellet disappears (~5 min).
15. Store the sample tube at 4 °C until use.

Part II. Genomic DNA isolation

Introduction

Bacteria, yeast, animal, and plant cells require different cell disruption methods. Once cells are broken open by treating with lysis buffer containing detergent, genomic DNA preparation usually involves removing proteins from the nucleic acid. Treatment with protein-digesting enzymes (proteinases) and/or extractions with the organic solvent phenol are two common methods of protein removal. The most frequently used enzyme is proteinase K, which is a very stable fungal protease that is active at high temperatures (65 °C) and in the presence of detergents and EDTA. However, its primary usefulness is that it rapidly inactivates endogenous RNases and DNases present in the cells. Phenol does not denature nucleic acids but proteins. Extracting cell lysate with buffer-saturated phenol separates the protein in the lower organic phenol phase from the polar DNA in the upper aqueous phase. The deproteinated DNA can be further purified by high-speed CsCl density centrifugation followed by dialysis and alcohol precipitation. The organic solvent extraction procedure does not remove polysaccharide contaminants, but much of the contaminants are left behind by the DNA spooling procedure. For the cell lysates containing large amounts of carbohydrates, cationic detergent cetyltrimethylammonium bromide (CTAB) is often included since it can bind to polysaccharides at high salt concentration and the CTAB–carbohydrates–proteins complex is removed by organic extraction.

In this lab exercise *E. coli* cell lysates are extracted with phenol and a small amount of alcohol is layered on top of the phenol-extracted nucleic acid solution. DNA is precipitated at the interface between the alcohol and aqueous phase. This precipitated DNA will be visible as long fibrous materials that can be spooled using a glass rod as nucleic acids adhere to SiO_2. It is recommended to watch the following YouTube video about genomic DNA extraction:

- http://www.youtube.com/watch?v=Q53NRh_KJ6w

Materials and equipment

- *E. coli* HB101 culture in LB broth medium
- TEL buffer (10 mM Tris (pH 8.0), 1 mM EDTA (pH 8.0) + 20 mg/mL of lysozyme)
- 10% (w/v) sodium dodecyl sulfate (SDS)
- 20 mg/mL of proteinase K in TE buffer
- Phenol/chloroform/isoamyl alcohol (25:24:1, v/v/v)
- CIA (chloroform/isoamyl alcohol, 24:1, v/v)
- 95% ethanol, 70% ethanol
- 3M sodium acetate pH 5.2 (adjust pH with glacial acetic acid)
- 15-mL conical centrifuge tube (sterile)
- 2.0-mL microcentrifuge tube (sterile)
- Sterile glass tube (10 × 75 mm)
- Sterile large-bore Pasteur pipette (or large-bore P1000 pipette tip)
- Sterile glass (SiO_2) rod
- Microcentrifuge, tabletop centrifuge

Procedure

Genomic DNA isolation

1. Grow 10 mL of *E. coli* culture overnight in LB broth (TA will do this step).
2. Harvest the cells in a 15-mL conical tube by centrifuging at 4000 rpm for 5 min.
3. Decant the supernatant and drain well on to a Kimwipe.
4. Suspend the pellet in 630 µL TEL buffer by repeated pipetting, transfer the cell suspension to a 2.0-mL microfuge tube, add 304 µL TEL buffer, mix well, and incubate for 10 min at 37 °C.
5. Add 60 µL of 10% SDS and 6 µL of 20 mg/mL proteinase K, mix well (DO NOT VORTEX), and incubate for 30 min at 50 °C.
6. Add an equal volume (1 mL) of phenol/chloroform/isoamyl-alcohol and mix very gently by inverting the tube several times until it becomes a homogeneous milky solution.

**Gentle mixing is important to avoid hydrodynamic shearing-induced fragmentation of very high molecular weight genomic DNA.*

7. Spin DNA/phenol mixture at 13 000 rpm for 10 min.
8. Pipette carefully off the upper aqueous phase using a sterile large-bore cut pipette tip and transfer to a new 2.0-mL microfuge tube (do not take the interface phase).

**The upper phase is very viscous, so do not insert the pipet tip deep into the solution but very slowly pipette out from a few millimeters from the surface level.*

9. Remove the interface layer, transfer to a new 2.0-mL microcentrifuge tube, add 100 µL of TE buffer and an equal volume of phenol/chloroform/isoamylalcohol, mix well, and spin at 13 000 rpm for 10 min (optional).
10. Pipette carefully off the upper aqueous phase using a sterile large-bore cut pipette tip and combine with the previous extraction phase (optional).
11. Transfer the upper aqueous phase to a sterile small glass tube (10 × 75 mm).
12. Add 5 M NaCl to a final concentration of 0.1 M; mix well.
13. Carefully layer two volumes of 95% ethanol on top of the aqueous phase by flowing down the inside wall of the tube. The alcohol should float on top and not mix with the aqueous phase. When soluble DNA comes into contact with alcohol, DNA becomes insoluble and white precipitates are visible at the interface.
14. Submerge the end of a glass rod just below the interface between the two layers and spool DNA on to a sterile glass rod (or bent Pasteur pipette with a heat-sealed end).

**A web-like mass of DNA will float at the junction of the two layers (the interface). Push a glass rod through the alcohol and slowly spin it between the fingers. Keep moving and rotating the rod in and out through the alcohol. Do not totally mix the two layers. As DNA adheres to the glass rod, gelatinous fibrous DNA becomes more compact. If you are not successful in spooling, you may transfer to microcentrifuge tubes and centrifuging at top speed for 5 min and then wash the pellet with 70% ethanol.*

15. Wash DNA by dipping the end of the spooling rod into 1 mL of 70% ethanol in the microcentrifuge tube for 10 s and air-dry.
16. Place 200 µL of TE buffer in a microcentrifuge tube and twirl the rod several times to dislodge the spooled DNA.
17. Seal the tube having the rod with parafilm or plastic wrap.
18. Let the spooled DNA rehydrate at room temperature. High molecular weight DNA can take several days to completely rehydrate and dissolve.

DNA quantification (NanoDrop 2000 and gel electrophoresis) (next week)

19. Determine the concentration by measuring the absorbance at 260 nm (DNA conc. = A_{260} × dilution factor × 50 µg/mL) and the purity by calculating the ratio of A_{260}/A_{280} (pure DNA has 1.8 to 2.0 of A_{260}/A_{280}).
20. Analyze DNA in 0.6% agarose gel to check for DNA status (TA will run the gel and take a photograph).
21. Store DNA at 4 °C for a short term and –20 °C for a long term.

Result

1. Observe the plasmid DNA pellet before and after air-dry.
2. Observe the genomic DNA after spooling.
3. Note any difference in DNA concentrations between the RNase-treated and untreated plasmid DNA.
4. Describe any unusual observations or problems.

Discussion

(Do not copy the discussion point. Write a paragraph in your own words.)

1. Did you successfully obtain plasmid DNA and spooled genomic DNA? If not, what might be the reason?
2. What are the major differences between the plasmid DNA and genomic DNA isolation procedures?

3. Describe the principle behind each step of DNA isolation including the modes of action of all the reagents used (listed in Materials).

Post-lab assignment

1. You set up a ligation reaction with a 2.8-kb plasmid and a 1.5-kb DNA fragment of your interest and then transformed the ligation mixture into competent *E. coli* cells. You obtained lots of putative transformant colonies from antibiotic selection plates. Unfortunately, the recombinant plasmid DNA after ligation has no blue/white screening marker to discriminate between the religated plasmid and the recombinant plasmid carrying the 1.5 kb DNA fragment. You prepared mini-prep plasmid DNAs from randomly selected 50 transformant colonies in order to fish out the transformants harboring the recombinant plasmid. How would you find such transformants without the use of restriction enzyme, labeled probe DNA, and PCR analysis?
2. You have a sample of purified plasmid DNA (~2 µg) in 50 µL TE buffer. To further concentrate the DNA, you added 100 µL of 95% ethanol, mixed well, incubated on ice for 10 min and centrifuged for 10 min at 4 °C but found no DNA precipitate in the bottom of the microcentrifuge tube. Why not? What should you have done to recover the DNA? Describe in detail what to do correctly.
3. For precise centrifugation conditions, certain procedures specify the relative centrifugal force (RCF) expressed in units of gravity (× g). Many microcentrifuges only have settings for speed (revolutions per minute, rpm), not RCF. Consequently, a formula for conversion is required to ensure that the appropriate setting is used in an experiment. The relationship between rpm and RCF is as follows: $g = (1.118 \times 10^{-5})\, RS^2$, where g is the relative centrifugal force, R is the radius of the rotor (from the center of the rotor to the sample) in centimeters, and S is the rpm speed of the centrifuge. If a lab protocol specifies 5000 × g for your experiment, what rpm should you use with the microcentrifuge you have in the lab? Show your calculation.
4. The isolation of genomic DNA utilizes reagents in the neutral pH range. What effects would you expect acidic or alkaline pH to have on the long fibrous DNA? Why?
5. Why is the spooling method not effective for recovering plasmid DNA?
6. Explain why EDTA is an important component of the buffer to isolate DNA?

Further reading

Birnboim, H.C. and Doly, J. (1979). A rapid alkaline extraction procedure for screening recombinant plasmid DNA. *Nucleic Acids Research*, **7** (6): 1513–1523.

Syn, C.K. and Swarup, S. (2000). A scalable protocol for the isolation of large-sized genomic DNA within an hour from several bacteria. *Analytical Biochemistry*, **278**: 86–90.

Wilson, K. (2000). Preparation of genomic DNA from bacteria. In *Current Protocols in Molecular Biology*, pp. 2.4.1–2.4.5. John Wiley & Sons, Inc. ISBN 0-471-50338-X.

10 MB experiment 10: PCR-based *Alu*-human DNA typing

Purpose: This is up to you to write down.

Introduction

Highly polymorphic (having many different forms) sequences are distributed across many regions of human chromosomes. Such sequences have been frequently used for a genetic disease marker and forensic and paternity testing analyses. Human chromosomes contain large amounts of DNA that does not code for proteins. Approximately 95% of polymorphisms in the human genome are known to be located in the non-coding regions. Thus, such polymorphic changes do not generally appear to affect the phenotypes of individuals but are likely to be passed on to their offspring. In humans, 10 to 15% of the chromosomal DNA constitutes tandemly repeated DNA in which a short nucleotide sequence is consecutively repeated numerous times.

The type of repeated DNA sequence we will be studying is the interspersed repeated DNA whose sequence occurs only once at any one location, but there are numerous copies of the same sequence on different chromosomes or in other regions of the same chromosome. Interspersed repeated DNA makes up 25 to 40% of the human genome, and the sequence we will be looking at belongs to the *Alu* family that is only found in primates.

Alu elements found in the human genome are typically about 300 bp containing a site (5′-AGCT-3′) recognized by the restriction enzyme *AluI* and are classified as short interspersed elements (SINEs). There are 14 known families of *Alu* elements, and approximately half a million *Alu* sequences interspersed throughout the haploid human genome, representing about 5.5% of the haploid genome. *Alu* is a defective retrotransposon because it does not encode reverse transcriptase and transposase enzymes that enable it to replicate and relocate. Therefore, *Alu* insertions are relatively stable once integrated; however, *Alu* may utilize the enzymes of other transposons to propagate itself to other genomic locations. The *Alu* element can integrate into the exon, intron, or intergenic region. Adjacent inverted *Alu* repeats can change mRNA by pairing and forming long stable stem-loop structures, which induces RNA editing by deaminating adenosine to inosine via an unknown mechanism. *Alu* insertion into an exon contributes to many inheritable diseases by either disrupting a gene function or a splice signal. For example, *Alu* Ya5 insertion into *BRCA1* or *BRCA2* locus is attributed to breast cancer.

In this lab exercise, each student extracts genomic DNA from his/her own cheek cells using Chelex 100 chelating resin, amplifies a short DNA sequence by the polymerase chain reaction (PCR), separates the amplicons by agarose gel electrophoresis, and analyzes allelic frequencies in the classroom student population. Exposure to both alkalinity of the Chelex suspension (pH 10–11) and boiling temperature brings about cell membrane disruption and DNA denaturation, making it possible to amplify DNA from forensic-type samples such as sperm, hair, and blood stains. The primers used in this experiment span a 292-bp region in chromosome 8. Since the *Alu* element is ~300 bp, a chromosome that lacks the *Alu* sequence will produce a 292 bp DNA fragment after PCR, while a chromosome that has a copy of the *Alu* sequence inserted into a gene will produce a 592 bp DNA fragment after PCR. It is recommended to participate in the virtual lab of DNA extraction:

- http://learn.genetics.utah.edu/content/labs/extraction/

Pre-lab assignment

(Typing and submission must be completed before lab work begins.)

1. What is a retrotransposon?
2. Explain the function of each component in the PCR setup in Part B.
3. Calculate the final concentrations of A25 primers, dNTP, and *Taq* enzyme in the Part B reaction setup.
4. What is the advantage of the use of a master mix in the PCR setup?
5. Describe the three main steps of the PCR amplification cycle. What happens to each step?

Materials and equipment

- Sterile saline (0.9% NaCl) solution
- 10% Chelex® suspension (10 mL): weigh out 1 g of Chelex 100 (100–200 mesh, sodium form), add 9 mL of 50 mM Tris-HCl (pH 8.0) and store at 4 °C.
- 2 mL screw-cap microcentrifuge tubes, sterile cotton swabs, thin-walled PCR tubes
- GoTaq® DNA polymerase (Promega), 5× PCR reaction buffer
- Forward A25 primer (10 μM): 5′-TATAATATGGCCTGGATTATA CCTGTGTTG-3′

Methods in Biotechnology, First Edition. Seung-Beom Hong, M. Bazlur Rashid and Lory Z Santiago-Vázquez.

Companion Website: www.wiley.com\go\hong\Methodsinbiotechnology

- Reverse A25 primer (10 μM): 5′-CCACAAATAGGCTCATGTA GAACTACAG-3′
- Horizontal mini-gel electrophoresis apparatus, power supply
- 10× lithium borate buffer (100 mM, pH 8.2)
- 10 mg/mL ethidium bromide
- 6× DNA sample loading dye (0.25% bromphenol blue, 0.25% xylene cynol, 30% glycrol in 1 M Tris-HCl, pH 8.0)
- Agarose
- DNA size marker (Quick-Load 50 bp DNA Ladder: NEB Cat. No. NO473S)
- Thermal cycler, boiling water beaker, floating microtube rack, ice bucket, microcentrifuge
- UV transilluminator/digital camera system

Procedure

Part A. Isolation of human DNA

1. Label a 2-mL screw-cap microcentrifuge tube containing 1.5 mL of saline solution with your name.
2. Use a sterile cotton swab to rub and twirl the inside of both of your cheeks, between the gum line and under your tongue. Alternatively, using a sterile pipet tip, gently scrape the inside of both cheeks 10 times each with the tip. You should see a small volume of white cells in the pipet tip.
3. Submerge the swab in the saline tube and twist vigorously for 30 s to detach the cells. Press the cotton head against the wall of the tube to squeeze out as much liquid as possible. Repeat if necessary.

**Your solution should be cloudy to get a detectable DNA band.*

4. Place your sample tube, along with other student sample tubes, in a balanced configuration in a microcentrifuge, and spin at 13 000 rpm for 90 s.
5. Pipet off the supernatant. Be careful not to disturb the cell pellet.

**Check to see that the supernatant is clear and that there is a visible white pellet in the bottom of the tube. If the supernatant is cloudy with little or no pellet, spin the tube(s) for an additional 1 min.*

6. Add 100 μL of Chelex to the tube containing the pellet.

**Chelex beads settle quickly. Therefore, before measuring 100 μL, cut off the pipet tip end to a few millimeters to make a larger opening and draw the Chelex in and out of the pipet tip several times to resuspend and take it.*

7. Resuspend the cells by pipetting up and down several times. Examine in front of a light source to check that no visible cell clumps remain.
8. Screw cap tight and place your sample tube on a floating microtube rack in a boiling water bath for 10 min.
9. Cool on ice for 30 s. Spin at 13 000 rpm for 2 min.
10. Transfer 5 μL of your denatured DNA to your name labeled thin-walled PCR tube.

**It is very important that no Chelex beads and cell debris are in the supernatant. Hold up the tube to a light source to look for any beads. Any carryover of Chelex to the PCR will not yield results.*

11. Place your sample on ice while you set up the PCR.

Part B. DNA amplification by PCR

12. Prepare the master mix for each 4-member group as follows (all volumes μL):

Reagent	– Control	Samples 1 to 4	Master mix (×5.5)
Nuclease-free H_2O	13.3 (+ 5.0*)	11.7	64.35
Green 5× GoTaq® reaction buffer	5.0	5.0	27.50
10 μM F-A25	0.5	1.3	7.15
10 μM R-A25	0.5	1.3	7.15
10 mM dNTPs	0.5	0.5	2.75
DNA	–*	5.0	–
Taq enzyme (5 U/μL)	0.2	0.2	1.10†
Total volume (μL)	25	25	110.00

*Add 5.0 μL of 50 mM tris-HCl (pH 8.0) instead of DNA.
†Pipette carefully and slowly because the high content (50%) of glycerol in the storage buffer may lead to pipetting errors.

13. Add 20 μL of the master mix and mix by pipetting up and down; try to avoid making bubbles. Keep on ice until PCR cycling is ready.

**At room temperature, primers bind non-specifically to your denatured DNA and Taq polymerase is still active even at room temperature, albeit at a slower rate. Once a few non-specific DNAs are synthesized, its T_m is high enough to re-anneal at high annealing temperatures, giving rise to non-specific amplification bands.*

**To mix, set the P20 pipettor to 10 μL, immerse the pipette tip fully into the solution, and gently pipette the solution up and down 3 to 4 times. Do not take out the pipette tip during mixing.*

14. Add one drop (~20 μL) of mineral oil to the tube. (This is unnecessary in a thermal cycler with a heated lid.)
15. Program the thermal cycler using the following parameters:

Cycle steps		Temperature (°C)	Time (min)
Initial denaturation		94	3
30 Cycles	Denaturation	94	0.5
	Annealing	51	0.5
	Extension	72	0.65
Final extension		72	10
Hold		8	∞

16. Begin the program cycle. Once the initial denaturation temperature reaches 90 °C, quickly spin your sample tube and load in the thermal cycler.
17. During the PCR cycle, prepare agarose gel (Part C) and set up the restriction enzyme digestion reaction (Part A of Experiment 11).

Part C. Preparation of 1.5% agarose gel

18. Add 0.9 g of agarose solution in a 250-mL Erlenmeyer flask.
19. Add 60 mL of 1× lithium borate buffer and swirl.

20. Microwave to boil: KEEP WATCHING! As soon as it boils, stop microwaving immediately, swirl solution gently, wait for ~1 min, and repeat this process until all the clear gel particles have disappeared (~5 min).

CAUTION: *hot molten agarose steam may escape when the flask is swirled and cause severe burns. Wear protective heat-resistant gloves.*

21. Cool agarose solution to ~60 °C.
22. Add ethidium bromide (EtBr) to a final concentration of 0.5 µg/mL.

CAUTION: *EtBr is a mutagen and must be handled as a hazardous chemical. Wear gloves while handling the gel.*

23. Place a gel casting tray into the horizontal electrophoresis chamber so that the rubber tubes at both ends of the tray can be fastened down to prevent the agarose solution from leaking.
24. Pour agarose solution to a thickness of about 1 cm. Get rid of any bubbles with a pipette tip if there are any, or push to the edge.
25. Insert a 10-well comb into the top notch of a casting tray and wait for 10 to 15 min for the gel solution until it becomes solidified. It is translucent once hardened.

**You may pour the gel at 4 °C (cold room or refrigerator) to speed up the hardening time.*

Part D. Loading samples and electrophoresis

26. Place the solidified gel in an electrophoresis chamber such that the sample wells are positioned toward a negative (black) electrode.
27. Add 1× lithium borate buffer, enough to cover the gel, and remove the comb carefully.
28. Add 4 µL of 6× loading dye to the PCR tube (25 µL reaction sample).

**If Green GoTaq® buffer is used for the PCR, do not add dye but use directly.*

29. Load 10-µL samples into wells, along with a DNA size marker (Lane 1).

**Use your free hand finger to help stabilize the pipette while you are ejecting a sample.*

**Do not insert the tip too deep into the well or you might puncture it.*

30. Put the lid on the gel electrophoresis chamber and plug black (negative) and red (positive) leads in the power pack.
31. Electrophorese at 300 V for ~30 min. Turn off the power supply.

Part E. Viewing and photography

32. Remove the gel from the tray and place in a plastic tray.
33. Place the gel on the UV transilluminator and align the gel in the center under the camera.
34. Turn the UV transilluminator on.
35. On the computer desktop, click on "PC Image" icon.
36. On File (upper toolbar), click on "Acquire Multiple Image."
37. Click on "Capture" on the bottom right-hand side of the screen.
38. Click on "Expose Series."
39. Select the best picture to save (the other can be closed without saving).
40. Save the picture file in the "Local Disk C" folder.
41. Click on "Print."
42. Turn the transilluminator off and wipe it off with a paper towel and wash bottle.
43. Dispose of the gel and contaminated gloves into a biohazard container.

Result

1. Label the photograph (lane numbers, marker band sizes, date).
2. Estimate the approximate size of individual bands.
3. Indicate which band is a + or – allele.
4. Record the number of each genotype (+/+, +/–, –/–) in all the students of all sections. What is my genotype based on the result?

Discussion

(Do not copy the discussion point. Write a paragraph in your own words.)

1. How and why does the + allele differ from the – allele in terms of band size and band intensity?
2. Did you see any additional fuzzy band (less than 50 bp) lower on the gel? If so, what are they?
3. Did you see any additional faint or strong bands? If so, what are they?
4. Calculate genotype and allele frequencies in your class. All groups need to share their data to fill out the chart below.

Genotype	Number of students	Observed genotype frequency*	Number of + allele	Number of – allele
+/+				0
+/–				
–/–			0	
Total		–		
		Allele frequency†		

*Genotype frequency = # students of each genotype/total # students.
†Allele frequency = # of each allele/total alleles.

An allele frequency is the ratio of the number of copies of a particular allele to the total number of alleles within a population. Imagine a class of 100 students with their genotype distribution as follows:

$$+/+\ 20,\quad +/-\ 50,\quad -/-\ 30$$

Since humans are diploid, the total number of alleles in the class is $2 \times 100 = 200$. The allele frequency for + is:

$$2 \times 20\,(\text{homozygotes}) + 50\,(\text{heterozygotes})/200 = 90/200 = 0.45$$

and the frequency for – is:

$$2 \times 30\,(\text{homozygotes}) + 50\,(\text{heterozygotes})/200 = 110/200 = 0.55$$

Therefore,

$$\text{"+" allele plus "−" allele} = 0.45 + 0.55 = 1$$

Using the genotypic distribution for your class, calculate the frequencies of the + and – alleles.

5. If the allele frequencies resulting from Mendelian inheritance of sexual reproduction remain constant generation after generation, such a genetically stable genotype frequency can be mathematically expressed as the Hardy–Weinberg equation, which is based on the assumption that the population undergoes no net change in allele frequencies over time. Once the allele frequencies are determined from the existing population, the distribution of expected genotypes can be described by:

$$p^2 + 2pq + q^2 = 1$$
$$+/+ \quad +/- \quad -/-$$

where p and q represent the frequencies of alleles; p^2 and q^2 are the homozygote frequencies; $2pq$ is the heterozygote frequency; and $p + q = 1$. Use the allele frequencies observed in your class to determine the expected genotype frequencies.

Make + allele = p, allele = q, and +/− = pq in the equation:

Expected " + / + " genotype frequency:

p^2 = ________

Expected " + / − " genotype frequency:

$2pq$ = ________

Expected " − / − " genotype frequency:

q^2 = ________

Use the table below to calculate how many students in your class should have each genotype.

Genotype	Expected genotype frequency	Total number of students in class	Expected number of students with ppecific genotype*
+/+			
+/−			
−/−			

*Expected # of students with specific genotype = expected genotype frequency × total number of students.

6. Are the observed genotypes frequencies significantly similar to or significantly different from the expected genotypes frequencies calculated from the Hardy–Weinberg equation?

To determine the statistical significance of the "goodness of fit" between observed and expected data, use the chi (χ^2)-square test as follows:

(a) Calculate χ^2 using the formula. Complete all calculations to three significant digits. Round off your answer to two significant digits.

$$\chi^2 = \Sigma[(O{-}E)^2/E]$$

where

O = observed number in a variable
E = expected number of a variable
Σ = sum of individual χ^2 values

(b) Calculate the p value by looking up the online chi-square calculator (http://www.fourmilab.ch/rpkp/experiments/analysis/chiCalc.html) or by using the chi square table below.

- Determine degree of freedom (DF) = the number of genotypes – the number of alleles. In this lab exercise, DF = 3 – 2 = 1.
- Locate the value closest to your calculated χ^2 on that DF row.
- Move up the column to determine the p value.

DF \ p value	0.95	0.9	0.7	0.5	0.3	0.2	0.1	0.05	0.01	0.001
1	0.004	0.016	0.15	0.46	1.07	1.64	2.71	3.84	6.64	10.83
2	0.1	0.21	0.71	1.39	2.41	3.22	4.61	5.99	9.21	13.82
3	0.35	0.58	1.42	2.37	3.67	4.64	6.25	7.82	11.35	16.27
4	0.71	1.06	2.20	3.36	4.88	5.99	7.78	9.49	13.28	18.47
5	1.15	1.61	3.00	4.35	6.06	7.29	9.24	11.07	15.09	20.52
6	1.64	2.20	3.83	5.35	7.23	8.56	10.65	12.59	16.81	22.46
7	2.17	2.83	4.67	6.35	8.38	9.8	12.02	14.07	18.48	24.32
8	2.73	3.49	5.53	7.34	9.52	11.03	13.36	15.51	20.09	26.13
9	3.33	4.17	6.39	8.34	10.66	12.24	14.68	16.92	21.67	27.88
10	3.94	4.87	7.27	9.34	11.78	13.44	15.99	18.31	23.21	29.59
	Non-significant; do not reject							Significant; reject		

χ^2 value

(c) State your conclusion in terms of your hypothesis.

- If the p value for the calculated χ^2 is larger than 0.05 (i.e., the calculated χ^2 is smaller than the χ^2 critical value of $p = 0.05$ at DF), data are close enough to the expected result; there is no statistically significant difference between the observed and expected values. The deviation is small enough that chance alone accounts for it. A p value of 0.6, for example, means that there is a 60% probability that any differences between the observed and expected values are due to chance only. Therefore, we accept the null hypothesis that the population is in Hardy–Weinberg equilibrium.
- If the p value for the calculated χ^2 is less than 0.05 (i.e., the calculated χ^2 is larger than the χ^2 critical value of $p = 0.05$ at DF), reject the experimental null hypothesis (data are not close enough to the expected result: significant difference), and conclude that some factor other than chance is operating for the deviation to be so great. For example, a p value of 0.01 means that there is only a 1% probability that this deviation is due to chance alone. Therefore, other factors must be involved in causing a significant difference.

7. If they are significantly different, what might cause the difference?

Post-lab assignment

1. Why did we use a 0.9% saline rinse to remove cheek cells? What would happen if we used just water?

2. Why did we add Chelex® and subsequently heat the cheek cell mixture for 10 min at 99 °C?

3. Cheek samples we used may be contaminated with lots of food particles and bacteria. Does this affect the result or not? Explain your choice.

4. Perform "Electronic PCR" by initiating a BLAST (Basic Local Alignment Search Tool) search to find DNA sequences in the database using the primers used in this experiment:

F(orward)-A25: 5′-TATAATATGGCCTGGATTATACCTGTGT TG-3′

R(everse)-A25: 5′-CCACAAATAGGCTCATGTAGAACTACA G-3′

Provide the answers to all the questions indicated by bold letter sentences.

4.1. • Copy the forward A25 primer sequence.
- Open the Internet site of NCBI BLAST (http://blast.ncbi .nlm.nih.gov/Blast.cgi).
- Click "nucleotide blast."
- Paste in the search window under "Enter Query Sequence."
- Select "Others (nr, etc.)" under "Choose Search Set."
- Click on BLAST:
  - Query sequences are sent to a server and BLAST algorithm will attempt to match the primer sequences to the millions of DNA sequences stored in the database. This may take a few seconds or more than a minute depending on the number of queued searches at the server.
  - BLAST search results are displayed in three ways: Graphical Summary, Descriptions (a list of sequences producing significant alignments), and Alignments (a detailed view of the query sequence (primer sequence) aligned to the subject sequence (database nucleotide sequence of the search hit).
  - In the Descriptions, notice the scores in the E-value column on the right. The E (Expectation) value is the number of alignments with the query sequence that would be expected to occur by chance in the database. Therefore, the lower the E value, the higher the probability that the hit is related to the query.

What does the E value of 5e-07 mean?

4.2. Click the sequence name with 100% Query cover in Description. This will show the display of sequence alignment between the query and subject sequences. Choose only Query 1 alignment that shows the best match, copy, and paste the alignment into a text document in a Courier New font size 10 and note the lowest nucleotide position in the subject sequence.

Show this sequence alignment.

4.3. Repeat steps 4.1 and 4.2 using the reverse A25 primer sequence. Note the highest nucleotide position in the same subject accession number sequence as the one noted by the BLAST hit of the forward A25 primer sequence.

Show this sequence alignment.

4.4. • Click Sequence ID: (denoted by letters followed by numbers).
- Click the "Change region shown ▾" menu on the upper right side.
- Click "selected region from_begin_ to _end_". Type the lowest (begin) and highest (end) nucleotide positions of the subject sequence based on the blast hit of forward and reverse primers query searches of steps 4.2 and 4.3.
- Click "Update View". The sequence between the forward and reverse primers will be displayed.

Copy and paste this sequence into a text document. Underline the nucleotide sequences of both forward and reverse primers.

4.5. • Return to the NCBI BLAST page. Click "nucleotide blast" and paste the entire sequence of the A25 amplicon into the search window of "Enter Query Sequence."
- Select "Human genomic plus transcript" under "Choose Search Set" and click on BLAST.
- Scroll down to Alignments and see the Features. Based on the description of all the Features of 100% alignments, answer the following question.

Where is the *Alu* A25 element located?
- **Copy and paste all two BLAST alignments into a Microsoft document as a Courier New font size 8, remove space before a paragraph, and note the position range of the subject sequence.**

4.6. • Click the "Map Viewer" of "Related Information" on the right side in the BLAST alignment. This will display the "Master Map: Contig." Red marks (■) indicate the "Blast Q1
hit" position to the assemblies of the human genome sequence (Contig), Genes seq, and RNA seq.
- Click "Map2: RefSeq Transcripts On Sequence Table View" under the Summary of Maps, which are shown below the Master Map. This will display the UTRs, exons, and introns.
- Repeat the above procedures for the other alignments.

(a) In which location is A25 *Alu* inserted on the basis of the start and stop positions of the transcript and BLAST alignment position range of the subject sequence (step 4.5)? Does your work agree with the NCBI described features of the BLAST alignment? Show your work.

(b) What is the expected phenotype caused by the *Alu* insertion based on the BLAST hit position? Explain why.

(c) Explain the difference between mRNA, CDS, exon, and ORF.

5. Distinguish between the following terminologies that are widely used in the scientific community.
- Positive and negative strands
- Plus and minus strands
- Coding and non-coding strands
- Sense and nonsense strands
- Template and non-template strands
- Forward and reverse strands
- Top and bottom strands
- Watson and Crick strands
- Complementary and reverse complementary strands

Which ones are identical to each other among them? Which ones are neutral to become either strand?

Further reading

Arcot, S.S., Fontius, J.J., Deininger, P.L., and Batzer, M.A. (1995). Identification and analysis of a "young" polymorphic *Alu* element. *Biochimica Biophysica Acta*, **1263**: 99–102.

Daniel, C., Silberberg, G., Behm, M., and Ohman, M. (2014). Alu elements shape the primate transcriptome by *cis*-regulation of RNA editing. *Genome Biology*, **15**: R28 (http://genomebiology.com/2014/15/2/R28).

Dehininger, P. (2011). *Alu* elements: know the SINEs. *Genome Biology*, **12**: 236 (http://genomebiology.com/2011/12/12/236).

Walsh, P.S., Metzger, D.A., and Higuchi, R. (2013). Chelex 100 as a medium for simple extraction of DNA for PCR-based typing from forensic material. *Biotechniques*, **54** (3): 134–139.

11 MB experiment 11: Restriction enzyme digestion

Purpose: This is up to you to write down.

Introduction

Many restriction enzymes have been isolated from diverse bacteria that use them as a defense against invading viral DNA. Among them, type II restriction endonuclease is useful since it recognizes short specific DNA sequences that are typically characterized by dyad symmetry (or palindromic sequence) and cleaves double stranded (ds) DNA phosphate backbone within the same recognition sequences. The resulting fragments have a blunt end or sticky ends with 3′-overhang or 5′-overhang depending on the enzymes. Such site-specific cleavages are very useful for the downstream recombinant DNA applications such as subcloning of fragments in vectors, probe DNA labeling for hybridization analysis, physical DNA mapping, and so on.

For each restriction enzyme, optimal reaction conditions are given by the manufacturer. Enzymes must be stored at −20 °C and should be carefully maintained on ice during the experiment. Manufacturers provide a panel of four 10× buffer systems for use with restriction enzymes. However, many restriction enzymes have been genetically engineered so that enzymes can work in a single universal buffer and shorter incubation times for digestion.

After the restriction digestion reaction, the reaction sample is mixed with loading tracking dye, and the mixture is electrophoresed in agarose gel to separate the DNA fragments. In addition, a DNA marker or DNA ladder must be run on the same gel, alongside the restriction digest samples, in order to determine the approximate sizes of the DNA fragments. After gel electrophoresis, DNA can be visualized using a fluorescent dye ethidium bromide, SYBR green dye, or methylene blue (0.025%). The DNA/ethidium bromide complex strongly absorbs UV light at 300 nm, excites the ethidium molecule which re-emits visible light at about 590 nm, and glows a bright orange-red.

In this lab exercise, the mini-prep pGLO plasmid DNA you isolated from transformant *E. coli* cells (Experiment 9) will be used for the restriction analysis. It is important to confirm that your putative transformants do indeed contain the transforming plasmid. The actual cleavage patterns and fragment sizes generated by restriction enzymes should agree with those expected from the restriction map of transforming plasmid. Once verified by restriction analysis, the transformant bacteria carrying the plasmid should be selected, grown, resuspended in 25% glycerol, and stored at − 80 °C for the permanent stock culture, though this step is not practiced in this lab.

Pre-lab assignment

(Typing and submission must be completed before lab work begins.)

1. What do 3′-overhang and 5′-overhang mean? Draw and indicate each example.
2. Why do we graph the log of each DNA fragment? Why do we not just graph the length of the fragment in base pairs?
3. What is the purpose of adding a tracking dye to the DNA sample?
4. What is a marker DNA fragment and why is it important in gel electrophoresis?

Materials and equipment

- HF restriction enzymes and 10× buffers (NEB)
- Plasmid pGLO, *E.coli* genome (obtained from Experiment 9)
- 10× lithium borate buffer (100 mM, pH 8.2)
- 10 mg/mL of ethidium bromide
- 10× DNA sample loading dye (0.25% bromphenol blue, 0.25% xylene cynol, 30% glycrol in 1 M Tris-HCl, pH 8.0)
- DNA size marker (λ DNA/*Hin*dIII + *Eco*RI Thermo Scientific or 1-kb DNA ladder NEB)
- Horizontal mini-gel electrophoresis apparatus, power supply
- 37 °C water bath

Procedure

Restriction digestion setup

1. Label nine sterile microcentrifuge tubes by numbering.
2. Pipet reagents into *the bottom* of each tube according to the table below, which shows the ingredients for the restriction enzyme digest (all volumes in μL).

Methods in Biotechnology, First Edition. Seung-Beom Hong, M. Bazlur Rashid and Lory Z Santiago-Vázquez.

Companion Website: www.wiley.com\go\hong\Methodsinbiotechnology

Tube	1	2	3	4	5	6	7	8	9
Component	**Uncut**	**RI**	**HIII**	**P**	**RV**	**RI+RV**	**P+RV**	**Uncut**	**RI**
Nuclease-free H_2O	13	11	11	11	11	10	10	8	7
10x NEB buffer 4	2	2	2	2	2	2	2	2	2
pGLO DNA	5	5	5	5	5	5	5	–	–
E. coli genome	–	–	–	–	–	–	–	10	10
*Eco*RI-HF (RI)	–	1	–	–	–	1	–		1
*Hind*III-HF (HIII)	–	–	1	–	–	–	–	–	–
*Pst*I-HF (P)	–	–	–	1	–	–	1	–	–
*Eco*RV-HF (RV)	–	–	–	–	1	1	1	–	–
Total volume (µL)	20	20	20	20	20	20	20	20	20

*Always keep the enzyme tubes in an ice bath and put them back in the −20 °C freezer as soon as you are finished with them.

- Add water first to all of the tubes using the same pipette tip.
- Add buffer second to all of the tubes using the same pipette tip.
- Addition of DNA and enzyme requires the use of a new, clean pipette tip.
- Add enzyme last. Please keep them on ice. Before you take 1 µL of enzyme, check the 1 µL volume in your pipette tip by taking 1 µL of dH_2O.
- DO NOT VORTEX to mix the content; mix gently by pipetting up and down three or four times while immersing the tip in the solution using a P20 micropipettor set at 10 µL volume. Do not take the pipette tip out of the solution during mixing.
- DO NOT MAKE A BUBBLE OR FOAM DURING PIPETTING.

Note: Restriction enzymes are stored in a 50% glycerol solution. Glycerol tends to stick to the outside of the pipette tip. If there is an excess material on the tip, gently wipe the tip on the inside rim of the tube. The best way to avoid this is to insert the pipette tip just below the top edge of the enzyme solution and slowly release the plunger button of the pipettor. Do not press your pipettor plunger button after inserting the pipette tip into the solution.

3. Place all the reaction tubes in a 37 °C water bath for at least 15 min.

12 MB experiment 12: Agarose gel electrophoresis

Introduction

DNA is a highly negatively charged molecule because of its phosphate backbone. Accordingly, DNA molecules are attracted toward the positive pole (anode) when placed in an electrical field. A porous agarose gel acts as a molecular sieve to sort DNA fragments by size. Because of the constant charge-to-mass ratio, DNA mobility in agarose gel is inversely proportional to the logarithm of size. Optimal separation of DNA fragments can be achieved by adjusting the concentration of agarose in the gel. A relatively low concentration of agarose matrix produces a more porous loose gel, which separates large fragments more effectively, as shown in the figure below. On the other hand, a high concentration produces a less porous rigid gel, which resolves small fragments more effectively. The table in the figure below gives the approximate concentrations of agarose to separate the indicated size ranges of linear DNA fragments.

Agarose (%)	Linear DNA (kb)
0.3	40 − 5.0
0.6	20 − 1.0
0.7	10 − 0.8
0.9	7 − 0.5
1.2	6 − 0.4
1.5	4 − 0.2
2.0	3 − 0.1

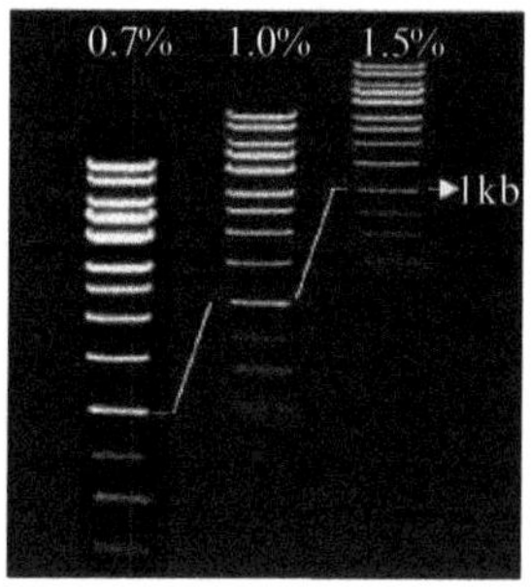

Reproduced from a 1 kb DNA Ladder (NEB) visualized by ethidium bromide staining. Mass values are for a 0.5 μg marker DNA.

The strength of electrical field also influences the resolution of DNA fragments. Two most widely used running buffers are TBE (Tris-borate-EDTA) and TAE (Tris-acetate-EDTA). TBE has a higher buffer capacity than TAE. However, TAE is the buffer of choice when DNA is to be extracted from the electrophoresed gel for a cloning experiment because borate is a strong inhibitor of many enzymes. Agarose gel must be prepared in the same ionic strength and the same type of buffer as the electrophoresis running buffer. In general, large DNA fragments are best resolved by running a lower percentage gel at a lower voltage for long times (1–5 V/cm, 12–16 h), whereas smaller fragments diffuse faster and therefore band sharpness is increased by running a higher percentage gel at a higher voltage. Use of Bionic™ buffer allows for fast agarose gel electrophoresis at 30 V/cm for 10–30 min without sacrificing DNA band resolution, as compared with 7--10 V/cm in TAE or TBE buffer. Agarose gels can separate DNA molecules ranging from 100 bp to about 20 kb, though it is difficult to resolve more than 30–40 kb and less than 100 bp DNA fragments. However, it is possible to separate DNA fragments ranging in size from 5 to 6000 kb by pulse-field gel electrophoresis and to resolve 1 bp difference in small DNA fragments by polyacrylamide-based DNA sequencing gel. It is recommended to participate in the following virtual lab:

- http://learn.genetics.utah.edu/content/labs/gel/

Procedure

Agarose gel electrophoresis of restriction enzyme digestion samples

1. Prepare a 60 mL of 0.5% agarose gel in 1× lithium borate buffer, pour into a mini-gel mould tray (see Experiment 10, protocol Part C), and solidify the gel in the refrigerator (4 °C).
2. Add 2 μL of 10× loading dye to each restriction digestion tube (20 μL).
3. Load the sample in the order of: lane 1, DNA marker (10 μL λ DNA/*Hin*dIII + *Eco*RI or 5 μL of 1-kb DNA ladder); lane 2, uncut pGLO; lane 3, pGLO/*Eco*RI; lane 4, pGLO/*Hin*dIII; lane 5, pGLO/*Pst*I; lane 6, pGLO/*Eco*RI+*Eco*RV; lane 7, *Pst*I + *Eco*RV; lane 8, uncut *E. coli* genome; lane 9, *E. coli* genome/*Eco*RI.
4. Electrophorese at 300 V for 30 min.
5. Take a photograph of the gel, along with a ruler placed on the side of the DNA MW marker.
6. Label all visible bands in the DNA ladder, indicate the gel lanes containing samples, and measure the distance from the bottom of the well to the bottom of each band.
7. Prepare a standard curve of with $\log_{10}$bp (y axis) versus distance migrated in cm (x axis) using marker DNA bands.

**If you have not measured the distances directly from the gel, measure the migration distance (cm) from the top well to each band of molecular weight marker DNA fragments on the photograph using a ruler. You may use the enlarged picture.*

Methods in Biotechnology, First Edition. Seung-Beom Hong, M. Bazlur Rashid and Lory Z Santiago-Vázquez.

Companion Website: www.wiley.com\go\hong\Methodsinbiotechnology

**If you use semi-log graph paper for this plot, you do not need to calculate the $\log_{10}$ (MW); simply plot the MW(bp) on the log scale y axis.*

8. Determine the apparent base pair length of the restriction fragments using the standard calibration curve.

Result

1. Show your work table for the distances of fragments of MW marker and digested pGLO DNA, plot the standard curve, and estimate the approximate size of fragments of pGLO DNA.
2. Fill in the following table based on the result.

pGLO sample	*Expected number of fragments	*Expected size of fragments	Actual number of fragments	Estimated size of fragments
pGLO/*Eco*RI				
pGLO/*Hin*dIII				
pGLO/*Pst*I				
pGLO/*Eco*RV				
pGLO/RI+RV				
pGLO/P+RV				

*Obtain this information from the post-lab assignment work.

Discussion

(Do not copy the number and discussion point. Write a paragraph in your own words.)

1. Explain the cleavage patterns at each lane. Do your predicted results match your actual results in terms of the number and size of the restriction fragments? If your results do not agree with your expected results, give a possible explanation as to why this might be so.
2. Did you see any high background intensity color on the top portion of gel? If so, what causes this background and why is the background displayed only in the top portion of gel?
3. Does your selected transformant *E. coli* cell harbor the pGLO plasmid on the basis of your result? Explain the reason.
4. Where is the expected position of the RNA band in your gel if undegraded RNAs are present in your DNA sample?
5. How many bands do you expect to see in an uncut pGLO DNA sample? If there is more than one, what are they and what possibly caused this to happen?
6. Was the DNA in your digestion completely cut? How did you know whether or not it was cut?
7. Are you able to see any specific bands in uncut and cut genomic DNAs? If so, what happened to the samples?

Post-lab assignment

1. How are restriction enzymes named? Use *Eco*RI and *Hin*dIII as an example.
2. What is a multiple cloning site (MCS) or a polylinker site? Why is an MCS useful to a molecular biologist and biotechnologist?
3. In order to obtain a pGLO restriction map and pGLO restriction fragment lengths, you must use the actual name of pGLO, pBAD-GFPuv in your NCBI database search.

- Visit NCBI GenBank database (http://www.ncbi.nlm.nih.gov/genbank/). On the main entry page, you will see a search menu bar at the very top. Select (*v*) "Nucleotide" from the drop-down search bar, type in "pBAD-GFPuv" in the blank box, and then click "Search." This will display the entire sequence with the GenBank number and annotation.
- Copy the entire sequence, paste as Courier New font size 10 into a Microsoft Word document and save it. Include sequence positions in the sequence data.
- Using the annotated information of sequence features, find the CDS sequence of the GFP gene, highlight the nucleotide sequence from start codon to stop codon, and indicate the start and stop codons with capital letters in the saved nucleotide sequence data.

4. Visit New England Biolab NEBcutter web site (http://tools.neb.com/NEBcutter2/).

- Paste the entire pGLO sequence in the blank box, or type in the GeneBank number.
- Select "The sequence is: circular" button.
- Select Enzyme to use: "All commercially available specificities."
- Click "Submit". You will get the restriction enzyme map of pGLO with ORFs.
- Click "Custom Digest" under the Main Options menu (bottom left).
- Choose *Bam*HI, *Eco*RI, *Eco*RV, *Hin*dIII, and *Pst*I. Click "Digest" located at the bottom of the window. It will display the restriction map with just these five enzymes. Click "Enzymes & sites" on the List menu. It will display the number of cuts and cut positions.
- Based on the above information, show how many and what sizes of fragments are produced as a result of *Eco*RI, *Hin*dII, *Pst*I, *Eco*RI+*Eco*RV, and *Pst*I+*Eco*RV digestions of pGLO plasmid. If you choose a set of enzymes in "New Custom Digest" under Main Options and click "View gel," fragment lengths will be displayed.
- Underline *Eco*RI and *Hin*dIII site positions on the pGLO sequence data obtained from your work of post-lab assignment 3.

5. Approximately how many base pairs are MCS? *Hint:* you must use the physical map, restriction enzyme sites, and DNA sequence information of pGLO.
6. How many single restriction enzyme sites are within the GFP coding sequence?

- Copy and paste the ORF sequence of GFP in NEB Cutter2, click the "Liner" button and submit.
- Click "1 cutters" on the list menu.

7. What does "star activity" of the restriction enzyme mean? What conditions cause the star activity?
8. The *E. coli* genome is about 4.6×10^6 bp. Assuming that all four nucleotides are randomly distributed in equal portions in the *E. coli* genome sequence, how many fragments on average will you generate by digesting the *E. coli* genome with a 6-bp recognition *Eco*RI enzyme?
9. Study the gel below and answer the questions. Lanes are numbered from left to right; each 10 μL of DNA was mixed with 2 μL of 6× loading dye, and the mix was added to each well.

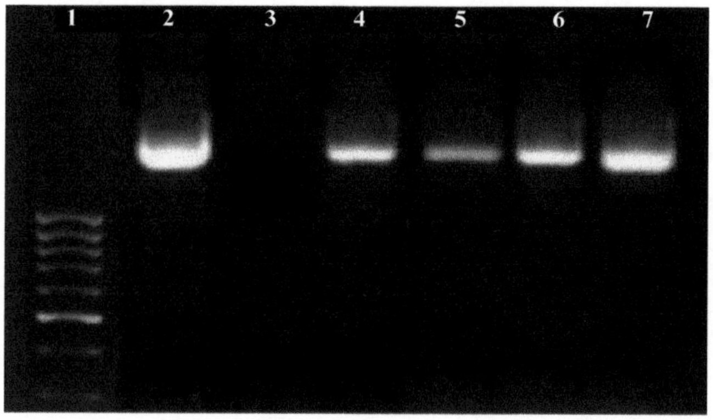

Lane 1: Molecular Maker DNA
Lane 2: sample A
Lane 3: sample B
Lane 4: sample C
Lane 5: sample D
Lane 6: 50 ng/μL DNA
Lane 7: 100 ng/μL DNA

Using the information of lanes 6 and 7, estimate the amount and concentration of DNA in the following lanes:

Lane 2: amount of DNA = _____ ng; concentration of DNA = ______ ng/uL

Lane 3: amount of DNA = _____ ng; concentration of DNA = ______ ng/uL

Lane 4: amount of DNA = _____ ng; concentration of DNA = ______ ng/uL

Lane 5: amount of DNA = _____ ng; concentration of DNA = ______ ng/uL

Further reading

Block, K.D. (2000). Unit 3.1. Digestion of DNA with restriction enconucleases. Unit 3.2. Restriction mapping. In *Current Protocols in Molecular Biology* (Edited by F.M. Ausubel). John Wiley & Sons Inc. ISBN 0-471-50338-X.

Maniatis, T., Fritcch, E.F., and Sambrook, J. (1982). Agarose gel electrophoresis, pp. 150–172. In *Molecular Cloning: A Laboratory Manual.* Cold Spring Harbor Laboratory. ISBN 0-87969-136-0.

13 MB experiment 13: Ouchterlony and ELISA immunoassays

Purpose: This is up to you to write down.

Introduction

All immunoassays are based on the unique ability of antibody to bind with high specificity to antigen in a complex mixture. These techniques are especially suited for detection of substances at low concentration and thus have been used in various applied and basic researches such as diseases diagnosis, pregnancy tests, and forensic identification of bloodstain origin, immunoblot, immunofluorescence, and immunoprecipitation. Immunoassays come in several different formats with semi-solid agar, solution, membrane, *in situ* tissue section, and whole cell. One of the widely used semi-solid agar and solution-based immunoassays are Ouchterlony and enzyme-linked immuno sorbent assay (ELISA), respectively.

Ouchterlony assay, also known as a double immunodiffusion technique, is based on the fact that when hundreds of antibodies (Ab) bind to hundreds of antigens (Ag), they produce aggregated complexes that are easily precipitated out of solution. In this method, wells are cut into agarose gel prepared in a Petri dish and are filled with Ag and Ab solutions. As the Ag and Ab diffuse outward toward each other at rates that are correlated to their concentration, size, and shape, a concentration gradient is established. When both Ag and Ab meet in an optimal ratio (equivalence), a milky white line of precipitate forms between the wells. The precipitin line is the result of large cross-linking lattices of Ag/Ab because polyclonal Ab in antiserum has two antigen binding sites and Ag is multivalent; that is, Ag has several antigenic determinants per molecule to which Ab can bind. Precipitation will not occur if excess Ag is present or if excess Ab is present. There is no one Ag/Ab ratio in which all different Ag/Ab reactions are at equivalence. The position of the precipitin band varies depending on the concentrations of Ag and Ab as well as the molecular weight of Ag. This technique can be used to determine an antigenic relationship between antigens based on the precipitin line patterns, as well as to detect viruses in infected sample extracts. Although this technique is simple and easy to perform and interpret for the presence of Ab for a particular Ag, as well as to determine its titer and identity, it requires a relatively large amount of Ag to detect a precipitation reaction. ELISA was developed to overcome this detection limitation.

ELISA is highly sensitive for a particular pathogen/antigen in a sample. In a basic ELISA procedure, a diluted series of protein mixture containing Ag is incubated in the wells of a microtiter plate. During incubation, the protein Ag is immobilized on to the surface of plastic wells because proteins adsorb to a non-modified polystyrene surface through hydrophobic interaction. After washing off the excess proteins and blocking any unoccupied plastic site in the wells, a specific Ab is added to the wells where the Ab–Ag interaction takes place during incubation. The Ab is covalently conjugated to an enzyme so that when a colorless substrate of the enzyme is added to the wells, the enzyme linked to the Ab converts the substrate to a colored product. The intensity of color is proportional to the amount of Ab bound and thus the amount of Ag in the original mixture. There are several variations in ELISA to choose the best assay format depending on the chemical properties of the analyte surface.

It is recommended to watch the following immunoassay animation and YouTube videos:

- http://www.sumanasinc.com/webcontent/animations/content/immunohistochemistry.html (Immunochemistry)
- http://www.youtube.com/watch?v=h-KptLVJpU0 (Antibody titration)
- http://www.youtube.com/watch?v=hmK7yYr2T54 (Antibody–antigen patterns)
- http://www.youtube.com/watch?v=RRbuz3VQ100 (ELISA)

In this lab exercise, you will carry out Ouchterlony and ELISA assays in order to study the specificity of the Ab–Ag reaction. In the Ouchterlony assay, you will use whole anti-BSA rabbit serum to test the polyclonal Ab for binding to a diluted series of BSA to determine the BSA antigen titer and for binding to the rabbit, pig, cow, and horse sera to determine the cross reactivity. For the ELISA assay, you will use goat anti-rabbit IgG-HRP (horseradish peroxidase) conjugate Ab and chromogenic substrate TMB (3,3′,5,5′-tetramethylbenzidine) that changes to a blue product upon reaction of HRP with H_2O_2. Addition of an acid stop solution converts blue to yellow, and the intensity of the yellow color is directly related to the amount of HRP activity and thus the amount of antigen in the sample. TMB is very sensitive, with a detection limit of 20 to 80 pg/mL, so it may produce a significant background if too much detection antibody protein is added.

Methods in Biotechnology, First Edition. Seung-Beom Hong, M. Bazlur Rashid and Lory Z Santiago-Vázquez.

Companion Website: www.wiley.com\go\hong\Methodsinbiotechnology

Pre-lab assignment

(Typing and submission must be completed before lab work begins.)

1. What is an antigen?
2. How is serum prepared and what proteins are present in the serum?
3. What is difference between antiserum and antibody? What is difference between polyclonal antibody and monoclonal antibody? Which antibody is used in this lab exercise?
4. What is the antigen used to produce goat anti-rabbit IgG-HRP (horse radish peroxidase) Ab for ELISA? How is this conjugated detection Ab produced?
5. What is the expected result of ELISA (Part E, step 16) in terms of color development?

Materials and equipment

- Pig serum, horse serum, cow serum, rabbit serum
- Anti-bovine serum albumin (BSA) antibody produced in rabbit (whole antiserum: Sigma B1520-2ML)
- Anti-horse serum antibody produced in rabbit (Sigma H8890-2ML)
- holo-Transferrin bovine (Sigma T1283-50 MG; 50 mg/mL in 1× PBS)
- 1% BSA (bovine serum albumin) in 1× PBS
- Goat anti-rabbit IgG (H+L)-HRP polyclonal secondary Ab (Life Technologies Cat No. G21234, lyophilized 1 mg): diluted at the optimal concentration (20–30 ng/mL) in blocking buffer immediately before use
- 10× PBS (phosphate buffered saline, pH 8.0): dissolve 80 g NaCl, 2 g KCl, 14.4 g Na_2HPO_4, and 2.4 g KH_2PO_4 in dH_2O to give final volume of 1 liter buffer; autoclave
- Washing buffer: 1× PBS + Tween 20 (0.05%): dilute 100 mL of 10× PBS to 895 mL with dH_2O; add 5 mL of 10% Tween 20 to buffer; store at room temperature
- Blocking buffer: 2% gelatin in washing buffer; prepare fresh before use
- Stop solution (0.1 M HCl)
- TMB substrate solution (TMB (3,3′,5,5′-tetramethyl benzidine) + H_2O_2):
    A. Fisher Scientific, Cat. No. PI-34021, TMB substrate kit (2 parts: 200 mL of 0.4 g/L TMB, 200 mL of 0.02% H_2O_2): mix equal volume just before use.
    B. Home recipe:
        1. Dissolve 1.46 g Na_2HPO_4 and 1.02 g citric acid (not sodium citrate) in water to final volume of 200 mL. This is citrate phosphate solution (0.05 M). This solution can be stored indefinitely in refrigerator.
        2. To prepare color reagent solution, add 1 mg of TMB (Sigma-Aldrich Product #860336-1G) to 10 mL of citrate phosphate solution. Next, add 2 μL of hydrogen peroxide (3%) to this solution. Make up immediately before use. Use this solution (0.1 g/L TMB, 0.06% H_2O_2) on the same day and store in the refrigerator.
- 1.25% agarose in 1× PBS
- Petri plates (100 × 15 mm), Pasteur pipette, microcentrifuge tubes
- 1/4 Microtiter (96 well) polystyrene plates
- Platform shaker
- Spectronic 20, cuvette

Procedure

Part I. Ouchterlony (double diffusion) assay

1. Pour melted agar solution (15 mL of 1.2% agarose in 1× PBS per Petri dish) into two Petri dishes (with no air bubbles trapped) and let stand for 10 min to solidify (TA will do this step).
2. Place the Petri plates over the sketch pattern below. Number the wells on the bottom side of the plate. Label your group number on the outer edge of the bottom side of the plate.

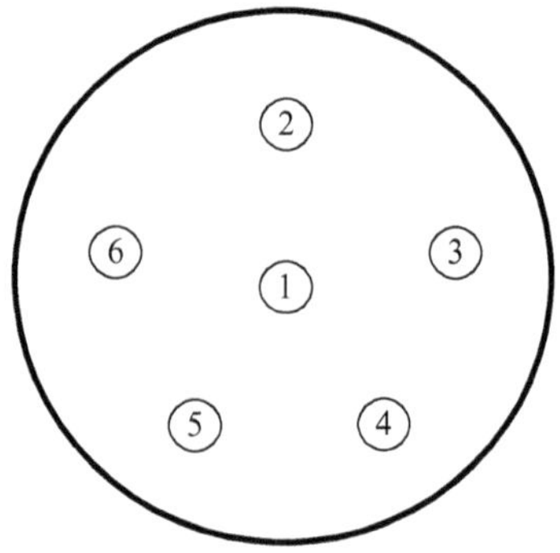

3. Push the larger end of the glass Pasteur pipette into the agarose to punch the wells. Take out the cut agar from the well using the narrow end of the pipette.
4. Make serial twofold dilutions in five microcentrifuge tubes using 1% (10 mg/mL) BSA so that the final BSA concentrations in each 50 μL final volume after dilution with 1× PBS should be 0.5%, 0.25%, 0.125%, 0.0625%, and 0.03125%. *Show your calculations as to how to dilute in the pre-lab notebook.*
5. Put 46 μL of anti-BSA rabbit whole antiserum in the center well #1 of plate A and 46 μL of the diluted BSA solutions in the assigned outer wells.
6. Put 46 μL anti-BSA rabbit whole antiserum in the center well #1 of plate B and 46 μL of each diluted serum in the assigned outer wells #2 to #6 as indicated in the table below.

Well no.*	Plate A	Plate B†	Plate C†
1	Anti-BSA rabbit whole serum	Anti-BSA rabbit whole serum	Anti-BSA rabbit whole serum: Anti-horse serum Mix (1:1)
2	0.50% BSA	Rabbit serum	Rabbit serum
3	0.25% BSA	Pig serum	Pig serum
4	0.125% BSA	Horse serum	Horse serum
5	0.0625% BSA	Cow serum	Cow serum
6	0.03125% BSA	Cow transferrin	Cow transferrin

*Wells should appear full, but be careful not to overfill the wells causing spillage on the surface.
†Group 1: 50% serum or 50% transferrin = 23 μL of 100% each serum or 100% cow transferrin + 23 μL of 1× PBS for plates B and C.
Group 2: 25% serum or 25% transferrin = 11.5 μL of 100% each serum or 100% cow transferrin + 34.5 μL of 1× PBS for plates B and C.
Group 3: 12.5% serum or 12.5% transferrin = 5.8 μL of 100% each serum or 100% cow transferrin + 40.2 μL of 1× PBS for plates B and C.
Group 4: 6.25% serum or 6.25% transferrin = 2.9 μL of 100% each serum or 100% cow transferrin + 43.1 μL of 1× PBS for plates B and C.

7. Put 23 μL each of anti-BSA rabbit whole serum and anti-horse serum in the center well #1 of plate C and 46 μL of each diluted serum in the assigned outer wells #2 to #6 as shown in the table.
8. Cover each plate and place upright inside the moist chamber (box or vinyl bag containing wet paper towels). Incubate for 48 to 72 hours at room temperature in a lab drawer.
9. Observe the precipitin lines appearing between the center and outer wells. Compare the intensity, length and width, and distance from the center well. An ideal line produced by a well-balanced reaction is narrow and dense with sharp margins, and is positioned away from the antibody well. Note down the highest antigen dilution at which the precipitin line is formed. This is the titer value of the antigen.

**The gel can be stained for proteins, especially if the precipitin line is very faint.*

Part II. ELISA assay

A. Preparation of materials

1. Prior to performing the ELISA assay, each group should have the following materials:

- One 1/4 section of a microtiter plate
- Sharpie marking pen
- Test tubes containing the indicated amounts of the following solutions:
  0.1 N HCl (~5 mL), 1× PBS (~25 mL), blocking buffer (~20 mL), washing buffer (~20 mL)
- Paper towel to be used to discard liquids
- 50 μL each of 0.5% pig, horse, cow, and rabbit serum
- 50 μL each of 0.1% pig, horse, cow, and rabbit serum

2. Label the plate section as shown below using a black marker pen. Each well can now be identified by a single alphabet letter and number.

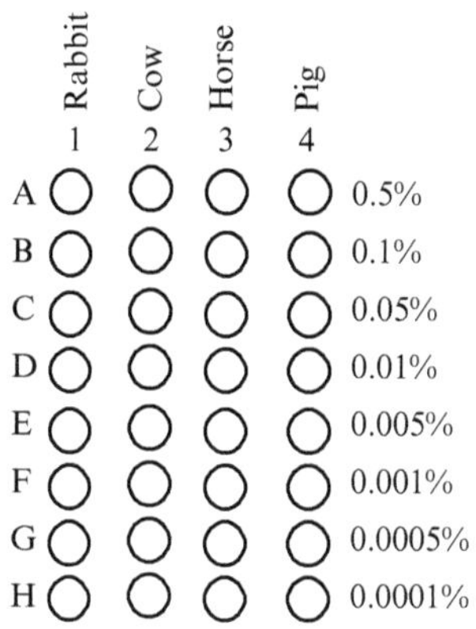

B. Coating antigen to microtiter plate

3. Place 45 μL of 1× PBS into wells 1C → 1H, 2C → 2H, 3C → 3H, and 4C → 4H.
4. Place 50 μL of 0.5% rabbit serum into well 1A, 50 μL of 0.5% cow serum into well 2A, 50 μL of 0.5% horse serum into well 3A, and 50 μL of 0.5% pig serum into well 4A.
5. Place 50 μL of 0.1% rabbit serum into well 1B, 50 μL of 0.1% cow serum into well 2B, 50 μL of 0.1% horse serum into well 3B, and 50 μL of 0.1% pig serum into well 4B.
6. Prepare and transfer serial tenfold dilutions so that the concentrations of rabbit serum in the wells after dilution should be as shown in the figure.

- To make the dilutions for 0.05 %, 0.005%, and 0.0005%, transfer 5 μL of the serum from well 1A to well 1C. Mix the contents in well 1C thoroughly by pipetting up and down several times. Work carefully to minimize bubbles. Transfer 5μL of the sample from well 1C to 1E, and 1E to 1G. Mix well between each transfer. Pipette 5 μL out from 1G.
- To make the dilutions for 0.01%, 0.001%, and 0.0001%, transfer 5 μL of the serum from well 1B to well 1D. Mix the contents in well 1D thoroughly by pipetting up and down several times. Transfer 5μL of the sample from well 1D to 1F. Transfer 5 μL of the sample from well 1F to 1H. Pipette 5 μL out from 1H.

7. Prepare and transfer 1:10 serial dilutions of cow, horse, and pig serum in the same way as described above.
8. Cover the plate and let the plate stand at room temperature for ~20 min in order to allow time for the proteins to be adsorbed to the well surfaces.

**It is important not to allow the plate to dry at any point. The most thorough adsorption and the lowest well-to-well variation occur overnight (16 to 18 hours) at 4 °C; the wells should be sealed to prevent evaporation.*

C. Blocking

9. Remove the coating solution by flicking the plate over a sink, tap out a residual fluid over the paper towel, and wash the plate by filling the wells with 200 μL of 1× PBS (do not touch the bottom of the well with the pipette tip). Remove the remaining drops by patting the plate on a paper towel.
10. Immediately add 300 μL of blocking buffer to each well. Cover the plate and gently rock on a platform shaker for 10 to 15 min at room temperature.
11. Remove the entire blocking buffer from each well by turning the plate upside down and tapping out residual fluid on to a paper towel.
12. Rinse the plate once with 300 μL of 1× PBS.

D. Antibody reaction and washing

13. Obtain one tube of goat anti-rabbit IgG-HRP antibody (~1.6 mL) from the instructor and add 50 μL of the antibody to each well.
14. Cover the plate and gently rock on a platform shaker for 20 min at room temperature to allow for the antibody to bind to the immobilized antigen.
15. Remove the entire antibody solution by turning the plate upside down and tapping out residual fluid on to a paper towel. Wash the plate four times with 200 μL of washing buffer per well each for 5 min at room temperature on a platform shaker.

**Do not let it dry between the washing times.*

E. Color development

16. Obtain one tube (~1.6 mL) of color substrate TMB solution from the instructor, add 50 μL of the solution to each well starting from the lowest dilution well H, and incubate for 5 to 20 min or until the desired color develops.

**The reaction rate by HRP will be most likely slow after 20 min.*

**High concentrations of HRP yield a greenish solution. A precipitated product indicates the presence of too much HRP and the need to optimize experimental conditions.*

17. Place the plate over a white sheet of plain paper and examine the intensity of the blue color. The blue product is the insoluble product of the peroxidase reaction.

**Optimal incubation times may vary depending on the amount of antigen present.*

18. Add 150 μL of 0.1 M HCl to each well. At a low pH, the insoluble blue product is converted to a soluble yellow product that should be easier to detect. This is the end point assay.
19. Place the plate over a white sheet of plain paper and record the relative intensity of yellow color in each well: No color (0), Low color (+), Medium color (++), High color (+++ lemon yellow).
20. Transfer only yellow color samples to the labeled test tubes containing 2.5 mL dH_2O and measure the intensity of the yellow product at A_{450} nm using a blank solution that consists of 2.5 mL of dH_2O, 50 μL of TMB solution, and 150 μL of 0.1M HCl. The dilution factor is 13.5 (2.7/0.2).
21. Draw an ELISA signal curve by plotting A_{450} values (y axis) versus % concentration (x axis).

Result

1. Take a picture of plate A and record the intensities (thick or thin; narrow and sharp or broad and blurry) and the distances of the precipitin lines from the center well in Ouchterlony plate A.

Well position	BSA concentration (%)	Distance (mm)	Precipitin line
2			
3			
4			
5			
6			

2. Obtain all four groups of Ouchterlony plates B and C, take a picture, and record precipitin line pattern results.

%	Rabbit serum		Pig serum		Horse serum		Cow serum		Cow transferrin	
	B	C	B	C	B	C	B	C	B	C
50										
25										
12.5										
6.25										

3. Plot the curve of A_{450} values versus antigen concentrations for the ELISA result.

Discussion

(Do not copy the discussion point. Write a paragraph in your own words.)

1. Which BSA antigen concentration forms the best precipitin lines in the Ouchterlony test? Explain your selection criteria.
2. Study the relationship between the antigen concentration and the position precipitin line as well as between the antigen concentration and intensity of the precipitin line from plate A.
3. Study the relationship between the pure antigen/antiserum (plate A) and the impure antigen/antiserum (plate B) reactions with respect to the intensities and sharpness of the precipitin lines.
4. Compare the results of plate B with different serum concentrations.
5. Describe the precipitation patterns of plate C. Determine the identity of relationship of antigens from plate C.
6. What is the detection sensitivity of ELISA (% serum concentration that is penultimate to the serum concentration yielding no detectable signal) based on the result?
7. Study the colors of the solutions of the positive and negative controls and serial dilutions in the ELISA assay. Are they the colors you expected them to be?
8. When the measured levels of antigen show a significantly lower absorbance than the actual level present in a sample, it is referred to as the "high dose hook effect." It is typically displayed as a plateau line in the ELSA signal concentration curve due to the loss of sensitivity at a high range of concentrations. Do you observe this effect in the ELISA result? If so, what possibly caused this phenomenon?
9. Are there any false positives (all the wells showing color) or false negatives (all the wells showing no color)? If so, discuss the possible errors in the ELISA experiment.

Post-lab assignment

1. Is your experiment a direct ELISA or an indirect ELISA assay? Explain your choice.
2. What type of ELISA would you choose to detect bacterial pathogens in a suspected sample? Explain your choice.
3. What type of ELISA should be used to detect a mycotoxin such as aflatoxin in suspected samples? Note that mycotoxins are secondary metabolites.
4. What is fluorescent antibody and what is the advantage of its use?
5. Is Ouchterlony assay precipitation or agglutination or both reactions? How does precipitation reaction differ from an agglutination reaction? Give examples of each.
6. How are polyclonal and monoclonal antibodies prepared? What is the difference between the polyclonal and monoclonal antibody? What type of diagnostic test are monoclonal antibodies used for?
7. What is a competitive ELISA and how does it work? Refer to https://www.youtube.com/watch?v=Kb26nQVMHds (Competitive ELISA Tutorial 1).
8. What is a test strip ELISA and how does it work? Refer to
- https://www.youtube.com/watch?v=riijHvCwHuI (Pregnancy Test Strip) and
- https://www.youtube.com/watch?v=rt0XGIz9gwQ (GMO Testing of Cotton Seed).

Further reading

Antigen Antibody Interaction. G-Biosciences (2012). Catalog No. BE-501.

Bailey, G.S. (1996). Ouchterlony bouble immunodiffusion. In *The Protein Protocols Handbook* (Edited by J.M. Walker), pp. 749–752. Human Press, Inc. DOI: 10.1007/978-1-60327-259-9_135; ISBN 978-0-89603-338-2.

Frey, A., Meckelein, B., Externest, D., and Schmidt, M.A. (2000). A stable and highly sensitive 3, 3′, 5, 5′-tetramethylbenzidine-based substrate reagent for enzyme-linked immunosorbent assays. *Journal of Immunological Methods*, **233**: 47–56.

Hornbeck, P., Winston, S.E., and Puller, S.A. (2000). Enzyme-linked immunosorbent assays. Unit 11.2.1–11.2.22. In *Current Protocols in Molecular Biology*. John Wiley & Sons, Inc. ISBN 0-471-50338-X.

14 MB experiment 14: Testing plant substances for antimicrobial activity

Purpose: This is up to you to write down.

Introduction

Nature is a very important source of new leads for drug development. Mostly plants and microorganisms have been screened for this purpose. Of all drugs used in Western medicine, about 25 to 50% is estimated to be derived from plants either as pure natural compounds or as natural product derivatives. A widespread use of alternative medicinal therapy mainly relies on traditional herbal plants, though in most case no scientific studies have been made to confirm their bioactivities. The number of plants studied so far is still limited, and thus there is a great potential in both terrestrial and aquatic plants for drug development. Plants produce a broad spectrum of so-called secondary metabolites, some of which are used for drugs, flavors, fragrances, and dyes. Plant secondary metabolites have numerous pharmacological activities; examples are anti-virus, anti-microbes, anti-insect, anti-parasite, anti-cancer, anti-oxidant, anti-cholesterol, anti-inflammation, anti-hypertension, anti-diabetes, anti-pyretic, analgesic, sedative, and antisepsis. The potential compounds of a medicinal value are usually screened using animal cell culture lines or pathogen samples before stage 3 of clinical trials for the safety and efficacy.

After the discovery of antibiotics, the indiscriminate use of commercial antimicrobial drugs, especially for farm animals, has brought about the widespread occurrence of multidrug-resistant bacteria and malaria pathogens. This is now one of the greatest threats to human health. Therefore, there is a constant need to develop new and effective antimicrobial drugs for the treatment of infectious diseases, and a sensitivity test is essential for effective chemotherapy against pathogens isolated from each patient. Plant-derived chemicals represent a vast untapped source of medicines and further exploration of antimicrobial phytochemicals needs to be made.

Phytochemicals are usually extracted by mixing pulverized tissues with a solvent for several hours or days followed by filtration and centrifugation. Water and/or ethanol are good for extracting highly polar compounds. For less polar compounds, one can use ethyl acetate, acetone, or chloroform. Petroleum ether or hexane is good for extracting highly non-polar compounds. The clarified extract is typically concentrated to dryness by means of a rotatory evaporator or a lyophilizer.

In this lab exercise, you will extract plant substances using water or methanol. Due to the time limitation, extraction will not be carried out for hours or days, and extracts will not be concentrated. You will then test the phytoextract for antibacterial activity using the disk diffusion sensitivity method, also called the "Kirby–Bauer" method. For this assay, either an agar overlay or cotton swab containing a test bacterial culture can be used. In the agar overlay technique, bacteria are added to a soft top agar, which has been melted and cooled to 45 °C. This temperature is warm enough for the top agar to remain liquid, but cool enough for the bacteria to survive. The melted agar/bacterial suspension mixture is poured evenly across the top of a Müeller–Hinton agar plate and allowed to solidify. In the cotton swab technique, a sterile cotton swab is immersed in a bacterial culture and is dabbed evenly on to the surface of an agar plate. Alternatively, the bacterial culture is spread on to the agar plate using a sterile glass rod spreader. Then, small round filter disks containing test compounds are placed on to the assigned positions of the agar surface, and the plates are incubated. Mueller–Hinton media contains beef infusion, casein hydrolysate, and starch. Beef infusion and casein hydrolysate are provided as a source of energy and nutrients, whereas starch acts as a colloid protector that absorbs toxic metabolites produced by bacteria that may interfere with the test compounds. Mueller–Hinton agar should always be used for disk diffusion susceptibility testing, according to NCCLS (National Committee for Clinical Laboratory Standards) and international guidelines.

It is recommended to watch the YouTube video regarding the Kirby–Bauer disk diffusion assay:

- http://www.youtube.com/watch?v=sx1uDYSfINA

After incubation, the bacteria distributed on the agar surface or through a thin layer of top agar will grow to produce a homogeneously turbid lawn. If the freshly seeded lawn is exposed to various antibacterial substances before incubation, any growth inhibition will give rise to a reduced turbidity of the lawn near the agent: the greater the antibacterial action, the wider the circular zone of inhibition called a "halo." Thus, the antibacterial strength of a substance can be judged by the width of the zone of inhibition around it. Minimum inhibitory concentration (MIC) is used to assess the antimicrobial susceptibility with regard to various concentrations. Aerobic or facultative bacteria that grow well on unsupplemented Müeller–Hinton agar can be tested. Test for certain fastidious bacteria requires additional supplements in Müeller–Hinton agar. Many conditions can affect a disk diffusion susceptibility test. When performing the tests, certain conditions

Methods in Biotechnology, First Edition. Seung-Beom Hong, M. Bazlur Rashid and Lory Z Santiago-Vázquez.

Companion Website: www.wiley.com\go\hong\Methodsinbiotechnology

are held constant from test to test so that only the halo size is variable. These include the depth of agar, the amount of organism (culture density), the concentration of chemicals, and incubation conditions (time, temperature, and atmosphere).

Pre-lab assignment

(Typing and submission must be completed before lab work begins.)

1. Good labeling is very important for every experiment. Draw how you label the eight different agar plates for the Part A experiment. Full details are required.
2. Draw the schematic flow diagram for the procedure of Parts B and C experiments.
3. *Klebsiella pneumonia, Pseudomonas aeuroginosa,* and *Staphylococcus aureus* bacteria used in this experiment are important human pathogens. What types of diseases are caused by them?
4. Two different solvents (aqueous and methanol) are used to prepare extracts. What general characteristic of chemical compounds is extracted by each solvent?

Materials and equipment

- *Allium sativum* (garlic) bulb, *Zingiber officinale* (ginger) rhizome, *Solanum tuberosum* (potato) tuber, *Camellia sinensis* (dried green tea powder)
- Overnight Mueller–Hinton broth cultures of *Escherichia coli, Klebsiella pneumonia, Pseudomonas aeuroginosa, Staphylococcus aureus*
- McFarland turbidity standard 0.5 (Fisher Scientific, Cat. No. R20410)
- Ampicillin disk (10 μg; BD Diagnostic Systems, Cat. No. 230705)
- Kanamycin disk (30 μg; BD Diagnostic Systems, Cat. No. 230829)
- Blank disk (6 mm diameter; BD Diagnostic Systems, Cat. No. 231039)
- Ice-cold sterile dH_2O, methanol
- Agar plates of Mueller–Hinton medium
- L-shaped Glass Rod Spreader, Bunsen burner
- Mortar and pestle, spatula, sterile cheesecloth, funnel, ice bucket
- Sterile capped 13 × 100 mm tubes, 2-mL microcentrifuges
- Fine-point forceps, Kimwipe
- 37 °C incubator

Procedure

Part A. Labeling Mueller–Hinton agar plates

1. Obtain 8 agar plates (each group). Label group #, four indicator bacteria (*E.c; K.p; P.a; S.a*) for water extracts (W) on the bottom edge of 4 Petri plates. Make the same labeling for methanol extracts (M) on the bottom edge of 4 Petri plates.

2. Mark the balanced positions for 6 filter disks on the bottom of 8 Petri plates: 4 bacteria Petri plates of each labeled 4 plant water extracts, water, and ampicillin disk positions; 4 bacteria Petri plates of each labeled 4 plant methanol extracts, methanol, and kanamycin disk positions.

**Each marked position will be used to place sample disks on to it in step 18.*

3. Warm all the agar plates at 37 °C (invert the plates during incubation).

Part B. Preparing plant extracts

4. Using a mortar and pestle, grind up ~4 g of plant tissue in liquid nitrogen, add 4 mL of sterile dH_2O, and grind again until the sample becomes a relatively uniform suspension. Use 1 g for dried green tea powder.

**Each group works on one plant tissue only (garlic, ginger, potato, green tea).*

5. Put three layers of cheesecloth in a funnel inserted into a 15-mL plastic conical centrifuge tube, pour the homogenate, and squeeze the cloth. Keep the juicy extract on ice.
6. Transfer ~1.8 mL of extract to each of two labeled 2-mL microcentrifuge tubes and spin at 7000 rpm, at 4 °C for 5 min. Transfer the supernatant to two new labeled microcentrifuge tubes and keep on ice. Discard the pellet.

**This supernatant is also used for the Experiment 15 exercise.*

7. Repeat steps 4 and 5 using methanol as the extracting solvent, transfer ~1.8 mL of extract to one labeled 2-mL microcentrifuge tube, and spin at 7000 rpm, at 4 °C for 5 min. Discard the pellet.
8. Repeat steps 4 to 7 for the other plant samples.

**Each group prepares water and methanol extracts from one plant tissue and shares with other groups for the subsequent lab exercises.*

9. Using forceps, drop 4 filter blank disks into each tube containing 100 μL of a water-extracted sample (garlic, ginger, potato, green tea).
10. Repeat step 9 for each tube containing 100 μL of methanol-extracted sample (garlic, ginger, potato, green tea).
11. Drop one filter disk into each tube containing 25 μL of sterile dH_2O or 25 μL of methanol.
12. Let them stand at room temperature until ready to use.

Part C. Bacterial inoculation and sample application

13. Pipette out 150 μL of a McFarland turbidity 0.5 culture of *E. coli*, place in the center of two agar plates labeled *E.c*, and spread evenly over the entire surface using a sterile L-shaped glass rod spreader.
14. Repeat step 13 for three other bacterial cultures of McFarland turbidity 0.5.
15. Open the lids of all eight plates and let them dry in the hood until there is no liquid on the surface.
16. Using sterile forceps (sterilized by alcohol flaming), take one filter disk out of the tube, blot any excess liquid on to Kimwipe, and place the disk on a marked position of a Petri plate. Keep

all the water-extracted samples on the same dish and all the methanol-extracted samples on the separate same dish.

17. Gently press down the disks to ensure complete contact of the disk with the agar surface.

18. Place ampicillin, kanamycin, water, methanol, and plant extract disks on the marked positions of each Petri plate.

19. Invert all the plates and incubate at 37 °C overnight (14 to 20 h).

20. Photograph the plates. Fill in the table in the Results section.

Result

The antimicrobial disk susceptibility tests are shown in the table below.

Disc no.	Control agents and plant extracts		Inhibition zone diameter (mm)*			
			E. coli	*P. aeuroginosae*	*K. pneumonia*	*S. aureus*
1	Garlic	Aqueous				
		MeOH				
2	Ginger	Aqueous				
		MeOH				
3	Potato	Aqueous				
		MeOH				
4	Green tea	Aqueous				
		MeOH				
5	Antibiotic	Ampicillin (aqueous extract only)				
		Kanamycin (MeOH extract only)				
6	Control	Aqueous				
		MeOH				

*Diameter in mm along with the disc diameter.

- Describe any unexpected observations and problems including fugal contamination that you found.

Discussion

(Do not copy the number and discussion point. Write a paragraph in your own words.)

1. Evaluate the efficacy of each plant extract as a source of potential antimicrobial medicine. Which bacterium is most resistant to all the chemicals used on the basis of the zone of inhibition?

2. Discuss the inhibition mechanism of antibiotics ampicillin and kanamycin in relation to the growth inhibition zones of the three different bacteria.

3. You tested both aqueous and methanol extracts. Which extract is more active? Does the zone of inhibition tests indicate that microorganisms have been killed by an antimicrobial product? How would you determine this?

4. Based on the results obtained, what combination of extract would be the best to inhibit or kill all four microbes tested?

5. Discuss the possible errors in the experiment that could give false data.

Post-lab assignment

1. What is the McFarland Standard? Why is it useful for the disk diffusion sensitivity test?

2. Plant extracts contain many different compounds. Therefore, the initial screening assay results are preliminary. What should be done to further identify the main bioactive ingredient?

3. If a phytochemical extract gives a negative reaction in the assay, does this mean that the extract does not contain any antimicrobial ingredient against the tested microbe? Explain your choice.

4. How would you test a phytochemical extract for anti-cancer activity at the pre-clinical stage?

5. How would you test a phytochemical extract for anti-viral activity at the pre-clinical stage?

Further reading

Cowan, M.M. (1999). Plant products as antimicrobial agents. *Clinical Microbiology Review*, **12** (4): 564–582.

Hudzicki, J. (2009). Kirby–Bauer Disk Diffusion Susceptibility Test Protocol. ASM Microbe Library (http://www.microbelibrary.org/component/resource/laboratory-test/3189-kirby-bauer-disk-diffusion-susceptibility-test-protocol).

15 MB experiment 15: Peroxidase enzyme activity assay

Purpose: This is up to you to write down.

Introduction

Under normal metabolic conditions, all aerobic organisms produce reactive oxygen species (ROS) as a byproduct of cellular respiration that uses O_2 as the final electron acceptor to metabolize sugar into CO_2 and H_2O and ATP. ROS are highly reactive oxygen-containing molecules that are either radicals such as superoxide ($\bullet O_2^-$), hydroxyl radical ($\bullet OH$), and peroxynitrite ($\bullet ONOO^-$) carrying an unpaired electron in the outermost shell or non-radicals capable of generating radicals such as hydrogen peroxide (H_2O_2) and singlet oxygen ($^1O_2$). Some ions such as hydroxyl ion (OH^-) and hypochlorite ion (OCl^-) also can be ROS. If ROS loses its own electron or gains an electron from a nearby molecule, its victim molecule now becomes a free radical, giving rise to a sequential chain reaction called "a free radical cascade."

It is recommended to watch the following YouTube video regarding free radical damage and antioxidant:

- http://www.youtube.com/watch?v=KCF6prDSrHE

Since ROS are highly toxic, cells must get rid of it rapidly. In doing so, cells use a variety of phase II detoxification enzymes including superoxide dismutase (SOD), catalase, glutathione S transferase, and peroxidase. In addition, cells may use antioxidant molecules having extra electrons to prevent a free radical cascade effect. Hydrogen peroxide is a common end product of oxidative metabolism and is not only a powerful oxidizing agent that is very damaging to a wide variety of molecules within cells but also a key regulator of cell signaling in a number of oxidative stress-related states. It is an important signal molecule involved in plant hypersentive response and apoptosis. H_2O_2 is generated by the mitochondrial respiratory chain as well as by a range of oxidase enzymes including glucose oxidase.

SOD catalyzes the dismutation of superoxide into oxygen and H_2O_2. Catalase directly breaks H_2O_2 down into water and oxygen as a member of the peroxidase family. Peroxidases split H_2O_2 into water and oxidized donor molecules. Some catalases are bifunctional in that they can also use a variety of organic substances as an electron donor. The overall reaction equations are:

$$2\cdot O_2^- + 2H^+ \xrightarrow{\text{Superoxide disumutase}} O_2 + H_2O_2$$

$$2H_2O_2 \xrightarrow{\text{Catalase}} 2H_2O + O_2$$

$$H_2O_2 + \text{Electron donar substrate} \xrightarrow{\text{Peroxidases}} 2H_2O + \text{Oxidized donar substrate}$$

Due to its implication of high accumulation levels in many disease states, there is much interest in sensitive assays for H_2O_2. Since oxygen gas makes bubbles in solution, catalase or peroxidase activity can be easily monitored by measuring the amount of oxygen bubbles produced as the reaction proceeds. The catalase activity can be quantitatively measured by direct reading at 240 nm, a maximum wavelength of H_2O_2. The peroxidase activity can be quantitatively measured using a chromogenic substrate indicator TMB (3,3′,5,5′-tetramethylbenzidine; MW 240.35 g/mole) as an electron donor reducing agent. As H_2O_2 is broken down by peroxidase enzymes, electrons are transferred from TMB to H_2O_2 (an oxidation-reduction reaction). As TMB

Lewis Structure of Reactive Oxygen Species (•unpaired electron that is unstable)

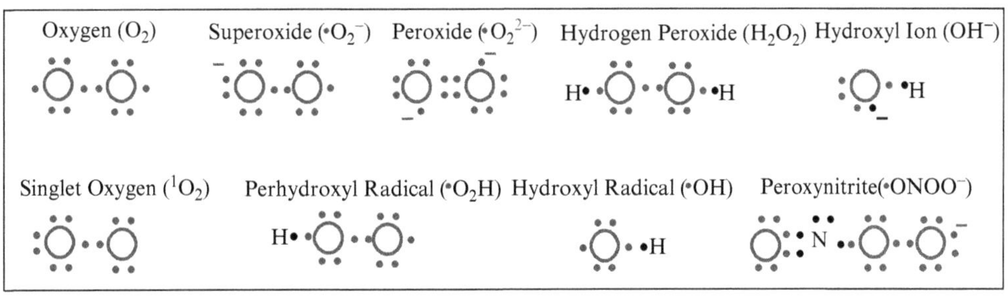

Methods in Biotechnology, First Edition. Seung-Beom Hong, M. Bazlur Rashid and Lory Z Santiago-Vázquez.

Companion Website: www.wiley.com\go\hong\Methodsinbiotechnology

donates one electron to H_2O_2, the oxidized TMB turns a blue color that can be read directly at 650 nm ($\varepsilon = 3.9 \times 10^4\ M^{-1}cm^{-1}$). When an acid stop solution is added, TMB (two-electron oxidations of TMB) becomes completely oxidized, turning from a blue to a yellow color that can be read at 450 nm ($\varepsilon = 5.9 \times 10^4\ M^{-1}cm^{-1}$) with two- to fourfold increased sensitivity. The degree of yellow color is directly related to the amount of peroxidases in the sample. Because the acid stop solution kills the enzyme, the blue color reading is more useful for kinetic assays for monitoring the rate of the peroxidase reaction as a function of time.

In this lab exercise, you will test the plant protein extracts obtained from Experiment 14 for peroxidase activity and study the effects of concentrations of enzyme peroxidases and substrate hydrogen peroxide on the rate of the activity. In order to determine the enzyme kinetics, you will add a fixed amount of crude enzyme extract to a series of different substrate concentrations, record the progress of the reaction rate (change in absorbance of the blue color product), and prepare Michaelis–Menten and Lineweaver–Burk plots where the reaction velocity is plotted against substrate concentration. All peroxidases are very stable and easy to assay, and thus are useful for comparisons among species and tissues. It is recommended to watch the following animation regarding enzyme kinetics: http://www.wiley.com/college/pratt/0471393878/student/animations/enzyme_kinetics/index.html

Pre-lab assignment

(Typing and submission must be completed before lab work begins.)

1. Describe how you would perform your serial dilutions so that you can end up with 0.1 mL in step 3 of the Part B experiment?
2. The TMB substrate solution in step 5 of the Part B experiment contains 0.2 g/L of TMB and 0.01% (v/v) H_2O_2. What is the final concentration of TMB and hydrogen peroxide (density 1.46 g/mL) in terms of molarity after dilution? Show your calculation.
3. What are the final molar concentrations of TMB and hydrogen peroxide after dilution in a total volume of 3.0 mL in each tube of step 10 of the Part C experiment? Show your work.
4. Draw the schematic flow diagram of Parts B and C experimental procedures.

Materials and equipment

- TMB substrate solution kit (Fischer Scientific, Cat. No. PI-34021; 0.4 g/L TMB and 0.02% hydrogen peroxide: 1:1 mix gives 0.2 g/L TMB and 0.01% H_2O_2: prepare just before use)
- MilliQ H_2O
- Spectronic 20, cuvette
- Glass test tubes (13 × 100 mm), ice bucket, water wash bottle.

Procedure

Part A. Preparation of crude protein extract

1. Each group uses the aqueous water extract prepared (step 6 of Part B in Experiment 14).

Group 1: Garlic bulb extract
Group 2: Ginger rhizome extract
Group 3: Potato tuber extract
Group 4: Any of the above extract

Part B. Effect of peroxidase concentration on the Rate of activity

2. Fill a colorimeter test tube with dH_2O. Place this test tube in a Spectronic 20 colorimeter and adjust the readout so that it displays zero absorbance at A_{650} nm.
3. Prepare serial dilutions of your plant protein extract as shown below using ice-cold sterile dH_2O in a total volume of 100 μL in each glass test tube. Keep them in an ice bucket.

- Tube A: 1/1 (undiluted) extract • Tube B: 1/2 dilution • Tube C: 1/4 dilution
- Tube D: 1/8 dilution • Tube E: 1/16 dilution • Tube F: sterile dH_2O

4. Transfer the contents of tube A to the colorimeter tube.
5. Add 2.8 mL of dH_2O at room temperature and 100 μL of TMB substrate solution (0.2 g/L TMB and 0.01% H_2O_2) to the colorimeter tube. Immediately place the tube in the colorimeter and start the timer. Record the absorbance (A_{650}) of the solution every 20 s for 2 min.

**If the concentration in the undiluted extract is too high, A_{650} values decrease over time. Then use the diluted sample.*

6. Pour off the solution into a waste beaker and wash the cuvette thoroughly using a water wash bottle.
7. Repeat steps 4 to 6 using the contents of Tube B, Tube C, Tube D, Tube E, and Tube F. Record all the absorbance data in the first table in Results.

Part C. Effect of substrate concentration on the rate of activity

8. Fill a colorimeter test tube with dH_2O. Place this test tube in the Spectronic 20 colorimeter and calibrate so that it displays zero absorbance at A_{650} nm.
9. Label 7 glass test tubes (S0, S1, S2, S3, S4, S5, S6) and add dH_2O, 0.02% H_2O_2, 50 μL of TMB solution (0.4 g/L) to each tube at room temperature, as shown in the table below.

Tube	dH_2O (mL)	0.02% H_2O_2 (μL)	0.4 g/L TMB[a] (μL)
S0	2.85	0	50
S1	2.84	10	50
S2	2.83	20	50
S3	2.81	40	50
S4	2.79	60	50
S5	2.77	80	50
S6	2.75	100	50

[a] TMB substrate is light sensitive (degraded by sunlight and by fluorescent lights).
Add last just before use.

10. Add 100 μL of undiluted or diluted extract that gives A_{650} change 0.1 to 0.2 per min (based on the Part B experiment) to the colorimeter tube cuvette.

**This ΔA_{650} is fast enough to be efficient for assay. However, it may vary widely depending on the plant and tissue type.*

11. Pour 3 mL of tube S6 content to the colorimeter cuvette. *Immediately* place the tube in the colorimeter and start the stopwatch. Record the absorbance of the solution every 20 s for 2 min.
12. Pour off the solution back into tube S6. Wash the colorimeter tube cuvette thoroughly with a water wash bottle.
13. Repeat steps 10 to 12 using the contents of tubes S5, S4, S3, S2, S1, and S0. Record all the absorbance data in the second table in Results.

Results

• The table below shows the effect of peroxidase concentration on the rate of activity (Part B experiment).

Protein extract (plant name)		$A_{650\,nm}$ incubation time (s)						
Dilution	μL	0	20	40	60	80	100	120
0	100							
1/2	50							
1/4	25							
1/8	12.5							
1/16	6.25							
dH_2O only	0							

Plot all data on a single graph labeled with the *y* axis as absorbance (A_{650}) and the *x* axis as incubation time. Among each series of data points, choose only the points in the LINEARLY increased portion and draw the best-fit line. Indicate each line with a protein extract amount (dilution or μL extract; 0 dilution = 100 μL). Note the slope of the graph line for each solution. This is an indicator of how fast the reaction proceeded.

• The table below shows the effect of substrate concentration on the rate of activity (Part C experiment).

Time (s)	$A_{650\,nm}$ Substrate consumed						
	S0	S1	S2	S3	S4	S5	S6
0							
20							
40							
60							
80							
100							
120							
Initial H_2O_2 concentration (mole/L)							

Plot all the data on a single graph labeling the *y*axis absorbance (A_{650}) and the *x* axis as incubation time. Draw the best-fit line to each series of data points. Indicate each line with substrate (H_2O_2) concentration.

Discussion

1. Identify any possible sources of error that may affect the results of the experiment.
2. What does the slope of line indicate in the plot of the first table data in Results?
3. Do the changes in substrate concentration in the second table in Results exhibit the same effect on the rate of activity as the changes in enzyme concentration in the first table? Explain, based on the plots.
4. Determine the V_{max} (maximum velocity) and K_m (substrate concentration at $\frac{1}{2}V_{max}$) from the obtained kinetic data. To do this, proceed to do the following steps.

(a) Correct each A_{650} values of S1 to S6 in the second table by subtracting a reading value of S0.
(b) Calculate the molar concentrations of the oxidized TMB blue products in the second table using Beer's Law ($A = \varepsilon cl$) and its molar extinction coefficient (ε) of $3.9 \times 10^4\ M^{-1}cm^{-1}$. This is equivalent to the molar concentration (M) of H_2O_2 consumed during the time of reaction.
(c) Convert all of the calculated molarities into μmoles (remember that the final reaction volume is 3 mL, so 1 mole/liter × 3 mL = 3×10^{-3} moles = 3×10^3 μmoles). This is equivalent to the amount of the substrate H_2O_2 consumed during the time of reaction.
(d) Plot μmoles of substrate (*y* axis) reacted at each concentration against time (*x* axis).
(e) From the plot of (d), determine which time point(s) are within the range of the LINEARLY increased portion of each curve S1 to S6.

**There may be several time points within the linearly increased portion of each curve.*

(f) Select a time point in the middle of the linear range where the measured activity is a linear function of enzyme (i.e., this is a steady-state first-order reaction stage in which the enzyme consumes a substrate and produces a product at a constant rate).
(g) Calculate the initial velocity V_0 (μmoles/min) at each substrate concentration in the time point of the liner range.

**If you select two or three time points instead of one time point, you must use the average V_0 value at each substrate concentration. Make sure to convert s to min.*

(h) Plot the kinetic data using initial velocity V_0 (μmoles/min) on the *y* axis and molar concentration of substrate hydrogen peroxide [S] on the *x* axis. This is the Michaelis–Menten plot.
(i) Determine the V_{max} (maximum velocity) and K_m (substrate concentration of $\frac{1}{2}V_{max}$) of peroxidases in the plant extract from the Michaelis–Menten plot.
(j) Prepare a Lineweaver–Burk plot of $1/V_0$ (*y* axis) versus $1/[S]$ (*x* axis), and determine K_m and V_{max}:

$$x \text{ intercept} = -1/K_m. \text{ Thus, } K_m = -1/x \text{ intercept value}$$

$$y \text{ intercept} = -1/K_{max}. \text{ Thus, } K_{max} = -1/y \text{ intercept value}$$

(k) Compare the values of V_{max} and K_m obtained from the Michaelis–Menten plot with those of the Lineweaver–Burk plot. Which values do you think are more correct?

Do not skip any one step for correct data processing, which is outlined below.

Correction of the second table data (effect of substrate Concentration on the rate of activity) by subtracting S0 reading value

↓

Calculation of H_2O_2 molar concentrations using Beer's Law ($A = \varepsilon cl$); $\varepsilon = 3.9 \times 10^4\ M^{-1}\ cm^{-1}$ in the second table data

↓

Conversion of molarities into μmoles in the second table: mole/liter × 10^3 × total Rx volume (mL)

↓

Plot the μmoles H_2O_2 (y axis) versus time (x axis)

↓

Select one or more time points within the linearly increased portion of the plotted curve

↓

Calculate average initial velocity (μmoles/min) at each substrate concentration

↓

Make a Michaelis–Menten plot: V_0 (μmoles/min) on y axis and $[H_2O_2]$ on x axis

↓

Determine the K_m and V_{max}

↓

Make a Lineweaver–Burk plot of $1/V_0$ (y axis) versus 1/[S] (x axis)

↓

Determine the K_m and V_{max}

Post-lab assignment

1. What are the damaging effects of ROS on cells at the molecular level?

2. Despite the fact that nitric oxide (•NO) is a free radical as a reactive nitrogen species, it is produced in almost all types of organisms including humans. What are the biological functions of •NO in humans?

3. Besides antioxidant enzymes, plants contain a variety of antioxidant compounds that protect the human body from oxidative stress. What are those from spinach, tomato fruit, red grape, green tea, carrot, broccoli, and cruciferous plants, respectively?

4. Even if hydrogen peroxide is toxic, it is widely used as a topical solution for an antiseptic and gargle, rinse, or discoloration. What percentage of the solution is used for that purpose?

5. The turnover number (K_{cat}) of an enzyme is defined by V_{max} in the presence of one mole of enzyme (the moles of substrate reacted per min per mole of enzyme per catalytic site per unit of time; the unit is reciprocal minutes, min^{-1}). If V_{max} is 3.0 μmole/min in the presence of 1.5 μg of tetrameric enzyme (monomer MW 7500) with each monomer carrying one catalytic site, what is the K_{cat}?

6. What is K_m?

- Is K_m constant for the particular enzyme and substrate?
- Is K_m dependent on or independent of enzyme concentration?
- What does it mean that K_m of a mutant enzyme is lower than K_m of a wild type enzyme?

7. What does it mean when K_m is unchanged between the wild type and the mutant, but K_{cat} is altered?

References

Elavarthi, S. and Martin, B. (2010). Spectrophotometric assays for antioxidant enzymes in plants. *Methods in Molecular Biology*, **639**: 273–280.

Georgescu, R. (2003). High-throughput tetramethylbenzidine (TMB) screen for peroxidases. *Methods in Molecular Biology*, **230**: 171–176.

Montogomery, R. and Swenson, C.A. (1976). *Quantitative Problems in the Biomedical Sciences*, 2nd edition. W.H. Freeman and Company. ISBN 0-7167-0178-2.

Tsutsui, H., Kinugawa, S., and Matsushima, S. (2011). Oxidative stress and heart failure. *American Journal of Physiology and Heart Circulation Physiology*, **301**: H2181–H2190. DOI: 10.1152/ajpheart.00554.2011.

Advanced Methods in Biotechnology (AMB) 1 Laboratory Exercises

16 AMB 1 experiment 16: Aseptic technique and culture handling

Purpose: This is up to you to write down.

Introduction

The aseptic technique is a very important laboratory skill for preventing contamination by unwanted microorganisms. This skill is required for preparing culture media, transferring cultures, inoculating media, isolation of pure cultures, and performing biological tests as a prerequisite for biotechnology. It is not limited to culture but must also be used for any other stock solutions. The passage of cultures to a fresh medium (subculturing) is typically performed when they are in the logarithmic phase before they reach the stationary phase. Both streaking and subculturing are the most widely used techniques for a short-term maintenance and use of a pure stock culture. Subculturing can be performed from solid to liquid medium, liquid to solid medium, liquid to liquid medium, or solid to solid medium. If subculturing is performed on a solid agar medium by either streaking or spreading dilutions, each isolated cell grows into a genetically identical colony as a result of amassed progeny cells derived from a single ancestor. However, even during normal growth in a nutrient-rich medium, some cells may develop into mutants due to DNA changes. Bacteria and eukaryotic microbes produce about one mutation per 300 chromosome replications. In a normal *Escherichia coli* population, the incidence of mutants is estimated to be one in 10^6 to one in 10^7 cells. Because the bacterial population in a rich medium can reach more than one billion in an overnight culture, a certain number of mutants may arise. Nevertheless, it is often necessary to grow a large number of bacteria from a single colony for molecular cloning since most of cloning is performed in a small circular plasmid DNA that replicates independently from the chromosome. In the case of plant and animal tissue cultures, somaclonal variation may arise over long periods of culture. This is why it is important to routinely examine and record the number of subcultures. For a long-term maintenance, a cryoprotectant such as glycerol or dimethylsulfoxide (DMSO) is typically used to store in a –80 °C freezer. It is advisable to scrape off a portion from the top of the frozen stock culture rather than to thaw the culture for inoculum because cells would die by the repeated freezing and thawing cycles.

Genetically engineered *E. coli* K-12 strains are most widely used for recombinant DNA studies. Examples of the engineered mutations are removal of restriction modification systems (*hsdR*-) and endonuclease activity (*endA*-), modification of DNA recombination systems (*recA*-) and deletion of lactose catabolic genes (Δ*lacZYA*). *E. coli* K-12 and its derivatives are not able to utilize sucrose as a sole carbon source. Unlike *E. coli* K-12 strains, *Saccharomyces cervisiae* is able to utilize sucrose but is not able to use lactose as a sole carbon source. In this lab exercise, you will streak cultures of two most widely used model organisms, *E. coli* K12 derivative MM294 and *S. cervisiae*, on to rich complex (LB and YPD) and selective differential (YM, EMB) agar media to observe the difference of their growth patterns on the respective media. In addition, you will set up overnight a culture in liquid broth using the aseptic technique.

It is recommended to watch the following YouTube videos regarding the aseptic techniques to inoculate, isolate, and transfer pure cultures:

- https://www.youtube.com/watch?v=AaG3Pt3nwLQ (Streak plate)
- https://www.youtube.com/watch?v=pg5ZEkHiZ-4 (Subculture)

Pre-lab assignment

(Typing and submission must be completed before lab work begins.)

1. What is the difference between the quadrant and the continuous streak method?
2. What is the main purpose of subculture?
3. Describe a step-by-step procedure on how to perform the subculture steps 13 and 14 in the Part B exercise.
4. What is the purpose of shaking at step 15 of the Part B exercise?

Materials and equipment

- *E. coli* MM294 strain genotype: F⁻ *glnX44(AS), λ⁻, rfbC1, spoT1, thiE1, hsdR17, creC510*
- *Saccharomyces cerevisiae* YNN 281 strain genotype: MATa, *trp1Δ, his3Δ200, ura3-52, lys2-801(amber mutation), ade2-1(ochre mutation), gal, mal, CUP(r)*
- LB (Luria-Bertani) agar plates, LB broth in culture tube
- YPDA (yeast extract, peptone, dextrose, adenine) agar plates, YPDA broth in culture tube
- YM (yeast mold) agar plates
- Levine EMB (eosin methylene blue) agar plates with lactose

Methods in Biotechnology, First Edition. Seung-Beom Hong, M. Bazlur Rashid and Lory Z Santiago-Vázquez.

Companion Website: www.wiley.com\go\hong\Methodsinbiotechnology

- Levine EMB agar plates with sucrose
- Levine EMB agar plates without carbon source
- Inoculating loops, Bunsen burner
- 37 °C and 30 °C incubators and shakers

Procedure

Part A. The streak plate (subculture in solid medium)

1. Spray your lab bench with 70% ethanol and wipe out with a clean paper towel.
2. Label your group number and *E. coli* or *S. cervisiae* with a Sharpie pen on the edge of the bottom of the following agar plates:

E. coli	*S. cervisiae*
(a) LB	LB
(b) YPDA	YPDA
(c) YM	YM
(d) EMB + lactose	EMB + lactose
(e) EMB + sucrose	EMB + sucrose
(f) EMB with no carbon source	EMB with no carbon source

3. Take *E. coli* and *S. cerevisiae* cells using a sterile loop and carry out a quadrant streak as shown below.

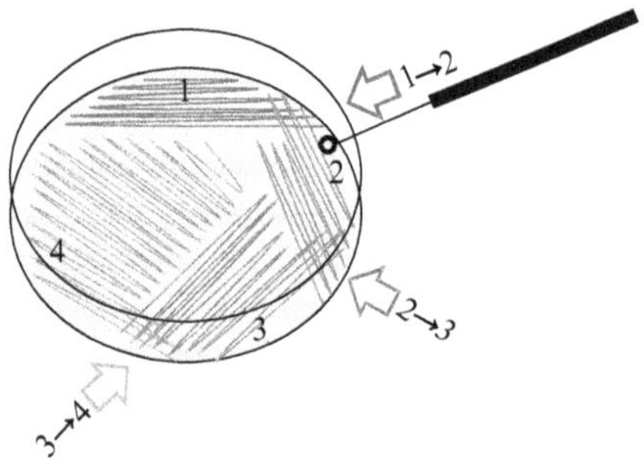

4. Flame an inoculating loop, cool down the loop, and touch a well-isolated single colony or a loopful of broth culture.
5. Lift the bottom of the Petri agar plate in your hand, leaving the lid on the clean bench.
6. Place the inoculating loop at the edge of the agar surface, and glide inoculum over the agar surface in quadrant section 1 so that cells can be evenly spread. Put the agar plate back on to the lid.
7. Flame, cool the loop (confirm by touching the loop on the edge of the agar), rotate the plate to the 1/4 turning left, open the plate lid, make 2 to 3 crossovers from section 2 into section 1, and do a continuous zigzag streak in section 2.
8. Repeat the above step 7 for the next two sections 3 and 4.

**Dot not push the loop on to agar surface; otherwise the loop will cut the agar.*

**If you use a heavy amount of single colony for inoculum, reflame and cool the loop for streaking on to every sector.*

**If you use a tiny amount of a single colony or liquid culture for inoculum, perform the above qudrant streaking without reflaming the loop (→ continuous streaking).*

**Make sure that the flamed loop is cooled down by touching the edge of the agar surface before touching the inoculum. If it sizzles, it is too hot.*

**Do not draw (slow movement) but streak (fast movement) on the surface.*

9. Incubate all *E. coli* plates at 37 °C overnight and all *S. cervisiae* plates at 30 °C for 2 days.

**Invert and stack all the Petri plates in the incubator.*

10. Tabulate the results of growth on each medium.
11. Evaluate your streaking skill based on the result. Examine the colonies of *E. coli* and yeast and note any differences between them.

**All used cultures must be placed into a biohazard bag for disposal.*

Part B. Culture transfer (subculturing in liquid medium)

12. Label your group number and *E. coli* or *S. cerevisiae* with a Sharpie pen on LB and YPDA broth culture test tubes, respectively.
13. Pick up a streaked single colony from the agar plate using a sterile inoculating loop and inoculate into a fresh liquid medium in a culture test tube.

**Do not put the cap on the bench but keep with your little finger.*

14. Remove a liquid culture medium using a sterile inoculating loop and inoculate into a fresh liquid medium in a culture test tube.
15. Shake-incubate *E. coli* at 37 °C, 200 rpm overnight and *S. cerevisiae* at 30 °C, 200 rpm.
16. Tabulate the results of growth on each medium.
17. Examine the turbidity in liquid media cultures and the growth of isolated colonies and any contaminants in the agar media cultures. Discuss in relation to the expected outcomes.

Post-lab assignment

1. How can you be certain that you have a pure bacterial culture on each plate?
2. What is the function of flaming the tube lip right after opening the culture tube cap?
3. Why do you cross over the second streak section back into the first section and from the third section back across the second section?
4. What is the purpose of using YM and Levine EMB with lactose or sucrose? What type of media are they?
5. Why are the inoculated agar plates incubated in an inverted position?

Further reading

Brent, R. and Lech, K. (2000). Unit 1.1. Media preparation and bacteriological tools. In *Current Protocols in Molecular Biology* (Edited by F.M. Ausubel), John Wiley & Sons, Inc. ISBN 0-471-50338-X.

Difco™ & BBL™ Manual of Microbiological Culture Media (2009). 2nd edition. Becton, Dickinson and Company. ISBN 0-9727207-1-5.

Drake, J.W., Charlesworth, B., Charlesworth, D., and Crow, J.F. (1998). Rates of spontaneous mutations. *Genetics*, **148**: 1667–1686.

Lee, J.W., Choi, S., Park, J.H., Vickers, C.E., Nielsen, L.K., and Lee, S.Y. (2010). Development of sucrose-utilizing *Escherichia coli* K-12 strain by cloning β-fructofuranosidases and its application for L-threonine production. *Applied Microbiology and Biotechnology*, **88** (4): 905–913.

Roth, J. and Anderson, D. (2004). Amplification–mutagenesis – how growth under selection contributes to the origin of genetic diversity and explains the phenomenon of adaptive mutation. *Research in Microbiology*, **155**: 342–351.

17 AMB 1 experiment 17: Yeast culture media preparation

Purpose: This is up to you to write down.

Introduction

To prepare a culture medium requires sterilization procedures to kill all unwanted organisms. A liquid medium can be sterilized by either filtration or autoclaving depending on the stability of the components in a pressurized moist heat condition. Agar-based semi-solid medium should be autoclaved because agar melts at high temperatures. A temperature of 121 °C for 15–20 minutes is the accepted standard condition for sterilizing up to one liter of culture medium. The pressure should not exceed 20 psi, as higher pressures may lead to the decomposition of carbohydrates and other components of the medium. For autoclaving, it is important to use a flask or bottle large enough to hold the volume (not more than 50% volume capacity of container) as well as to loosen a screw cap of the bottle in order to avoid boiling over and exploding caused by the internal pressure buildup. The sterilized media should also not be left in the closed autoclave chamber for a prolonged time after autoclaving is finished because the inside hot temperatures will change the pH and the medium composition as manifested by carmelization or darkening of the medium. If heat-labile media supplements are needed, then filter-sterilized solutions should be added aseptically to the autoclaved medium that had been cooled down to about 50 °C, since liquid agar begins to solidify at around 45 °C. Use of high-quality MilliQ H_2O is unnecessary to prepare the culture medium; single-deionized water (dH_2O) is sufficient to prepare the culture medium.

In this lab exercise, you will aseptically prepare a complex and a synthetic defined (SD) yeast medium to use in the next week's lab exercise of a growth curve. It is recommended to watch YouTube video regarding how to make an agar plate:

- http://www.youtube.com/watch?v=OljTdYH_Wtg (Agar plate pouring)
- http://www.youtube.com/watch?v=MsDpyZwpsoY (Removing bubbles from agar plates)

Pre-lab assignment

(Typing and submission must be completed before lab work begins.)

1. List all the components in the SD medium (www.mpbio.com/index.php?cPath=1_7_45).
2. List all the components in the Ura DO (dropout) supplement (www.mpbio.com/index.php?cPath=1_7_45).

Materials and equipment

- YPD (Clontech Cat. No. 630409): 10 g/L yeast extract, 20 g/L peptone, 20 g/L dextrose
- Adenine hemisulfate
- YNB base with ammonium sulfate (Life Technologies Cat No. Q300-09)
- Ura Dropout supplement (ClonTech Cat. No. 630416)
- Uracil
- Bacto-Agar (Difco)
- Petri plates (100 × 15 mm)
- Weigh boats, spatula, labeling tape, aluminum foil
- Analytical balance, magnetic stir plate, magnetic stir bar
- Autoclave
- Water baths (55 °C)
- Laminar floor hood cabinet

Procedure

Part A. YPDA complex medium preparation

1. Prepare 400 mL with 20 g of pre-mixed YPD powder (Clontech) and 0.012 g adenine hemisulfate, or pre-mixed YPDA powder (Clontech). Mix the powder into the water as homogeneously as possible.

Methods in Biotechnology, First Edition. Seung-Beom Hong, M. Bazlur Rashid and Lory Z Santiago-Vázquez.

Companion Website: www.wiley.com\go\hong\Methodsinbiotechnology

2. Dispense 50 mL into a 200-mL flask A and 350 mL into a 1-L flask B. Add 7 g of agar to flask B.

Part B. Synthetic defined medium preparation

3. Prepare 50 mL of liquid in a 200-mL flask C and 350 mL of solid agar media in a 1-L flask D using the components as below.

Components	YNB-glucose (per 100 mL)		Final concentration
	Liquid (g)	Solid (g)	
YNB with $(NH_4)_2SO_4$*	0.67	0.67	0.67%
Glucose*	2	2	2%
Ura DO supplement† (Clontech)	0.077	0.077	0.077%
Uracil	0.20	0.2	0.002%
Agar	—	2.0	2%

*Clontech minimal SD base contains 2% glucose plus 0.67% YNB with $(NH_4)_2SO_4$.
†Ura DO supplement chemical is not dissolved at room temperature. Prepare the liquid medium separately.

4. Cover the opening top with aluminum foil and autoclave at 121 °C, 15 psi for 15 min in a slow exhaust mode.
5. Label YPDA, date, and group number on the edge of bottom side of 15 Petri plates using a Sharpi pen.

**A standard size Petri plate (100 × 15 mm) will hold 20 to 25 mL.*

6. After autoclaving, place the agar media flasks in a 60 °C water bath for 20 min.
7. Turn on the laminar flow hood, disinfect the surface by spraying 70% ethanol, place all the plates, and open the plate lids.
8. Swirl YPDA agar medium in Flask B gently.

**A dense agar solution settles down in the bottom of the flask. However, do not shake vigorously as this will make lots of air bubbles.*

**Remove your flask from the water bath only when you are ready to pour your plates.*

9. Wrap a clean paper towel around the neck of the flask, and hold it to pour the molten agar medium into pre-labeled Petri plates. Remove any air bubbles on the surface of the agar using a sterile Pasteur pipette or push air bubbles to the edge of the plate.

**Wipe out the drenched water outside the flask with a clean paper towel before pouring. This prevents contaminated water from dripping into your sterile Petri plate while pouring.*

**As soon as the liquid medium touches the circled edge of the plate, stop pouring and pour into the next plate.*

**Immediately after pouring, rinse the empty flask with tap water.*

10. Do the same as above (steps 5 to 9) for YNB/Glc medium in Flask D.
11. Allow all the plates to stand with the lid open until they are completely solidified.
12. Close the cover lid, place the agar plate upside down into a vinyl bag, and store at 4 °C.

**You may stack up all the plates upside down into a vinyl bag and incubate at room temperature until next week to check for contamination.*

Post-lab assignment

1. Why is the requirement for amino acids much lower than that for glucose in the defined medium?
2. Explain how a complex medium differs from a synthetic defined medium and describe the usage of complex and synthetic defined media.
3. List the ways to contaminate the stock media and stock solutions.
4. Contamination occurs most frequently in a student lab exercise. List six ways you can contaminate a culture during inoculation and transformation.

Further reading

Bergman, L.W. (2001). Growth and maintenance of yeast. *Methods in Molecular Biology*, **177**: 9–14.

Treco, D.A. and Winston, F. (2000). Unit 13.1. Preparation of yeast media. In *Current Protocols in Molecular Biology* (Edited by F.M. Ausubel). John Wiley & Sons, Inc. ISBN 0-471-50338-X.

18 AMB 1 experiment 18: Growth curve

Purpose: This is up to you to write down.

Introduction

Cell counting is useful for physiological, biochemical, or fermentation studies. For example, if one knows the number of cells present in a culture or wet weight of cells, then one can calculate the amount of protein or metabolite products that can be isolated from that population based on the quantitative yield data. Microbial enumeration is also routinely used in the area of public health. Food or water microbiologists test food, milk, or drinking water for the numbers of microbial pathogens to determine if they are safe for human consumption. Coliform bacteria are referred to as indicator organisms because their presence often indicates the presence of feces from humans and other warm-blooded animals and the concurrent existence of more serious pathogens. The acceptable limits range from 50 to 1000 coliforms per gram of food. This means that the mere presence of contaminating bacteria does not much matter but their concentrations (cell population densities) matter, since many pathogens produce toxic products in a cell density-dependent manner known as the quorum sensing mechanism.

The number of cells that are present in a given population can be determined (i) by using the spectrophotometer to measure the optical density of the population, (ii) by directly counting the microorganisms using a hemocytometer (counting chamber) under a microscope, or (iii) by serially diluting the cells and plating the diluted cells on media that support the growth. The latter method is called viable cell counting because it identifies the living cells in that population. Cell counting by absorbance (A_{600}) measurement is usually limited to 10^7 cells/mL or greater and 10^5 cells/mL or greater in bacterial and yeast cell suspensions, respectively. On the other hand, a few numbers of cells can be determined by direct counting using a microscope or by plate counting.

The cell numbers are typically plotted on a graph as a function of time to make a growth curve. The growth curve has four distinct phases, lag, exponential or logarithmic (log), stationary, and death phases. When a microbe is introduced into a fresh medium, there is no detectable growth in order to adapt to a new environment. This phase is termed the lag phase, and the length of lag phase depends on the previous growth condition of the organism. Following this phase, cells enter the log phase in which they grow rapidly at a constant rate as long as cells have adequate nutrients in favorable conditions. As cells continue to grow and nutrients are depleted, they enter a stationary phase in which growth rate is gradually decreased to no further growth. The subsequent accumulation of toxic waste metabolites in media accelerates the cell death. The rate of mutations significantly increases during the late stationary and death phases and some microbes produce spores to resist this adverse condition.

Growth kinetics during the log phase can be mathematically expressed as $N = (N_0)2^n$ or $n = 3.3\ (\log N - \log N_0)$, where N is the final total number of cells at the end of time interval (t); N_0 is the initial number of cells at the beginning of the time interval (t_0); n is the number of generations determined from the incubation time divided by the doubling time. In the growth curve, the slope of line in the log phase represents a growth rate constant (i.e., the number of doublings per unit of time) K, which can expressed as $K = \Delta Y/\Delta X = (\log N - \log N_0)/(t - t_0)$. Because N would be 2 $\times N_0$ for the generation time or doubling time (dt), the formula becomes: $K = (\log 2N_0 - \log N_0)/\text{dt} = \log 2/\text{dt} = 0.301/\text{dt}$. As a result, $\text{dt} = 0.301/K$.

In this lab exercise, you will use serial dilutions based on the optical density of yeast cell culture, spread on to agar medium, and determine the number of live yeast cells in a time-point culture population. Normal laboratory haploid yeast strains have a doubling time of approximately 90 min in YPDA medium and approximately 140 min in synthetically defined (SD) media at the optimum temperature of 30 °C. However, strains with greatly reduced growth rates in synthetic media are often encountered. Usually yeast strains reach a maximum density of 2×10^8 cells/mL in YPDA medium. Yeast YNN281 has approximately 3×10^7 cells/mL for the optical density at A_{600} of 1.0. For optimal aeration and growth, the medium should constitute no more than 20% of the total volume of the flask in a shaking incubator at 250 to 300 rpm.

Pre-lab assignment

(Typing and submission must be completed before lab work begins.)

1. What environmental conditions affect the growth of microbes?
2. You obtained $A_{600} = 0.08$ for a 1/10 diluted yeast culture. How would you dilute the cell culture if you want to get about 100 colonies after 0.1 mL of the last dilution is plated on to a Petri plate?
3. How would you dilute a 0.5-mL culture at step 8 of the Part C experiment using 1.5-mL microcentrifuge tubes?

Methods in Biotechnology, First Edition. Seung-Beom Hong, M. Bazlur Rashid and Lory Z Santiago-Vázquez.

Companion Website: www.wiley.com\go\hong\Methodsinbiotechnology

4. Six tubes are in a rack. You mix 1 mL of cell culture with 0.5 mL of water in tube #1, transfer 0.1 mL of tube #1 to tube #2 containing 0.9 mL of water and mix well, and the same procedure (0.1 mL + 0.9 mL water) is repeated for the remaining 4 tubes. After plating 0.1 mL of the last dilution tube #6, you obtain 100 colonies. What is the cell density (cells/mL) of the original cell culture?
5. How would you experimentally determine the growth rate of cells?

Materials and equipment

- Calculator, Sharpie pen
- YPDA and SD +Ura liquid media (Experiment 17, Part B)
- Sterile water (~10 mL) in a conical tube for dilution of yeast culture cells
- Autoclaved 1.5 mL microfuge tubes for dilution
- 10 YPDA and SD +Ura agar plates (Experiment 17, Part B)
- Spectronic 20 and Cuvettes
- Denatured ethanol in beaker, glass spreader, and Bunsen burner

Procedure

A. Harvest of time point cultures (TA will perform this)

1. Inoculate *Saccharomyces cerevisiae* YNN281 into 2 mL of both YPDA and SD +Ura liquid media.
2. Shake-incubate overnight (16 to 18 h) at 30 °C, 300 rpm (~10^8 to 5×10^7 cells/mL).

**Doubling time is ~90 to 100 min in YPDA medium and ~200 min in YNB/glucose medium. A_{600} = 1.0 for ~3×10^7 cells/mL.*

**YNN281 growth at 30 °C, 250 rpm for 16 h (6:00 pm to 10:00 am) gave ~2.35 of A_{600} (=7×10^7 cells/mL) when the single colony cells are inoculated into 50 mL of YPD medium in a 250 mL flask.*

3. Inoculate 0.2 mL of each overnight culture into each 100 mL of YPDA and SD +Ura liquid media in a 500-mL sterile Erlenmeyer culture flask.

**YPD and YNB/glucose overnight cultures should be used to inoculate into YPD and YNB/glucose media, respectively.*

4. Swirl the cultures to mix, take 4 mL aliquots of the YPDA and SD +Ura culture into sterile test tubes (label the time point and medium type), and store at 4 °C (→0 h time point).
5. Shake-incubate the culture flasks, and take 4 mL aliquots at 4, 6, 8, 12, and 24 h from a YPDA culture and 4 mL aliquots from 6, 8, 12, and 24 h from an SD +Ura culture. Store all aliquots of cultures at 4 °C.

B. Cell density determination

6. Calibrate Spectronic 20 at A_{600} using 3 mL of YPDA medium as a blank.
7. Vortex all time point cultures to get a uniform cell suspension.
8. Read 3 mL of YPDA cultures at 0 h, 4 h, 6 h, 8 h, and 12 h time points at A_{600}.

**Use YPD medium as a blank to set zero. Read 0 h first, 4 h next, and 8 h last; then you do not need to wash the cuvette when you read each time point culture. After reading, do not put back into the original culture tubes but trash.*

9. If the A_{600} reading is above 1.0, dilute the cells to 1/10 in YPDA medium in a total volume of 3 mL (i.e., 0.3 mL culture + 2.7 mL YPDA) and read the absorbance again.

**The linear range for a spectrophotometer is between 0.05 and 0.7 at A_{600}, so dilution is necessary to obtain good readings. Always mix your culture well before withdrawing a sample.*

10. Calculate the cell density (cells/mL) of each time point culture by $A_{600} \times 3 \times 10^7 \times$ dilution factor (DF).
11. Repeat steps 6 to 10 using YNB cultures (0 h, 6 h, 8 h, 12 h, and 24 h) and YNB medium as a blank and diluent.

C. Serial dilution

12. Dilute the cell culture of each time point in sterile water so that the last dilution tube contains about 500 cells/mL. An example of dilution is shown below.

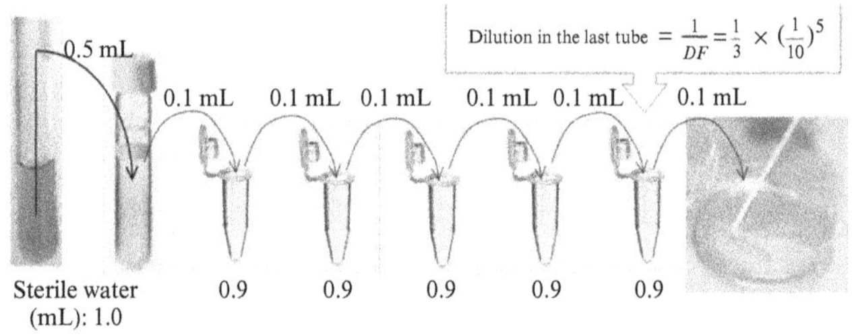

For A_{600} read 0.5 of 1/10 diluted culture, cell density of undiluted yeast culture = $0.5 \times (3 \times 10^7) \times 10 = 1.5 \times 10^8$ cells/mL. The last dilution tube should have 500 cells/mL if 0.1 mL plating yields 50 colonies. Thus, the dilution factor (DF) in the last dilution tube will be (1.5×10^8 cells/mL) ÷ (500 cells/mL) = 3×10^5. Dilution =1/DF = $(1/3) \times (1/10)^5$. This means that for a 1/3 dilution, 0.5 mL of undiluted culture is mixed with 1.0 mL of sterile H_2O and for $(1/10)^5$ dilutions, 0.1 mL of a 1/3 dilution is mixed with 0.9 mL sterile H_2O, and this 1/10 dilution is performed in the same way four more times.
13. Dilute a 0 h time point culture to 1/200 in sterile water for plating if the A_{600} reading is 0.
14. Record the dilution factors (DF) on the plate, group #, and section # on the bottom edge of YPDA and YNB agar plates.

**DF on the plate = DF in the last dilution tube × 10 for a 0.1-mL plating.*

15. Spread 0.1 mL each of the final dilution at each time point evenly on to agar plates.

**Perform duplicate plating at each time point culture. Good spreading is important; sterilize the glass spreader by flaming and cooling down before plating each time point culture.*

16. Incubate all plates upside down at 30 °C; check YPDA plates within 2 days and YNB plates after 2 to 4 days.
17. Count the number of colonies, and calculate cells/mL in each time point of culture based on the dilution factor on the plate.
18. Plot the following data on linear paper:

- Log_{10} cell density (cfu/mL) versus time
- Log_{10} A_{600} versus time

Note: your discussion must include your estimation of the sensitivity of the spectrophotometer based on the A_{600} value versus the viable cell count and the doubling time in YPD and YNB media. Refer to the Post-Lab Assignment 3 to find how to determine the doubling time.

Incubation time	A_{600}	Expected no. of cells/mL calculated	Actual no. of cells/mL observed

Post-lab assignment

1. At 4 p.m. A_{600} of your yeast culture is 1.0. You want to harvest 10 mL culture of yeast cells at A_{600} of 1.0 at noon on the next day. You know that yeast doubles every 2.5 h under the same growth conditions used. How much volume of a 4 p.m. culture would you need to inoculate into a fresh 10 mL medium, assuming that 100% of the 4 p.m. culture cells are alive?
2. Given that a typical *E. coli* culture contains 3×10^9 cfu/mL/A_{600} and that the sensitivity of a spectrophotometer is about 0.005 of A_{600}, what is the smallest bacterial population that can be studied with a spectrophotometer?
3. **(a)** Prepare two graphs: a semi-log graph of A_{600} (y axis) as a function of time (x axis) and a standard linear graph of $log_{10}A_{600}$ as a function of time from the growth data below. Indicate lag, exponential, and stationary phases.

Time	Hours	A_{600}	log A_{600}
10:00	0.00	0.045	
10:30	0.50	0.050	
11:00	1.00	0.070	
11:30	1.50	0.119	
12:00	2.00	0.215	
12:30	2.50	0.375	
01:00	3.00	0.485	
02:15	4.25	0.660	
02:45	4.75	0.740	
03:15	5.25	0.750	

*To prepare a semi-log graph using Microsoft Excel 2010, select and drag over all the data in the two columns to highlight → select "Insert" menu → select "Scatter with only Markers" → select "Layout:" menu → select "Axes" submenu → choose "Primary Vertical Axis ▸" → choose "Show Axis with Log Scale" → Select "Gridlines" submenu → Choose "Primary Horizontal Gridlines ▸" → Choose "Major & Minor Gridlines" → Select "Gridlines" submenu again → Choose "Primary Vertical Gridlines ▸" → Choose "Major & Minor Gridlines" → Label X and Y titles using "Axis Titles" submenu.

(b) Determine the doubling time from the graph by taking the time elapsed between the two points in which A_{600} or log A_{600} has doubled in the exponential growth phase.

**Use a ruler to draw a line through the growing linear phases of the curve. This represents the exponential or log phase. Pick an A_{600} or log A_{600} value on the y axis; note its corresponding time on the x axis. Then, find the point at which the A_{600} or log A_{600} has doubled and note that time. The time elapsed between the two points is the doubling time.*

(c) What is the growth rate constant (K)? Determine the doubling time using the K value.
4. What is bioreactor and how is it useful?
5. As soon as you are hired at the Center for Food Safety of USDA, you are asked to assess the extent of *Salmonella typhimurium* contamination in a tomato. You obtained the tomato from a store, prepared 0.25 g of the skin tissue sample, macerated the tissue in sterile water, filtered through cheesecloth, and brought the final volume up to 50 mL. You then diluted 1 mL of the solution by adding 1 mL to 19 mL sterile water, plated 100 μL each of the diluted solution on to three LB agar plates, and incubated for 24 h at 37 °C. You observed 300, 290, and 250 colonies in the three independent plates.
(a) Relative to the starting tissue sample, what was the final dilution factor (per gram of sample) that was achieved on the plate?
(b) Calculate the number of bacteria per gram of the tissue sample.
6. What is the quorum sensing (QS) mechanism? Provide examples resulting from the physiological process of QS.

Further reading

Boten, E., Little, C., Aird, H., Greewood, M., McLauchlin, J., Meldrum, R., Surman-Lee, S., Tebbutt, G., and Grant, K. (2009). Guidelines for assessing the microbiological safety of ready-to-eat foods placed on the market. Health Protection Agency (https://www.gov.uk/government/uploads/system/uploads/attachment_data/file/363146/Guidelines_for_assessing_the_microbiological_safety_of_ready-to-eat_foods_on_the_market.pdf).

Madigan, M.T., Martinko, J.M., and Clark, D.P. (2009). Microbial growth, Chapter 6. In *Brock Biology of Microorganism*, 12th edition. Pearson Benjamin Cummings. ISBN 0-132-32460-1.

Martins, S.B. and Selby, M.J. (1980). Evaluation of a rapid method for the quantitative estimation of coliforms in meat by impedimetric procedures. *Applied and Environmental Microbiology*, **39**: 518–524.

19 AMB 1 experiment 19: Mini plasmid prep

Purpose: This is up to you to write down.

Introduction

Plasmids are an essential vector tool for genetic manipulation in biotechnology, and nearly all plasmids being used are a genetically engineered form of naturally occurring plasmids that have a wide range of sizes ranging from one to several hundred kb. They typically contain a host-specific origin of replication (*ori*), a multiple cloning site (MCS) for insertion of a foreign DNA, and an antibiotic and/or nutritional selection marker gene. Most engineered plasmid vectors are of a small size less than 20 kb and thus can be easily manipulated and isolated. Several mini-prep procedures have been developed to purify plasmid DNA from small amounts of bacterial culture for rapid screening of a large number of transformants. The purified DNA can then be used for restriction digestion, PCR, and DNA sequencing to determine the size, orientation, and sequence of the insert fragments. Depending on the type of usage, plasmid DNA will be the expression vector for protein expression, cloning vector for propagation of recombinant DNA, shuttle vector for replication in two different host cells, integration vector for targeting a gene of interest to a specific locus of genome, and binary vector for separation of a *cis*-acting DNA element from a *trans*-acting DNA element for transkingdom gene transfer in two separate compatible vectors. The copy number of plasmid greatly influences the yield. Plasmid pUC vector is a copy number mutant of ColE1 type plasmid pBR322 in which the *rop* gene is deleted; rop protein downregulates initiation of plasmid replication to maintain 15 to 20 copies per cell. As a result of this mutation, pUC plasmid replicates in 500 to 700 copies per cell. pRY121 is the expression plasmid that carries the *E. coli* β-galactosidase gene in the shuttle vector that has replication origins of both yeast 2 μm plasmid and ColE1 plasmid. In this lab exercise, you will use a QIAprep Spin Miniprep Kit to isolate two different plasmids, pUC19 and pRY121, which differ in size and copy number. The whole procedure is based on alkaline lysis of *E. coli* cells followed by adsorption of DNA on to silica in the presence of high chaotropic salt. It consists of three steps: (1) preparation and clearing of a bacterial lysate, (2) adsorption of DNA on to the QIAprep silica membrane, and (3) washing and elution of plasmid DNA. For more information, refer to the Introduction (pp. 10–13) and Background (pp. 39–44) information of the QIAprep Miniprep Kit Handbook; it is important to understand what each step accomplishes. To prepare plasmid DNA, *E. coli* is typically grown overnight in LB or TB medium. TB medium is preferred when low copy number plasmid is prepared since it is known to give higher yields of plasmid DNA than LB medium.

It is recommended to watch the following YouTube video regarding plasmid isolation:

- http://www.youtube.com/watch?v=nsv8py5ujA0 (Plasmid mini-prep)
- http://www.youtube.com/watch?v=8xEDEJ0DHFA (Qiagen mini-prep kit)
- http://www.youtube.com/watch?v=04cHNr1rG-w (Mini-prep tips)

After plasmid isolation, you will run an aliquot on to agarose gel in order to check the status of plasmid DNA. Native plasmid is covalently closed circular DNA, which is a compact form in a tightly twisted supercoiled status. However, a relaxed form of nicked open circular DNA is usually found in plasmid prep as a result of shear force-mediated DNA nicks during plasmid DNA preparation. This open circular form of plasmid DNA migrates much more slowly than the supercoiled form of plasmid DNA despite the identical size. The amount of open circular DNA varies depending on the way the samples are handled. The appearance of smearing above and below a distinct band or absence of distinct bands on agarose gel indicates the degradation of DNA.

Pre-lab assignment

(Typing and submission must be completed before lab work begins.)

1. Find out what step of the original protocol is omitted or added in the protocol we use. Why is this step omitted or added? (Refer to QIAprep Miniprep Kit Handbook.)

2. What step must be modified to isolate low-copy plasmids and cosmids, large plasmids (>10 kb)? (Refer to QIAprep Miniprep Kit Handbook.)

3. What plasmid DNAs do we isolate today? What are the sizes of the plasmids? How do two plasmids differ in terms of copy number, replication in host, and screening marker?

4. What is the typical yield of high copy number pUC18 plasmid from a 1.5-mL LB overnight culture when DNA is eluted with 50-μL EB in the QIAprep miniprep system? (Refer to QIAprep Miniprep Kit Handbook.)

Methods in Biotechnology, First Edition. Seung-Beom Hong, M. Bazlur Rashid and Lory Z Santiago-Vázquez.

Companion Website: www.wiley.com\go\hong\Methodsinbiotechnology

Materials and equipment

- QIAprep Miniprep Kit (Qiagen, Cat. No. 27104)
- LE392 (pRY121) (ATCC 37658)
- DH5α (pRY121) overnight culture grown in LB-Amp or TB-Amp media
- DH5α (pUC19) overnight culture grown in LB-Amp or TB-Amp media
- 10× lithium borate buffer (100 mM, pH 8.2)
- 0.5% agarose gel in 1× lithium borate buffer containing 0.5 µg/mL ethidium bromide
- 1 kb DNA ladder (NEB Cat. No. 3232S)
- Microcentrifuge tube rack, 2-mL microcentrifuge tubes
- Horizontal electrophoresis apparatus and power supply

Procedure (adapted from QIAprep miniprep kit)

1. Pick a single colony of DH5α (pUC19) and DH5α (pRY121) from streaked selective plates and inoculate a culture of 2 to 4 ml of LB or TB medium containing 100 µg/mL of ampicillin. Incubate overnight (12 to 16 h) at 37 °C with shaking at 200 rpm.

**Use a tube or flask with a volume of at least 4 times the volume of the culture. TA will do this.*

Before proceeding, each group must label two 2-mL microcentrifuge tubes (pU19) and two 2-mL microcentrifuge tubes (pRY12).

2. Harvest a 2-mL culture of DH5α (pUC19) and a 4-mL culture of DH5α (pRY121) in labeled 2-mL microcentrifuge tubes by centrifuging at 13 000 rpm for 1 min.

**A 4-mL cell culture can be harvested by transferring another 2 mL of culture to the first 2-mL harvest tube, centrifuging, and removing the supernatant again.*

3. Aspirate off the culture supernatant. Repeat steps 2 and 3 for a 4-mL culture harvest.

**Remove as much of the supernatant as possible without disturbing the pellet.*

4. Suspend the bacterial pellet in a 250-µL buffer P1.

**To suspend cells, cap the tube and vortex on the highest setting or strike the bottom of the tube against the tube rack. Look very closely for any undispersed pellet before proceeding to the next step. It is essential that the pellet be completely re-suspended in buffer P1.*

5. Add 250 µL of buffer P2 and mix gently by inverting 4 to 6 times.

**During this step, a viscous bacterial lysate forms. No incubation is necessary. Proceed immediately to do the next step.*

6. Add 350 µL of buffer N3 and mix gently by inverting 4 to 6 times.

**Do not vortex to mix. A white precipitate consisting of cell debris and SDS will form. No incubation is necessary. Proceed immediately to do the next step.*

7. Spin in the microcentrifuge at 13 000 rpm for 10 min.

**A white pellet will form on the bottom and side of the tube after centrifugation. During this centrifugation step, place the necessary number of Qiagen spin columns into the respective number of 2-mL microcentrifuge collection tubes and label each Qiagen spin column appropriately.*

8. Use a P1000 micropipettor set at about 850 µL to transfer all the supernatant to the labeled spin column that had been inserted into the collection tube.

**It is important not to contaminate the spin column with too much white precipitate, which will clog up the column bed.*

9. Spin at 13 000 rpm for 1 min, pour off the flow-through, and place the centrifuged Qiagen spin column back into its 2-mL collection tube.

**Flow-through is the liquid that passed through in the bottom of the collection tube.*

10. Add 750 µL of buffer PE to the above column, and spin at 13 000 rpm for 1 min.

11. Remove the flow-through liquid, place the centrifuged spin column back into its 2-mL collection tube, and spin again at 13 000 rpm for 1 min.

**This step is important to remove any residual ethanol wash solution that might still be trapped in the column.*

12. Place the spin column into a new sterile, labeled 1.5-mL microcentrifuge tube and carefully add 50 µL of buffer EB (pre-heated to 70 °C) to the *center* of the column.

**Do not puncture the spin column membrane using a pipette tip during loading on to the column.*

13. Stand at room temperature for 1 min. Centrifuge at 13 000 rpm for 1 min.

14. Remove the spin column, tightly close the lid* on the tube containing the eluted DNA, and save the spin column and collection tubes for regeneration and re-use.

**Do not touch the inside cap of the tube with your bare finger.*

15. Spot 5 µL of TE buffer plus 2 µL of 6× DNA loading dye to a clean parafilm, and add 5 µL of each mini-prep plasmid DNA to each spot.

16. Load the samples, along with a 5 µL (0.5 µg) of 1 kb DNA ladder on to 0.5% agarose gel in 1× lithium borate buffer (prepared by TA).

17. Run the gel at 300 V for 25–30 min and take a photograph under UV light (see the figure on page 85).

18. Store the plasmid DNA samples at –20 °C.

Post-lab assignment

1. Explain the functional significance of the following reagents for a mini-plasmid prep:

- EDTA
- SDS
- NaOH
- Potassium acetate (K^-Ac^+)
- Guanidine HCl

2. Describe in detail how you would prepare the following reagents:

(a) 100 mL each of 0.5M EDTA (pH 8.0) and 1 M Tris-Cl (pH 8.0) stock solutions

(b) 100 mL of buffer P1 using the above stock solutions

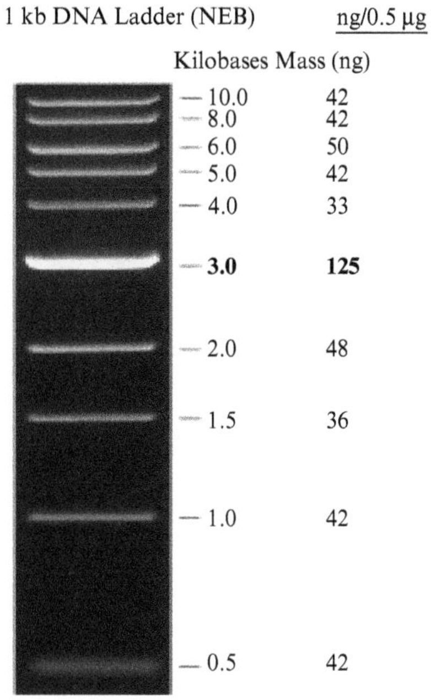

A 1 kb DNA ladder visualized by ethidium bromide staining. Mass values of individual bands are per 0.5 µg marker DNA. (Copyright by NEB. Reprinted with permission)

3. Are your mini-prep DNAs good or bad in terms of DNA status (integrity) based on the agarose gel electrophoresis? If so, how do you know? Do the plasmids migrate to the expected positions as judged by a 1-kb DNA ladder MW marker? If not, what is the reason? What do you have to do if you want to more accurately estimate their sizes (DNA sequencing is not the answer)?

4. When you run the uncut plasmid DNA sample prepared from the purification kit, you had an extra band that migrated faster than supercoiled DNA on an agarose gel. This extra band that migrated just below the supercoiled DNA was still present in the same amount as in the uncut sample when the same amount of uncut plasmid DNA sample was digested by restriction enzymes, indicating that this extra band was resistant to digestion by restriction enzymes. What is the reason? What would you do to troubleshoot this problem?

5. Assuming that you recover 70% of pUC18 plasmid DNA (2686 bp), calculate the total DNA yield (in µg) and DNA concentration (in ng/µL) using the following information:

• Cell density of an overnight *E. coli* culture	2×10^9 cells/mL
• Average MW of double-strand DNA	660 g/mole
• pUC19 plasmid copy number	700 copies per cell
• Overnight culture volume harvested	1.5 mL
• Final DNA sample volume after plasmid mini-prep	50 µL

Additional information

Recycling Qiagen column

1. Save the collection tubes and the columns after elution of DNA.
2. Fill the column with 700 µL of 1 M HCl.
3. Cap and store in an airtight container for at least 24 h (less than a month).
4. Wash the columns and collection tubes in a large beaker of water.
5. Assemble the column and collection tube, add 700 µL of sterile H_2O to the column, spin for 10 s at 13 000 rpm, and discard the water.
6. Repeat the washing step 5.
7. Fill the column with 700 µL of buffer QBT (750 mM NaCl, 50 mM MOPS, pH 7.0, 15% isopropanol, 0.15% Triton X-100), spin down, and discard the buffer.
8. Place the columns in an airtight plastic bag for storage (mark the number of times used; the column can be regenerated up to 20 times).
9. Wash the collection tubes, air dry, and store them for reuse.

Chemical contents of Qiagen reagents for spin column

Buffer	Type	Composition
P1	Cell suspension buffer	50 mM Tris-HCl, pH 8.0 10 mM EDTA, pH 8.0 → Autoclave and cool down 50-100 µg/mL RNaseA* (store at 4 °C after adding RNase A)
P2	Alkaline lysis buffer	0.2 M NaOH 1% SDS (dissolve at 37 °C if precipitates form)
N3	Neutralization buffer	4.2 M guanidine-HCl† 0.9 M potassium acetate pH 4.8 with acetic acid (dissolve at 37 °C if precipitates form)
PB	Wash buffer I	5.0 M guanidine-HCl 20 mM Tris-HCl, pH 6.6 Add ethanol (molecular biology grade) to a final 38%
PE	Wash buffer II	20 mM NaCl (sterile) 10 mM Tris-HCl pH 7.5 (sterile) Add ethanol (molecular biology grade) to a final 80%
EB	Elution buffer	10 mM Tris-HCl, pH 8.5 (sterile)

*DNase-free RNase A (Sigma, for molecular biology grade, R6513-10MG).
†Guanidine-HCl: (Sigma, for molecular biology, ≥99% product # G3272-100G).

Buffer notes

- Add the provided RNase A solution to buffer P1, mix, and store at 4 °C.
- Add ethanol (molecular biology grade 100%) to buffer PE before use.
- Check buffers P2 and N3 before use for salt precipitation. Redissolve any precipitate by warming to 37 °C. Do not shake buffer P2 vigorously.
- Close the bottle containing buffer P2 immediately after use to avoid acidification of buffer P2 from CO_2 in the air.

References

Buffer composition by Qiagen MiniPrep. http://www.my-whiteboard.com/buffer-composition-for-qiagen-miniprep-midiprep-maxprep-pcr-clean-up-and-gel-extraction-kits/.

Formulas for Qiagen Kit Buffers. http://openwetware.org/wiki/Qiagen_Buffers.

QIAprep Spin Miniprep Kit (Catalog No. 27104) Handbook, 12/2006.

Further reading

Chromatographic Purification and Separation Process for Mixtures of Nucleic Acids. May 7, 2002. US Patent 6,383,393 B1.

Siddappa, N.B., Avinash, A., Venkatramanan, M., and Ranga, U. (2007). Regeneration of commercial nucleic acid extraction columns without the risk of carryover contamination. *BioTechniques*, **42**(2): 186–192.

Tartoff, K.D. and Hobbs, C.A. (1987). Improved media for growing plasmid and cosmid clones. *Bethesda Research Laboratory Focus*, **9**: 12.

Yanisch-Perron, C., Vieira, J., and Messing, J. (1985). Improved M13 phage cloning vectors and host strains: nucleotide sequences of the M13mp18 and pUC19 vectors. *Gene*, **33**, 103–119.

20 AMB 1 experiment 20: Restriction digestion, purification, concentration, and quantification of DNA

Purpose: This is up to you to write down.

Introduction

DNA and RNA absorb light maximally at a wavelength of 260 nm. Because of this property, they can be quantified spectrophotometrically using the Beer–Lambert equation: $A = \varepsilon cl$, where A = absorbance, ε = extinction coefficient, c = concentration of DNA or RNA in solution, and l = the length (cm) of the light path through the solution. Extinction coefficients of nucleic acids are 0.020 $(\mu g/mL)^{-1}cm^{-1}$ for double-stranded DNA, 0.027 $(\mu g/mL)^{-1}cm^{-1}$ for single-stranded (ss) DNA, 0.025 $(\mu g/mL)^{-1}cm^{-1}$ for ssRNA, 0.030 $(\mu g/mL)^{-1}cm^{-1}$ for single-stranded oligonucleotides of 60–100 bp long, and 0.040 $(\mu g/mL)^{-1}cm^{-1}$ for single-stranded oligonucleotides of less than 40 bp long. These values mean that A_{260} of 1 equals 50 µg/mL dsDNA, 37 µg/mL ssDNA, 40 µg/mL ssRNA, 33 µg/mL ss oligonucleotides of 60–100 bp, and 25 µg/mL ss oligonucleotides of less than 40 bp. Low salt buffers such as TE, Tris, and phosphate buffers (pH 7.0–8.0) give very stable readings, whereas water gives up to ± 14% fluctuated readings.

The purity of a nucleic acid solution can also be determined by calculating the A_{260}/A_{280} ratio. Pure DNA and RNA have an A_{260}/A_{280} ratio of 1.8 to 2.0 and 2.0, respectively. The ratio of a nucleic acid sample contaminated with protein or phenol is less than the values given above. If there is a significant protein contamination as judged by the A_{260}/A_{280} ratio, protein can be removed by phenol extraction, and this must be followed by ethanol precipitation. It is recommended to watch the following YouTube video as to how to extract DNA with phenol:

- http://www.youtube.com/watch?v=ZjbG1efem2M

Microvolume nucleic acid samples can be directly quantified without the use of cuvettes using NanoDrop 2000c spectrophotometer:

- http://www.jove.com/video/2565/nanodrop-microvolume-quantitation-of-nucleic-acids

Due to the simple, rapid, and non-destructive measurement, absorption spectroscopy has been most frequently used to measure the amount of DNA and RNA. However, it is relatively insensitive since it requires at least 2 µg/mL to obtain reliable estimates of A_{260}. Moreover, absorption spectroscopy cannot readily distinguish between DNA and RNA, and it cannot be used with crude nucleic acid preparations due to the interference. Electrophoresis through mini-agarose gel provides a rapid and convenient way not only to estimate the quantity of DNA but also to analyze its physical state and integrity, and to check if the sample contains detectable amounts of RNA at the same time.

DNA precipitation and washing by ethanol is most widely used to concentrate DNA from dilute solution and to remove salt. The lowest DNA concentration that can be precipitated in the absence of carrier after a 15-min incubation on ice is known to be 20 ng/mL. For more information on ethanol precipitation, read an article (https://tools.lifetechnologies.com/Content/Focus/Focus%20Volume%207%20Issue%204.PDF). This study is based on a 3.2 kb DNA fragment. However, small DNA molecules (<200 bp) are inefficiently precipitated by ethanol; a long incubation time is critical for precipitation of small DNA fragments (incubate at least 1 h at −80 °C and wash the pellet in 95% ethanol). Adjusting the solution of small DNA molecules to a final concentration of 0.01 M $MgCl_2$ prior to addition of ethanol also greatly improves its recovery efficiency.

To clean up a DNA sample, either the silica method or the phenol/chloroform/isoamylalcohol extraction method can be used. DNA purification by the silica method is superior to the organic extraction method, because unwanted oligonucleotide primers (<50 bp) and impurities such as salts, proteins, RNA, oils, carbohydrates, ethidium bromide, dyes, unincorporated dNTP, detergents, residual phenol, and chloroform do not bind to the microparticulate silica. In the presence of high chaotropic salts, DNA binds to silica particles with great affinity. The chaotropic salts, such as guanidine thiocynate and sodium iodide denture proteins and solubilize agarose impurities including denatured proteins and agarose monomers, can be washed away from the silica with ethanol-containing buffer. Contrary to DNA adsorption, DNA elution is most efficient under basic pH and low salt concentrations. As a result, the silica method is efficient and convenient for purifying DNA fragments as short as 100 bp after many enzymatic reactions and gel extraction. Many silica column-based commercial kits are available for a DNA clean-up.

In this lab exercise, you will cut pUC19 plasmid vector DNA with restriction enzymes, purify by either organic extraction

Methods in Biotechnology, First Edition. Seung-Beom Hong, M. Bazlur Rashid and Lory Z Santiago-Vázquez.

Companion Website: www.wiley.com\go\hong\Methodsinbiotechnology

followed by ethanol precipitation or by the silica method, and quantify the purified DNA by a band intensity comparison.

Pre-lab assignment

(Typing and submission must be completed before lab work begins.)

You will perform both this experiment and Experiment 21 today. Describe a flow diagram of the overall procedure in the quickest way to finish these experiments. You do not need to describe the detailed procedures. However, your description should include which part is the first, second, third, fourth, etc., with an approximate time needed for performing each part of the experiments.

Materials and equipment

- 10× lithium borate buffer (200 mM, pH 8.2)
- 10× DNA loading dye
- pUC19 plasmid mini-prep (0.07 to 0.1 μg/μL; Experiment 19)
- *Sma*I, *Eco*RI-HF, *Hin*dIII-HF (NEB)
- MW marker standard DNA (λDNA/*Hin*dIII+*Eco*RI or 1 kb DNA Ladder, NEB N3232S)
- Phenol/chloroform-isoamylalcohol (25:24:1 (v/v/v))
- CIA (chloroform-isoamylalcohol; 24:1, v/v)
- Isopropanol, 75% ethanol
- Finely ground amorphous silica particles (Sigma S-5631) suspension (100 mg/mL in PB buffer)
- PB buffer (5 M guanidine-HCl, 20 mM Tris-HCl pH 6.6, 30% isopropanol)
- PE buffer (10 mM Tris-HCl pH 7.5, 80% ethanol)
- Elution buffer (10 mM Tris-HCl pH 8.5), TE buffer (pH 8.0)
- Agarose (low EEO type), ethidium bromide (10 mg/mL)
- Horizontal electrophoresis apparatus, power supply
- 25 °C and 37 °C water bath

Procedure

Part A. Restriction enzyme digestion

1. **(a)** Digest two sets of 1 μg of pUC19 (2686 bp) with *Sma*I in two sterile 1.5-mL microtubes as follows. One set is used for Parts B and C, and the other for Part D.

Addition order	Component	Volume (μL)
1	Sterile dH_2O	*x*
2	10× NEB buffer 4	2
3	pUC19 DNA	*y*
4	*Sma*I	1.5
	Total volume	20

Mix gently (avoid bubbles!) and incubate at *25 °C* for 15 min.

(b) Digest 1 μg of pUC19 with *Eco*RI + *Hin*dIII in a sterile 1.5-mL microtube as follows. This will be used for Part D.

Addition order	Component	Volume (μL)
1	Sterile dH_2O	*x*
2	10× NEB buffer 4	2
3	pUC19 DNA	*y*
4	*Eco*RI-HF	1
4	*Hin*dIII-HF	1
	Total volume	20

Mix gently (avoid foams!) and incubate at 37 °C for 15 min.

2. During incubation time, prepare 0.5% agarose mini-gel in 1× lithium borate buffer containing 0.5 μg/mL of ethidium bromide.

**This gel is used for Part E of this experiment and PCR DNA samples for Experiment 21.*

Part B. DNA extraction by phenol/chloroform/isoamylalcohol

3. Add 180 μL TE buffer (pH 8.0) to one pUC19/*Sma*I digestion sample (total volume 200 μL).
4. Add 1/10 volume (20 μL) of 3 M sodium acetate (pH 5.2).
5. Add an equal volume of the phenol–CIA mixture and vortex.
6. Spin at the maximum speed for 3 min in a microcentrifuge.
7. Remove the upper phase carefully and transfer to a new microfuge tube.
8. Add an equal volume of CIA and vortex.
9. Spin at the maximum for 1 min in a microcentrifuge.
10. Remove the upper phase carefully and transfer to a new microfuge tube.

**Do not dispose of the extracted phenol CIA wastes in the regular waste but in the fume hood.*

Part C. DNA Precipitation by ethanol

11. Add an equal volume of isopropanol, mix well by inverting the tube several times, and stand at room temperature for 5 min. Alternatively, you may add two volumes of cold 95% ethanol and incubate at 0 °C for 10 min.
12. Spin at the maximum for 10 min at room temperature for isopropanol precipitation.
13. Decant the supernatant and tap on to a paper towel to remove the residual fluid.
14. Rinse the DNA pellet by adding 500 μL of 75% ethanol and decanting as above.
15. Allow the pellet to air dry in the laminar-flow hood until no fluid is present.
16. Add an appropriate volume (~20 μL for 1 μg DNA) of elution buffer, mix gently, and pulse spin.

Part D. DNA purification by silica

17. Add 5 volumes of PB buffer to pU19/*Sma*I and pUC19/*Eco*RI+*Hin*dIII (Part A).
18. Add 10 μL (~1 mg) of silica suspension and vortex gently.

**Mix very well before use; 1 mg silica binds ~3 to 4 μg of DNA.*

19. Stand at room temperature for 5 min with occasional mixing to keep the silica powder from sedimenting.

**If a large amount of DNA is purified or if the volume of the binding reaction is greater than 1.5 mL, increase the incubation time of the binding step to 15 min.*

20. Spin at the maximum for 10 s in a microcentrifuge and aspirate off the supernatant.
21. Pulse spin and remove the residual liquid carefully.
22. Suspend the pellet in 500 μL of PE buffer.

**For DNA fragments ≥5 kb, suspend the pellet by inverting the tube. Vortexing can cause shearing of large DNA molecules.*

23. Spin at the maximum for 10 s in a microcentrifuge and aspirate off the supernatant.
24. Repeat steps 22 and 23 twice more.
25. Allow the pellet to air dry in the laminar-flow hood for ~10 min.
26. Add 32 μL of elution buffer and vortex gently.

**Elution efficiency depends on pH and temperature. The maximum elution efficiency is achieved between pH 7.0 and 8.5 at 50–80 °C.*

27. Incubate at 55 °C for 5 min with occasional mixing. Spin at the maximum for 30 s in a microcentrifuge.
28. Transfer the supernatant to a new microfuge tube.

**Typically yields are >80%.*

Part E. DNA quantification by EtBr fluorescence-based "band-brightness" comparison

29. Spot 8 μL of TE buffer and 1 μL of 10× loading dye on to the clean side of Parafilm.
30. Take 2 μL of the DNA sample out from steps 16 and 28 and mix with loading dye on the parafilm. Store the remaining DNA samples at −20°C until use (Experiment 22).
31. Load the above DNA sample on to a 0.5% agarose gel in 1× lithium borate buffer containing 0.5 μg/mL ethidium bromide, along with 10 μL (0.5 μg) of phage *λ*DNA/*Hind*III *Eco*RI MW marker standard DNA or 1 kb ladder DNA (0.5 μg).

**Load PCR DNA samples (Experiment 21) on to the same gel.*

32. Electrophoresis at 300 V (30 V/cm) for 20–25 min.
33. Visualize DNA bands under a UV transilluminator and photograph the gel.
34. Estimate the DNA concentration by a band brightness comparison and record the result. (These data are required for Experiment 22).

**See an example of the estimation below:*

Lane 1: 2 μL DNA sample
Lane 2: 2 μL DNA sample
Lane 3: 5 μL (0.1 μg/μL) of 1-kb DNA ladder (reproduced with permission of NEB)

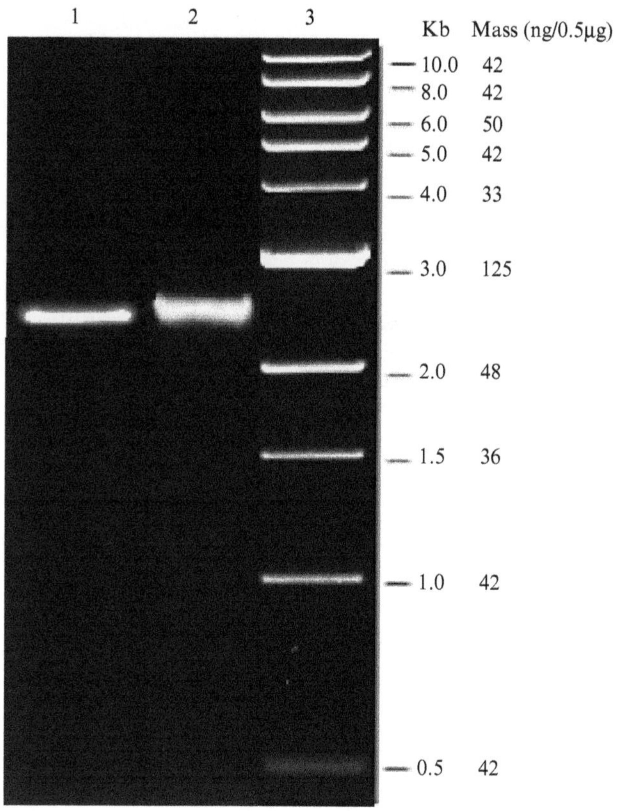

The intensity of the DNA band in Lane 1 is approximately the same as that of a 6.0-kb marker band that contains 50 ng of DNA. Thus, 2 μL of DNA ≈ 50 ng; the concentration of Lane 1 DNA = 25 ng/μL.

The intensity of the DNA band in Lane 2 is approximately the same as that of a 3.0-kb marker band that contains 125 ng of DNA. Thus, 2 μL of DNA ≈ 125 ng; the concentration of Lane 1 DNA = 62.5 ng/μL.

35. Compare the yield of the silica method with that of the phenol/CIA extraction/ethanol precipitation method. Calculate the % recovery yield of each method: (initial DNA amount used for reaction ÷ final DNA amount obtained after purification) × 100.

Post-lab assignment

1. What are the major advantages of the ethidium bromide-visualized DNA quantification method over the NanoDrop spectrometric quantification method?
2. What steps would you take if your DNA preparation had an A_{260}/A_{280} ratio of 1.40 or 1.98?
3. You have 1 mL of plasmid DNA sample with a concentration of 0.01 mg/mL. Using 100% and 70% cold ethanol and 3 M sodium acetate (pH 5.2), you want to concentrate the DNA sample to a final concentration of 0.1 μg/μL in TE buffer. Describe the detailed step-by-step procedure including volumes of each solution to be used as to how you concentrate your DNA sample.
4. What are the roles of salt and ethanol in DNA precipitation?
5. What are the effects of incubation temperature, incubation time, and centrifugation time in ethanol precipitation of DNA? (Read a short article: https://tools.lifetechnologies.com/Content/Focus/Focus%20Volume%207%20Issue%204.PDF)
6. The GenBank sequence accession number of pUC19 is L09137.

- Visit NCBI GenBank database (http://www.ncbi.nlm.nih.gov/genbank/). On the main entry page, you will see a search menu bar at the very top. Select (v) "Nucleotide" from the drop-down search bar, type in the accession number in the blank box, and then click "Go."

- Copy and paste the numbered sequences of pUC19 in Courier New font (size 10) into the Microsoft Word document.
- Note the position of polylinker *Hin*dIII-*Sph*I-*Pst*I-*Sal*I-*Xba*I-*Bam*HI-*Sma*I-*Kpn*I-*Sac*I-*Eco*RI in the annotation and locate the *EcoRI* and *Hin*dIII sites.
- On the basis of the following restriction cut ends produced, remove the polylinker sequence *starting from each 5′ cut end sequence*. Remember that the pUC19 sequence is shown in the 5′ to 3′ direction in only one strand and that pUC19 is a circular dsDNA.

*Eco*RI (5' overhang 5'-AATT-3')

```
5′· · · G▼A A T T C · · ·3′
3′· · · C T T A A▲G · · ·5′
```

*Hin*dIII (5' overhang 5'-AGCT-3')

```
5′· · · A▼A G C T T · · ·3′
3′· · · T T C G A▲A · · ·5′
```

- (a) What is the sequence of the smaller fragment that is removed by *HindIII* + *EcoRI* double digestion? Write *both top and bottom strand nucleotides.*
- (b) What is the linearized pUC19 sequence after *HindIII* + *EcoRI double* digestion? When you describe the linearized sequence, begin with the *cut 5′ end* sequence of *Hin*dIII site and terminate with *the same strand cut 5′ end* sequence of *Eco*RI site to assemble.

To re-format the DNA sequence, visit online DNA Tools software:

- http://www-bimas.cit.nih.gov/molbio/readseq/

- Type ">pUC19/H3+RI" and copy and paste the entire DNA sequence on the next line in the box (→ FASTA format). You do not need to remove the position numbers on the pasted sequence; they will be automatically removed and renumbered after formatting.
- Select "Genbank/GB" in the output format drop-down menu and click "submit."

Or:

- http://www.ebi.ac.uk/Tools/sfc/emboss_seqret/

- Type ">pUC19/H3+RI" and copy and paste the entire DNA sequence on the next line in the box (→ FASTA format). You do not need to remove the position numbers on the pasted sequence; they will be automatically removed after formatting.
- Choose "DNA" from the drop-down menu of sequences.
- Select "FASTA format including NCBI-style IDs" in the INPUT FORMAT and "Genbank entry format" in the OUTPUT FORMAT.
- Click "Submit" and save the FASTA format sequence that will be used for Post-Lab Assignment 5 of Experiment 21.

Further reading

Boyle, J.S. and Lew, A.M. (1995). An inexpensive alternative to glassmilk for DNA purification. *Trends in Genetics*, **11** (1): 8.

Desjardins, P. and Conklin, D. (2010). NanoDrop microvolume quantitation of nucleic acids. *Journal of Visual Experiments*, (**45**), e2565. DOI:10.3791/2565.

Moore, D. (2003). Unit 2.1A. Purification and concentration of DNA from aqueous solutions. In *Current Protocols in Molecular Biology* (Edited by F.M. Ausubel). John Wiley & Sons, Inc. ISBN 0-471-50338-X.

Zeugin, J.A. and Hartley, J.L. (1985). Ethanol Precipitation of DNA. *Focus* Vol. **7**, No 4. Bethesda Research Laboratories Life Technologies Inc. (https://tools.lifetechnologies.com/Content/Focus/Focus%20Volume%207%20Issue%204.PDF).

http://www.ogt.com/resources/literature/483_understanding_and_measuring_variations_in_dna_sample_quality

21 AMB 1 experiment 21: Polymerase chain reaction (PCR)

Purpose: This is up to you to write down.

Introduction

PCR is used for a wide variety of applications including gene cloning, assembly and fusion, colony screening, molecular genotyping, mutagenesis, sequencing, forensics, and gene expression analyses. PCR is routinely performed using a thermostable polymerase in a thermal cycler that works like a molecular photocopier to produce billions of the same copies of DNA from tiny amounts of initial template DNA during the repeated cycles of heat denaturation, annealing, and extension. The exponential phase of PCR amplification can be mathematically expressed as $N_f = N_i \times (1 + E)^n$, where N_f is the final number of amplified molecules, N_i is the initial number of template molecules, n is the number of amplification cycles, and E is the amplification efficiency expressed as a decimal fraction. E is 1 when the efficiency is 100%. Assuming overall 70% efficiency and 10^5 template molecules, about 10^{12} amplified molecules will be produced after 30 cycles. The efficiency, sensitivity, specificity, and fidelity of PCR are all dependent on the conditions of PCR components and cycling parameters. More background information on the optimization of PCR components and cycling conditions is given at the end of the protocol. Inclusion of positive and negative control reactions is essential to check whether the PCR conditions used can successfully amplify the target sequence.

In this lab exercise, you will amplify a small portion of *lacZ* DNA from *Escherichia coli* chromosome and plasmid pRY121. You will also amplify the GFP_{uv} coding sequence from pGLO plasmid. You will then analyze 1/10th volume of PCR sample by agarose gel electrophoresis in order to check for the specificity and efficiency of PCR. The specificity is assessed by the number and size of band, whereas the efficiency is assessed by the quantity (intensity) of the target band. If the PCR band is of a single expected size, then the PCR reaction sample can be directly purified using the spin column method. However, if the PCR band is not specific, you have to cut out a specific band from the gel and extract. All of your PCR fragments will be purified by either method and cloned into pUC19 vector in Experiment 22 next week.

Pre-lab assignment

(Typing and submission must be completed before lab work begins.)

1. What factors affect the efficiency of the PCR process if an optimized reaction buffer is used?
2. What factors affect the fidelity of the PCR process if an optimized reaction buffer is used?
3. What factors affect the specificity of the PCR process if an optimized reaction buffer is used?
4. What are the differences between the GoTaq® and Phusion® DNA polymerase in terms of the efficiency and fidelity of the PCR process?
5. Why should the *Taq* polymerase PCR reaction (step 1) be set up on ice?

Materials and equipment

- Plasmid pRY121 (~25 ng/μL)
- *E. coli* genomic DNA
- Plasmid pGLO (~50 ng/μL)
- LacZ primer 1: GTTGTTGCAGTGCACGGCAG
- LacZ primer 2: GCTGGAATTCCGCCGATACTG
- GFP SLIC primer 1: AAAACGACGGCCAGTGAATTGGTACCGAGCTCGAATTC
- GFP SLIC primer 2: GACCATGATTACGCCAAGCTTGGCTAGCAAAGGAGAAG
- 10 mM dNTP mix (Promega, Cat. No. U151)
- GoTaq® DNA polymerase (Promega, Cat. No. M3001; 5 units/μL) and 5× buffer
- Phusion® High-Fidelity DNA polymerase (NEB M0530S; 2 units/μL) and 5× buffer
- Three thermal cyclers or one thermal cycler with three independently controlled blocks
- Thin-walled 0.2-mL PCR tube
- Agarose, 10× lithium borate buffer, ethidium bromide (10 mg/mL)
- 10× or 6× DNA loading dye

Methods in Biotechnology, First Edition. Seung-Beom Hong, M. Bazlur Rashid and Lory Z Santiago-Vázquez.

Companion Website: www.wiley.com\go\hong\Methodsinbiotechnology

- DNA MW marker (1 kb DNA ladder, NEB Cat. No. N3232S)
- Ice box

Procedure

1. Add the following components to your labeled thin-walled PCR tube *on ice* in this order.

Taq-LacZ/pRY121

Component (for plasmid pRY121 PCR)	Volume (μL)	Final
Nuclease-free sterile dH_2O	35.25	
5× Green GoTaq buffer	10.0	1×
10 mM dNTPs	1.0	200 μM each
10 μM LacZ primer 1 + 2 mix	2.5	0.5 μM each
pRY121 plasmid (1 ng/μL)	1.0	1 ng
GoTaq® DNA polymerase (5U/μL)	0.25*	0.025 U/μL

Taq-LacZ/genome

Component (for genomic PCR)	Volume (μL)	Final
Nuclease-free sterile dH_2O	35.25	
5× Green GoTaq buffer	10.0	1×
10 mM dNTPs	1.0	200 μM each
10 μM LacZ primer 1 + 2 mix	2.5	0.5 μM each
E. coli genome DNA	1.0	50 ng
GoTaq® DNA polymerase (5U/μL)	0.25*	0.025 U/μL

*Pipette carefully and slowly because the high content (50%) of glycerol in the storage buffer may lead to pipetting errors.

2. Program the thermal cycler I using the following parameters.

Cycle	Steps	Temperature (°C)	Time
1	Initial denaturation	94	4 min
25	Denaturation	94	30 s
	Annealing	56	30 s
	Extension	72	36 s
1	Final extension	72	5 min
1	Hold	4	∞

3. Begin the program cycle. Once the initial denaturation temperature reaches to 90 °C, quickly spin your sample tube and place in the thermal cycler.

4. Add the following components to your labeled thin-walled PCR tube *on ice* in this order.

Phusion-LacZ/pRY121

Component (for plasmid pRY121 PCR)	Volume (μL)	Final
Nuclease-free sterile dH_2O	35.0	
5× Phusion HF buffer	10.0	1×
10 mM dNTPs	1.0	200 μM each
10 μM LacZ primer 1 +2 mix	2.5	0.5 μM each
pRY121 plasmid (1 ng/μL)	1.0	1 ng
Phusion® DNA polymerase (2U/μL)	0.5*	0.02 U/μL

*Pipette carefully and slowly because the high content (50%) of glycerol in the storage buffer may lead to pipetting errors. Set up in ice because 3′→5′ exo activity degrades primers and template.

5. Program the thermal cycler II using the following parameters.

Cycle	Steps	Temperature (°C)	Time
1	Initial denaturation	98	30 s
30	Denaturation	98	10 s
	Annealing	70	10 s
	Extension	72	9 s
1	Final extension	72	5 min
1	Hold	4	∞

6. Begin the program cycle. Once the initial denaturation temperature reaches 90 °C, quickly spin your sample tube and place in the thermal cycler.

7. Add the following components to your labeled thin-walled PCR tube *on ice* in this order.

Phusion-GFP/pGLO

Component (for plasmid pGLO PCR)	Volume (μL)	Final
Nuclease-free sterile dH_2O	35.0	
5× Phusion HF buffer	10.0	1×
10 mM dNTPs	1.0	200 μM each
10 μM GFP SLIC primer 1 +2 Mix	2.5	0.5 μM each
pGLO plasmid (1 ng/μL)	1.0	1 ng
Phusion® DNA polymerase (2U/μL)	0.5*	0.02 U/μL

*Pipette carefully and slowly because the high content (50%) of glycerol in the storage buffer may lead to pipetting errors. Set up in ice because 3′→5′ exo activity degrades primers and template.

8. Program the thermal cycler III using the following parameters.

Cycle	Steps	Temperature (°C)	Time
1	Initial denaturation	98	30 s
30	Denaturation	98	10 s
	Annealing	58	10 s
	Extension	72	12 s
1	Final extension	72	5 min
1	Hold	4	∞

9. Begin the program cycle. Once the initial denaturation temperature reaches 90 °C, quickly spin your sample tube and place in the thermal cycler.

10. Prepare a 0.5% agarose gel containing 0.5 μg/mL of ethidium bromide in 1× lithium borate buffer.

**The same gel prepared in Experiment 20 (Part A, step 2) is used to load PCR DNA samples.*

11. Spot 6 μL of TE buffer on to a clean Parafilm and add 2 μL of 6× DNA loading dye and 4 μL of PCR sample.

12. Load the 4 PCR samples and 1 kb DNA ladder 5 μL (0.5 μg) on to a 0.5% agarose gel.

13. Carry out electrophoresis at 300 V (30 V/cm) for 20–25 min.

14. Visualize DNA bands under a UV transilluminator and photograph the gel.

15. Determine the specificity of PCR based on the size and number of PCR product.

**Ideally only a single band of expected size should be present.*

16. Estimate DNA concentration by band brightness comparison (see an example in Part E of Experiment 20).

**These data are required for Experiment 22.*

**A nanodrop assay of PCR sample does not accurately quantify DNA unless you use PicoGreen assay.*

17. Label Group #, sample name, and concentration on each tube: Taq_LacZ/pRY121, Taq_LacZ/genome, Phusion_LacZ/pRY121, Phusion_GFP/pGLO.

18. Store all the PCR samples at –20 °C.

**All these samples will be used for PCR cloning in Experiment 22.*

Post-lab assignment

1. Suppose your PCR reaction yielded multiple products in addition to the one you desired.What changes would you make in the reaction to increase specificity without using different primers? Assume that you used the optimized PCR mater mix.

2. If a yeast cell contained a nucleotide base mutation on an allele of a chromosome and the region flanking the mutation was PCR-amplified using DNA isolated from the yeast mutant cells, approximately what percentage of amplicons would contain this mutation?

3. If your PCR did not yield any product, suggest four things you could alter to obtain a product. Explain the rationale behind each suggestion, assuming that you used the optimized PCR mater mix.

4. A GenBank accession number of *E. coli* gene *lacZ* is V00296. The nucleotides of *lacZ* primers 1 and 2 are given in Materials and Equipment.

- Visit the NCBI GenBank database (http://www.ncbi.nlm.nih.gov/genbank/). On the main entry page, you will see a search menu bar at the very top. Select (*v*) "Nucleotide" from the drop-down search bar and type the accession number in the blank box; then click "Search."
- Copy and paste the sequence of *lacZ* as Courier New font size 10 into a Microsoft Word document.
- Visit NCBI Align Sequences Nucleotide BLAST (blastn): http://blast.ncbi.nlm.nih.gov/Blast.cgi?PAGE_TYPE=BlastSearch&PROG_DEF=blastn&BLAST_PROG_DEF=megaBlast&BLAST_SPEC=blast2seq.
- Copy the *lacZ* primer 1 sequence and paste on to the "Enter Query Sequence" box.
- Click "Align two or more sequences" to display "Enter Subject Sequence" box, type V00296 in the box, and click "BLAST."
- Copy the highest match in the Alignments and paste as HTML format into Microsoft Word.
- Underline the matching sequence in the primer 1 sequence to indicate the *lacZ* overlap sequence.
- Repeat the procedure using the *lacZ* primer 2 sequence and V00296.

Based on the obtained information, write the sequence of *lacZ* amplicon and underline the *lacZ* primer 1 and 2 sequences.

5. The GenBank sequence accession number of pGLO (= pBAD-GFPuv) is U62637. The sequences of GFP SLIC primers 1 and 2 are given in Materials and Equipment.

- Have your FASTA format sequence of the linearized pUC19 cut with *Hind*III + *Eco*RI from the Post-Lab Assignment 6 of Experiment 20.
- Visit NCBI GenBank database (http://www.ncbi.nlm.nih.gov/genbank/). On the main entry page, you will see a search menu bar at the very top. Select (*v*) "Nucleotide" from the drop-down search bar and type in U62637 in the blank box; then click "Search."
- Copy and paste the numbered sequences of pGLO in Courier New font (size 10) into a Microsoft Word document and save.
- Visit NCBI Align Sequences Nucleotide BLAST (blastn): http://blast.ncbi.nlm.nih.gov/Blast.cgi?PAGE_TYPE=BlastSearch&PROG_DEF=blastn&BLAST_PROG_DEF=megaBlast&BLAST_SPEC=blast2seq.

(a) Copy the GFP SLIC primer 1 sequence and paste on to the "Enter Query Sequence" box.

Copy the GFP SLIC primer 1 sequence and paste on to the "Enter Query Sequence" box.

Copy and paste the linearized pUC19/*Hind*III + *Eco*RI sequence (obtained from Post-Lab Assignment 6 of Experiment 20) in FASTA format on to the "Enter Subject Sequence" box (displayed by clicking "Align two or more sequence").

Click "BLAST."

**Note: "Strand = Plus/Plus" or "Strand = Plus/Minus." The default query sequence is Plus.*

- Copy the highest match in the Alignments and paste in HTML format (Courier New font size 10) into a Microsoft Word document.

(b) Repeat the above (a) procedure using the GFP SLIC primer 2 sequence.

(c) Copy the GFP SLIC primer 1 sequence and paste on to the "Enter Query Sequence" box.

Type GenBank accession numbers of pGLO, U62637 in the "Enter Subject Sequence" box.

Click "BLAST."

Copy the highest match in the Alignments and paste in the HTML format (Courier New font size 10) into a Microsoft Word document.

Underline the matching sequence in the primer 1 to indicate the pGLO overlap sequence.

(d) Repeat the above (c) procedure using the GFP SLIC primer 2 sequence.

(e) From (a) and (c) alignments, write the entire primer 1 in which the pUC19 and pGLO sequences are clearly distinguished by underlining GFP sequence.

(f) From (b) and (d) alignments, write the entire primer 2 in which the pUC19 and pGLO sequences are clearly distinguished by underlining the GFP sequence.

**In your descriptions, indicate 5′-overhang sequence (4 bp) of EcoRI and HindIII sites of the linearized pUC19 as a boxed bold sequence in primers 1 and 2.*

6. **(a)** Find the pGLO DNA sequence (GFP_{uv}) flanked by the primers 1 and 2 using the pGLO overlap primer sequences. To do this:

- Visit http://www.ncbi.nlm.nih.gov/nucleotide/.

- Type pGLO Genbank accession number, "U62637" in the Search window and click "Search."
- Click "Change region shown ▼" menu on the upper right side.
- Click "selected region from_begin_ to _end_." Type the lowest (begin) and highest (end) nucleotide positions of the subject sequence based on the blast hit of primers 1 and 2 query searches against pGLO in steps (e) and (f) of assignment 5 above.
- Click "Update View." The sequence between the primers 1 and 2 will be displayed.
- **Copy and paste this sequence into a Microsoft Word document. Underline the overlap primer 1 and primer 2 sequence. What happens to a start codon? Where is a stop codon?**
- **Locate and indicate a stop codon of the GFPuv sequence as a boxed bold sequence.** To get answers, you need to compare with the GFP_{uv} CDS (1342 ... 2061) sequence of U62637 (http://www.ncbi.nlm.nih.gov/nuccore/U62637).

(b) Convert the above GFP_{uv} sequence to reverse the complementary sequence. To do this:

- Visit the online Sequence Format Converter (http://www.bioinformatics.org/sms/rev_comp.html) or (http://www.cellbiol.com/scripts/complement/dna_sequence_reverse_complement.php).
- Copy the GFP_{uv} sequence (1341..2080) and paste in the submission window.
- Select "reverse-complement." Click "Submit."
- **Copy this sequence and paste into a Microsoft Word document. Underline the overlap primer 1 and primer 2 sequence. Type ">GFP CDS Rev-complement" on the top of the sequence to display as a FASTA format. Remove unmatched end sequences.**

(c) Find the primers 1 and 2 overlap sequences in the linearized pUC19/*Hin*dIII + *Eco*RI sequence (obtained from Post-Lab Assignment 6 of Experiment 20).

- Locate and indicate a start codon of *lacZ* α of pUC19 as a boxed bold sequence, referring to the pUC19 map (https://www.neb.com/products/n3041-puc19-vector).

(d) Based on primers 1 and 2 overlap sequences, assemble both pUC19/*Hin*dIII + *Eco*RI DNA (lower case letter nucleotides) and GFP CDS Rev-complement DNA (upper case letters) nucleotides.

**Note that the assembled sequence is actually a circular DNA rather than a linear DNA.*

(e) After assembling, copy and paste the manually assembled sequence (FASTA format with >pUC19-GFP) into the submission window of the online Sequence Format Converter (http://www.ebi.ac.uk/Tools/sfc/emboss_seqret/).

- Select "Unknown format" in the INPUT FORMAT and "Genbank entry format" in the OUTPUT FORMAT, and click "Submit."
- **Copy and paste this sequence into a Microsoft Word document. How many base pairs does the assembled pUC19-GFP have?**

(f) Locate primers 1 and 2 sequences by underlining in the assembled sequence. How many and what amino acid sequences LacZα of pUC19 are fused to the GFP_{uv} sequence?

7. Draw the map of an assembled circular plasmid using the assembled entire sequence and online plasmid mapping software:

- http://wishart.biology.ualberta.ca/PlasMapper/
- http://mgcheo3.med.uottawa.ca:8080/PlasMapper/

Simply paste the FASTA format DNA sequence into the text window below and click "Graphic Map." Copy image and paste as a HTML format into Microsoft Word document.

Additional background information

Part A. PCR primer design

In PCR, DNA between the oligonucleotide primers is amplified. For this to be specific, careful attention should be made to primer design, because false priming to other sites can lead to false products and/or contaminating co-products. PCR primers are normally designed such that their 3′ ends point toward each other with a forward primer being a top strand sequence to the left of the region to be amplified and a reverse primer being a bottom strand sequence to the right of that region.

There are a few basic rules for the design of primers:

1. Primers should be at least 18 bases long, which theoretically corresponds to a unique sequence in 6.9×10^{10} ($= 4^{18}$) bp DNA, but preferably be less than 40 bases. The longer the oligonucleotide, the more sequence errors may occur due to the increasing risk of depurination during the synthesis process.

2. Unless it is desired, primers should not be designed from within repetitive sequences.

3. The primers should be preferably in the range 40 to 60% G/C.

4. There should be no runs of any one base longer than four bases; this is particularly important for runs of G and C residues. In addition dinucleotide repeats should be avoided to prevent a hairpin loop.

5. The sequence and length of each primer will determine its Tm. The Tm is defined as the temperature at which 50% of a sequence is dissociated from its complementary sequence. Tm can be calculated using an online Tm calculator:

- http://tmcalculator.neb.com/#!/
- https://www.idtdna.com/calc/analyzer

The primers should be designed such that their Tm values are as closely matched as possible to ensure efficient annealing, preferably within 1 °C. Preferable oligos should be chosen with Tm values between 52 and 65 °C. Annealing temperature (Ta) of PCR is normally at 5 °C below the lower Tm value if the GC content is less than 50%. If the GC content is more than 50%, the lower Tm value of the primer set is recommended to be used for Ta.

6. The primers should be free of internal self-complementarity (homo-dimer) and should not show significant complementarity to each other (hetero-dimer). As a rule, primers should have no more than three bases in a row that are complementary to either of the two primers used.

7. Because DNA polymerization occurs starting from the 3′ end of each primer and the G/C base pair is more stable, whenever possible the 3′ end of each primer should be a single G or C.

8. Non-complementary bases for introducing mutations and restriction sites can be incorporated into the primers. In this case, they should be at or close to the 5′ end of the primers.

Although primers can be chosen manually following the rules, inspection of complementarity and Tm calculations can be tedious and difficult to do manually. Therefore, the use of free primer design software such as Primer 3 and Oligo 6 is recommended. These effortlessly calculate Tm and search for complementary sequences and give the lists of recommended primer pairs. Primer 3 or NCBI PCR Primers Design can be used online via a web browser.

- NCBI BLAST – PCR Primers Design: http://www.youtube.com/watch?v=6thmaB0liXo
- Primer 3: WWW primer tool: http://biotools.umassmed.edu/bioapps/primer3_www.cgi
- Web implementation of Primer 3Plus: http://biotools.umassmed.edu/

Part B. Setting up a PCR reaction

PCR is very sensitive to trace contamination. Therefore, it is very important that care is taken to avoid contaminating all the reagents to be used for PCR reactions. Negative control of no template DNA should always be included in any PCR experiment to ensure that your product band is due to the presence of both primers, and to check for contamination of the reagents.

For standard PCR using *Taq* polymerase, reactions are set up in a thin-walled 0.2-mL tube with the following components of stock solutions in ice:

Template DNA	
10 μM forward primer	
10 μM reverse primer	
10 mM dNTP mix	
10× Reaction buffer	
Taq polymerase	
Nuclease-free sterile water	*Always add dH_2O and buffer first and enzyme last. *Always set up a reaction in ice bath.

Primers

PCR uses a pair of forward and reverse primers that allow amplification of DNA between specific sites. The two primers base-paired to the top and bottom strand and the 3′-OH ends of oligonucleotides primers are in convergent directions. The primers are usually delivered in a dehydrated form in nmoles quantity by manufacturers. To resuspend your oligonucleotides to make a 100 μM stock, simply multiply the number of nmoles by 10 to get the microliter volume of water to be added. For example, assuming you have an oligonucleotide of 20 nmole of a lyophilized form of oligonucleotide, 20 × 10 = 200 μL of sterile high-quality water is required to get a 100 μM stock. The final concentration of each primer in reaction is 0.1 to 2 μM. To calculate micrograms (μg) of an oligo when nmoles and molecular weight (g/mole) are known, you can use the formula: μg = nmoles × (MW/1000). If oligo primers are suspended in water, it is advised to make aliquots and store at −20 °C because repeated freezing and thawing causes the primers to degrade.

Template DNA

PCR can be used to amplify DNA from the impure sources such as bacterial and yeast colonies, but it is generally safe to use a pure template. This is because contaminating DNA can give false products, and some chemical contaminants can either chelate magnesium ions that are required by polymerase and inhibit the enzyme activity or alter annealing conditions. The integrity of the DNA is also important. Thus, it is advisable to check DNA status by agarose gel electrophoresis rather than relying upon a spectrophotometric reading.

For a standard PCR of 25–30 cycles, template DNA amount is 0.1 to 10 ng of plasmid DNA, 5 to 20 ng of bacterial genome, and 100 to 500 ng of human genomic DNA per 50 μL. Use of too much DNA may lead to non-specific amplicons and smearing. Conversely, use of too little DNA will not give rise to an appreciable amount of amplicon. In general, the higher the DNA concentration and number of cycles, the lower is the primer specificity. To calculate the number of copies of your DNA sequence when moles are known, copy number = moles × (6×10^{23}). To convert μg/μL concentration into μM or ng/pmole concentration when MW or bp is known, μM = (μg/μL ÷ MW) × 10^6, or ng/pmole = bp × 0.66.

dNTP

Avoid repeated thawing and freezing by making aliquots. Use a stock solution of high-quality dNTP that has all four nucleotides at an equal concentration. A default final concentration is 200 μM. If you add more dNTP, you may need to add more $MgCl_2$ because dNTP binds magnesium ions.

Buffer

Buffer is normally supplied with each enzyme as a 10× or 5× concentrate. It is recommended to use only the buffer supplied with the particular enzyme.

Enzyme

Enzymes are typically supplied in stabilized formulations at a concentration of around 1 to 5 units/μL. Each PCR will require between 2 and 5 units of enzyme. Although PCR enzymes are stable for many rounds of temperature cycling, they can perish during normal usage and are sensitive to oxidation. Therefore, care must be taken to keep the enzyme at −20 °C when not in use and to avoid introducing air bubbles in the reaction setup. Take the enzyme out only when required, keeping it on ice at the bench. If you want to clone a PCR product for protein expression, use of a high-fidelity proofreading Phusion or Q5 polymerase enzyme (NEB) is recommended.

H_2O

Nuclease-free, sterile high-quality pure water must always be used.

$MgCl_2$

DNA polymerases require magnesium ions as a cofactor. The optimal concentration of Mg^{+2}, as magnesium chloride or magnesium sulfate, is provided in the buffer supplied with each enzyme. A default concentration is 1.5 mM. However, since all forms of nucleotides including dNTPs and even some chemicals such as EDTA bind magnesium ions, their level can be suboptimal. In such a circumstance, as an impure DNA sample, supplementation of the PCR reaction with more magnesium could be necessary. However, addition of too much magnesium ions will result in either non-specific or no PCR product.

Master mix

To expedite setting up multiple reactions and reduce the chance of pipetting errors, it is preferred to prepare a master mix that contains either all the reaction components except for DNA template when different individual DNA templates are used for genotyping or all the reaction components lacking the primer set when the same DNA template is used for amplifying different regions. The total volume of master mix should be the volume of single reaction minus the volume of a missing component multiplied by the total number of reactions plus one in order to make up for slight errors in dispensing volumes.

PCR tubes/microtiter plates

Thin-walled 0.2-mL tubes are recommended since heat transfer during temperature cycling is more efficient. Alternatively thin-walled micro-tire plates can be used, provided that they are such a good design that the wells fit into the heat block.

Part C. Thermal cycling condition

Number of cycles	Step	Temperature (°C)	Time
1	Initial denaturation	95	2 min
25–35	Denaturation	95	0.5–1 min
	Annealing	42–65	0.5–1 min
	Extension	72	1 min/kb (Taq) 15–30 s/kb (Phusion or Ampli- Taq)
1	Final extension	72	5 min
1	Hold	4	Indefinite

*In all cases it is best to use a so-called "hot start" technique; this simply means pre-cycling heat denaturation while the tubes are in the thermal cycle, before the first cycle of denaturing, annealing, and extension step. This minimizes the possibility of insufficient denaturation and getting junk priming (including primer dimers) during the setup of the reactions.

*The annealing temperature (Ta) should be optimized for each primer set based on the primer. Ta can be calculated using an online calculator (http://tmcalculator.neb.com/#!/).

*The secondary structures such as a primer dimer, self-dimer, and hairpin loop can be detected using the online IDT OligoAnalyzer (https://www.idtdna.com/calc/analyzer).

*As the amount of DNA increases in later cycles, the number of *Taq* DNA polymerase molecules may become limiting for the extension time allotted. Increasing the extension times in later cycles may be needed to maintain efficiency of amplification.

Further reading

AmpliTaq® DNA Polymerase Manual (LifeTechnologies, Cat. No. N8080171).

Viljoen, G.J., Nel, L.H., and Crowther, J.R. (2005). PCR – the basic reaction. In *Molecular Diagnostic PCR Handbook*, pp. 20–49. ISBN 978-1-4020-3403-9.

22 AMB 1 experiment 22: TA, blunt end, SLIC, and CPEC cloning of PCR product

Purpose: This is up to you to write down.

Introduction

The ends of PCR DNA can be either a 3′-A overhang or blunt end depending on the type of DNA polymerases used. TA cloning is a direct cloning method in which the PCR product with a 3′-A overhang tail end is directly ligated to the 3-T overhang tailed vector called the T-vector. Any plasmid vector can be used for a T-vector if the plasmid contains a unique blunt-end restriction site in the polylinker cloning sites. For example, the *Eco*RV site of pBluescript plasmid and the *Sma*I site in pUC19 plasmid, both of which carry a *lacZα* fragment for blue/white color selection, can be used for TA cloning. However, sticky end fragments must be blunt-ended by filling in using Klenow enzyme for DNA fragments with 5′ overhangs or by trimming with T4 DNA polymerase for DNA fragments with 3′ overhangs. Blunt-ended plasmid DNA is then tailed by adding 3′-T overhangs to generate T-vector.

There are two methods for adding T overhangs to the cut end of vector DNA. One method utilizes a terminal deoxynucleotidyl-transferase (TdT) that transfers a dideoxythymidine (ddT) residue from dideoxythymidine triphosphate (ddTTP) to the 3′ end of blunt-ended cut plasmid. The other method uses *Taq* polymerase and dTTP for 3′-T tailing of blunt-end cut plasmid. However, *Taq* polymerase adds the 3′-T residues inefficiently, which results in only a portion of the blunt-ended plasmids being T-tailed at both ends. On the other hand, terminal transferase adds ddT to 3′ ends very efficiently. Therefore, T-vectors prepared with terminal transferase give more reliable results than those obtained from *Taq* polymerase.

For TA cloning, it is best if the 5′ end of PCR primers is designed to have G because the template-independent 3′-A overhang terminal transferase activity of *Taq* polymerase is most efficient at C and less efficient at A. If a high-fidelity proofreading polymerase such as *Pfu* and *Phusion* is used for PCR cloning, blunt-end ligation to the vector must be performed. Alternatively, a bit of normal *Taq* polymerase is added to the purified PCR DNA because *Taq* polymerase preferentially adds A to the 3′ end in the presence of all four dNTPs. To facilitate A addition, *Taq* polymerase is added after high-fidelity PCR is completed and the mixture is further incubated for 10 min at 72 °C.

In addition to the PCR product, any restriction fragments can also be TA cloned. In the presence of all four dNTPs, *Taq* polymerase will first fill in the 3′-recessive sticky ends of DNA fragments with 5′ overhang ends and then add A to each 3′ end. *Taq* polymerase also adds A directly to each 3′ end of a blunt-end fragment. DNA fragments with 3′ overhang sticky ends can be trimmed with T4 DNA polymerase to make them blunt-ended before treatment with *Taq* enzyme.

For TA cloning, the PCR reaction should be checked by running 1/10 volume of reaction sample on to agarose gel to see if the product is of a single band. If there are smaller fragments in the PCR sample, they will be cloned preferentially, decreasing the probability of getting the larger fragments cloned. In this case, a correct sized product should be purified from the gel. Restriction fragments of a correct size also should be gel-purified. Fresh PCR products are recommended for TA cloning ligation, since 3′-A overhangs will gradually be lost during storage. The universal TA cloning method is useful for subcloning a restriction DNA fragment of interest when compatible restriction sites are absent in a destination vector.

Another simple, rapid, and versatile cloning method is the sequence and ligation-independent cloning (SLIC) method, which utilizes T4 DNA polymerase that has a 3′ to 5′ exonuclease activity. In this method, both linearized vector and PCR fragments carrying a flanking overlapping vector DNA sequence (15 to 25 bp) are treated with T4 DNA polymerase in the absence of dNTPs to produce complementary ssDNA 5′ overhang ends. The assembled fragments are then directly transformed into *E. coli* where ssDNA gaps between the fragments are repaired *in vivo*.

The circular polymerase extension cloning (CPEC) method is also very simple, directional, and sequence and ligation independent, requiring both linearized vector and PCR fragments that have a flanking overlapping vector DNA sequence of 15 to 25 bp. In this method, linear dsDNA insert(s) and a single vector are first heat-denatured together in a typical high-fidelity PCR reaction mixture without any PCR oligo primers. The resulting single strands anneal with their complementary overlapping ends and extend using each other as a template to form double-strand circular recombinant plasmids. The assembled fragments are then directly transformed into *E. coli* cells where nicks are repaired *in vivo*. Unlike SLIC, CPEC uses double-stranded overlapping insert(s) and a linearized vector directly.

There are also several other PCR cloning methods: TOPO TA cloning (Invitrogen), Gateway cloning (Invitrogen), In-Fusion cloning (Clontech), Expressioneering (Lucigen), Gibson assembly cloning, and overlap extension (OE) cloning. It is recommended to watch the following videos about the different PCR cloning methods.

Methods in Biotechnology, First Edition. Seung-Beom Hong, M. Bazlur Rashid and Lory Z Santiago-Vázquez.

Companion Website: www.wiley.com\go\hong\Methodsinbiotechnology

- http://media.invitrogen.com.edgesuite.net/presentations/gateway/index.html
- https://www.youtube.com/watch?v=Mjbf0DeYtxk
- https://www.youtube.com/watch?v=5bAmqzkS8pw
- https://www.youtube.com/watch?v=MuQcjO6t3eo

In this lab exercise, you will prepare T-vector from pUC19 using the terminal transferase method, purify PCR product using the silica method, ligate *Taq* polymerase PCR products to pUC19 T-vector, transform the ligated vector into *E. coli*, and identify positive colonies. The PCR product of Phusion polymerase will be blunt-end ligated to a dephosphorylated vector. The insert DNA of TA and blunt-end cloning is ligated to the vector in either direction. The orientation of the insert in the vector can be determined by either DNA sequencing or restriction mapping. In addition, you will clone GFP PCR product into pUC19 vector using the SLIC and CPEC methods. Unlike TA cloning, both SLIC and CPEC cloning allow for directional seamless assembly of multiple fragments as well as in-frame fusion cloning into protein expression vector in a single-tube, one-step reaction. In this lab exercise, directional in-frame fusion cloning of GFP gene into pUC19 vector can be directly confirmed by visual GFP expression on an agar plate. To conduct and finish the experiments efficiently requires your correct answers to pre-lab assignment questions.

Pre-lab assignment

(Typing and submission must be completed before lab work begins.)

1. Show your work as to how to estimate the concentration of the purified pUC19/*Sma*I and pUC19/*Eco*RI + *Hin*dIII in the Experiment 20 exercise. Calculate the volume of pUC19/*Eco*RI + *Hin*dIII in step 32 of Part H (SLIC cloning) and step 34 of Part I (CPEC cloning).
2. Using *ng/pmole = bp × 0.66*, calculate the total pmole DNA ends of pUC19/*Sma*I plasmid purified from organic extraction/ethanol precipitation and silica method (Experiment 20).
3. Calculate the amount (ng) of *lacZ* PCR insert to be ligated to a 50 ng vector DNA in step 28 (a) and (b) of Part G (TA and blunt-end cloning).
4. If you use 10 μL of pUC19/*Sma*I vector at dephosphorylation, steps 3 to 7 of Part C, what concentration should it be in order to use 2 μL of dephosphorylated vector for step 28(b) of Part G?
5. Describe a flow diagram of overall major steps in the quickest way to finish up today's experiment. Your description should be based on the results of Experiment 21 and include which step is the first, second, third, and so on, with an approximate time of each major step (A to K).

Materials and equipment

- pUC19/*Sma*I (Experiment 20)
- pUC19/*Eco*RI + *Hin*dIII (Experiment 20)
- PCR samples (Experiment 21): Taq_LacZ/pRY121, Taq_LacZ/Genome, Phusion_LacZ/pRY121, Phusion_GFP/pGLO
- Silica Membrane Econo Spin Column (Epoch Life Science, Cat. No. 1920-050)
- PB buffer (5 M guanidine-HCl, 20 mM Tris-HCl pH 6.6, 30% isopropanol)
- PE buffer (10 mM Tris-HCl pH 7.5, 80% ethanol)
- QG buffer (6 M guanidine thiocynate, 50 mM Tris-HCl, pH 6.6, 20 mM EDTA, pH 8.0, 0.006% cresol red)
- Elution buffer (10 mM Tris-HCl pH 8.5)
- 0.5 M EDTA (pH 8.0)
- Terminal transferase (NEB, Cat. No. M0315S, 20 units/μL)
- 2′,3′-Dideoxythymidine-5′-triphosphate (ddTTP: USB/Affy metrix Product # 77116 1 UM, 10 mM)
- T4 DNA polymerase (NEB, Cat. No. M0203S, 3 units/μL)
- Antarctic phosphatase (NEB, Cat. No. M0289S, 5 units/μL); phosphatase buffer (10×)
- Quick Ligation™ Kit (NEB Cat. No. M2200S, 2000 units/μL)
- 20× lithium borate buffer (200 mM, pH 8.2)
- 0.5% agarose gel containing 0.5 μg/ml ethidium bromide in 1× lithium borate buffer
- 10× BSA (NEB), 10× buffer 2 (NEB)
- 10× DNA loading dye
- Standard DNA reference (pUC19/*Sma*I and 1 kb DNA ladder NEB, Cat. No. N3232S)
- LB/Amp (100 μg/mL)-X-β-Gal (40 μg/mL)–IPTG (0.1 mM) agar plates
- $CaCl_2$ solution I (100 mM $CaCl_2$; autoclave or filter-sterilize; store at 4 °C)
- $CaCl_2$ solution II (100 mM $CaCl_2$, 15% glycerol, 10 mM HEPES, pH 7.5; filter-sterilize)
- *E. coli* DH5α or TOP10
- 25 °C, 37 °C, 42 °C, and 65 °C water baths
- Bio-Rad Thermocycler, thin-walled 0.2-mL PCR tube
- NanoDrop 2000 spectrophotometer

Procedure (day 1)

A. Blunt end digestion and purification of vector plasmid

1. Determine the concentration of the purified pUC19/*Sma*I DNA in the Experiment 20 exercise and calculate the total pmole DNA ends to set up the terminal transferase reaction (step 2).

B. T-vector preparation

2. The terminal transferase method is given below.

Addition order	Component	Volume (μL)
1	Nuclease-free sterile H_2O	*x*
2	10× TdT buffer	5
3	2.5 mM $CoCl_2$	5
4	10 mM ddTTP	0.5
5	Up to 1 pmole DNA ends* of purified pUC19/*Sma*I	*y*
6	Terminal transferase (20 units/μL)	0.1 (2 units/pmole DNA ends)
	Total volume	50

*To determine the ng/pmole, multiply the number of base pairs by 0.66.

- Incubate at 37 °C for 30 minutes.

**During incubation, prepare 0.5% agarose gel in 1× lithium borate buffer with 0.5 µg/mL ethidium bromide for step 24 of Part F (TA prepares a gel with a 14-well comb).*

- Stop the reaction by adding 4 µL of 0.5 M EDTA (pH 8.0).
- Proceed to do the silica column purification steps of Part E.

C. Dephosphorylation of vector (for blunt-end ligation)

3. Add 1/10 volume of 10× Antarctic phosphatase reaction buffer to the pUC19/*Sma*I (purified from Experiment 20; up to 5 µg of DNA) cut with any restriction endonuclease in any buffer.
4. Add 1 µL of Antarctic phosphatase (5 units) and mix.
5. Incubate for 15 min at 37 °C for 5′ overhangs or blunt ends, 60 min for 3′ overhangs.
6. Heat to inactivate for 5 min at 65 °C and keep on ice until use.
7. Use directly for the ligation at step 28(b) of Part G.

D. Gel extraction of PCR product (*skip if the PCR produces a single band in experiment 21*)

8. Run the *Taq*-LacZ, *Phusion*-LacZ, and *Phusion*-GFP PCR samples (from Experiment 21) on to 0.5% agarose gel in 1× lithium borate buffer at 30 V/cm for 10 min.
9. Weigh an empty sterile 1.5-mL microcentrifuge tube.
10. Place the ethidium bromide-stained gel on to a UV transilluminator covered with plastic wrap.
11. Cut out the gel bands of a correct size with a clean razor blade as quickly as possible and place into the empty microcentrifuge tube.

**Take special care to avoid sources of nuclease contamination.*

**Minimize the time of UV exposure as much as possible. Exposure of DNA to UV radiation for as little as 45 seconds can damage the DNA.*

**Minimize the size of the gel slice by removing extra agarose.*

12. Weigh the gel-containing microcentrifuge tube and calculate the gram quantity of the gel slice.

**Put the gel slice up to 0.25 g in a 1.5-mL microcentrifuge tube.*

13. Add 3 volumes of buffer QG to 1 volume of gel (1 g equals about 1 mL).
14. Incubate at 50 to 57 °C for 10 min (or until the gel slice has completely dissolved). Mix by vortexing the tube every 2 min during the incubation.

**After the gel slice has dissolved completely, check the color of the mixture. If the color is changed, add 10 µL of 3 M sodium acetate (pH 5.2) and mix to turn the color of the mixture to its original color.*

E. Silica spin column purification of PCR products and T-vector

15. Label 4 spin columns: Taq-LacZ, Phusion-lacZ, Phusion-GFP, and T-vector.

16. Transfer PCR samples (previously in Experiment 21) to fresh 1.5-mL tubes and add 5 volumes of PB buffer to 1 volume of sample and mix.

DNA sample	PB buffer volume to be added (µL)
Taq-LacZ/pRY121 PCR DNA (~46 µL)*	460
Taq-LacZ/*E. coli* genome PCR DNA (~46 µL)*	
Phusion-LacZ/pRY121 PCR DNA (~46 µL)	230
Phusion-GFP/pGLO PCR DNA (~46 µL)	230
T-vector reaction of pUC19/*Sma*I (54 µL from step 2b of Part B)	270

*Combine the two samples.

17. Place a spin column in a provided 2-mL collection tube.
18. To bind the DNA, apply the sample to the spin column and centrifuge for 30 to 60 s.
19. Discard the flow-through. Place the spin column back into the same tube.

**Collection tubes are reused for recycling.*

20. To wash, add 750 µL of PE buffer to the spin column and centrifuge for 30 to 60 s.
21. Discard the flow-through and place the spin column back in the same tube. Centrifuge the column for an additional 1 min at maximum speed.

**Residual ethanol from buffer PE will not be completely removed unless the flow-through is discarded before this additional centrifugation. Any residual buffer PE may interfere with subsequent enzymatic reactions.*

22. Place the spin column in a clean 1.5 mL microcentrifuge tube.
23. To elute DNA, add 32 µL of buffer EB (10 mM Tris-Cl, pH 8.5) to the *center* of the spin column membrane, let the column stand for 1 min, and centrifuge the column for 1 min.

**Ensure that the elution buffer is dispensed directly on to the QIAquick membrane for complete elution of bound DNA. Typical recovery of the yield is 60 to 80%.*

F. DNA quantification (NanoDrop 2000 and gel electrophoresis)

24. Add 3 µL of each purified sample to 7 µL of TE buffer plus 2 µL of 6× DNA loading dye.

- Load on to 0.5% agarose gel in 1× lithium borate buffer.
    - Lane 1: 1 kb DNA ladder
    - Lane 2: Silica purified Taq-lacZ PCR (G#1)
    - Lane 3: Silica purified Phusion-lacZ PCR (G#1)
    - Lane 4: Silica purified Phusion-GFP PCR (G#1)
    - Lane 5: Silica purified T-vector (G#1)
    - Lane 6: Silica purified Taq-lacZ PCR (G#2)
    - Lane 7: Silica purified Phusion-lacZ PCR (G#2)
    - Lane 8: Silica purified Phusion-GFP PCR (G#2)
    - Lane 9: Silica purified T-vector (G#2)
    - Lane 10: Silica purified Taq-lacZ PCR (G#3)
    - Lane 11: Silica purified Phusion-lacZ PCR (G#3)

- Lane 12: Silica purified Phusion-GFP PCR (G#3)
- Lane 13: Silica purified T-vector (G#3)
- Run at 300 V for 20 min.

25. Compare the band brightness to determine the approximate concentration of each sample. Store the samples at −20 °C (label each concentration) if not used immediately.

**These data are also required for Experiment 24, Southern Blot Analysis.*

26. During electrophoresis, spot 2 μL of each purified sample on to NanoDrop 2000 and measure for DNA quantification using EB as a blank. Record the ng/μL and A_{260}/A_{280} ratios.

27. Compare each sample concentration estimated by band brightness with that of NanoDrop 2000. Are the values similar to each other? If considerably different, which one do you think is more accurate and why?

G. TA and blunt-end cloning of LacZ PCR fragment

28. Combine 50 ng of T-vector or 50 ng of blunt-end vector with a threefold molar excess of insert. Adjust the volume to 10 μL with dH_2O.

*To calculate the insert amount, use the following formula:

$$\text{ng of insert} = \text{ng of vector} \times \left(\frac{\text{bp of insert}}{\text{bp of vector}}\right) \times \text{fold molar excess of insert}$$

(a) TA ligation

Component	Volume (μL)
Nuclease-free sterile H_2O	*x*
Silica-purified *Taq*-LacZ PCR insert (563 bp)	*y*
Silica-purified T-vector (pUC19/*Sma*I 2686 bp)	*z* (~50 ng)
2x Quick ligation buffer	5
T4 Ligase	0.5
Total volume	10

(b) Blunt-end ligation

Component	Volume (μL)
Nuclease-free sterile H_2O	*x*
Silica-purified Phusion-LacZ PCR insert (563 bp)	*y*
Dephosphorylated pUC19/*Sma*I from step 7 of Part C	*z* (~50 ng)
2x Quick ligation buffer	5
T4 Ligase	0.5
Total volume	10

29. Mix gently by pipetting up and down, and incubate at 25 °C for 5 min.

**Do not make foam! Set the P20 micropipettor to 5 μL and use it to pipette up and down while immersing the pipette tip in the solution.*

**Do not rely on ambient air temperature but place the tube in a room temperature water bath.*

30. Chill on ice and then transform (Part K) or store at −20 °C.

H. SLIC cloning of GFP PCR fragment

31. Mix the linearized vector and insert at a molar ratio of 1:2 in a final total volume of 10 μL in a 0.2-mL microtube.

Component	Volume (μL)	Final concentration	Total amount
Nuclease-free sterile H_2O	*x*	–	–
10x BSA	1	1x	10x
10x NEB buffer 2	1	1x	10x
pUC19/*Eco*RI+*Hind*III (2631 bp)	*y*	10 ng/μL	100 ng
GFP PCR (783 bp)	*z*	4 ng/μL	40 ng
Total volume	20		

32. Add 0.2–0.5 μL of T4 DNA polymerase (3 U/μL) to the mixture and incubate at 25–30 °C for 2.5 min; do not incubate longer.

33. Put the reaction mixture on ice immediately to stop the reaction and incubate on ice for 10 min. Add 5 or 10 μL of the reaction mixture to 50 or 100 μL competent *E. coli* cells for transformation (Part K).

I. CPEC cloning of GFP PCR fragment

34. Add the following components to your labeled thin-walled PCR tube *on ice* in this order.

Component	Volume (μL)	Final concentration	Total amount
Nuclease-free sterile H_2O	*x*	–	–
5x Phusion HF buffer	4.0	1x	20x
10 mM dNTPs	0.4	0.2 mM	4×10^{-6} mmole
pUC19/*Eco*RI+*Hind*III (2631 bp)*	*y*	5–10 ng/μL	100–200 ng
GFP PCR (783 bp)*	*z*	2–4 ng/μL	40–80 ng
Phusion® DNA polymerase (2U/μL)	0.2	0.02 U/μL	0.4 U
Total volume	20		

*The insert: the vector molar ratio should be between 1:1 to 2:1.

35. Program the thermal cycler using the following parameters.

Cycle	Steps	Temperature (°C)	Time
1	Initial denaturation	98	1 min
15	Denaturation	98	10 s
	Annealing	55	30 s
	Extension	72	56 s*
1	Final extension	72	5 min
1	Hold	4	∞

*The total length of the assembled length (kb) × 15.

36. Begin the program cycle. Once the initial denaturation temperature reaches 90 °C, quickly spin your sample tube and place in the thermal cycler.

J. Preparing competent cells (this part will be performed by TA)

37. Pick up a well-isolated single colony from the streaked LB agar plates of *E. coli* DH5α or TOPO10, inoculate into a 2-mL LB medium, and incubate at 37 °C, 200 rpm overnight.
38. Next day, inoculate 50 mL of LB medium in a 250-mL flask with 0.5 mL of the overnight culture (1:100 dilution).
39. Grow the culture at 37 °C, 250 rpm, for 2 to 3 h to a mid-log phase (A_{600} = 0.7 to 0.8).
40. Aliquot the culture into two sterile centrifuge tubes.
41. Centrifuge at 4500 rpm for 10 min at 4 °C.
42. Pour off the culture supernatant and suspend gently in 10 mL ice-cold $CaCl_2$ solution I.

**Do not vortex the cell suspension and it must be kept on ice from now on.*

43. Keep the suspended cells on ice for 30 min at least (up to 1 h).
44. Centrifuge the cells at 3000 rpm for 5 min at 4 °C.

**Do not increase the rpm and time of centrifugation. After centrifugation, cell pellet is loose; take care not to disrupt cell pellet.*

45. Decant supernatant and resuspend gently in 2 mL of ice-cold $CaCl_2$ solution II (final volume of 0.04 of the starting culture volume).

**Now it is ready for transformation, and the competent cells can be used immediately.*

46. Keep the suspended cells on an ice box and place in a cold room (4 °C) overnight.

**Competency increases with time on ice and reaches a maximum at 12 to 24 h.*

47. Next day, swirl the tube gently to resuspend the precipitated cells, dispense suspended cells into *pre-chilled sterile* microcentrifuge tubes (100 μL aliquots), and freeze immediately at −70 °C. When you use frozen competent cell later, thaw quickly between the thumb and then keep on ice before use.

K. Transformation

48. Label six sterile 1.5-mL microtubes (Exp-TA, Exp-Blunt, Exp-SLIC, Exp-CPEC, + Control, − Control) and put into ice.
49. Aliquot a 100 μL of the competent *E. coli* cells into *pre-chilled* microtubes on ice.

Transformation tube	Reaction components		Volume added (μL)
	Vector	Insert	
Exp. (TA cloning)	pUC19/*SmaI*$_{\text{T overhang}}$	LacZ3′	10
Exp. (blunt end cloning)	pUC19/*SmaI*	LacZ3′	10
Exp. (SLIC cloning)	pUC19/*EcoRI*+*HindIII*	GFP_{uv}	10
Exp. (CPEC cloning)	pUC19/*EcoRI*+*HindIII*	GFP_{uv}	10
+ Control (+ DNA)	pUC19 (0.1 μg/μL)	–	1
– Control (– DNA)	TE buffer	–	5

**Be very gentle when working with competent cells. Competent cells are very sensitive to changes in temperature or mechanical lysis caused by pipetting. Transformation should be started immediately as soon as the last bit of ice disappears. Mix by swirling or tapping the tube gently, not by pipetting.*
**The volume of the reaction mixture should not exceed 10% of the competent cell volume in the transformation process.*

50. Add the ligation or reaction mixtures to each test tube containing competent cells and swirl gently.
51. Incubate tubes on ice for 30 min; do not disturb.
52. Heat shock to 42 °C for 1 min.

**The length of time may differ, depending on the bacterial strain and composition of chemicals used to prepare competent cells.*

53. Immediately chill the tubes on ice for 1 min and add 500 μL of sterile LB or SOC broth to each tube.
54. Incubate at 37 °C, 200 rpm for 1 h.
55. While the tubes are being shaken in the incubator, label your 10 agar plates on the edge of the bottom as follows:
- Label two LB/Amp/X-β-Gal/IPTG plates: pUClacZ3′$_{TA}$; Group#, Section#
- Label two LB/Amp/X-β-Gal/IPTG plates: pUClacZ3′$_{blunt}$; Group#, Section#
- Label two LB/Amp/IPTG plates: SLIC; Group#, Section#
- Label two LB/Amp/IPTG plates: CPEC; Group#, Section#
- Label one LB/Amp/X-β-Gal/IPTG plate: + Control pUC19
- Label one LB/Amp/X-β-Gal/IPTG plate: – Control

56. Plate 100 μL of the transformation reaction (using a sterile spreader) on to LB/Amp/X-β-Gal/IPTG agar plates.

**Make sure that the alcohol-flamed glass spreader is completely cooled down before use.*

**Plate on to the – Control plate first, then the Exp plate, and the + Control plate last.*

57. Incubate the plates upside down at 37 °C overnight (16–18 h).

**Do not incubate for more than 18 h.*

Procedure (day 2)

L. Screening of transformants

58. Count the number of white colonies on TA and blunt-end cloning plates and compare the cloning efficiency based on the number of white colonies.
59. Count the total number of colonies on SLIC and CPEC plates, and then place them under a UV transilluminator to count the number of green glowing colonies; compare cloning efficiency.
60. Pick colonies for colony PCR or mini-prep plasmid DNA analysis (optional).

Post-lab assignment

1. Calculate the molarity of a linearized 5-kb vector DNA if you put 50 ng of the vector in a 20 μL of ligation reaction. MW of 1 bp dsDNA is 650 g/mole. Show your calculations.
2. Determine the mass of a 1-kb insert needed to achieve a 3:1 molar ratio of insert:vector using 50 ng of the above 5-kb vector in a 20 μL ligation reaction. Show your calculations.
3. Why is ddTTP used instead of dTTP in terminal transferase reaction for T-vector construction?
4. What is the purpose of dephosphorylation of a linearized vector in Part C?
5. Why are white colonies analyzed for colony PCR or mini-prep DNA analysis?

6. You picked several white colonies, grew the cultures, and prepared mini-prep plasmid DNAs in the Part L experiment. In order to determine (a) the size and (b) the orientation of the TA-cloned PCR insert (*Taq*-LacZ PCR obtained from Experiment 21), what restriction enzymes would you use? Show your work to explain your choice and the expected cleavage patterns using the web sites below.

- To obtain a restriction map and sites of the LacZ PCR insert, use the NEB cut tool program (http://tools.neb.com/NEBcutter2/).
- To obtain a pUC19 map visit the web site (https://www.neb.com/products/n3041-puc19-vector).

7. **(a)** What does "inframe fusion cloning" mean?
(b) Show the first 10 amino acid sequences translated from the start codon of the LacZα-GFP fusion sequence in the assembled plasmid sequence of pUC19-GFP (Post-Lab Assignment 6 of Experiment 21). In your description, distinguish between the amino acid sequences of LacZα and GFP.

Further reading

Holton, T.A. and Graham, M.W. (1991). A simple and efficient method for direct cloning of PCR products using ddT-tailed vectors. *Nucleic Acids Research,* **19** (5): 1156.

Jeong, J.-Y., Yim, H.-S., Ryu, J.-Y., Lee, H.S., Lee, J.-H., Seen, D.-S., and Kang, S.G. (2012). One-step sequence- and ligation-independent cloning as a rapid and versatile cloning method for functional genomics studies. *Applied and Environmental Microbiology,* **78**: 5440–5443.

Li, M.Z. and Elledge, S.J. (2007). Harnessing homologous recombination *in vitro* to generate recombinant DNA via SLIC. *Nature Methods,* **4**: 251–256.

Marchuk, D., Drumm, M., Saulino, A., and Collins, F.S. (1991). Construction of T-vectors, a rapid and general system for direct cloning of unmodified PCR products. *Nucleic Acids Research,* **19** (5): 1154.

Mead, D.A., Pey, N.K., Herrnstadt, C., Marcil, R.A., and Smith, L.M. (1991). A universal method for the direct cloning of PCR amplified nucleic acid. *Nature Biotechnology,* **9**: 657–663.

Quan, J. and Tian, J. (2011). Circular polymerase extension cloning for high-throughput cloning of complex and combinatorial DNA libraries. *Nature Protocols,* **6**: 242–251.

23 AMB 1 experiment 23: One-step multifragment assembly cloning

Purpose: This is up to you to write down.

Introduction

Bioengineers often need to put multiple DNA fragments together to construct revised biosynthetic pathways or reprogram genetic regulatory circuits. Methods such as BioBrick™, Golden Gate, In-Fusion™, SLIC, CPEC, isothermal assembly, successive hybridization assembly (SHA), and *in vivo* recombination in yeast have been developed to implement directional assembly. BioBrick™ and Golden Gate methods do not require overlap sequences but all the other methods require overlaps between the assembly fragments.

Isothermal assembly is commonly referred to as "Gibson Assembly" because it was developed by Daniel G. Gibson. This method allows the efficient directional assembly of multiple overlapping restriction or PCR DNA fragments in one reaction and utilizes three enzymatic activities: (1) T5 exonuclease removes nucleotides from the 5′ ends of double-stranded DNA molecules to create single-stranded 3′ overhangs that facilitate annealing of fragments that share complementarity at one end of the overlap region, (2) Phusion DNA polymerase fills in gaps, and (3) *Taq* DNA ligase seals the nicks. This method was used to assemble 101 DNA fragments into four quarter segments of synthetic *Mycoplasma genitalium* genome, each 136–166 kb (Gibson *et al.*, 2008). If a region of DNA interest is too large to be amplified by PCR, it can be divided into several overlapping PCR products that can be joined together into one piece. This approach eliminates the need to remove or add restriction enzyme sites within DNA fragments to join together. The only requirement is a 20–40 bp overlapping sequence between the DNA fragments.

If the final construct is to be linear rather than a fragment to be subcloned into a plasmid vector, the ends of the DNA will be retracted at each 5′ end. If desired, these ends can be filled in a separate reaction using a short oligonucleotide primer, complementary to the 3′ end. However, this is not necessary for circular constructs, since all gaps are filled and the complementary ends are ligated during the isothermal reaction.

A major limitation to Gibson assembly is that the termini of the DNA sequence fragments to be joined must not have a stable secondary structure of single-stranded DNA, such as a hairpin stem loop and repeated sequences, since this would interfere with the required single-stranded annealing and priming of adjacent assembly fragments. This is also true for In-Fusion™, SLIC, and CPEC assembly. The Gibson assembly method was used to simultaneously assemble and clone up to nine of 275 to 980-bp DNA fragments into a plasmid vector in a single tube one-step reaction (Gibson *et al.*, 2009).

The Gibson assembly is a powerful tool to produce different combinations of genetic building blocks in order to design and build novel biological systems. It is recommended to watch the following videos regarding synthetic biology.

- http://www.youtube.com/watch?v=rD5uNAMbDaQ
- http://www.youtube.com/watch?v=Ywy5b_Om0Q8

In this lab exercise, you will amplify the *lacZ* alpha and *ori* fragments from pUC19 plasmid and kanamycin resistance (Km^R) gene cassette from pGBKT7 plasmid. You will also isolate the DNA fragment containing *araC*-P_{BAD} promoter by restriction enzyme digestion of pGLO plasmid. All these four fragments are gel-purified and are assembled using CPEC and Gibson assembly methods in order to express *lacZ* alpha under the control of arabinose-inducible promoter in a plasmid backbone carrying the pUC19 origin of replication and Km^R selection marker. You will transform the assembled construct into *E. coli* TOP10 or HB101 strain that is capable of transporting L-arabinose but not metabolizing it so that the level of L-arabinose remains constant inside cells for the maximum GFP expression.

Pre-lab assignment

(Typing and submission must be completed before lab work begins.)

1. Calculate the volumes of the four DNA fragments at the step 26 of Part G, assuming that each DNA concentration is 50 ng/µL. Show your work.
2. Calculate the extension rate (time/kb) of Phusion® High-Fidelity DNA polymerase on the basis of the expected amplicon size and the cycling condition used in Parts A and B. Show your work.
3. The speed of ramping of Bio-Rad MyCycler is a heating rate of 2.0 °C/s and cooling rate of 1.0 °C/s. Based on this information, how long does it take to finish (a) PCR of *lacZ* alpha fragment and (b) PCR of Km^R gene and pUC19 *ori* starting at a 98 °C ready mode? Show your work.
4. What are the expected results of transformation (step 38) on each plate in terms of color of the colony when the experiment is successful?

Methods in Biotechnology, First Edition. Seung-Beom Hong, M. Bazlur Rashid and Lory Z Santiago-Vázquez.

Companion Website: www.wiley.com\go\hong\Methodsinbiotechnology

5. Describe the outline of the experiments to finish up in the shortest time you will spend. You are to work using two thermal cyclers without a break time.

Materials and equipment

- Primer 1: GAAGGAGATATACATATGGCCATGATTACGC CAAGCTTGC
- Primer 2: CCACCTGAAC GAAGCATCTG CGGTGATGAC GGT-GAAAACC
- Primer 3: CAGATGCTTC GTTCAGGTGG
- Primer 4: CCATAGTTGCCTGACTCCCCGCCCCTAGGG TGGGC-GAAGAAC
- Primer 5: CGGGGAGTCAGGCAACTATGG
- Primer 6: TTGACAGGCACATTATGCATTGCGTATTGGGCGC TCTTCC
- Phusion® High-Fidelity DNA polymerase (NEB Cat. No. M0530S)
- PB buffer (5 M guanidine-HCl, 20 mM Tris-HCl pH 6.6, 30% iso-propanol)
- PE buffer (10 mM Tris-HCl pH 7.5, 80% ethanol)
- QG buffer (6 M guanidine thiocyanate, 50 mM Tris-HCl, pH 6.6, 20 mM EDTA, pH 8.0, 0.006% cresol red)
- EB buffer (10 mM Tris-HCl pH 8.5)
- Isopropanol
- EconoSpin™ Mini Spin Column (Epoch Life Science, Cat. No. 1920-050)
- 10× Lithium borate buffer
- Agarose (low EEO type)
- 10× DNA loading dye, ethidium bromide (10 mg/mL)
- DNA marker standards (pUC19/*Sma*I, 50-bp DNA ladder, NEB Cat. No. N0473G)
- Plasmid DNAs: pUC19, pGLO, pGBKT7
- *Nhe*I-HF™ (NEB Cat. No. R3131S; 20 U/μL), *Cla*I (NEB Cat. No. R0197S; 5 U/μL)
- Gibson Assembly Master Mix (NEB Cat. No. E2611S)
- *E. coli* TOP10 culture in LB (0.4 to 0.8 of A_{600}): F^{-} *mcr*A (*mrr-hsd*RMS-*mcr*BC) *ϕ80lac*Z Δ M15 Δ *lac*X74 *deo*R *rec*A1 *ara*D139 Δ (*ara*A-*leu*) 7697 *gal*U *gal*K *rps*L (Str^{R}) *end*A1 *nup*G λ^{-}
- Chemically competent TOPO10 or HB101 cells
- LB/Km (30 μg/mL)-X-β-Gal (40 μg/mL)/L-arabinose (2 × 10^{-3} M) agar plates
- LB/Amp (100 μg/mL)-X-β-Gal (40 μg/mL)/IPTG (0.1 mM) agar plates
- LB or SOC medium
- Ice box, 37 °C, 42 °C, and 50 °C water bath
- Nuclease-free sterile H_2O
- Razor blade, thin-walled PCR tubes
- Horizontal agarose gel electrophoresis system and power supply
- Two thermal cyclers, NanoDrop 2000 spectrophotometer

Procedure (day 1)

A. PCR of *lacZ* alpha fragment

1. Add the following components to your labeled thin-walled PCR tube *on ice* in this order.

	Component (for *lacZ* α)	Volume (μL)	Final concentration
1	Nuclease-free sterile dH_2O	33.5	
2	5× Phusion HF buffer	10.0	1×
3	10 mM dNTPs	1.0	200 μM each
4	25 μM primer 1 +2 mix	1.0	0.5 μM each
5	pUC19 plasmid (1 ng/μL)	4.0	4 ng
6	Phusion® Hot Start II DNA polymerase (2U/μL)	0.5*	0.02 U/μL

[a] Pipette carefully and slowly because the high content (50%) of glycerol in the storage buffer may lead to pipetting errors. It is critical that the Phusion DNA polymerase is the last component added to the PCR mixture, since the enzyme exhibits 3′→5′ exonuclease activity that can degrade primers in the absence of dNTPs.

Program the thermal cycler I using the following parameters.

Cycle	Steps	Temperature (°C)	Time
1	Initial denaturation	98	30 s
20	Denaturation	98	10 s
	Annealing	68	10 s
	Extension	72	15 s
1	Final extension	72	5 min
1	Hold	4	∞

Begin the program cycle. Once the initial denaturation temperature reaches to 90 °C, quickly spin your sample tube and load in the thermal cycler.

B. PCR of Km® gene cassette and pUC19 *ori* fragment

2. Add the following components to your labeled thin-walled PCR tube *in ice*.

	Component (for Km^{R})	Volume (μL)	Final concentration
1	Nuclease-free sterile dH_2O	33.5	
2	5× Phusion HF buffer	10.0	1×
3	10 mM dNTPs	1.0	200 μM each
4	25 μM primer 3+4 Mix	1.0	0.5 μM each
5	pGBKT7 plasmid (1 ng/μL)	4.0	4 ng
6	Phusion® Hot Start II DNA polymerase (2U/μL)	0.5*	0.02/μL

*Pipette carefully and slowly because the high content (50%) of glycerol in the storage buffer may lead to pipetting errors. It is critical that the Phusion DNA polymerase is the last component added to the PCR mixture, since the enzyme exhibits 3′→5′ exonuclease activity that can degrade primers in the absence of dNTPs.

3. Add the following components to your labeled thin-walled PCR tube *in ice*.

	Component (for *ori*)	Volume (μL)	Final concentration
1	Nuclease-free sterile dH_2O	33.5	
2	5× Phusion HF buffer	10.0	1×
3	10 mM dNTPs	1.0	200 μM each
4	25 μM primer 5 + 6 Mix	1.0	0.5 μM each
5	pUC19 plasmid (1 ng/μL)	4.0	4 ng
6	Phusion® Hot Start II DNA polymerase (2U/μL)	0.5*	0.02/μL

*Pipette carefully and slowly because the high content (50%) of glycerol in the storage buffer may lead to pipetting errors. It is critical that the Phusion DNA polymerase is the last component added to the PCR mixture, since the enzyme exhibits 3′→5′ exonuclease activity that can degrade primers in the absence of dNTPs.

4. Program the thermal cycler II using the following parameters.

Cycle	Steps	Temperature (°C)	Time
1	Initial denaturation	98	30 s
20	Denaturation	98	10 s
	Annealing	67	10 s
	Extension	72	30 s
1	Final extension	72	5 min
1	Hold	4	∞

Begin the program cycle. Once the initial denaturation temperature reaches 90 °C, quickly spin your sample tubes (steps 2 and 3), and load in the thermal cycler.

C. Double digestion and purification of vector plasmid pGLO

5. Set up 1 μg of pGLO DNA with *Nhe*I-HF™ and *Cla*I in a sterile 1.5-mL microtube in the following order.

Sterile dH_2O (as needed)	x μL
10× NEB buffer 4	4 μL
pGLO DNA	y μL
*Nhe*I-HF™	1 μL
*Cla*I	1 μL
Total volume	40 μL

Mix gently by pipetting up and down a few times and incubate at 37 °C for 15 min or until the PCR is finished.

6. During PCR cycling and incubation time, prepare 0.5% agarose min-gel (60 mL) in 1× lithium borate buffer containing 0.5 μg/mL of ethidium bromide. Use a large tooth comb.

D. Agarose gel electrophoresis

7. Add 5 μL of 10× DNA loading dye to each PCR sample and restriction digestion sample.

8. Load on to the gel, along with 5 μL (0.5 μg) of a 50-bp DNA ladder and run at 300 V for 15 min.

E. Silica spin column gel extraction of PCR products and pGLO/ClaI + NheI (QIAquick Gel Extraction Kit Protocol)

**This step is to prevent carryover of intact or undigested vector templates after PCR or restriction digestion.*

9. Label four sterile 1.5-mL microcentrifufge tubes per each group as below and weigh each of all the empty tubes:

Group 1: LacZα; Group 2: Km^R; Group 3: pUC_Ori; Group 4: pGLO/CN

10. Place the ethidium bromide stained gel on to a UV transilluminator covered with plastic wrap.

11. Cut out the gel bands of a correct size (see the figure below and steps 28 and 30) with a clean razor blade as quickly as possible and place into the empty microcentrifuge tube.

**Minimize the time of UV exposure as much as possible.*

**Minimize the size of gel slice by removing extra agarose that has no DNA band.*

**Each group collect the other three group's gel bands and proceed to the next step using a single spin column in order to increase concentration of the eluent sample.*

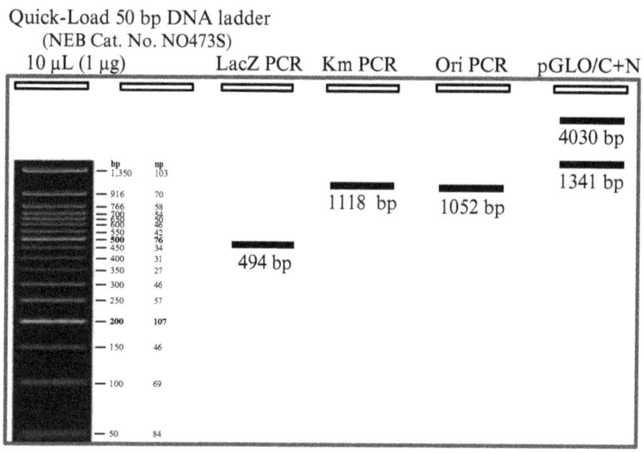

The Quick-Load 50 bp DNA ladder visualized by ethidium bromide staining. Mass values of individual bands are per 1.0 μg marker DNA. (Copyright by NEB. Reprinted with permission.)

12. Weigh the gel-containing microcentrifuge tube and calculate gram quantity of the gel slice.

**Put the gel slice up to 0.25 g in a 1.5-mL microcentrifuge tube.*

13. Add 3 volumes of buffer QG to 1 volume of gel (1 g equals about 1 mL).

14. Incubate at 50–55 °C for ~10 min (or until the gel slice has completely dissolved). Mix by vortexing the tube every 2 min during the incubation.

**If the color of the mixture is orange or violet after the gel slice has dissolved, add 10 μL of 3 M sodium acetate (pH 5.2) and mix to turn the color of the mixture to yellow.*

15. Add one gel volume of isopropanol to the sample and mix.

**0.1 mL of isopropanol is added for 0.1 g of gel slice.*

16. Place a QIAquick spin column in a provided 2-mL collection tube.
17. Apply the sample to the spin column and spin at 13 000 rpm for 1 min.

**For sample volumes of more than 800 μL, simply load and spin again.*

18. Discard flow-through and place the spin column back in the same collection tube.
19. Add 500 μL of PE buffer and spin at 13 000 rpm for 1 min.
20. Discard flow-through and spin at 13 000 rpm for an additional 1 min.
21. Place the spin column in a fresh labeled 1.5-mL microtube.
22. Add 25 μL of EB buffer (warmed at 50–55 °C) to the *center* of the spin column membrane, let the column stand for 1 min, and then centrifuge for 1 min.

**Elution efficiency depends on pH and temperature and typically yields >80%.*

F. DNA quantification of the purified PCR and restriction fragments

23. Spot 2 μL of each purified sample on to NanoDrop 2000 and measure for DNA quantification using EB as a blank. Record ng/μL and A_{260}/A_{280} ratios.
24. Add 4 μL of each purified sample to 6 μL of TE buffer plus 2 μL of 6× DNA loading dye and run on to 0.5% agarose gel in 1× lithium borate buffer at 300 V for 25 min (TA will do this step).
25. Compare each sample concentration estimated by band brightness with that of NanoDrop 2000.

G. Gibson assembly reaction

26. Calculate the amounts (ng) that correspond to 0.1 pmole each of pGLO/*Cla*I-*Nhe*I, *lacZ* α, Km^R, and pUC19 ori fragments, based on the fragment length and weight using the following formula:

$$\text{pmole} = \frac{\text{weight in ng}}{\text{bp} \times 650\,\text{g/mole}} \times 1000$$

27. Calculate the volumes (μL) that contain the above-determined amounts of the vector and insert using the estimated DNA concentrations (step 23).
28. Set up the following experimental and control reactions in a PCR tube on ice using the determined volumes (step 27).

By each group:

Experimental component	Volume (μL)
pGLO/*ClaI-NheI* (1341 bp)*	
lacZ α fragment (494 bp)*	
Km^R gene cassette (1118 bp)*	
pUC19 *ori* (1052 bp)*	
Nuclease-free sterile ddH_2O	
Gibson Assembly Master Mix (2×)	5
Total olume	10

*All are equimolar with a total of 0.2–1.0 pmole of DNA fragments assembled.

By one group:

+ Control component	Volume (μL)
2 overlapping dsDNA fragments	5
Gibson Assembly Master Mix (2×)	5
Nuclease-free sterile ddH_2O	0
Total volume	10

29. Incubate samples in a thermal cycler at 50 °C for 60 min. Store samples on ice or at −20 °C for the subsequent transformation.

H. CPEC assembly reaction

30. Set up the following reaction in a PCR tube on ice (each group).

Experimental component	Volume (μL)
Nuclease-free sterile ddH_2O	
5× Phusion HF buffer	4
10 mM dNTPs	0.4
pGLO/*ClaI-NheI* (1341 bp)*	
lacZ α fragment (494 bp)*	
Km^R gene cassette (1118 bp)*	
pUC19 *ori* (1052 bp)*	
Phusion® DNA polymerase (2U/μL)	0.2
Total volume	20

*All are equimolar with a total of 0.2–1.0 pmole of DNA fragments.

31. Program the thermal cycler using the following parameters.

Cycle	Steps	Temperature (°C)	Time
1	Initial denaturation	98	1 min
15	Denaturation	98	10 s
	Annealing	55	30 s
	Extension	72	56 s*
1	Final extension	72	5 min
1	Hold	4	∞

*The total length of the assembled length (kb) × 15.

32. Begin the program cycle. Once the initial denaturation temperature reaches to 90 °C, quickly spin your sample tube and place in the thermal cycler.
33. Store samples on ice or at −20 °C for subsequent transformation.

Procedure (day 2)

I. Chemical transformation

34. Obtain chemically competent *E. coli* TOP10 cells and add 4 μL of Gibson and 10 μL of CPEC reaction sample to 100 μL chemically competent *E. coli* TOP10 cells on ice. Add 4 μL of positive Gibson assembly control reaction sample to 100 µL chemically competent *E. coli* TOP10 cells. For positive and negative control, add 1 µL of pUC19 sample and 1 µL of TE buffer, respectively.

**Be very gentle when working with competent cells. Competent cells are very sensitive to changes in temperature or mechanical lysis caused by pipetting. Transformation should be started immediately as soon as the last bit of ice disappears. Mix by swirling or tapping the tube gently, not by pipetting.*

**The volume of Gibson reaction mixture should not exceed 5% of the competent cell volume.*

35. Incubate the tubes on ice for 30 min.
36. Heat-shock the cells for 1 min in a 42 °C water bath and keep on ice for 1 min.
37. Add 0.5 mL SOC medium and shake-incubate at 37 °C, 200 rpm for 1 h.
38. While the tubes are being shaken in the incubator, label your agar plates on the edge of the bottom as follows:

- Label two LB/Km/X-β-Gal/L-arabinose plates: Gibson Expt (pAraCZK); Group#
- Label two LB/Km/X-β-Gal/L-arabinose plates: CPEC Expt (pAraCZK); Group#
- Label one LB/Km/X-β-Gal/L-arabinose plate: – Control (TE): share by all groups
- Label one LB/Amp/X-β-Gal/IPTG plate: + Control (Gibson): share by all groups
- Label one LB/Amp/X-β-Gal/IPTG plate: + Control (pUC19): share by all groups

39. Plate 100 µL of the transformation reaction on to the agar plates.

**Make sure that the alcohol-flamed glass spreader is completely cooled down before use.*

**Plate on to – Control plate first, flame-cool, then Expt plate, flame-cool and + Control plate last.*

40. Incubate the plates upside down at 37 °C overnight (16–18 h).

Procedure (day 3)

J. Screening of transformants

41. Count the number of blue and white colonies in the experimental and control plates.
42. Pick blue colonies for colony PCR or mini-prep plasmid DNA analysis (optional).

Post-lab assignment

1. What are the differences among the (a) BioBrick assembly, (b) Golden Gate assembly, (c) Gibson assembly, (d) SLIC assembly, (e) CPEC assembly, and (f) SHA?

2. (a) Describe how to make Gibson Assembly Master Mix based on the article by Gibson *et al.* (2009).
(b) The Master Mix described in the paper is 1.33×. What volumes of the Master Mix and the combined DNA fragments should be added in a total 20 µL of the isothermal assembly reaction?

3. GenBank accession numbers of pUC19 and pGLO (= pBAD-GFPuv) are L09137 and U62637, respectively. The sequence of pGBKT7 is found on the Clontech web site as.gb file. Do not skip anyone of the steps below for correct data processing.

- To obtain the sequence of pGBKT7, perform the following steps:
  - Go to the Clontech web site http://www.clontech.com/xxclt_searchResults.jsp.
  - Type pGBKT7 in the blank box and click "Search."
  - Double click "pGBKT7 vector sequence file in.gb format" to download the pGBKT7.gb file.
  - Open the downloaded.gb file using a text editor (Microsoft Wordpad, Notepad, or TextEdit).
  - The sequence will be displayed in Courier New font size 11.
  - Copy and paste all the text into a Microsoft word document and convert to Courier New font size 9.
  - Save the file as rich text format (RTF).
- To obtain the sequence of pUC19 and pGLO, perform the following steps:
  - Visit the NCBI GenBank database (http://www.ncbi.nlm.nih.gov/genbank/). On the main entry page, you will see a search menu bar at the very top. Select (*v*) "Nucleotide" from the drop-down search bar and type in the accession number in the blank box; then click "Search." Alternatively, you can get the sequences directly from:
    http://www.ncbi.nlm.nih.gov/nuccore/L09137
    http://www.ncbi.nlm.nih.gov/nuccore/U62637
  - Copy and paste the *numbered* sequences of pUC19 and pGLO as HTML Format (Courier New font size 9). Save the all three pGBKT7, pUC19, and pGLO sequences in a single Microsoft Word document.
- To find the overlap sequences in the primers, perform the following steps:
  - Visit NCBI Align Sequences Nucleotide BLAST (blastn): http://blast.ncbi.nlm.nih.gov/Blast.cgi?PAGE_TYPE=BlastSearch&PROG_DEF=blastn&BLAST_PROG_DEF=megaBlast&BLAST_SPEC=blast2seq
  - Copy the primer 1 sequence and paste on to the "Enter Query Sequence" box.
  - Type L09137 in the "Enter Subject Sequence" box and click "BLAST."
  - (a) Copy the highest match in the alignments and paste as HTML format into Microsoft Word. Indicate a pUC19 overlap matching sequence in primer 1 by underlining.

**Note that the Query sequence is always the "Plus" strand.*

  - (b) Repeat the procedure using the primer 2 sequence and L09137.
  - (c) Repeat the procedure using the primer 1 sequence and U62637.
  - (d) Repeat the procedure using the primer 2 sequence and the entire sequence of pGBKT7.
  - (e) Repeat the procedure using the primer 3 sequence and the entire sequence of pGBKT7.
  - (f) Repeat the procedure using the primer 4 sequence and the entire sequence of pGBKT7.

- (g) Repeat the procedure using the primer 4 sequence and L09137.
- (h) Repeat the procedure using the primer 5 sequence and L09137.
- (i) Repeat the procedure using the primer 6 sequence and L09137.
- (j) Repeat the procedure using the primer 6 sequence and U62637.
- (k) Write each primer and distinguish between the pUC19, pGLO, and pGBKT7 sequences with a start–end position number and a strand of subject sequences. An example is shown below.

Primer #: NNNNNNNNNNNNNNNNNNNNNNNNNNNNNNNNNNNNNNNNNNN
Sbjct 1 (24…5) Minus Sbjct 2 (670…690) Plus

- (l) Based on the result, **what primers have overlaps with pGLO sequences?**

4. **(a)** Based on the overlap primer sequences, sketch a scheme of four merged DNA fragments and the location of six primers. Remember that a merger results in a circular DNA.

Experiment	DNA fragment	Primers or restriction enzymes/template DNA
Part C	*araC*-P_{BAD}	*NheI* + *ClaI*/pGLO (overlap ends with primers)
Part A	*lacZα*	Primer 1 + 2/pUC19
Part B2	*Km^R*	Primers 3 + 4/pGBKT7
Part B3	*ori*	Primers 5+ 6/pUC19

(b) Based on the result of primer overlap sequences and pUC19, pGLO, and pGBKT7 sequences and the reaction setup experimental design used in Parts A, B, and C (summarized as above), copy each DNA fragment sequence and paste as HTML Format.

Note: if the aligned subject sequence is the "Minus" strand, use a reverse complementary sequence to locate the overlap position.

(c) Add the remaining primer sequence at each end of the DNA fragment if there is one. Pay attention to the overlap primer sequence and strand.

(d) Assemble the four DNA fragments based on the primer overlap sequences. To do this, first convert *lacZα*, Km^R, and *ori* DNA fragment sequences into a reverse complementary sequence, and then begin with the *araC*-P_{BAD} DNA fragment for merging the ends of two fragments.

- To convert into a reverse complementary sequence, copy the sequence and paste into the "Input Sequence format" window of the online Sequence Format Converter (http://www.bioinformatics.org/sms/rev_comp.html) or (http://www.cellbiol.com/scripts/complement/dna_sequence_reverse_complement.php).
- Select "reverse-complement." and click "Submit."
- Copy and paste this sequence into a Microsoft Word document and put "revc fragment name" on the top of sequence.

**Do not convert the araC-P_{BAD} DNA fragment into a reverse complementary sequence.*

- Merge the fragment based on your sketch.

**You need to assemble only three times to finish a final construct. In your description of the assembled sequence, indicate all six primer sequences by underlining.*

(e) Remove the overlap primer sequence with *araC-P_{BAD}* from the tail end of the assembled fragment.

(f) After assembly, copy and paste the manually assembled sequence into the "*Input Sequence format*" window of the online Sequence Format Converter (http://www.ebi.ac.uk/Tools/sfc/emboss_seqret/).

Select "Unknown format" in the INPUT FORMAT and "Genbank entry format" in the OUTPUT FORMAT, and click "Submit." Copy and paste this sequence into a Microsoft Word document and put ">pAraCZK" on the top of the sequence (→FASTA Format).

How many base pairs does the assembled pAraCZK plasmid have?

5. **(a)** Visit the following online EMBOSS 6.3.1: merger site (http://mobyle.pasteur.fr/cgi-bin/portal.py?#forms::merger).
(b) Paste a pair of DNA fragment sequences into two separate input section boxes (a and b section option "Enter your data below").
(c) Click "Run", enter your e-mail address, and click OK; loading takes some time. If it takes too long, click "back to form" and try again.
(d) Click "full screen" in the **outseq_out option** (*Sequence*).
(e) Copy the assembled sequence and paste in HTML Format into a Microsoft Word document.
(f) Close the window, click "back to form" and "Reset."
(g) Copy the merged sequence and paste into the input section box (asequence option). Copy the next fragment to be merged and paste into a second input section box (bsequence option).

**You need to assemble only three times. It is important to remove the overlap primer sequence with araC-P_{BAD} from the last fragment to be merged.*

**Correct assembly by the EMBOSS merger program is entirely dependent on the correct overlap sequences at each end of the DNA fragment at step (c) of Post-lab assignment 4.*

(h) Copy the entire sequence and reformat using the online Sequence Format Converter.

How many base pairs does the assembled pAraCZK plasmid have? Is the total size of your manually assembled sequence identical to that of the online assembler?

(i) If both are different, you may use the online alignment Clustal omega software (http://www.ebi.ac.uk/Tools/msa/clustalo/) to find any mismatch.

6. Draw the map of the assembled circular plasmid using the assembled entire sequence and online plasmid mapper software (http://wishart.biology.ualberta.ca/PlasMapper/). Simply paste the DNA sequence into the text window below and click "Graphic Map."

References

Gibson, D.G, Benders, G.A., Andew-Pfannkoch, A., Dennisova, E.A., Baden-Tillson, H., Zaveri, J., Stockwell, T.B., Brownley, A., Thomas, D.W., Algire, M.A., Merryman, C., Young, L., Noskov, V.N., Glass, J.I., Venter, J.G., Hutchison, C.A., and Smith, H.O. (2008). Complete C chemical synthesis, assembly, and cloning of a *Mycoplasma genitalium* genome. *Science*, **319**: 1215–1220.

Gibson, D.G., Young, L., Chuang, R.-Y., Venter, J.C., Hutchison, C.A., and Smith, H.O. (2009). Enzymatic assembly of DNA molecules up to several hundred kilobases. *Nature Methods*, **6** (5): 343–345.

Further reading

Benders, G. A., Noskov, V.N., Denisova, E.A., Lartigue, C., Gibson, D.G., Assad-Garcia, N., Chuang, R.-Y, Carrera, W., Moodie, M., Algire, M.A., Phan, Q., Alperovich, N., Vashee, S., Merryman, C., Venter, J.C., Smith, H.O., Glass, J.I., and Hutchison, C.A. (2010). Cloning whole bacterial genomes in yeast. *Nucleic Acids Research*, **38** (8): 2558–2569.

Gibson, D.G., Glass,J.I., Lartigue, C., Noskov, V.N., Chuang,R.-Y., Algire, M.A., Benders, G.A., Montague, M.G., Ma, L., Moodie, M.M., Merryman, C., Vashee, S., Krishnakumar, R., Assad-Garcia, N., Andrews-Pfannkoch, C., Denisova, E.A., Young, L., Qi, Z.-Q., Segall-Shapiro,T.-H., Calvey, C.H., Parmar, P.P., Hutchison, C.A., Smith, H.O., and Venter, J.C. (2010). Creation of a bacterial cell controlled by a chemically synthesized genome. *Science*, **329**: 52–56.

Jiang, X., Yang, J., Zhang, H., Zou, H., Wang, C., and Xian, M. (2012). *In vitro* assembly of multiple DNA fragments using successive hybridization. *PLoS ONE*, **7** (1): e30627.

Kouprina, N. and Larionov, V. (2003). Exploiting the yeast *Saccharomyces cerevisiae* for the study of the organization and evolution of complex genomes. *FEMS Microbiology Reviews*, **27**: 629–649.

Quan, J. and Tia, J. (2009). Circular polymerase extension cloning of complex gene libraries and pathways. *PLoS ONE*, **4** (7): e6441.

24

AMB 1 experiment 24: Restriction enzyme digestion and fast agarose gel electrophoresis

Purpose: This is up to you to write down.

Introduction

Restriction enzymes work like molecular scissors that recognizes and cuts a very specific set of nucleotides in double-strand DNA. However, these endonucleases cannot work on single-stranded DNA. There are five known types of the enzymes, type I, II, IIS, III, and intron-encoded homing endonucleases, which differ in recognition binding and cutting sites. Type 1 enzyme is not useful because its cleavage sites are neither specific nor identical to its recognition sites. However, types II, IIS, and homing endonucleases are very useful and thus frequently used for molecular cloning. Type II enzymes typically recognize palindromic sequences and cut within the recognition site, whereas type IIS enzymes recognize non-palindromic sequences and cut at the outside of the recognition site. Unlike types II and IIS, which typically recognize 4 to 8 nucleotides, homing endonucleases recognize very rare asymmetric 14 to 40 nucleotides.

Depending on the type of enzymes, endonuclease action results in a blunt end or cohesive end with either 5′ or 3′ overhang. The specificity and efficiency of restriction cleavage depend on the reaction conditions of components, incubation time, and temperature, as well as DNA methylation. Almost all *E. coli* strains used for cloning have two site-specific DNA methylases: Dam methylase adds a methyl group to adenine residue of GA*TC, whereas Dcm methylase transfers a methyl group to cytosine residue of C*CAGG and C*CTGG. Many enzymes do not cleave methylated DNA. More background information on the optimization of restriction enzyme digestion is provided at the end of the protocol.

In this lab exercise, you will use four different type II enzymes to digest pRY121 plasmid DNA and separate the cleaved fragments by fast agarose gel electrophoresis in order to conduct the Southern blot experiment (Experiment 25). The physical map of plasmid pRY121 is shown below.

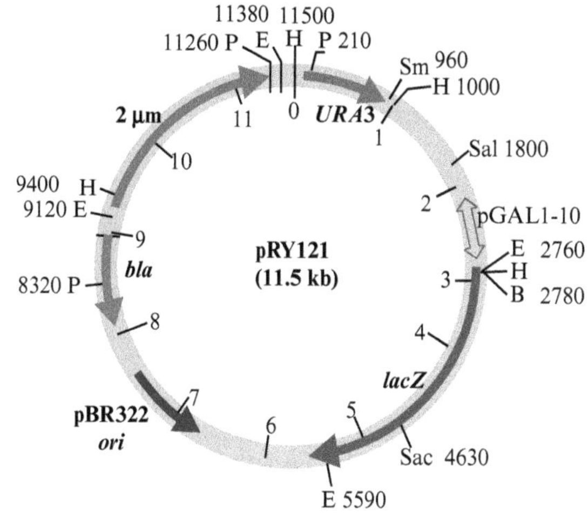

The full names of abbreviations in the figure are as follows: H, *Hin*dIII; E, *Eco*RI; P, *Pst*I; B, *Bam*HI; Sm, *Sma*I; Sal, *Sal*I; Sac, *Sac*I; *bla*, beta-lactamase gene conferring ampicillin resistance; *URA3*, orotidine-5′-phosphate decarboxylase gene involved in uracil synthesis; *lacZ*, a 3069-bp *E. coli* β-galactosidase coding sequence; pGAL1-10, a 680-bp galactose inducible promoter derived from yeast *GAL1* and *GAL10* genes; while the numbers inside and outside the circle denote approximate lengths in kb and bp, respectively. This plasmid is able to replicate in both yeast and *E. coli* cells because the presence of replication origins of both yeast 2-μm circle and bacterial pBR322 plasmids. *URA3* is a nutritional selection marker for yeast transformant and *bla* is an antibiotic selection marker for *E. coli* transformant. For more detailed information of pRY121 plasmid, refer to the references (Yocum *et al.*, 1984; West *et al.*, 1984).

Pre-lab assignment

(Typing and submission must be completed before lab work begins.)

1. Calculate the expected numbers and sizes of the DNA fragments after complete digestion of plasmid pRY121 DNA with (a) *Bam*HI, (b) *Hin*dIII, (c) *Pst*I, and (d) *Eco*RI.
2. Based on the expected size of the above DNA fragments, (a) what DNA fragments would be least separated and thus appear as a doublet band on agarose gel? (b) What DNA fragments would be hard to detect in the ethidium bromide stained gel?
3. If you want the smallest band of *Pst*I enzyme-digested pRY121 DNA to contain 20 ng DNA, how much pRY121 DNA do you need to use? Show your work.
4. Describe in detail how to set up *Pst*I-HF enzyme digestions of pRY121 DNA (stock 25 ng/μL) so that the smallest band contains 20 ng of DNA. Your description also must include the order of addition of each component including the use of a total 10 units of enzyme in a total 25 μL reaction.

Methods in Biotechnology, First Edition. Seung-Beom Hong, M. Bazlur Rashid and Lory Z Santiago-Vázquez.

Companion Website: www.wiley.com\go\hong\Methodsinbiotechnology

Materials and equipment

- Restriction enzymes *Bam*HI-HF, *Eco*RI-HF, *Hin*dIII-HF, *Pst*I-HF (NEB, 20 units/μL)
- 10× CutSmart™ buffer (NEB)
- Sterile nuclease-free H_2O
- pRY121 plasmid DNA (mini-prep from Experiment 19), LacZ PCR DNA (Experiment 22)
- *E. coli* genomic DNA
- Ethidium bromide (10 mg/mL)
- 10× Lithium borate buffer
- Agarose (low EEO)
- 10× DNA loading dye
- 1 kb DNA ladder marker (NEB Cat. No. N3232S)
- Sterile 1.5-mL microcentrifuge tubes
- Erlenmeyer flasks (250 mL)
- 37 °C water bath, microwave oven, ice bucket
- Horizontal mini-gel electrophoresis apparatus (10 cm length mini-gel), power supply

Procedure

1. Label five sterile microcentrifuge tubes as shown in the table below and pipet reagents into the bottom of each tube.

Order of addition	Component \ Tube	1	2	3	4	5
		B	H	P	E	E
1	Nuclease-free H_2O	16.5	0.5	0.5	0.5	0.5
2	10× CutSmart™ buffer	2.5	2.5	2.5	2.5	2.5
3	pRY121 DNA	5	20	20	20	–
3	*E. coli* genome	–	–	–	–	20
4	*Eco*RI-HF (E)*	–	–	–	2	2
4	*Hin*dIII-HF (H)*	–	2	–	–	–
4	*Pst*I-HF (P)*	–	–	2	–	–
4	*Bam*HI-HF (B)*	1	–	–	–	–
	Total volume (μL)	25	25	25	25	25

*Always keep the enzyme tubes in an ice bath and put them back in the −20 °C freezer as soon as you have finished with them.

2. Mix gently by pipetting up and down twice, and incubate the reaction tubes at 37 °C for at least 15 min.
3. Prepare 60 mL of 0.5% agarose in 1× lithium borate buffer for a mini-gel of 10 cm length, add 3 μL of ethidium bromide (10 mg/mL), swirl the flask, and pour the gel solution at 4 °C.
4. Add 2.5 μL of 10× DNA loading dye to each reaction tube.
5. Prepare ~10 ng of LacZ-PCR sample (from Part F of Experiment 22) in 10 μL of TE buffer and add 1 μL of 10× DNA loading dye.
6. Load all the samples on to agarose gel, along with a 1-kb DNA ladder.
   Lane 1: 1-kb DNA ladder
   Lane 2: LacZ-PCR DNA
   Lane 3: pRY121/*Bam*HI
   Lane 4: pRY121/*Hin*dIII
   Lane 5: pRY121/*Pst*II
   Lane 6: pRY121/*Eco*RI
   Lane 7: *E. coli* genomic DNA/*Eco*RI
7. Run the gel at 300 V for 30 min.
8. Take a photograph under a UV transilluminator.

**The photograph will be used to measure the distance between the loading well and each band of DNA.*

Post-lab assignment

1. Restriction enzymes *Xho*I and *Sal*I cut C↓TCGAG and G↓TCGAC, respectively.
   (a) Explain why *Xho*I cleavage fragments can be inserted into a *Sal*I site of plasmid vector.
   (b) If a plasmid vector has a single *Sal*I site, how could you select plasmids containing *Xho*I fragment inserts?
2. How can you tell that the restriction enzyme digestion is complete? Provide at least two independent ways to explain.
3. *Dpn*I, *Mbo*I, and *Sau3A*1 are isoschizomers. What is an isoschezomer and how do the three enzymes differ from each other?
4. You have 10 different DNA samples and want to carry out the same restriction enzyme digestions on 2 μL DNA, each with 1.0 unit of *Eco*RI enzyme in a total reaction volume of 10 μL. Therefore, you want to make an 11× pre-mix using 10× buffer, dH_2O, and *Eco*RI (11 units/μL) in order to save on pipetting. Describe how to make up 11× pre-mix and set up the restriction enzyme reaction.
5. In a previous experiment, a 2-kb PCR product from human genome was cloned into a TA cloning vector pNBR that was cut with *Sma*I and then treated with terminal deoxynucleotidyl transferase. Below is a diagram of the gel patterns from each digest (U, uncut; Sa, *Sal*I; Sm, *Sma*I; R, *Eco*RI; B, *Bam*HI; P, *Pst*I) after agarose gel electrophoresis.

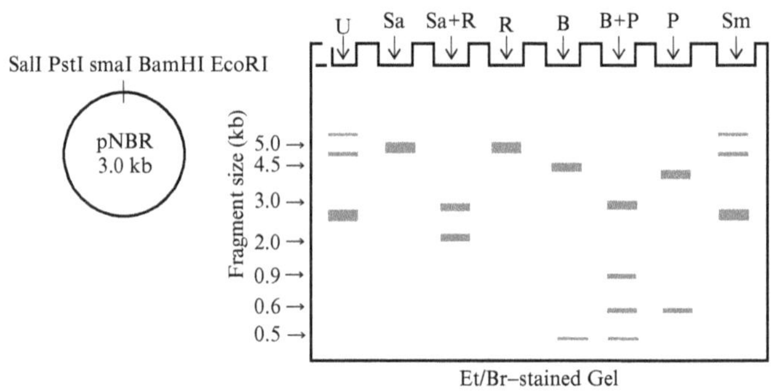

Determine a map of the plasmid. In your map, distinguish between the human DNA fragment insert and the plasmid vector. Indicate all the restriction sites and the distances between the restriction sites.

Additional background information

(1) DNA amount for digestion

The amount of DNA needed to visualize DNA may vary depending on the size of the sample well, the type of staining dye, and the number of digestion fragments. Suppose that you cut a 5-kb plasmid carrying a 2-kb insert with *Eco*RI enzyme and anticipate three bands of 3 kb, 1.5 kb, and 0.5 kb cleavage fragments. If you want to make the smallest 0.5-kb band contain about 10 ng of DNA for easy detection, you need to use 100 ng of plasmid DNA because the smallest band is 10 times smaller than the plasmid DNA prior to digestion. The resulting three cleavage bands will contain 60 ng (100 × 3/5), 30 ng (100 × 1.5/5), and 10 ng (100 × 0.5/5) of DNA, respectively. The largest band would be about six times more glowing fluorescent than the smallest band, assuming that ethidium bromide molecule has no sequence specificity in binding to DNA molecules. However, if you digest 100 ng of the above same plasmid with a different enzyme that cuts only once, then the single band of 100 ng of DNA will be very bright.

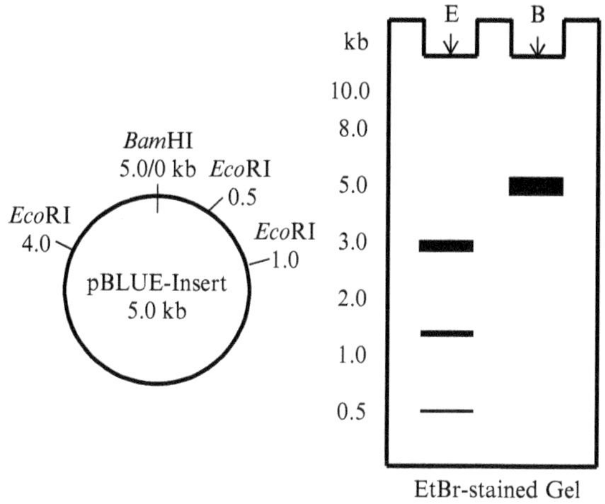

E (*Eco*RI): 3 kb (60 ng), 1.5 kb (30 ng), 0.5 kb (10 ng)
B (*Bam*HI): 5 kb (100 ng)

Overloading, which depends on the sample amount and well size, typically results in trailing and smearing of DNA bands. They appear to run fast as if they are smaller than they actually are. On the other hand, with too little DNA loading the smaller bands are too faint to detect. Usually digestion of 2–5 μL of a 50 μL of high copy plasmid DNA prepared by a mini-prep kit will give a satisfactory result. However, it depends on the number and size of the bands expected.

(2) Typical protocol

An example of a reaction setup in a total volume of 10 μL is shown below. The order of addition matters:

(1st) *X* μL H_2O (*X* volume may vary depending on the *Y* volume used)
(2nd) 1.0 μL 10× buffer
(3rd) *Y* μL DNA
(4th) 0.5 μL enzyme

Mix gently by pipetting up and down a few times while immersing the pipette tip in the solution.

Buffer

Usually 10× buffer comes with the restriction enzyme when purchased. Most companies provide three kinds of basic buffers and occasionally a uniquely different buffer for a particular enzyme. You may check in the company catalogue (NEB, Promega, Life Technologies, Roche, etc.) page displaying the list of the enzyme/buffer compatibility. The most important parameter is salt cation concentration (high: 100 mM, medium: 50 mM, and low: 0 mM NaCl or KCl), and the relevant 10× buffer is added at a final concentration of 1×. Mostly the buffer maintains pH 8.0. Sometimes bovine serum albumin (BSA) is added to the reaction mix at a final concentration of 0.1 mg/mL to stabilize the enzyme as well as bind some impurities.

H_2O

Sterile nuclease-free distilled water must be used. It is always safe to add the water and buffer into the tube first, DNA second, and enzyme last in the reaction tube. Addition of enzyme directly to water or concentrated buffer may cause irreversible denaturation of enzyme.

DNA

DNA preparations may have impurities (phenol, chloroform, alcohol, EDTA, detergents, or excessive salts), which can inhibit restriction enzyme digestion activity. Although DNA is more prone to nuclease digestion the absence of EDTA, high concentrations of EDTA can inhibit restriction digestion. Thus, one would prefer to use $T_{10}E_{0.1}$ (10 mM Tris-Cl, pH 8.0, 0.1 mM EDTA) rather than $T_{10}E_1$ buffer to dissolve the DNA pellet. This is especially true for PCR or a DNA sequencing reaction. As a general rule, restriction enzymes require six nucleotides on either side of their recognition site to cut efficiently. The recognition site located very close to the ends of linear DNA segments would not be cleaved. For more information on the cleavage efficiency for the end of DNA fragments, refer to https://www.neb.com/~/media/NebUs/Files/Chart%20image/cleavage_olignucleotides_old.pdf

Restriction enzyme

Enzymes are supplied in concentrated units, usually between 5 and 100 units depending on the enzyme and manufacturer. Usually 1 to 0.5 μL of enzyme is sufficient to cut mini-prep plasmid DNA in 20 to 25 μL reaction volume, though you can use less if you cut longer. One unit is expressed as the amount of enzyme that cuts 1 μg in 1 h under the optimum assay conditions. *Do not use enzyme more than 10% of the final reaction volume* (i.e., not more than 1 μL of enzyme in 10 μL of the total reaction volume). This is because the enzyme storage buffer contains a high concentration of anti-freezing compound glycerol to maintain the enzyme activity at –20 °C. At concentrations above 10%, glycerol not only inhibits the digestion but also can cause star activity, leading to aberrant, non-specific cuts of the DNA. All manufacturers of restriction enzymes supply specific buffers with the enzymes, and these should also be stored in the –20 °C freezer.

The optimum reaction temperature of most enzymes is 37 °C; check the optimum reaction temperature for the individual enzymes because certain enzymes require either a higher or a lower temperature (e.g., 25 °C for *Sma*I, 50 °C for *Bsa*I, 65 °C for *Bsm*I). The incubation time of standard enzymes is 1 h, but 5 to 15 min incubation is enough when FastDigest enzymes (Thermo Scientific) or HF™ enzymes (NEB) are used.

Pre-master mix

If you plan to perform many digestions, then it is better to prepare a pre-master mix in order to reduce pipetting. Suppose that you are doing 25 restriction digestions with the same enzyme but with 25 different DNA samples. You should set up a 26× pre-mix containing a mixture of buffer, water, and enzyme rather than

25× pre-mix in order to make up for any volume deficiency due to pipetting inaccuracies, especially for allocating to the last reaction tube. To prepare a 26× pre-mix for a total 10-μL reaction tube each containing 2 μL of DNA and 0.5 μL of enzyme, you add 192.4 μL of H_2O, 2.6 μL of 10 × buffer, and 13 μL of enzyme in order (total reaction volume = 26 × 10 = 260 μL; total premix volume = 260 − (26 × 2) = 208 μL; 10 × buffer needed = (26 × 1)/10 = 2.6 μL; enzyme needed = 26 × 0.5 = 13 μL; water volume needed = 208 − 2.6 − 13 = 192.4 μL). Mix gently by pipetting up and down after addition of enzyme and keep the pre-mix on ice. *Avoid air bubbles* during mixing; the enzyme exposed to air pockets is prone to be denatured. Place 2 μL of DNA first into each of 25 microcentrifuge tubes. Then, add 8 μL of pre-mix and mix gently by pipetting up and down a few times while immersing the pipette tip in the solution. Use a fresh pipette tip when you add DNA and pre-mix to each reaction tube.

Double digestion

You may want to digest DNA with two or more enzymes. In this case, you have to make sure that the buffer to be used is compatible with all the enzymes. Companies often provide a universal buffer that can be used for several enzymes at the same time (NEB Double Digest Finder). If two enzymes are not compatible for a single buffer, determine which enzyme requires a lower salt (NaCl) buffer. Then, set up restriction digestion using the enzyme with the lower salt requirement first, adjust the first reaction buffer to the optimal salt concentration of the second enzyme using a concentrated salt solution after the first digestion, and conduct the second enzyme digestion by adding the enzyme and incubating at an appropriate temperature. However, double digestions can be simplified by using a single buffer if FastDigest enzymes (Thermo Scientific) or HF™ enzymes (NEB) are used. However, remember the *10% rule of total enzyme amount* added to avoid the glycerol inhibition effect.

DNA size marker

Many size markers are available commercially including 1 kb DNA ladder and lambda DNA cut with restriction enzyme(s).

10× DNA loading dye (20 mL)

Dissolve in 6.25 mL of H_2O:

- 0.025g of xylene cynol (optional)
- 0.025g of bromophenol blue (optional: 0.025 g of orange G)
- 1.25mL of 10% SDS
- 12.5mL of glycerol

Add 1/10 volume of dye to the restriction-digested sample before loading on to agarose gel.

Agarose gel electrophoresis

Gels of 0.7 to 0.8% (w/v) agarose in 1× TBE or 1× TAE buffer are suitable for separating linear DNA molecules of 0.5–10 kb in size. However, its running time takes much longer than 1× LB (lithium borate) or 1× bionic buffer (Sigma Aldrich, Product # B6185). The maximum resolution of DNA fragments larger than 10 kb is obtained when the gel is run overnight at a low voltage of less than 1 to 5 V/cm. However, running at a low voltage is not necessary to increase resolution of small fragments less than 1 kb.

Visualization of DNA bands

Ethidium bromide (EtBr) can be added to H_2O for a post-run gel staining or to a gel solution for pre-run staining at a final concentration of 0.5 μg/mL. Adding unnecessary excess amounts of EtBr does not increase the detection limit but rather causes a background staining problem. Always wear gloves to handle ethidium bromide because it is a potent mutagen. Ethidium bromide (EtBr) has a detection limit of about 2 to 5 ng per lane for a 4–5 mm sample well in agarose gel. Staining of denatured ssDNA or RNA is less sensitive than dsDNA, requiring some tenfold more nucleic acid. The fluorescence of EtBr is quenched by polyacrylamide, and thus its sensitivity is reduced ten- to twentyfold in polyacrylamide gel. Any excess of unbound EtBr molecule, which is positively charged, migrates to the cathode. This is often attributed to a problem of uneven background staining, especially with high background in the top portion of the gel. Detection can be improved by soaking the stained gel in 1 mM of $MgSO_4$ for ~ 30 min to reduce background staining. Unlike EtBr, GelRed™ is not hazardous and its staining is known to be ~10 times more sensitive than EtBr. However, GelRed™ is recommended for a post-run gel staining at a final concentration of 3× in H_2O. Destaining is not required, but the ethidium bromide or GelRed™ gel can be washed in water to reduce background if necessary. DNA bands in both types of gels can be visualized in a standard UV transilluminator. Staining solutions can be reused at least 2–3 times. Store staining solution at room temperature protected from light.

References

Yocum, R.R., Hanley, S., West, R., and Ptshne, M. (1984). Use of lacZ fusions to delimit regulatory elements of the inducible divergent GAL1–GAL10 promoter in *Saccharomyces cerevisiae*. *Molecular Cell Biology*, **4** (10): 1985–1998.

West, R.W., Yocum, R.R., and Ptshne, M. (1984). *Saccharomyces cervisiae GAL1-GAL10* divergent promoter region: location and function of the upstream activating sequence UAS_G. *Molecular and Cellular Biology*, **4**: 2467–2478.

Further reading

Block, K.D. (2000). Unit 3.1. Digestion of DNA with restriction endonucleases. Unit 3.2. Restriction mapping. In *Current Protocols in Molecular Biology* (Edited by F.M. Ausubel). John Wiley & Sons, Inc. ISBN 0-471-50338-X.

Brody, J.R., Calhoun, E.S., Gallmeier, E., Creavalle, T.D., and Kern, S.E. (2004). Ultra-fast high-resolution agarose electrophoresis of DNA and RNA using low-molarity conductive media. *BioTechniques*, **37**: 598–602.

Detection of DNA/RNA Using Ethidium Bromide. https://www.nationaldiagnostics.com/electrophoresis/article/ethidium-bromide-staining.

25 AMB 1 experiment 25: Southern blot transfer

Purpose: This is up to you to write down.

Introduction

The Southern blot technique was developed by Edwin M. Southern in 1975 shortly after many type II restriction enzymes had been isolated from different bacterial species since the first discovery of *Hin*dIII enzyme from *Haemophilus influenza* in 1970 (Southern, 1975). It enables detection of a specific DNA sequence in a complex genomic DNA. Based on the specificity of hybridization between the probe and target DNA sequences, it has been used for genetic fingerprinting known as restriction fragment length polymorphism (RFLP) analysis, detection of DNA insertion, deletion or point mutations, and identification of homologous genes in different species (orthologs) and within a species (paralogs).

Southern blotting involves the following sequential steps: (1) restriction enzyme digestion of DNA (genomic or plasmid), (2) agarose gel electrophoresis of DNA fragments, (3) ethidium bromide staining and photograph, (4) treatment of gel by acid depurination (optional), alkaline denaturation, neutralization, and equilibration, (5) blot transfer of denatured DNA fragments to nylon or nitrocellulose membrane in a high salt solution by upward capillary action, (6) immobilization of the DNA fragment on to membrane by UV cross-linking or baking in a hot oven, (7) pre-hybridization of the membrane to a blocking buffer, (8) hybridization of a labeled probe to the membrane, (9) washing the membrane at an appropriate stringency, and (10) radioactive or non-radioactive (chromogenic or chemiluscent) detection of hybridized bands. A choice must be made between the radioactive and non-radioactive labeling of probe DNA, as well as between nylon and nitrocellulose membrane. If non-radioactively labeled probe DNA is used, use of nitrocellulose membrane is preferred to reduce background hybridization signals.

The Southern blot experiment is routinely employed to verify the stable integration of transgene in a host genome, as well as to determine whether the insertion takes place in multiple loci or a single locus. Selection of restriction enzyme with respect to a transgene is important for this detection. The enzymes that cut only once within a transgene should be chosen. This is because multiple insertions are often clustered at a locus and two insertion events are most likely to be indistinguishable from a single insertion event when the enzymes used do not cut a transgene at all. To determine the exact position of a single insertion event, one should prepare a genomic library from the transgenic organism, screen the library using the Southern blot probe, and sequence the hybridized clone. Alternatively, the inverse PCR approach can be adopted. For the principle of this method, refer to the YouTube website:

- https://www.youtube.com/watch?v=n0Zn89AHE04 (Inverse PCR animation)

In this lab exercise, you will transfer restriction cleavage fragments obtained from Experiment 24 to nitrocellulose membrane (blotting) and fix the transferred DNA on to the membrane prior to use for hybridization to probe DNA.

Pre-lab assignment

(Typing and submission must be completed before lab work begins.)

1. Why is DNA cut with restriction enzymes prior to Southern blotting?

2. Why is the blotted DNA subjected to alkaline denaturation followed by neutralization? What happens if either step is omitted in Southern blot hybridization? Explain.

Materials and equipment

- 20× SSC (3 M NaCl, 0.3 M sodium acetate, pH 7.0)
- Denaturing solution (0.5 M NaOH, 1.5 M NaCl)
- Neutralizing solution (1 M Tris-HCl, pH 7.5, 1.5 M NaCl)
- Equilibrating solution (10× SSC)
- Protran® BA85 nitrocellulose membrane (Whatman)
- Whatman 3MM paper
- Paper towels
- Paper cutter, ruler, forceps, overhead projector transparency sheets, tape
- Pyrex glass dish
- Plate shaker, glass plate, weight
- UV cross-linker (Stratalinker 2400)

Methods in Biotechnology, First Edition. Seung-Beom Hong, M. Bazlur Rashid and Lory Z Santiago-Vázquez.

Companion Website: www.wiley.com\go\hong\Methodsinbiotechnology

Procedure (day 1)

1. After taking a photograph under UV illumination (Experiment 24), measure the dimension (width × length cm) of the gel.
2. Submerge the gel in denaturing solution (100 mL) and gently shake at room temperature for 30 minutes. Rinse briefly in distilled water.

**The gel can be soaked in 0.25 M HCl for 10 min prior to the denaturing step when a target DNA is >5 kb. This depurination step facilitates DNA fragmentation for efficient transfer.*

**During this time, set up a nick translation reaction (Experiment 26). Cut paper towels, 4 sheets of Whatman papers, and one nitrocellulose membrane fitted to the same dimension as the gel.*

3. Immerse and shake gently in neutralizing solution (100 mL) twice each for 10 min.
4. Immerse and shake gently in equilibration solution (100 mL 10× SSC) for 15 min.
5. Wet two sheets of paper bridges in 10× SSC and place them on to the support box in the Pyrex dish containing 10× SSC solutions.

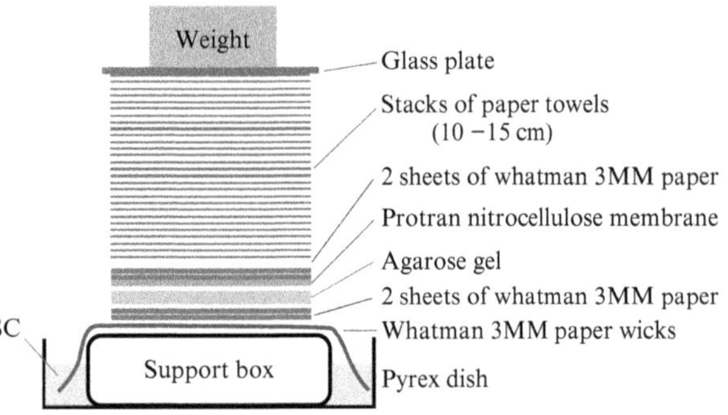

6. Place two sheets of Whatman paper on to the wet paper bridge and smooth out the surface using a clean glass test tube.
7. Flip the gel upside down by putting the gel between two overhead transparency sheets and place it on to the wet Whatman 3MM paper.

**DNA is usually on the bottom side of the gel.*

**Be careful when handling the 0.5% agarose gel; it is fragile.*

8. Cut one corner of your nitrocellulose to help you orient your blot, and lay a membrane exactly on to the gel. Make sure that the membrane does not hang over the agarose gel.

**When touching the blot membrane, always wear gloves and handle the membrane only at the edges with forceps.*

**Using a clean glass test tube, gently smooth out the surface to remove air bubbles trapped between the gel and Whatman 3MM paper, between the gel and membrane, and between the membrane and Whatman 3MM paper.*

9. Place the first Whatman 3MM paper on to the membrane and smooth it out to wet before adding the second Whatman 3MM paper. The second piece does not have to wet completely.

**Make sure that the Whatman 3MM paper and paper towels on the top of the membrane do not touch the gel or buffer-saturated Whatman 3MM paper bridge. You may put Parafilm strips on the paper bridge along the edge of the gel to prevent circuiting of buffer from the bridge to the stacks of paper towels.*

10. Put stacks of paper towels on the second Whatman 3MM paper.
11. Place a flat glass or plastic plate on the top of the stacked paper.
12. Put a weight on the plate, and leave the blot transfer overnight (12 to 16 h).

**Tape the whole sandwich assembly to the mini-gel size.*

Procedure (day 2)

13. Peel off the gel carefully. Remember which side of the membrane is the DNA binding side! Do not discard the gel, which is to be used for step 17.

**Before peeling the membrane off the gel, draw the line positions of each lane on the membrane using a blunt pencil. This helps to identify the DNA binding side and to determine the size of hybridized bands after probe detection. To draw the sample well lines, place the gel/membrane on to clean paper or plastic wrap with the membrane facing down and draw the lines directly on to the sample well lane positions.*

**You can write down the first and last lane numbers on the top of the membrane with a BLUNT pencil. Write your group and section # on the edge of the membrane.*

14. Place the membrane with *the DNA side facing up* on the paper towel, place into the UV cross-linker, and close the door.
15. Select the *Autocrosslink* mode; the bleeper will sound and a pre-set exposure of 1200 microjoules (×100) will be displayed; and press the "Start" button. It should count down to zero. The cross-linking will be complete in ~1 min.
16. Wrap with plastic Saran wrap, label the group and section #, and store at room temperature until use or at 4 °C for a long duration.
17. Examine the peeled-off gel under a UV transilluminator to check if the DNA is completely transferred.

Post-lab assignment

1. What are the advantage and disadvantage of using nylon and nitrocellulose membrane for blot transfer?
2. List the practical applications of the Southern blot technique.
3. How does the Northern blot (NB) differ from Southern blot (SB) in terms of procedure?
4. What are the colony and plaque lift hybridizations? How do they differ from the Southern blot in terms of procedure?

Reference

Southern, E.M. (1975). Detection of specific sequences among DNA fragments separated by gel electrophoresis. *Journal of Molecular Biology*, **98**: 503–517.

Further reading

Brown, T. (2003). Unit 2.9. A Southern blotting. In *Current Protocols in Molecular Biology* (Edited by F.M. Ausubel), pp. 2.9.1–2.9.15. John Wiley & Sons, Inc. ISBN 0-471-50338-X.

26 AMB 1 experiment 26: Probe labeling and purification

Purpose: This is up to you to write down.

Introduction

Currently there are at least five different probe DNA labeling methods: direct oligonucleotide synthesis, end labeling, random primer labeling, nick translation labeling, and PCR labeling. Probe DNA can be either radioactively or non-radioactively labeled. Radioactive labeling is typically carried out using ^{32}P-labeled dNTP in the reaction. Non-radioactive labeling can be accomplished by direct labeling with fluorophore-tagged dNTP or by indirect labeling with biotin- or digoxigenin (DIG)-labeled dNTP. For the respective non-radioactive labeling type, target-probe hybrids are detected through laser light excitation to emit a characteristic fluorescence, conversion of colorless chromogenic substrate to a color compound, and chemiluminscent substrate emission to light by receptor protein conjugated enzymes. Most widely used conjugated enzymes are horseradish peroxidase and alkaline phosphatase because of their high stable enzyme activity.

In this lab exercise, you will non-radioactively label the entire pRY121 DNA by the nick translation method using BioNick™ labeling kit and label a portion of *lacZ* DNA by the PCR labeling method. Nick translation yields heterogeneous DNA probes of different sizes and nucleotide sequences, whereas PCR labeling produces a defined size and sequence of DNA probe. PCR labeling can be carried out using a 10× dNTP mix of BioNick™ labeling kit components used for nick translation, since a 1× dNTP mix of BioNick™ labeling kit has 0.02 mM dNTP, which is a default concentration of the PCR reaction. After labeling, you will purify the labeled probes by desalting spin column chromatography. You will also prepare the spin column using Sephadex G50, which has an exclusion limit of 72 bases. After centrifuging the spin column loaded with the labeling reaction sample, all unincorporated dNTPs including labeled dNTP are retained in the column, but both labeled and unlabeled probe DNAs larger than 72 bases are eluted.

Pre-lab assignment

(Typing and submission must be completed before lab work begins.)

1. What enzymes are used for the nick translation reaction? Explain their functions.

2. What happens if the incubation time of reaction at step 2 of Part I is more than 1 h?

3. How many nucleotides of DNA are excluded from Sephadex G50? What other methods can be used to purify labeled oligo primers smaller than the DNA eluted from Sephadex G50?

Materials and equipment

- Bionick™ Labeling System Kit (Life Technologies, Cat. No. 18247-015)
- Swollen Sephadex G-50 resin equilibrated SDS column buffer
- SDS column buffer (50 mM Tris-Cl, pH 8.0, 2 mM EDTA, 250 mM NaCl, 0.1% SDS)
- Sterile glass wool
- Sterile 1.5-mL microcentrifuge tubes
- 15-mL conical tubes (used)
- Disposable 1-mL syringes
- Tabletop centrifuge
- LacZ primer 1: GTTGTTGCAGTGCACGGCAG
- LacZ primer 2: GCTGGAATTCCGCCGATACTG
- Phusion® high-Ffdelity DNA polymerase (NEB M0530S; 2 units/μL) and 5× buffer
- Thermocyclers

Procedure

Part I: Nick translation (groups 1 and 2 only)

1. Pipet the following components into a 0.5-ml microcentrifuge tube *on ice*:

Groups 1 and 2	Only one group
x μL nuclease-free H_2O	31 μL nuclease-free H_2O
5 μL 10× dNTP mix	5 μL 10× dNTP mix
y μL (1 μg) pRY121 DNA	4 μL (1μg) pBR322 DNA (0.25 μg/μL)
5 μL 10× enzyme mix	5 μL 10× enzyme mix
50 μL total volume	50 μL total volume

2. Mix well, pulse spin, and incubate at 16 °C for 1 h.

3. Add 5 μL of stop buffer (0.5 M EDTA, pH 8.0) and a 50 μL SDS column buffer.

**SDS prevents the biotinylated DNA from sticking non-specifically to the resin and glass wool plug in the Sephadex spin column.*

Methods in Biotechnology, First Edition. Seung-Beom Hong, M. Bazlur Rashid and Lory Z Santiago-Vázquez.

Companion Website: www.wiley.com\go\hong\Methodsinbiotechnology

4. During the reaction incubation time, prepare the Sephadex G50 spin column assembly as shown in the figure below.

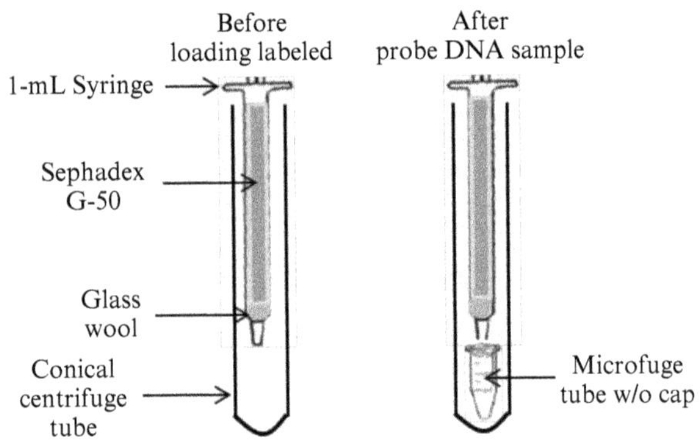

(a) Remove the plunger from a 1-mL disposable syringe.
(b) Soak a small ball of sterile glass wool in the SDS column buffer and press it down to the bottom of the syringe using the plunger.
(c) Fill up the glass wool syringe with the swollen Sephadex G50 resin and put into a 15-mL conical tube.
(d) Spin the assembly at 1600 × g for 4 min in a tabletop centrifuge. Aspirate off the column effluent in the bottom of the conical tube.
(e) Repeat steps (c) and (d) until the resin is packed to the 0.9-mL mark position. If the column is overpacked, add a small volume of column buffer and remove.
(f) Apply 100 μL of the SDS column buffer, spin at 3000 rpm for 4 min, and remove the effluent. Make sure that the column is packed to around the 0.9-mL position.
(g) Cut off the cap of a 1.5-mL sterile microcentrifuge tube, put into the conical tube, and insert the packed column syringe into the conical tube.
(h) Apply the labeled probe sample to the column and spin at 1600 × g for 4 min.
(i) Transfer the eluent to a new sterile labeled microcentrifuge tube.

5. Spot the 2-μL control DNA only on to NanoDrop 2000 and measure for ssDNA quantification using the SDS column buffer as a blank. Record the ng/μL and A_{260}/A_{280} ratios.

**Biotin-labeled DNA can be directly quantified at A_{260} because biotin has no absorbance at 260 nm.*

6. Store at −20 °C for hybridization experiments (step 2 of Experiment 27 and step 6 in Experiment 28) next week.

Part II. PCR labeling (groups 3 and 4 only)

- *10× dNTP mix of of BioNick™ labeling system:*
  0.2 mM each dCTP, dGTP, dTTP
  0.1 mM dATP
  0.1 mM biotin-14-dATP
  500 mM Tris-HCl (pH 7.8)
  50 mM $MgCl_2$
  100 mM β-mercaptoethanol
  100 μg/mL nuclease-free BSA

1. Set up the PCR reaction using a 10× dNTP mix of BioNick™ labeling system instead of a 10 mM dNTP to a final concentration of 1× dNTP mix and perform a PCR cycle as described in Experiment 21 protocol (Phusion-LacZ/pRY121).

Component (for plasmid pRY121 PCR)	Volume (μL)	Final
Nuclease-free sterile dH_2O	35.0	
5× Phusion HF buffer	10.0	1×
10× dNTP mix of BioNick™ labeling	1.0	1×
10 μM LacZ primer 1 + 2 mix	2.5	0.5 μM each
pRY121 plasmid (10 ng/μL)	1.0	10 ng
Phusion® DNA polymerase (2U/μL)	0.5*	0.02 U/μL

*Pipette carefully and slowly because the high content (50%) of glycerol in the storage buffer may lead to pipetting errors. Set up in ice because 3′→5′ exo activity degrades primers and template.

2. Program the thermal cycler II using the following parameters.

Cycle	Steps	Temperature (°C)	Time
1	Initial denaturation	98	30 s
	Denaturation	98	10 s
25	Annealing	70	10 s
	Extension	72	9 s
1	Final extension	72	5 min
1	Hold	4	∞

3. Begin the program cycle. Once the initial denaturation temperature reaches 90 °C, quickly spin your sample tube and place in the thermal cycler.
4. After the PCR cycle, add 50 μL of SDS column buffer to the PCR reaction tube and proceed to do step 4 of Part I.
5. Store at −20 °C for hybridization experiments (Experiment 27 and 28) next week.

Post-lab assignment

1. Describe the advantages and disadvantages of radioactive labeling and non-radioactive labeling methods.
2. We used the entire pRY121 plasmid DNA to prepare biotinylated probe DNA by the nick translation and PCR methods. What are your expected results if you hybridize each of the probes to the yeast *S. cervisiae* and *E. coli* genomic DNA restriction fragments that are immobilized on the blotted membrane?
3. There are five different labeling methods as briefly described in the introduction. If you want to label multiple DNA fragments of a series of differing sizes in order to make labeled standard molecular markers, what labeling method should you use? Explain your choice.
4. The genome size of *E. coli* K12 is 4.6397 × 10^6 bp. The coding sequence of the *lacZ* gene is 3054 bp. The MW of 1-bp dsDNA is 660 g/mole. You load 10 μg of genomic DNA cut with restriction enzyme and separated the DNA fragments on agarose gel to detect the *lacZ* gene by Southern blot analysis. In addition, you want to load a positive control sample of a 563-bp PCR-amplified fragment (Taq-LacZ) obtained from Experiment 21 on the gel.

(a) If you want to detect the amount corresponding to a single copy of the *lacZ* gene in the gel, how much of the purified Taq-LacZ PCR product should you use to load on agarose gel?

(b) If you want to detect a single copy amount of the *lacZ* gene using a purified 20 ng/μL Taq-LacZ PCR product, how would you prepare it?

Further reading

Becker, J.M., Caldwell, G.A., and Zachgo, E.A. (1990). Preparation, purification, and hybridization of probe. In *Biotechnology: A Laboratory Course*, 2nd edition, pp. 85–96. Academic Press, UK. ISBN 0-12-084562-8.

Extinction Coefficients and Fluorescence Data. http://www.glenresearch.com/Technical/Extinctions.html.

Leary, J.J., Brigati, D.J., and Ward, D.C. (1983). Rapid and sensitive colorimetric method for visualizing biotin-labeled DNA probes hybridized to DNA or RNA immobilized on nitrocellulose: Bio-blots. *Proceedings of the National Academy of Science USA*, **80**: 4045–4049.

Rigby, P.W., Dieckmann, M., Rhodes, C., and Berg, P. (1977). Labeling deoxyribonucleic acid to high specific activity *in vitro* by nick translation with DNA polymerase. *International Journal of Molecular Biology*, **113** (1): 237–251.

27 AMB 1 experiment 27: Dot blot assay

Purpose: This is up to you to write down.

Introduction

The dot blot (or slot blot) technique was originally developed as a semi-qualitative method to estimate the relative amount of target sequences in a series of DNA samples. A large number of samples can be applied simultaneously on the surface of the membrane, making it possible that many DNAs can be screened in a single hybridization experiment. Because of the sensitivity, rapidity, and simplicity, dot blot hybridization assay has been often used for detecting specific viruses in the environmental samples. This assay procedure is also employed for immunoassay to detect, analyze, and identify specific proteins from a lot of samples. In this case, primary and secondary antibodies are used as probes to detect specific proteins. The technique, however, provides no information on the size of target molecule because sample separation such as electrophoresis and chromatography is not required.

Biotinylation of DNA is widely used as a safe and convenient alternative to radioactive labeling, which entails hazardous handling issues. Non-radioactive biotin-labeled probe-target hybrids are detected with alkaline phosphatase (AP)-conjugated streptavidin, which has a high specific affinity for biotin. AP binds and cleaves the colorless chromogenic substrate BCIP (5-bromo-4-chloro-3-indoylphosphate) supplemented with the enhancer chromogen NBT (nitro blue tetrazolium). This results in the formation of an insoluble dark blue purple precipitate, which appears as a visible spot or band at the reaction site on the membrane. This system has a high sensitivity to detect as low as 0.03 pg of homologous DNA in dot blot hybridization after 16 hours of color development. As milk contains biotin, use of milk power as a blocking reagent is incompatible with a biotin-streptavidin detection system. Accordingly, other blocking reagents such as bovine serum albumin and gelatin must be used.

Application of multiple samples to a single membrane strip can be facilitated using a commercial dot/slot blotting manifold attached to a suction device. Individual spot densities can be densitometrically scanned for quantification by comparing with a series of diluted standard DNA amounts. In this lab exercise, you will manually spot a series of dilutions of the biotin-labeled probe directly on a membrane strip and compare with hybridization intensities of known standard concentrations. The labeling efficiency is defined as the lowest detectable concentration of the labeled probe.

Pre-lab assignment

(Typing and submission must be completed before lab work begins.)

Describe the roles of (a) streptavidin-AP, (b) biotin, (c) NBT, and (d) BCIP in detection of hybridized DNA bands.

Materials and equipment

- Washing buffer 1: 0.1 M Tris-Cl (pH 7.5), 0.1M NaCl, 2 mM $MgCl_2$ (autoclave)
- Blocking buffer 2: 3 % BSA (or gelatin) in wash buffer 1 (filter-sterilize)
- Detection buffer 3: 0.1 M Tris-Cl (pH 9.5), 0.1M NaCl, 50 mM $MgCl_2$ (autoclave)
- 6× SSC solution (diluted from 20× SSC)
- Nylon or Protran® nitrocellulose membrane strip (8 × 4 cm)
- Streptavidin-AP conjugate (2 mg/mL; Life Technologies)
- 1-Step™ NBT/BCIP (Life Technologies, Cat. No. 34042B)
- TE buffer (pH 8.0)
- Biotin-labeled probe DNAs (Experiment 26)
- Biotin-labeled control pBR322 DNA in Bionick™ Labeling Kit (Experiment 26)
- Safety cap lock for microcentrifuge tube
- Standard Petri plate (100 diameters × 15 mm height)
- UV cross-linker (Stratalinker 2400), scissors, microplate shaker

Procedure

All are performed at room temperature except for step 7.

1. Wear gloves, clean a scissor cutting edge with 70% ethanol tissue, and cut the nylon membrane to a strip of 8 × 3 cm, as exactly shown below. Mark with a blunt pencil. This figure size of strip fits into a standard Petri plate (100 × 15 mm).
2. Heat (95 °C) the biotinylated pRY121 (Groups 1 and 2) and *lacZ* probe (Groups 3 and 4) DNA samples (Experiment 26) and biotinylated pBR322 DNA in safety cap microcentrifuge tubes for 5 min, quickly cool on ice/water bath, and pulse-spin the tubes.
3. Make 5 series of serial 1/2 dilutions (2 µL + 2 µL of 6× SSC) of the biotinylated pRY121 and *lacZ* probe DNAs using 6× SSC. Also do the same for the biotin-labeled control pBR322 probe DNA sample. Record control probe DNA quantities in serial dilutions in the Lab notebook on the basis of NanoDrop 2000 quantification.

Methods in Biotechnology, First Edition. Seung-Beom Hong, M. Bazlur Rashid and Lory Z Santiago-Vázquez.

Companion Website: www.wiley.com\go\hong\Methodsinbiotechnology

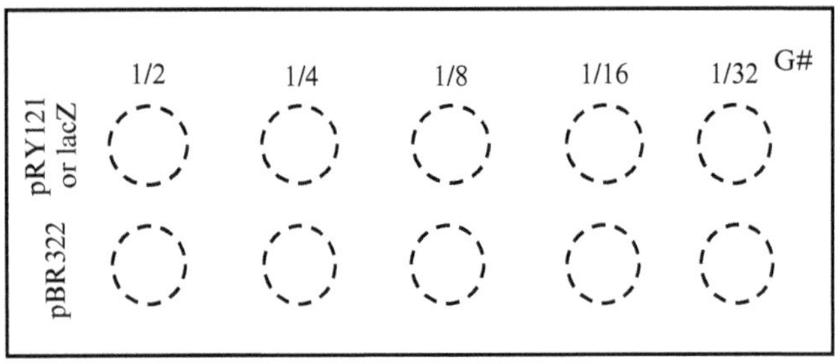

6× SSC in dilution buffer prevents the DNA spot from spreading on the membrane, which diminishes the color intensity of the spot.

4. Spot 1 μL of each dilution and 1 μL of 6× SSC (negative control) on to a test strip of nylon membrane and air dry (or spot under vacuum using a "slot blot" system).

**Do not touch or scratch the membrane with a micropipette tip.*

5. Expose to a short-wave UV light for 1 min under the "Autocrosslink" mode in a UV cross-linker.
6. Float the membrane in 10 mL of *washing buffer 1* for 1 min in a container (Petri dish 100 × 15 mm) to rehydrate.

**The membrane should float freely in solution.*

7. Incubate in *blocking buffer 2* (0.3 mL/cm^2 of membrane = 9.6 mL for 8 × 4 cm strip) for 30 min at *37 °C* with moderate shaking.

**Warm blocker buffer 2 to 50 to 60 °C to facilitate dissolution of the blocking reagent.*

**The membrane should float freely in solution.*

8. Prepare the *streptavidin-AP conjugate* (0.2 mL/cm^2 of membrane, diluted to 1 to 2 μg/mL in *washing buffer 1*).

**Dilute just prior to use.*

**The membrane should float freely in a Petri dish.*

9. Drain the *blocking buffer 2* from a tray and incubate the membrane in diluted streptoavidin-AP conjugate for 30 min on a platform shaker.
10. Transfer the membrane to a shallow dish.
11. Wash with *washing buffer 1* (0.6 mL/cm^2 of membrane) for 15 min with moderate shaking and discard the solution.
12. Wash with *washing buffer 1* (0.6 mL/cm^2 of membrane) for 15 min with moderate shaking and discard the solution.
13. Wash with *detection buffer 3* (0.2 mL/cm^2 of membrane) for 10 min with moderate shaking.
14. Incubate the membrane in substrate solution 1-Step™ NBT/BCIP (no dilution is necessary; use ~0.1 mL per cm^2 of membrane) in the dark.

**Mix well 1-Step™ NBT/BCIP bottle before use and add directly.*

**The blue-purple precipitate may become visible after 10 to 60 min of incubation.*

**Detection sensitivity increases with incubation time.*

15. Discard the solution, rinse with TE buffer (pH 8.0) or dH_2O for 30 s, and air-dry. The dried membranes can then be stored for several months in the dark to preserve the color.

Post-lab assignment

1. What are allele-specific oligonucleotide (ASO) dot blot and reverse ASO dot blot assays? (a) Briefly outline the procedure. (b) Distinguish between them. What useful information can be obtained from the assays?
2. The data below show the ASO probe results of testing patients for four mutant alleles (A, B, C, and D) in a gene. The letters N and M represent dot blotting using probes for the normal and mutant sequences, respectively. The symbol "•"denotes hybridization observed; the symbol "O" denotes no hybridization.

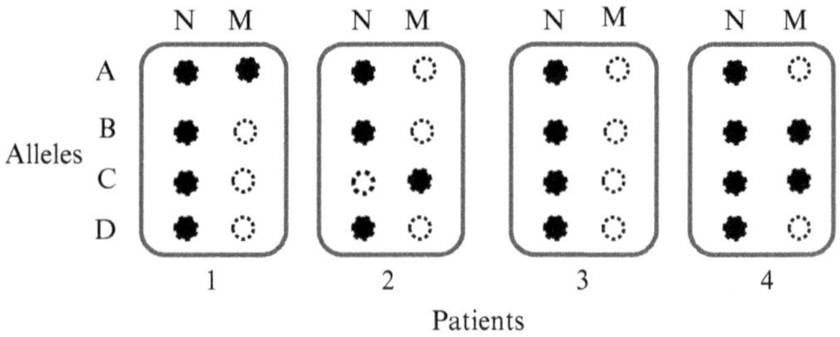

What type of dot blot assay is used for the above data? Explain your choice.
3. List at least four practical applications of the dot blot assay technique.
4. What is an *in situ* RNA or *in situ* DNA hybridization assay? Briefly outline the procedure. What useful information can be obtained from the assay?

Further reading

Biotin Chromogenic Detection Kit. Life Technologies Catalog No. KO661.
Brown, T. (2003). Unit 2.9B. Dot and slot blotting of DNA. In *Current Protocols in Molecular Biology* (Edited by F.M. Ausubel), pp. 2.9.15–2.9.20. John Wiley & Sons, Inc. ISBN 0-471-50338-X.

28 AMB 1 experiment 28: Pre-hybridization, hybridization, and detection

Purpose: This is up to you to write down.

Introduction

In a membrane hybridization, a specific probe DNA base-pairs to its complementary target DNA that is immobilized on a membrane support. Hybridization signal strength is influenced by a number of factors such as the amounts of DNA fixed on to membrane, GC content and complementarity of the probe and target DNA, length, concentration, and labeling efficiency of the probe, and stringency of hybridization and washing conditions. Generally, lower salt, higher formamide amount, shorter probe, lower target concentration, and higher temperature increase stringency, rendering unfavorable to hybridization by reducing thermal stability of probe-target hybrids. Both the length of probe and the ionic strength of hybridization solution are important for determining the rate of annealing. The DNA concentration and annealing time are also important. In theory, high probe concentrations will give maximum signal strength because probe-target hybrid formation occurs purely by chance. In practice, however, it requires a balance between maximizing the specific signal and minimizing the non-specific background. By controlling the stringency of hybridization and washing conditions, target sequences that are either a 100% match or some mismatch (>70%) to the probe can be detected. Optimal hybridization conditions should be empirically determined since high stringency parameters typically increase signal specificity but decrease signal intensity.

In order to prevent non-specific binding of the probe to the unoccupied sites of membrane, it must be pre-incubated in the pre-hybridization buffer that contains blocking agents, Denhardt's solution and salmon sperm DNA. Hybridization requires the use of single-stranded probe DNA. If DNA probes are double-stranded after the labeling reaction, they must be denatured. For denaturation, a suitable quantity of probe is placed in a safety cap microcentrifuge tube and heated for 5 minutes in a boiling water bath and then quickly chilled by placing the tube in an ice bucket. For hybridization, the used pre-hybridization buffer is discarded and replaced with a new pre-hybridization buffer to which the denatured probe is added. Pre-hybridization and hybridization steps can be performed in any container that can be tightly sealed, such as temperature-resistant roller bottles and sealable plastic bags. The container must be sealed during the procedure to prevent the hybridization buffer from releasing NH_4 and changing the buffer pH during incubation. Incubation is usually performed with gentle agitation overnight at an appropriate temperature. Hybridization stringency can be varied by manipulation of three factors: temperature, salt concentration, and formamide concentration; adding 1% formamide lowers the melting temperature (T_m) by 0.72 °C. If the probe is larger than 100 bp, hybridization is usually carried out at 68 °C in a low-salt buffer. However, if a short oligonucleotide probe of PCR primer length is used, the hybridization step is usually carried out at a temperature 5 to 10 °C below the calculated T_m for the desired probe-target hybrid.

After hybridization, Denhardt's solution and salmon sperm DNA are replaced with SDS. This step washes off unhybridized probe as well as unstably hybridized bound probe. Washing stringency can be varied by changing the temperature and salt concentration. Detection of hybridized bands can be accomplished radioactively or non-radioactively. Both radioactive and non-radioactive chemiluscent detection can be performed by autoradiography or phosphor imager scanning. Nonradioactive chromogenic bands are directly visualized by characteristic colors on the membrane. The intensity of both radioactive and non-radioactive bands can be densitometrically scanned for quantification.

It is recommended to visit the following animation or video web sites regarding membrane hybridization-based analysis applications:

- Southern blot application:
  http://www.youtube.com/watch?v=oR-TmnF0_es
- DNA microarray virtual lab:
  http://learn.genetics.utah.edu/content/labs/microarray/
- Dot blot application:
  http://www.youtube.com/watch?v=tLmaQEgQxOA
  http://www.youtube.com/watch?v=z1c7YsXq4bs
- Northern blot application:
  http://www.youtube.com/watch?v=Yiif0BJX0VE

Pre-lab assignment

(Typing and submission must be completed before lab work begins.)

1. What is the stringency and why is it important for the hybridization experiment? How is the stringency controlled?
2. What happens if you do not boil the probe before starting Southern blot hybridization?
3. Calculate the volume of salmon sperm DNA to be added at step 2 of Part A.

Methods in Biotechnology, First Edition. Seung-Beom Hong, M. Bazlur Rashid and Lory Z Santiago-Vázquez.

Companion Website: www.wiley.com\go\hong\Methodsinbiotechnology

Materials and equipment

- Biotin-Labeled pRY121 and *lacZ* probes (Experiments 26)
- 50× Denhardt's solution (1 g Ficoll, 1 g polyvinylpyrrolidine, 1 g BSA in 100 mL; filter-sterilize)
- 1.5× Pre-hybridization buffer (29.25 g NaCl, 3.0 g Tris, 0.175 g EDTA, 0.5 g SDS, 0.5 g sodium pyrophosphate in 250 ml ultrapure water, adjust pH to 7.9 with HCl, add ultrapure water to 375 mL total; autoclave)
- Salmon sperm DNA (sheared by sonication, heat-denatured; 10 mg/mL)
- 20× SSC (3 M NaCl, 0.3 M sodium acetate, pH 7.0)
- 10% SDS
- MilliQ H_2O
- Hybridization oven, glass roller bottle
- Washing buffer 1 (Experiment 27)
- Blocking buffer 2 (Experiment 27)
- Detection buffer 3 (Experiment 27)
- TE buffer (pH 8.0)

Procedure

Pre-hybridization and hybridization step (day 1)

1. Pipette 15 mL of 1.5× pre-hybridization buffer (warmed at 65 °C) and 4 mL of 50× Denhardt's solution into a glass roller bottle warmed at 65 °C.

2. Denature aqueous solution of sonicated salmon sperm DNA at 100 °C for 5 min and then chill on ice. Add the denatured DNA to the pre-hybridization solution to obtain a final DNA concentration of 50 µg/mL.

3. Place the blotted membrane *with the DNA side facing up* in a glass roller bottle containing the above and incubate at 65 °C for at least 2 h in a hybridization oven.

**If multiple blots are in a single container, make sure there is enough liquid in the container to keep the blots from sticking together during the incubation. They should move freely when agitated.*

**Take care to avoid trapping any air bubbles.*

**Do not allow the membrane to dry at any time from now on. If the membrane dries or sticks to a second membrane, the assay will have a high background.*

4. Discard the pre-hybridization solution after incubation and add a fresh pre-hybridization solution (15 mL of 1.5× pre-hybridization buffer + 4 mL of 50× Denhardt's solution + 0.1 mL of 10 mg/mL of denatured ssDNA) to the membrane (60 µL/cm^2) in a glass roller bottle.

5. Denature the biotin-labeled probe at 100 °C for 5 min and chill on ice.

6. Add the denatured probe to the glass roller bottle containing the membrane and pre-hybridization solution to obtain a final probe concentration of 25 to 100 ng/mL.

**In this lab experiment, you add the entire labeled probe you prepared in Experiment 26.*

7. Incubate at 65 °C overnight.

Washing step (day 2; total time ~1 h)

8. Wash the membrane twice in a clean glass baking dish containing a 100 mL of 2× SSC plus 0.1% SDS for 5 min at room temperature.

9. Wash the membrane twice with a 100 mL of 0.2× SSC plus 0.1% SDS for 5 min at room temperature.

10. Wash the membrane twice in a glass roller bottle containing a 100 mL of 0.15× SSC plus 0.1% SDS for 15 min at 50 °C.

11. Remove excess liquid from the membrane by blotting on to Whatman 3MM paper.

12. Do not dry out the membrane completely.

13. Wrap the membrane in plastic wrap, label, and store at room temperature.

Detection step (day 3)

14. Float the membrane in *washing buffer 1* for 1 min in a container to rehydrate.

15. Incubate in *blocking buffer 2* (0.3 mL/cm^2 of membrane) for 30 min at 37 °C with moderate shaking.

16. Prepare the *streptavidin-AP conjugate* (0.2 mL/cm^2 of membrane; diluted to 1 to 2 µg/mL in *washing buffer 1*).

17. Drain the *blocking buffer 2* from a tray and incubate the membrane in diluted streptavidin-AP conjugate for 30 min on a platform shaker.

18. Wash with *washing buffer 1* (0.6 mL/cm^2 of membrane) for 15 min with moderate shaking and discard the solution.

19. Wash with *blocking buffer 2* (0.6 mL/cm^2 of membrane) for 15 min with moderate shaking and discard the solution.

20. Wash with *detection buffer 3* (0.2 mL/cm^2 of membrane) for 10 min with moderate shaking.

21. Incubate the membrane in substrate solution 1-Step™ NBT/BCIP (no dilution is necessary; use ~0.1 mL per cm^2 of membrane) in the dark.

22. Discard the solution, rinse with TE buffer (pH 8.0) or dH_2O, for 30 s and air-dry.

23. Compare the pRY121-probed bands (Groups 1 and 2) with the *lacZ*-probed bands (Groups 3 and 4) on membranes.

Post-lab assignment

1. Why do you think molecular weight marker DNA (NEB 1 kb DNA ladder marker) bands are hybridized to the nick-translated pRY121 plasmid probe in your experiment? (*Hint:* refer to the following web site's FAQs: https://www.neb.com/products/markers-and-ladders/dna-markers-and-ladders/dna-markers-and-ladders.)

2. How would your final Southern blot appear under the following circumstances? Explain why it is in each case.

- **A.** Your hybridization temperature was too high.
- **B.** You omitted the pre-hybridization step.
- **C.** You omitted the neutralization step of the gel treatment.
- **D.** You neglected to wash off unbound probe after hybridization.

3. What would you do if your hybridized bands yielded very low signal intensities? Suggest as many possible causes as you can.

4. Southern blot analysis is frequently used to detect the presence of transgene integration in host genome. However, transfection into mammalian cell or transformation into plant often results in integration of multiple copies of transgenes. The integrated copy number can be determined by comparing the intensities of hybridized bands with those of copy number standards. Suppose that you obtained a green fluorescent transgenic *Arabidopsis thaliana* plant (the total number of DNA base pairs in a haploid genome: ~157 × 10^6 bp) that contains a 714-bp GFP coding sequence under the control of a plant promoter. To determine the copy number integration, you conducted Southern blot analysis using a 10 µg genomic DNA isolated from the T_0 line of *A. thaliana* (i.e., non-selfing transgenic *A. thaliana* that is hemizygous for the GFP gene) and a series of GFP copy number standard. What amounts (mass) of copy number standard are required for 1, 5, and 10 copies of GFP?

5. A DNA microarray is a microscale probe-target hybridization involving an arrayed series of many microscopic spots of DNA. In terms of probe and target DNA preparations, what are the major differences between the microarray and blot hybridization?

Further reading

Biotin Chromogenic Detection Kit. Life Technologies Catalog No. KO661.

Brown, T. (2003). Unit 2.10. Hybridization analysis of DNA blots. In *Current Protocols in* Molecular Biology (Edited by F.M. Ausubel), pp. 2.10.1–2.10.16. John Wiley & Sons, Inc. ISBN 0-471-50338-X.

29 AMB 1 experiment 29: Total yeast RNA isolation and RT-PCR

Purpose: This is up to you to write down.

Introduction

The mRNA component of total RNA is often the focus of gene expression studies since each mRNA molecule carries the coding information from a gene to the site of protein synthesis. There are several conventional techniques used to analyze the steady-state level of mRNA; examples are Northern blot, ribonuclease protection, primer extension, and nuclear run-off assays. Another technique called reverse transcription (RT) that produces cDNA from mRNA is also widely used. Because cDNAs represent all the genes that are expressed in the cells of interest and cDNAs can be amplified by PCR, PCR-based differential display (DD-PCR) and cDNA-amplified fragment length polymorphism (AFLP) have been used to discover differentially expressed genes. Alternatively, cDNA samples can be labeled with radioactive or fluorescent tags, and then either hybridized to an array of tens of thousands of PCR-amplified gene-specific cDNA sequences that have been immobilized on to a glass microscope slide in a grid pattern of spots (DNA microarray) or hybridized to an array of hundreds of thousands of short (25 ntd) oligonucleotides directly synthesized on the chip (DNA chip or gene chip). Comparison of the hybridization patterns of labeled cDNA probes from different RNA sources can clearly reveal which genes are expressed differently in the two samples. This opens the possibility of identifying co-regulated genes, which in turn reflects underlying regulatory mechanisms and functional interrelationships. With the advent of the low-cost next generation sequencing technologies, whole transcriptome shotgun sequencing, called RNA-seq, has been developed. RNA-seq provides far more accurate and comprehensive transcriptome data than microarray and is now the most dominant transcriptomic tool.

Whatever techniques are used, an undegraded intact RNA sample is prerequisite for gene expression analysis. The difficulty in RNA isolation is that RNA is susceptible to nearly ubiquitous ribonucleases, extremely stable enzymes requiring no cofactors to function. Moreover, RNA is chemically unstable due to spontaneous cleavage of phosphodiester backbone via intramolecular transesterification. An RNA sample must be free of protein, including nucleases, genomic DNA, and enzymatic inhibitors for RT-PCR. It is highly recommended that RNA purity and integrity of the sample is checked by a spectrophotometer reading ($A_{260}/A_{280} \geq 1.8$) and denaturing agarose-formaldehyde gel electrophoresis prior to the subsequent experiments. However, even when DNA is not visible on an agarose gel, trace amounts may still remain depending on the amount and nature of the sample.

Before you set up the RT-PCR reaction, you need to design the primers for the gene sequences of interest. There are several excellent sites for designing PCR primers. Gene-specific PCR primers can be designed by doing "Primer-BLSAT" your sequence (http://www.ncbi.nlm.nih.gov/tools/primer-blast/index.cgi?LINK_LOC=BlastHome) or using the Primer3 tool (http://biotools.umassmed.edu/bioapps/primer3_www.cgi), or can be obtained from the PrimerBank database (http://pga.mgh.harvard.edu/primerbank/). Whenever possible, it is desirable to design primers that anneal to the sequences that span the junction of two exons in order to distinguish between amplification of cDNA and amplification of contaminating genomic DNA.

Once high-quality RNA and well-designed primers are obtained, the conversion efficiency of mRNA to cDNA by reverse transcriptase is the single most variable step in the whole quantitative transcript abundance analysis procedure. RNA transcripts having a strong secondary structure of intramolecular base pairing must be disrupted by heat denaturation for efficient reverse transcription. Among several commercial reverse transcriptase enzymes, most of which are stable up to 45 °C, Verso™ reverse transcriptase is stable up to 57 °C and is able to synthesize long cDNA strands of up to 11 kb. The variability of conversion efficiency by reverse transcriptase can also be minimized by using a primer set that will generate short-length cDNA rather than a full-length cDNA.

Following the reverse transcription, heating at 95 °C denatures the RNA/cDNA hybrid, inactivates the reverse transcriptase, and dissociates the enzyme from the cDNA. This cDNA is then used for the template of end point PCR or real-time qPCR to detect and quantify the gene expression level. Because of its simplicity and sensitivity, RT-PCR is now widely used for evaluation of gene expression and alternative exon splicing levels, as well as for gene cloning. Moreover, RT-PCR validates gene expression data obtained from microarray and RNA-Seq experiments and gene silencing by RNA interference. It is recommended to watch the following videos regarding relative and absolute quantification of a copy number:

- http://www.youtube.com/watch?v=GQOnX1-SUrI (Real Time QPCR Data Analysis, Part 1)
- http://www.youtube.com/watch?v=tgp4bbnj-ng (Real Time QPCR Data Analysis, Part 2)

Methods in Biotechnology, First Edition. Seung-Beom Hong, M. Bazlur Rashid and Lory Z Santiago-Vázquez.

Companion Website: www.wiley.com\go\hong\Methodsinbiotechnology

- http://www.youtube.com/watch?v=GB4wcQsCawU (Droplet Digital PCR, Part 1)
- http://www.youtube.com/watch?v=wBlV-MBvftc (Droplet Digital PCR, Part 2)

In this lab exercise, you will isolate total RNA from the glucose-induced and uninduced yeast cells that were pre-grown aerobically on ethanol as the sole carbon source. Due to the time limitation, yeast RNA preps obtained from the SV Total RNA Isolation Kit will not be checked for integrity by running on to denaturing formaldehyde agarose gel. This kit is easy to use with reagents that have been tested for effectiveness. You will be using a constitutively expressed yeast ubiquitin-protein ligase mRNA (UBC6) as an internal, endogenous control (reference gene) to normalize the expression levels of pyruvate decarboxylase (PDC5) mRNA (target gene) in glucose-treated (test sample) and untreated control (calibrator sample) yeast. UBC6 is a housekeeping gene that is expressed at the same level irrespective of conditions and is thus used to correct the signal intensity data for any differences in the amount of cellular RNA placed in RT-PCR reactions. You will check the specificity of RT-PCR by agarose gel electrophoresis of PCR samples and the quantity of RT-PCR products of three different cycles (Semi-End Point qPCR) by ImageJ scan analysis of agarose gel stained with GelRed™ Dye. Last, you will also quantify the RT-PCR samples using PicoGreen assay.

Pre-lab assignment

(Typing and submission must be completed before lab work begins.)

1. Perform the nucleotide BLAST (http://blast.ncbi.nlm.nih.gov/Blast.cgi) using PDC5 and UBC6 primer sequences as queries and identify each GenBank accession number and gene description.
2. What are the expected sizes of amplification products when PDC5 and UBC6 primers are used for RT-PCR? Show your work using the information obtained from the above question.
3. What is the component of Verso™ Enzyme Mix?
4. Draw the flow diagram of today's experiment.

Materials and equipment

- YPDA cultures of *S. cerevisiae* BY4741 (MATa, *his3Δ1, leu2Δ0, met15Δ0, ura3Δ0*)
- Ethanol (molecular biology grade)
- YNB with $(NH_4)_2SO_4$ + 2% EtOH + 0.08% Ura Dropout supplement + 0.2% uracil
- Yeast suspension buffer (1M Sorbitol, 0.1 M EDTA, pH 8.0; add β-mercaptoethanol and lyticase to a final 0.1% and 0.5 units/μL, respectively, and prepare just before use)
- SV Total RNA Isolation Kit (Promega Z3100)
- Verso™ 1-Step RT-PCR Kit (Thermo Fisher Scientific, Cat. No AB 1454LDA)
- Yeast PDC5 primer:
  Forward 5′-GTTCCAATTACCAAGTCTACTCCA-3′
  Reverse 5′-GGACGATAGCGTATACATCTGTT-3′
- Yeast UBC6 primer (internal reference control):
  Forward: 5′- GATACTTGGAATCCTGGCTGGTCTGTCTC-3′
  Reverse: 5′- AAAGGGTCTTCTGTTTCATCACCTGTATTTGC-3′
- Quick-load 50 bp DNA ladder (0.1 μg/μL; NEB Cat. No. NO473S)
- Agarose, 10× lithium borate buffer, 6× DNA loading dye
- GelRed® (10 000X; VWR Cat. No. 89139-134)
- Ice box, sterile microcentrifuge tubes, thin-walled PCR tubes
- RNase-free sterile pipette tips
- NanoDrop 2000 spectrophotometer
- ImageJ software (download from http://rsb.info.nih.gov/ij/download.html and install according to the instructions; see AMB 1 Appendix 4, "Band Quantification by ImageJ Program")
- Quant-iT PicoGreen® dsDNA assay kit (Thermo Fisher Scientific, Cat. No. P7589)
- 96-well Microflour 2 plates (Fisher Scientific Cat. No. 14-245-176)
- Tecan Infinite M200 Spectrafluor Microplate Reader
- DNA standard (λ DNA: 100 μg/mL) in Quant-iT PicoGreen® dsDNA assay kit
- 1× TE buffer (pH 8.0)
- Sterile nuclease-free dH_2O

Procedure (adapted from SV total isolation kit, Promega and verso one-step RT PCR Kit, Life technologies)

Part A. Induction of yeast cells (TA will perform this part)

1. Grow yeast BY4741 cells in YPDA medium at 30 °C, 250 rpm, overnight from an isolated colony on a freshly streaked YPDA plate.
2. In the morning, inoculate the overnight culture into two 100-mL flasks, each containing a fresh 10-mL of YNB with $(NH_4)_2SO_4$ + 2% EtOH + 0.08% Ura Dropout supplement + 0.02% uracil medium to a final cell density of 0.1 at A_{600}. Grow the culture at 30 °C, 250 rpm for ~30 h.
3. Add 2.5 mL of 20% glucose to one flask (final 4%) and grow for 30 min.
4. Measure cell density of both flask cultures by reading at A_{600} (A_{600} = 1.0 for ~3×10^7 cells/mL); read 1/10 dilution in YNB (0.3 mL culture + 2.7 mL YNB).
5. Based on the A_{600} value, calculate the volume that contains 2×10^7 cells:

$$V(\text{mL}) = (2 \times 10^7) \div [A_{600} \times \text{DF}(10) \times (3 \times 10^7)]$$

Part B. Total yeast RNA isolation

Wipe the bench space and all micropipettors with 70% ethanol and wear gloves!

1. Harvest an equal number ($\geq 2 \times 10^7$) of each glucose-induced and uninduced yeast cells in a 1.5-mL tube by centrifuging at 8000 rpm for 30 s. Decant the supernatant and carefully remove any remaining media by aspiration.

**Repeat this step when the volume to be harvested for 2×10^7 cells is larger than 1.5 mL.*

2. Resuspend the cell pellet in 100 μL of *yeast suspension buffer*.

3. Incubate at 30 °C with gentle shaking for 15 to 30 min until the solution appears clear.

4. Add 75 µL of *RNA lysis buffer* and mix gently.

5. Add 350 µL of the *RNA dilution buffer* (blue). Mix by inversion and spin at 13 000 rpm, room temperature for 10 min. Transfer the cleared lysate to a fresh tube.

6. Add 200 µL of 95% ethanol to the lysate, mix by pipetting 3 to 4 times, transfer to the spin column assembly, and spin at 13 000 rpm,at room temperature for 1 min.

7. Take the spin basket from the spin column assembly, discard the liquid in the collection tube, and put the spin basket back into the collection tube.

8. Add 600 µL of *RNA wash solution* (ethanol added to a final 59.8%) to the spin column assembly, spin at 13 000 rpm, at room temperature for 1 min, empty the collection tube as before, and place it in a rack.

9. For each RNA sample prep, make the *DNase incubation mix* fresh by combining 40 µL of yellow core buffer, 5 µL of 0.09 M $MnCl_2$, and 5 µL of DNase I enzyme in a sterile tube kept on ice.

**Mix by gentle pipetting; do not vortex; DNase I is very sensitive to physical denaturation.*

10. Apply 50 µL of the freshly-prepared *DNase incubation mix* directly to the membrane inside the spin basket and incubate at room temperature for 15 min.

**Make sure that the solution is in contact with and thoroughly covering the membrane.*

11. Add 200 µL of *DNase stop solution* (ethanol added to a final 57.1%) to the spin basket and centrifuge at 13 000 rpm, at room temperature for 1 min (no need to empty the collection tube after the spin).

12. Add 600 µL of *RNA wash solution* (ethanol added) and spin at 13 000 rpm, at room temperature for 1 min.

13. Empty the collection tube, add 250 µL of *RNA wash solution* (ethanol added), and spin at 13 000 rpm, at room temperature for *2 min*.

**The long centrifugation dries the spin column membrane, ensuring that no ethanol is carried over during RNA elution.*

14. Remove the cap from the spin basket, transfer to the elution tube, add 100 µL of *nuclease-free water* to the membrane, and spin at 13 000 rpm, at room temperature for 1 min.

15. Cap the elution tube containing the purified RNA, label the tube, and store at –70 °C until use.

16. Determine the concentration and purity by pipetting 2 µL of RNA sample on to the bottom pedestal of the NanoDrop 2000 spectrophotometer using water as a blank.

**Pure RNA has 1.8 to 2.1 of A_{260}/A_{280}; A_{260} reading 1 = 40 µg/mL RNA.*

Part C. Setting up reactions

17. Add the reaction components for uninduced yeast RNA to the bottom of a thin-walled PCR tube on ice in order, starting from top to bottom in the table (one per each group).

18. Add the reaction components for glucose-induced yeast RNA to the bottom of a thin-walled PCR tube on ice in order, starting from top to bottom in the table (one per each group).

Components (tube 1/U-PDC5)	Volume (µL)
RNase-free H_2O	x*
PDC5 primer mix (10 µM each)	0.5
Uninduced yeast RNA	y* (100 ng)
Verso Enzyme Mix†	0.5
1-Step Master Mix (2×)†	12.5
Total volume	25
Components (tube 2 /U-UBC6)	**Volume (µL)**
RNase-free H_2O	x*
UBC6 primer mix (10 µM each)	0.5
Uninduced yeast RNA	y* (100 ng)
Verso Enzyme Mix†	0.5
1-Step Master Mix (2×)†	12.5
Total volume	25

*Volumes of water and RNA can be adjusted based on the concentrations of both induced and uninduced samples.
†Do not vortex the Verso Enzyme Mix or the 1-Step PCR Master Mix.

Components (tube 3/I-PDC5)	Volume (µL)
RNase-free H_2O	x*
PDC5 primer mix (10 µM each)	0.50
Glucose-induced yeast RNA	y* (100 ng)
Verso Enzyme Mix	0.5
1-Step Master Mix (2×)	12.5
Total volume	25
Components (tube 4/I-UBC6)	**Volume (µL)***
RNase-free H_2O	x*
UBC6 primer mix (10 µM each)	0.50
Glucose-induced yeast RNA	y* (100 ng)
Verso Enzyme Mix	0.5
1-Step Master Mix (2×)	12.5
Total volume	25

*Volumes of water and RNA can be adjusted based on the concentrations of both induced and uninduced samples.

Part D. Running RT-PCR

19. Program the thermal cycler using the following parameters.

Cycle	Steps	Temperature (°C)	Time
1	1st strand cDNA Synthesis	50	15 min
1	Initial denaturation	95	2 min
	Denaturation	95	20 s
35*	Annealing	50	30 s
	Extension†	72	17 s
1	Hold	4	∞

*Number of cycle depends on transcript abundance and template complexity.
†1 min/kb.

20. Transfer 1 µL of PCR sample to 99 µL of 1× TE buffer in the assigned microplate well (each well was already filled with TE buffer by TA), as shown below. Wash each pipette tip by pipetting up and down twice.

- Groups 1, 2, 3, and 4 use A, B, C, and D rows, respectively.
- Each group uses the following column for each PCR cycle sample:

Columns 1, 2, and 3:	0th, 25th, and 30[t]h cycle of the induced PDC5
Columns 4, 5, and 6:	0th, 25th, and 30th cycle of the uninduced PDC5
Columns 7, 8, and 9:	0th, 25th, and 30th cycle of the induced UBC6
Columns 10, 11, and 12:	0th, 25th, and 30th cycle of the uninduced UBC6

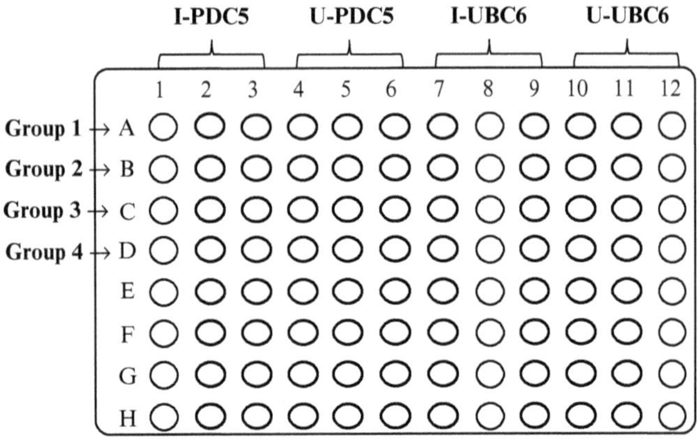

21. Begin the program cycle. Once the initial denaturation temperature reaches to 50 °C, quickly spin your sample tubes and load in the thermal cycler.

22. Press the "Pause Run" button on the thermal cycler to stall when the 25th PCR cycle extension step is completed, transfer 6 µL each of PDC5 and UBC6 samples to new labeled 0.5-mL microtubes (induced PDC5/25th cycle, induced UBC6/25th cycle; uninduced PDC5/25th cycle, uninduced UBC6/25th cycle), and keep them on ice.

23. Press the "Resume Run" button to re-start the PCR cycle.

24. Repeat the steps 22 to 23 at the 30th PCR cycle.

25. Transfer 1 µL each of the 25th PCR (step 22) and 30th PCR (step 24) samples to the assigned microplate wells.

Part E. Gel electrophoresis

26. During the PCR cycle, prepare 60 mL of 1.5% agarose gel in 1× lithium borate buffer using a 10-well comb.

27. After the 35th PCR cycle, add 5 µL each of PDC5 and UBC6 PCR sample to a labeled microtube.

28. Add 2 µL of 6× DNA loading dye to each of 6 tubes containing 10 µL of PCR DNA mix (5 µL PDC5 + 5 µL UBC6): induced 25th, 30th, 35th; uninduced 25th, 30th, 35th.

29. Load the samples and electrophoresis at 300 V for 30 min (or 25 V for 14 h).

Lane 1: 50 bp DNA ladder : 5 µL (0.1 µg/µL) direct loading
Lane 2: 25th cycle of the uninduced PDC5 + UBC6 RT-PCR
Lane 3: 25th cycle of the induced PDC5 + UBC6 RT-PCR
Lane 4: 30th cycle of the uninduced PDC5 + UBC6 RT-PCR
Lane 5: 30th cycle of the induced PDC5 + UBC6 RT-PCR
Lane 6: 35th cycle of the uninduced PDC5 + UBC6 RT-PCR
Lane 7: 35th cycle of the induced PDC5 + UBC6 RT-PCR

30. Stain in 2.5× GelRed solution for 10 min (fresh solution) to 30 min (old solution).

31. View the gel under UV, take a photograph, and save as an image format.

DNA size marker (NEB Quick-Load 50 bp DNA ladder):

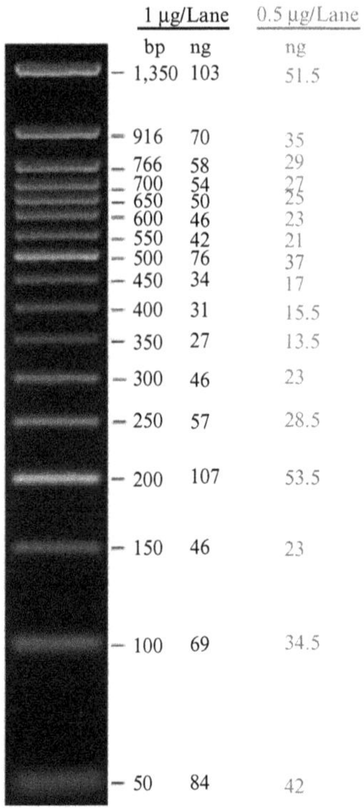

A Quick-Load 50 bp DNA ladder visualized by ethidium bromide staining. (Copyright by NEB. Reprinted with permission.)

Part F. Densitometric quantification by ImageJ processing analysis program

32. Open the ImageJ program and follow the instruction of Chapter 30.13 Gels ▶ (pages 144–145) in ImageJ User Guide Book or "Band Quantification by ImageJ Program" in Appendix 4.

33. Calculate the amounts of the RT-PCR products (refer to the example of the "Data Analysis" part of "Band Quantification by ImageJ Program") using a 200-bp marker (53.5 ng/0.5 µg/lane) as a standard for your calculation. Show your work.

34. Determine the fold changes in PDC5 and UBC6 gene expressions from uninduced and induced samples.

35. Determine the fold increase in the PDC5 gene expression, that is, the fold change in PDC5 divided by the fold change in UBC6.

Part G. PicoGreen assay

The PicoGreen reagent is a proprietary, asymmetrical fluorescent cyanine dye, 2-(*n*-bis-(3-dimethylaminopropyl)-amino)-4-(2,3-

dihydro-3-methyl-(benzo-1,3-thiazol-2-yl)-methylidene)-1-phenyl-quinolinium.

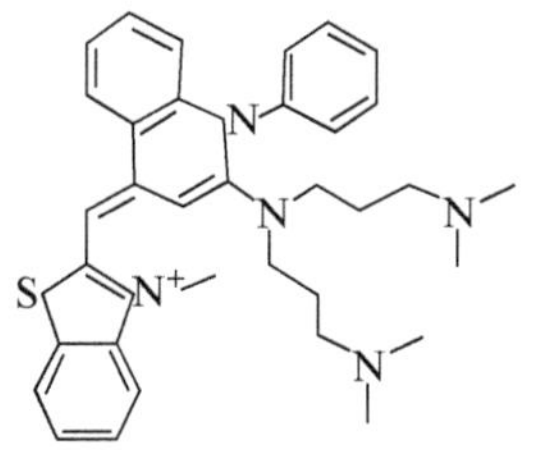

Free dye does not fluoresce, but upon binding to dsDNA it exhibits a >1000-fold fluorescence enhancement. PicoGreen is able to bind strongly not only to highly polymeric DNA but also to short duplexes of <20 bp. Hence, accurate quantification of the PCR product by PicoGreen assay requires the presence of single-band specificity and the absence of a primer dimer. PicoGreen assay is 10 000-fold more sensitive than the UV absorbance method and is highly selective for dsDNA over ssDNA and RNA. Fluorescence intensity is not affected by the GC content of DNA. Using a fluorospectrometer, a linear detection range for dsDNA bound with PicoGreen® reagent is 25 pg/mL to 1 μg/mL. This linearity is maintained even in the presence of several compounds commonly found to contaminate nucleic acid preparations, including salts, urea, ethanol, chloroform, detergents, proteins, and agarose, though the signal intensity may be affected.

1. Turn on the fluorescence plate reader at least 10 min before reading. Use the following settings to read the PicoGreen results:

	Wavelength/bandwidth
Excitation:	~485 nm/20 nm
Emission:	~535 nm/25 nm

2. Prepare 500 μL of 2 μg/mL (2 ng/μL) DNA standard for a high range by mixing 10 μL of 100 μg/mL stock with 5 μL of master mix (Part C) and 485 μL of 1× TE.

3. Prepare 500 μL of 50 ng/mL (50 pg/μL) DNA standard for a low range by mixing 10 μL of 2 μg/mL DNA standard with 5 μL of master mix (Part C) and 485 μL 1× TE.

4. ABCD rows/1–12 wells are already filled with each group's 12 RT-PCR samples from the Part D experiment as follows.

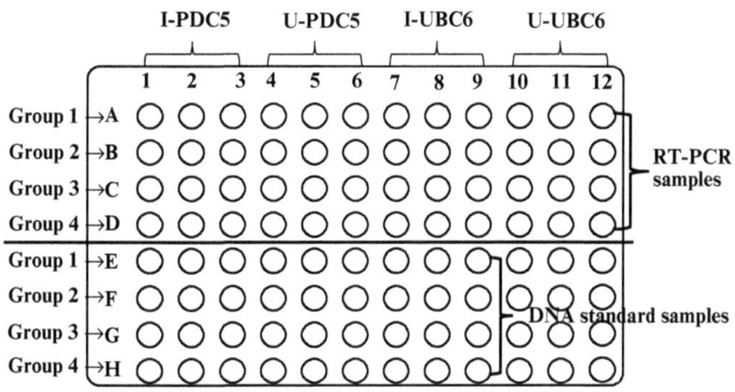

Prepare the two different ranges of standard samples in labeled 0.5-mL microtubes, as shown in the tables below, and transfer them to each well, as shown in the above figure E, F, G, and H wells.

Low range standard

Tube	Plate column well*	1× TE buffer (μL)	50 ng/mL DNA standard (μL)	Final total DNA (pg)/well	Final [DNA] (pg/μL) after mixing with 100 μL PicoGreen reagent
1	1 (blank)	100	0	0	0
2	2	99.5	0.5	25	0.125
3	3	99	1	50	0.25
4	4	90	10	500	2.5
5	5	0	100	5000	25

*Groups 1, 2, 3, and 4 use E, F, G, and H rows (see the figure above), respectively.

High range standard

Tube	Plate column well*	1× TE buffer (μL)	2 μg/mL DNA standard (μL)	Final total DNA (ng)/well	Final [DNA] (pg/μL) after mixing with 100 μL PicoGreen reagent
6	6	99.5	0.5	1	5
7	7	99	1	2	10
8	8	90	10	20	100
9	9	0	100	200	1000

*Groups 1, 2, 3, and 4 use E, F, G, and H rows (see figure above), respectively.

5. Prepare 9 mL of 1/200 dilution of the PicoGreen reagent in 1× TE (8.955 mL of 1× TE + 45 μL of PicoGreen reagent). Mix well (TA will do this).

**The PicoGreen reagent is light-sensitive and should be kept wrapped in foil while thawing and in the diluted state.*

6. Add 100 μL of diluted PicoGreen to every well. Mix by pipetting up and down. Briefly spin the plate (1000 rpm for 30 s) to remove any air bubbles.

7. Cover the microplate with foil and incubate at room temperature for 5 min.

**Protect the PicoGreen reagent and plate from light as much as possible until read.*

8. Perform Tecan plate analysis following operating procedures of the Tecan Infinite M200 microplate reader for DNA quantification.

9. Copy and paste the data into an Excel file.

10. Prepare *separate* low- and high-range standard curves. To graph a standard curve (the best-fit trend line) in Excel 2010:

- **(a)** Subtract out the blank value from each sample value.
- **(b)** List DNA standard (X) and intensity (Y) values in side-by-side columns.

(c) Click and drag a box over the data with a mouse to highlight all the data.

(d) Click the "Insert" menu and select "Scatter with only Markers."

(e) Click "Layout 9" in the Chart Layouts. This will display the regression line equation and correlation coefficient (R^2), along with unspecified titles.

(f) Type *x* and *y* axes and chart title on the linear regression graph.

11. Subtract the 0th PCR cycle value from the 25th and 30th PCR cycle values.

12. Calculate the DNA quantity (ng) of each PCR sample using the equation of the regression line plotting the standard curve.

**Use low- and high-range standard curves to determine DNA quantities from low and high intensities, respectively.*

13. Compare the amounts of PCR products obtained by PicoGreen assay with those of ImageJ analysis. If the difference is significant, which assay do you think is more correct and why?

**Note: the DNA quantity estimated by ImageJ and PicoGreen analysis in the lab protocol is based on 5 μL and 1 μL RT-PCR samples, respectively.*

Note: although any purified dsDNA preparation may be used, it is preferable to prepare the standard curve with DNA similar to the type being assayed; plasmid for quantitating plasmid DNA; linear DNA for quantitating linearized DNA fragments. Most linear dsDNA molecules have been found to yield approximately equivalent signals, regardless of fragment length. The PicoGreen assay remains linear in the presence of several compounds that contaminate nucleic acid preparations, though the signal intensity may be affected. Thus, to serve as an effective control, the dsDNA solution used to prepare the standard curve should be treated in the same way as the experimental samples and should contain similar levels of such compounds.

Post-lab assignment

1. (a) What is single-cell transcriptomics? (b) Why is it important? (c) What techniques are used for single-cell transcriptomics?

2. Briefly describe the overall procedure, advantage, and disadvantage of (a) Northern blot, (b) ribonuclease protection, (c) primer extension, and (d) nuclear run-off assays.

3. Describe the overall procedure and advantage and disadvantage of (a) differential display (DD-PCR) and (b) cDNA-AFLP techniques.

4. You set up the RT-PCR reaction in a total volume of 25 μL. Agarose gel analysis of the RT-PCR sample showed a single unique 250-bp DNA band only. PicoGreen assay for 1 μL of the RT-PCR sample at the 10th PCR cycle gave 0.2 pg.

(a) How many molecules of a 250-bp amplicon are produced as a result of the RT-PCR? Average MW of 1-bp ds DNA is 660 g/mole.

(b) Assuming that both conversion efficiency by reverse transcriptase and amplification efficiency by DNA polymerase are 100%, how many mRNA molecules are initially present?

5. What is the droplet digital (dd) PCR? Briefly describe the principle.

6. The following cycle threshold (C_T) values for the PDC5 target and UBC6 reference gene expression were obtained from a real-time qPCR experiment.

	Test (treated) sample C_T	Calibrator (untreated) sample C_T
PDC5 target	23.4284	29.4584
UBC6 reference	15.4642	18.3542

If expression of a reference gene is not affected by treatment conditions and amplification efficiency of a reference gene is equal to that of a target gene, a relative fold change in expression (X) in real-time PCR is given by the equation: Fold change X = $2^{-\Delta\Delta C_T}$, where $\Delta\Delta C_T = \Delta C_T$ target in test − ΔC_T target in calibrator; ΔC_T target in test = C_T target − C_T reference in test sample; ΔC_T target in calibrator = C_T target − C_T reference in calibrator sample.

Assuming that treatment does not affect the reference gene expression and that the slopes of four samples' regression lines on the graph are the same running parallel to each other, calculate the fold change in the PDC5 target gene expression level.

References

Quant-iT™ PicoGreen® dsDNA Assay Kit Manual (Thermo Fisher Scientific, Cat. No. P7589).

SV Total RNA Isolation System Manual Promega, Cat. No. Z3100).

Further reading

Livas, D., Almering, M.J.H., Daran, J.M., Pronk, J.T., and Gancedo, J.M. (2011). Transcriptional responses to glucose in *Saccharomyces cerevisiae* strains lacking a functional protein kinase A. *BMC Genomics*, **12**: 405.

Rasband, W.S. (2013). *ImageJ*. U.S. National Institute of Health, Bethesda, Maryland, USA (http://imagej.nih.gov/ij/).

Singer, V.L., Jones L.J., Yue, S.T., and Haugland, R.P. (1997). Characterization of PicoGreen reagent and development of a fluorescence-based solution assay for double-stranded DNA quantitation. *Analytical Biochemistry*, **249**: 228–238.

Tang, F., Lao, K., and Surani, M.A. (2011). Development and applications of single-cell transcriptome analysis. *Nature Methods*, **8** (4 Suppl.): S6–S11.

Teste, M.-A., Duquenne, M., François, J.M., and Parrou, J.-L. (2009). Validation of reference genes for quantitative expression analysis by real-time RT-PCR in *Saccharomyces cerevisiae*. *BMC Molecular Biology*, **10**: 99, DOI:10.1186/1471-2199-10-99.

30 AMB 1 experiment 30: Yeast-based *in vivo* recombination cloning

Purpose: This is up to you to write down.

Introduction

Saccharomyces cerevisiae, a scientific name for baker's yeast or brewer's yeast, is a single-cell eukaryotic model organism that has been most comprehensively studied for all aspects of cell and molecular biology. Yeast has many functions common to higher eukaryotic cells such as subcellular organelles and post-translational modifications that are absent in prokaryotic cells. Many other plant and animal genes are able to functionally transcomplement the auxotrophic yeast mutations. Yeast has a genome size of approximately 12.5 megabase containing about 5400 protein-coding genes, which are arranged on 16 linear chromosomes. Collective deletion mutants of almost all coding and non-coding open reading frames and complete genome sequence data are open to the public and are available in the *Saccharomyces* Genome Database (SGD).

Yeast has a highly efficient homologous recombination system that is much more proficient than bacteria and other eukaryotes. Unlike in *E. coli*, there is also little need for special yeast genotypes for *in vivo* homologous joining. Accordingly, this has been widely exploited for gene cloning, gene synthesis, plasmid construction, site-specific mutagenesis, and targeted gene disruption, replacement, and deletion. Yeast cells can take up multiple DNA fragments and correctly put the pieces whose ends are overlapped together into a single recombinant molecule. For example, a 592-kb *Mycoplasma genitalium* genome was assembled from 25 overlapping fragments, each 17 to 32 kb with at least 80-bp sequence overlaps on either side. Even two different double-stranded DNA fragments can be seamlessly fused by using single-stranded oligonucleotides that overlap the end nucleotide sequences of the two fragments. In one transformation event, yeast can take up and correctly assemble as least 38 single-stranded oligonucleotides that overlap as few as 20 bp, along with a linear double-stranded vector that overlaps the terminal oligonucleotides.

Yeast cells can take up easily both linear and circular DNA molecules. Introduction of these DNA molecules can be accomplished by spheroplast, lithium acetate, or electroporation transformation methods. The spheroplast method is more laborious and technically demanding than the other two methods. The electroporation technique requires the use of an expensive electric pulsing apparatus. On the other hand, the lithium acetate transformation method is technically easier, less expensive, and thus is the most commonly used protocol that requires no special equipment. The principle of the lithium acetate method is that alkali cations make yeast cells competent to take up DNA. Addition of carrier DNA and polyethylene glycol in combination with heat shock facilitates the uptake of transforming DNA.

In this lab exercise, you will cut pRY121 plasmid with restriction enzymes, isolate two overlapping linearized DNA fragments, and transform the DNA fragments into yeast using the lithium acetate method in order to reassemble pRY121 plasmid. This procedure is similar to the *E. coli* chemical transformation method despite a few differences, since heat shock treatment of a competent cell/DNA mixture at 42 °C is critical for DNA uptake.

Pre-lab assignment

(Typing and submission must be completed before lab work begins.)

1. What are the approximate sizes of the two bands to be cut out at step 6 of Part I, B?

2. What are the sizes of the nucleotide sequence overlap at the terminal ends of the two DNA fragments isolated?

3. How do you prepare a 10-mL PEG-TEL solution using 1 M Tris-HCl (pH 8.0), 0.5 M EDTA (pH 8.0), 1 M LiAc (pH 7.5), and 50% PEG stock solutions?

4. Your yeast cell culture is about 10^7 cells/mL based on an A_{600} reading. How much of the cell culture do you have to inoculate into 100 mL of fresh YPDA medium so that a final cell density can be 2×10^6 cells/mL? Show your calculation.

5. Calculate the volume of 1× TEL or 1× TELG at step 7 of Part II, C when you harvest a 10-mL culture whose 1/10 diluted culture has 0.08 of A_{600}.

Materials and equipment

- *S. scerevisiae* YNN281 (*MAT*a, *ura3-52*, *trp1-Δ*, *his3-200*, *lys2-801*, *ade2-1*)
- YPDA (1% yeast extract, 2% peptone, 2% glucose, 0.003% adenine hemisulfate)
- Uracil dropout YNB agar medium with glucose (0.077% uracil dropout supplement, 0.67% YNB, 2% glucose, 2% agar).

Methods in Biotechnology, First Edition. Seung-Beom Hong, M. Bazlur Rashid and Lory Z Santiago-Vázquez.

Companion Website: www.wiley.com\go\hong\Methodsinbiotechnology

- Uracil dropout YNB agar medium with galactose and raffinose (0.077% uracil dropout supplement, 0.67% YNB, 1% galactose, 1% raffinose, 2% agar)
- Uracil dropout YNB agar medium with galactose and glucose (0.077% uracil dropout supplement, 0.67% YNB, 1% galactose, 1% glucose, 2% agar)
- Plasmid DNA pRY121
- *Sma*I, *Sac*I-HF, and *Pst*I-HF (NEB) enzymes, 10× CutSmart™ buffer (NEB)
- 0.5% agarose gel in 1× bionic buffer (pre-made by TA)
- QG buffer (6 M guanidine thiocynate, 50 mM bionic Tris-HCl, pH 6.6, 20 mM EDTA, pH 8.0, 0.006% cresol red)
- Econo Spin™ Column (Epoch Life Science, Cat. No. 1920-050)
- 50% polyethylene glycol 4000 or 3500 (warm to dissolve; filter sterilize)
- 1 M Tris-HCl pH 8.0 (sterile)
- 1 M lithium acetate pH 7.5 (adjust pH with dilute acetic acid; filter sterilize)
- 0.5 M EDTA pH 8.0 (sterile)
- Salmon sperm DNA (sonicated and denatured, 10 mg/ml; store at −20 °C)
- High-quality sterile dH_2O
- Glycerol, DMSO
- 1× TEL (10 mM Tris-HCl, pH 8.0, 1 mM EDTA. 0.1 M lithium acetate; prepare before use)
- 1× TELG (10 mM Tris-HCl, pH 8.0, 1 mM EDTA, 0.1 M lithium acetate, 15% glycerol; prepare fresh before use)
- PEG-TEL (40% PEG-4000 or 3350, 10 mM Tris-HCl, pH 8.0, 1 mM EDTA, 0.1 M lithium acetate; prepare fresh before use)
- 60 °C water bath
- Speed vacuum concentrator
- NanoDrop 2000 spectrophotometer

Procedure

Part I. Isolation of overlapping DNA fragments (day 1)

A. Restriction digestion of pRY121 plasmid

1. Set up the single- and double-digestion reactions as follows:

Tube 1		Tube 2	
Component	Volume (μL)	Component	Volume (μL)
Nuclease free H_2O	1	Nuclease free H_2O	2
10× CutSmart™ buffer	2	10× CutSmart™ buffer	2
pRY121 plasmid	15	pRY121 plasmid	15
*Sma*I	1		
*Sac*I-HF	1	*Pst*I-HF	1
Total volume	20	Total volume	20

2. Mix the reaction mixtures gently and incubate at least for 15 min at 37 °C.

B. Gel purification of DNA fragments

3. Run the restriction samples on to a pre-made 0.5% agarose gel in 1× bionic buffer at 30 V/cm for 10 min.
4. Weigh an empty sterile 1.5-mL microcentrifuge tube.
5. Place the ethidium bromide stained gel on to a UV transilluminator covered with plastic wrap.
6. Cut out the gel band of the highest MW (the uppermost band) from each single and double digestion samples with a clean razor blade as quickly as possible and place into the empty microcentrifuge tube.

**Take special care to avoid sources of nuclease contamination.*

**Minimize the time of UV exposure as much as possible. Exposure of DNA to UV radiation for as little as 45 seconds can damage the DNA.*

**Minimize the size of the gel slice by removing extra agarose.*

7. Weigh the gel-containing microcentrifuge tube and calculate the gram quantity of gel.

**Put the gel slice up to 0.25 g in a 1.5-mL microcentrifuge tube.*

8. Add 3 volumes of buffer QG to 1 volume of gel (1 g equals about 1 mL).
9. Incubate at 50 to 57 °C for 10 min (or until the gel slice has completely dissolved). Mix by vortexing the tube every 2 min during the incubation.
10. Place a spin column in a provided 2 mL collection tube and apply the first DNA sample to the spin column and centrifuge for 30–60 s.
11. Discard the flow-through. Place the spin column back into the same tube and apply the second DNA sample to the same spin column and centrifuge for 30 to 60 s.

**Collection tubes are re-used for recycling.*

12. To wash, add 0.75 mL of buffer PE to the spin column and centrifuge for 30 to 60 s.
13. Discard the flow-through and place the spin column back in the same tube. Centrifuge the column for an additional 1 min at the maximum speed.

**Residual ethanol from buffer PE will not be completely removed unless the flow-through is discarded before this additional centrifugation. Any residual buffer PE may interfere with subsequent enzymatic reactions.*

14. Place the spin column in a clean 1.5 mL microcentrifuge tube.
15. To elute DNA, add 25 μL of buffer EB (10 mM Tris-Cl, pH 8.5) to the center of the spin column membrane, let the column stand for 1 min, and centrifuge the column for 1 min. Store DNA at −20 °C.

**Ensure that the elution buffer is dispensed directly on to the QIAquick membrane for complete elution of bound DNA. Typically the recovery of yield is 60–80%.*

16. Estimate the DNA concentration by spotting 2 μL of the eluted DNA sample on to NanoDrop 2000 using EB buffer as a blank.
17. Evaporate the eluted DNA by speed vacuum to reduce to the volume of about 5 μL.

**Calculate the total amount (μg) of DNA based on the NanoDrop quantification and the volume of the eluted DNA before concentrating by evaporation.*

18. Store the concentrated DNA at −20 °C until use.

Part II. Yeast transformation (day 2)

C. Preparing competent cells

1. Inoculate 0.4 mL YPDA with a single colony (2 to 3 mm in diameter), vortex vigorously to disperse any clumps, and transfer 0.4 mL cell suspension into a 100-mL flask containing a 20 mL of YPDA.
2. Incubate at 30 °C at 250 rpm overnight (16–18 h).
3. Next morning, transfer 2 mL overnight culture to a 250-mL flask containing a 50 mL YPDA and shake-incubate as above for 4 to 6 h.

**Cells must be freshly prepared. Do not store the cells in a refrigerator.*

4. Read A_{600} of 1/10 diluted culture (TA will do this step).
5. Obtain 10–15 mL of YNN281 yeast culture, place cells into the tubes, and centrifuge at 4000 rpm for 5 min at room temperature. Discard the culture supernatant.
6. Suspend in 5 mL sterile distilled water and centrifuge as above.
7. Discard the supernatant and resuspend the cell pellet in 1× TEL or 1× TELG to a final cell density of 10^9 cells/mL, assuming that 1 of A_{600} is equivalent to 3×10^7 cells/mL.
8. Use immediately or rotate gently for 1 h at room temperature before transformation (this incubation is not required).
9. Aliquot 100 μL cell suspensions in 1× TELG in sterile microcentrifuge tubes and freeze for future use by wrapping the tube in several layers of Kimwipe and/or paper towels and placing them in a cardboard box in a −80 °C freezer. It is important to wrap up the tube so that the cells *freeze slowly*. Unlike *E. coli* competent cells, flash freezing in liquid nitrogen will severely reduce their competency. Thaw cells at room temperature for transformation.

D. Transformation

10. Add 5 μL of salmon sperm carrier DNA (10 mg/mL) to a sterile microcentrifuge tube.
11. Add 5 μL of two linearized DNA fragments mix (obtained from step 15 of Part I) to the carrier DNA. As negative and positive controls, add 5 μL of sterile water and *x* μL pRY121 plasmid (the same amount of 5 μL as the linearized DNA fragments mix) to the carrier DNA in separate tubes, respectively.

Tube 1 (Expt): 5 μL (50 μg) of carrier DNA + 5 μL of linearized DNA fragments mix

Tube 2 (− Control): 5 μL (50 μg) of carrier DNA + 5 μL of sterile water

Tube 3 (+ Control): 5 μL (50 μg) of carrier DNA + *x* μL of pRY121plasmid

**Carrier DNA must be completely denatured to be single-stranded.*

12. Add 50 μL competent cells to each tube above.

**The volume of competent cell should be less than 1/10 the volume of transforming DNA.*

13. Add 500 μL PEG-TEL solution (made fresh from the stock solutions) to each tube and mix gently by pipetting up and down.

**It is critical to use PEG with MW of 4000 or 3500, as the use of a higher molecular weight dramatically decreases transformation efficiency.*

14. Incubate the tubes at 30 °C for 30 min. Mix gently every 10 min.
15. Add 25 μL of DMSO to each tube; mix well.
16. Place in a 42 °C water bath for 15 min. Mix gently every 5 min.

**The optimum time may vary for different yeast strains.*

17a. Spin in a microcentrifuge for 15 s at 8000 rpm at room temperature and proceed to the next step, step 18.

**The fragile cells may be damaged and die if you spin too long at high speed.*

17b. Add 1 mL of room temperature YPDA medium, transfer to a sterile 1.5-mL tube, incubate at 30 °C, 200 rpm for 1 h, and spin as in step 17a (optional step).

**This is not usually necessary. If transformation efficiency of a certain yeast strain is very low, this option may be helpful.*

18. Remove all of the supernatant. Resuspend the pellet by pipetting up and down gently in 0.5 mL of sterile water.
19. Plate 100 μL each of the experimental transformation mixture directly on to two YNB −Ura/+glucose selection plates. Plate 100 μL each of the positive and negative control mixtures on to the same selection plate.
20. Incubate at 30 °C until transformants appear. Transformants will be visible on the agar surface about 3 days after incubation.
21. Count the number of transformants and calculate transformation efficiency:

$$\frac{\text{cfu} \times \text{total suspension volume (mL)}}{\text{Volume plated (mL)} \times \text{amount of DNA used (μg)}} \times \text{dilution factor} = \frac{\text{cfu}}{\text{μg DNA}}$$

Example of calculation assuming:

- 100 colonies (step 20)
- Plating volume (step 19): 0.1 mL (no dilution = dilution factor = 1)
- Total suspension volume (step 18) = 0.5 mL
- Amount of plasmid DNA used (step 11) = 0.1 μg

$$\frac{100 \text{ cfu} \times 0.5 \text{ mL}}{0.1 \text{ mL} \times 0.1 \text{ μg}} \times 1 = \frac{5 \times 10^3 \text{ cfu}}{\text{μg DNA}}$$

22. Using a Sharpie pen, circle and number as many well-isolated transformant colonies as you can on the bottom of the experimental plate; pick up each colony using a sterile toothpick or micropipettor tip and make a cross (X) streak on to one [YNB −Ura/+galactose/+raffinose] plate, one [YNB −Ura/+galactose/+glucose] plate, and one [YNB −Ura/+glucose] plate, as shown below.

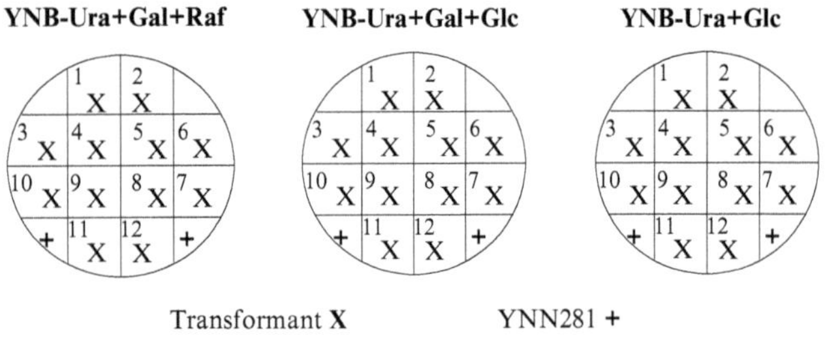

Note: each numbered streak on the above three plates should be made using the same colony from the transformant plates.

23. Incubate at 30 °C for 2–3 days. Store the plates at 4 °C.

**These plates will be used for X-β-gal filter lift assay of Experiment 33 next week.*

Further reading

Gibson, D.G, Benders, G.A., Andew-Pfannkoch, A., Dennisova, E.A., Baden-Tillson, H., Zaveri, J., Stockwell, T.B., Brownley, A., Thomas, D.W., Algire, M.A., Merryman, C., Young, L., Noskov, V.N., Glass, J.I., Venter, J.G., Hutchison, C.A., and Smith, H.O. (2008). Complete chemical synthesis, assembly, and cloning of a *Mycoplasma genitalium* genome. *Science*, **319**: 1215–1220.

Gibson, D.G., Benders, G.A., Axelrod, K.C., Zaveri, J., Algire, M.A., Moodie, M., Montague, M.J., Venter, J.C., Smith, H.O., and Hutchison, C.A. (2008). One-step assembly in yeast of 25 overlapping DNA fragments to form a complete synthetic *Mycoplasma genitalium* genome. *Proceedings of the National Academy of Science USA*, **105**: 20404–20409.

Gietz, R.D. and Schiestl, R.H. (2007). Frozen competent yeast cells that can be transformed with high efficiency using the LiAc/SS carrier DNA/PEG method. *Nature Protocols*, **2** (1): 1–4.

Iizasa, E. and Nagano, Y. (2006). Highly efficient yeast-based *in vivo* DNA cloning of multiple DNA fragments and the simultaneous construction of yeast/*Escherichia coli* shuttle vectors. *BioTechniques*, **40**: 79–83.

31 AMB 1 experiment 31: Plasmid DNA isolation from yeast

Purpose: This is up to you to write down.

Introduction

Most plasmids used for yeast studies are shuttle vectors that contain the sequence elements enabling them to be selected and propagated in *E. coli*. This allows for convenient amplification and subsequent manipulation *in vitro*. Most yeast vectors contain an origin of replication (*ori*) from pBR322 plasmid controlling 15 to 20 copy number maintenance in an *E. coli* cell and an antibiotic selection marker, β-lactamase gene *bla* (or Amp^R) or neomycin phosphotransferase II gene *npt*II (or Km^R), which confers resistance to ampicillin and kanamycin, respectively.

In addition, all yeast vectors contain nutritional markers that permit selection of transformants containing the desired plasmid in yeast. The most commonly used yeast markers are *URA3, HIS3, LEU2, TRP1,* and *LYS2,* which complement specific auxotrophic mutations in yeast, such as *ura3-52, his3-D1, leu2-D1, trp1-D1,* and *lys2-201*. These complementable yeast mutations have been chosen because of their low-reversion rate. Yeast markers *URA3, HIS3, LEU2, TRP1,* and *LYS2* can also complement specific *E. coli* auxotrophic mutations.

Because of the low copy number of most yeast plasmids, a tough cell wall and large chromosomal DNA, the yield and quality of plasmid DNA from yeast cells are very low. The yield of plasmid DNA from yeast varies widely depending on the yeast strain, plasmid size, growth conditions, and the methods employed. From a 5 mL yeast culture, E.Z.N.A.® Yeast Plasmid Kit (Omega Bio-Tek) was reported to give about 0.5 to 1 μg in 100 μL elution volume. Yeast 2-μ plasmid DNA recovered from a 1.5 mL overnight culture using EZ Yeast™ Plasmid Prep Kit (G-Biosciences) ranges from 0.1 to 0.3 ng, which is below the detection limit of ethidium bromide fluorescent DNA (1 to 5 ng/band). Accordingly, plasmid DNA recovered from yeast cells is not suitable for direct molecular analyses such as restriction digestion and yeast transformation.

In this lab exercise, you will use the Qiagen Miniprep Kit to prepare plasmid DNA from yeast cells, which can be directly used for transformation into *E. coli*, PCR, and DNA sequencing. The method was developed by Singh and Weil (2002). In this method, the yeast cell wall is first removed either by a physical method using glass beads or by an enzymatic method using lyticase, and then yeast cells are treated just like *E. coli*, following the Qiagen miniprep procedure. The third method involves vortexing yeast cells with glass beads in the presence of detergents and organic solvents followed by alcohol precipitation of DNA. Whatever method is used, yeast plasmid preparations are contaminated with mitochondria and genomic DNAs.

Materials and equipment

- YNN281 (pRY121) culture in YPD medium
- QIAprep® Spin Miniprep Kit (Qiagen Cat. No. 27104)
- Acid-washed glass beads (0.45–0.53 mm; Sigma G-8772)
- Lyticase stock solution (5 mg/mL lyticase in 1M Sorbitol, 0.1 M EDTA, pH 8.0; aliquots stored in at −20 °C; add β-mercaptoethanol to a final 0.1% just before use)
- Breaking buffer (2% Triton X-100; 1% SDS; 100 mM NaCl; 10 mM Tris-HCl pH 8; 1 mM EDTA)
- Buffers P1, P2, N3, and EB (QIAprep Miniprep Kit; Qiagen, Cat. No. 27104)
- Phenol/chloroform/isoamylalcohol (25:24:1, v/v/v) mixture
- TE (10 mM Tris-HCl pH 8.0, 1 mM EDTA), sterile H_2O
- GelRed® (10,000×; VWR Cat. No. 89139-134)
- Isopropanol
- Plasmid pRY121
- 0.5% agarose gel in 1× bionic buffer or 1× lithium borate buffer
- 37 °C water bath

Procedure

A. Lyticase lysis protocol

1. Grow yeast culture at 30 °C overnight at 30 °C, 200 rpm in 5 to 10 mL YPD medium.

**Do not incubate for more than 20 hours or store the culture for extended periods.*

2. Fill a 2-mL microtube with YPD yeast culture, spin at 8000 rpm for 30 s, and pour off the supernatant. Repeat this step to harvest 5 to 10 mL of culture.

3. Suspend the cell pellet in 500 μL of sterile H_2O, spin at 8000 rpm for 30 s, and remove the supernatant.

4. Resuspend the cell pellet in 200 μL of suspension buffer P1 by pipetting up and down gently several times until the mixture is homogeneous.

5. Add a 100 μL of lyticase stock solution, mix well, and incubate at 37 °C for at least 25 min.

Methods in Biotechnology, First Edition. Seung-Beom Hong, M. Bazlur Rashid and Lory Z Santiago-Vázquez.

Companion Website: www.wiley.com\go\hong\Methodsinbiotechnology

6. Add 300 µL of lysis buffer P2, mix by inverting the tube several times, and incubate at room temperature for 10 min.
7. Add 420 µL of neutralization buffer N3, mix by gentle inversion several times, and spin at 13 000 rpm for 10 min.
8. Place a Qiagen-Spin column into a 2-mL microfuge tube and apply the clarified supernatant.
9. Spin for 1 min at 13 000 rpm and discard the flow-through.
10. Wash the spin column by adding 500 µL of wash buffer PB and centrifuging for 1 min. Discard the flow-through.
11. Wash the spin column by adding 750 µL of wash buffer PE and centrifuging for 1 min.
12. Discard the flow-through and spin for 1 min to completely remove the residual wash buffer.
13. Place the spin column in a sterile 1.5-mL microfuge tube.
14. Add 50 µL of TE or EB buffer (pre-warmed to 65 °C) to the center of the spin column, let it stand at room temperature for 1 min, and elute plasmid DNA by spinning for 1 min.

**Use 10 and 2 µL of eluant to transform 100 µL chemically competent E. coli cells and electro-competent cells, respectively.*

B. Glass bead lysis protocol

15. Follow steps 1, 2, and 3 of procedure A.
16. Resuspend the cell pellet in 300 µL suspension buffer P1.
17. Add 180 µL acid-washed glass beads and vortex vigorously for 2 min. Let it stand to allow the beads to settle.
18. Transfer the supernatant to a new 1.5-mL microcentrifuge tube.
19. Add 250 µL of lysis buffer P2 and invert gently several times to mix; incubate at room temperature for 10 min.
20. Add 350 µL of neutralization buffer N3, mix by gentle inversion, and spin at 13 000 rpm for 10 min.
21. Follow steps 8 to 14 of procedure A.

C. Organic extraction protocol

22. Follow steps 1, 2, and 3 of procedure A.
23. Resuspend the yeast cell pellet in a 300-µL breaking buffer.
24. Add 300 µL phenol/chloroform/isoamylalcohol mixture and 0.4 g of acid-washed glass beads.
25. Vortex vigorously for 2 min and spin at 13 000 rpm for 5 min.
26. Transfer supernatant to a new tube, add 500 µL of isopropanol, and mix well.
27. Centrifuge at 13 000 rpm for 5 min.
28. Remove the supernatant, add 500 µL of 70% ethanol to the tube, mix gently by inverting the tube twice, spin at the top speed for 1 min, and pipette off the supernatant.
29. Spin briefly and remove any residual liquid in the tube.
30. Air-dry the pellet in the laminar floor hood for 10 min and dissolve in 30 µL of TE buffer.

D. Agarose gel electrophoresis

31. Load 10 µL of yeast plasmid prep plus 2 µL of 6× loading dye on to 0.5% agarose gel in 1× bionic buffer or 1× lithium borate buffer (prepared by TA).

Lane 1: 1 µL of pRY121 plasmid + 9 µL of TE + 2 µL of 6× loading dye
Lane 2: 10 µL of lyticase DNA + 2 µL of 6× loading dye
Lane 3: 10 µL of glass bead DNA + 2 µL of 6× loading dye
Lane 4: 10 µL of organic DNA + 2 µL of 6× loading dye

32. Run at 300 V for 30 min and stain in 2.5× GelRed® solution for 10–30 min.
33. Take a photograph under UV light.

References

E.Z.N.A.® Yeast Plasmid Mini Kit – Omega Bio-Tek (http://omegabiotek.com/store/wp-content/uploads/2013/07/D3376-Yeast-Plasmid-Mini-Kit-Combo-Online.pdf).

EZ Yeast™ Plasmid Miniprep Kit – Gbiosciences (http://www.gbiosciences.com/PDF/Protocol/EZ_Yeast_Plasmid_Prep.pdf).

Isolation of plasmid DNA from yeast using the QIAprep® Spin Miniprep Kit (User-Developed Protocol PR04.doc. Oct-01).

Singh, M.V. and Weil, P.A. (2002). A method for plasmid purification directly from yeast. *Analytical Biochemistry*, **307**: 13–17.

32 AMB 1 experiment 32: *E. coli* transformation with yeast plasmid DNA

Purpose: This is up to you to write down.

Introduction

To verify that a particular phenotype in yeast is conferred by plasmid or to obtain enough yeast plasmid for restriction cleavage analysis, transformation of yeast plasmid into *E. coli* followed by plasmid isolation is necessary because of the low quality and poor yield of plasmid from yeast. When plasmid isolated from yeast is used to transform into *E. coli*, transformation efficiency is so low that only a few colonies are usually obtained due to the low yield and/or high impurity of yeast plasmid preparation. *E. coli* is typically transformed by either a chemical or an electroporation method. Electroporation has a much higher transformation efficiency than the chemical method and allows up to a 200-kb plasmid to be transformed. Nevertheless, the $CaCl_2$ chemical method is most widely used because it is a very simple and inexpensive procedure with good transformation efficiency. Moreover, most transformation experiments involve a small size of recombinant plasmid carrying less than a 10-kb insert.

In the chemical transformation method, *E. coli* cells in the exponential phase are concentrated by centrifugation and treated with a high concentration of ice-cold $CaCl_2$ solution to make *E. coli* competent for DNA uptake. Plasmid DNA is then mixed with the cells and incubated in ice. A subsequent heat shock treatment of a cell and DNA mixture significantly increases the permeability of the cell wall and membrane, and DNA transport across the cell envelope is completed after heat shock. Heat shock treatment must be brief because extended heat shock time greatly decreases cell viability. Immediately after heat shock, cells are grown in a non-selective medium to allow a plasmid-borne antibiotic resistance marker gene to be expressed, and then plated on to antibiotic-containing medium to select transformant colonies. Most *E. coli* strains are amenable to transformation by the calcium chloride protocol, though transformation efficiency varies widely depending on strains.

Competent *E. coli* cells with high transformation efficiency are commercially available. The high efficiency is important when very small amounts of DNA are used or when nearly all of the DNA need to be transformed to prepare a library of plasmids carrying different inserts. However, low transformation efficiency is good enough for the purpose of subcloning since you need only one transformant for storage and propagating plasmid. The protocol described here is mostly sufficient for gene cloning. In this lab exercise, you will use yeast plasmid preparations isolated by three different protocols (Experiment 31) to transform into *E. coli* using the $CaCl_2$ chemical method and compare the transformation results. Many factors influence the efficiency of transformation. If one follows the well-established procedures faithfully, success of transformation will depend on the use of good competent cells, aseptic technique, and right amounts of antibiotic.

Materials and equipment

- Ice-cold $CaCl_2$ solution I (100 mM $CaCl_2$; autoclave or filter-sterilize; store at 4 °C)
- Ice-cold $CaCl_2$ solution II (100 mM $CaCl_2$, 15% glycerol, 10 mM HEPES, pH 7.5; filter-sterilize; store at 4 °C)
- Pre-chilled 1.5-mL sterile microcentrifuge tubes
- Ice-cold sterile distilled water
- *E. coli* cell (single isolated colonies on a streaked plate): HB101 K-12
- Plasmid DNA: pRY121 (35 ng/μL) from *E. coli* (Experiment 19) and from yeast (Experiment 31)
- LB or SOC broth
- LB agar-Amp (100 μg/mL) plates
- Absolute ethanol (molecular biology grade)
- Spreader in 99% denatured ethanol beaker, Bunsen burner
- Sterile 50-mL conical centrifuge tubes
- Ice bucket
- 42 °C water bath, 37 °C shake incubator

Procedure

A. Preparing competent cells (this part will be done by TA)

1. Pick up a well-isolated single colony from the streaked LB agar plates of *E. coli* HB101, inoculate into 2 mL of LB medium, and incubate at 37 °C, 200 rpm overnight.

2. Next day, inoculate a 0.5-mL overnight culture into a 50-mL LB medium in a 250-mL flask.

3. Grow a culture at 37 °C, 250 rpm, for 2 to 3 h to a mid-log phase (A_{600} = 0.7 to 0.8).

4. Dispense the culture into two sterile 50-mL conical centrifuge tubes.

5. Centrifuge at 4500 rpm for 10 min at 4 °C.

Methods in Biotechnology, First Edition. Seung-Beom Hong, M. Bazlur Rashid and Lory Z Santiago-Vázquez.

Companion Website: www.wiley.com\go\hong\Methodsinbiotechnology

6. Pour off the culture supernatant and suspend gently in 10 mL of an ice-cold $CaCl_2$ solution I.

**Do not vortex to suspend the pelleted cells.*

7. Keep the suspended cells on ice for at least 30 min (up to 1 h).
8. Centrifuge cells at 3000 rpm for 5 min at 4 °C.

**Do not increase the rpm and time of centrifugation. After centrifugation, the cell pellet is loose; take care not to disrupt the cell pellet. Cells are fragile in ice-cold $CaCl_2$ solution.*

9. Decant the supernatant and resuspend gently in 2 mL of ice-cold $CaCl_2$ solution II (0.04× of starting culture volume).

**Now it is ready for transformation, and the competent cells can be used immediately.*

**If not used immediately, make aliquots and store in a −80 °C freezer. Competent cells can also be used on the next day with higher transformation efficiency if stored on ice in a 4 °C refrigerator overnight.*

**The same procedure can be used to prepare competent cells for other E. coli strains.*

B. Transformation

10. To each 100 µL competent cell in a *pre-chilled* 1.5-mL tube, add the following DNAs:

x = 10 (Group 1), 7.5 (Group 2), 5.0 (Group 3), 2.5 (Group 4)

- x µL plasmid DNA isolated by lyticase protocol (each group)
- x µL plasmid DNA isolated by glass bead protocol (each group)
- x µL plasmid DNA isolated by organic extraction protocol (each group)
- 2 µL pRY121 plasmid DNA (one positive control shared by all groups)
- 2 µL TE buffer (one negative control shared by all groups)

11. Keep all the tubes on ice for at least 30 min.

**A 10-min incubation can be used but transformation efficiency drops.*

12. Transfer all the tubes to a 42 °C water bath, incubate for 1 min, and transfer them to an ice bath.

13. Add a 0.5 mL SOC or LB medium to each tube and incubate at 37 °C, 200 rpm for 1 h.

**A 30-min incubation can be used but transformation efficiency drops.*

14. Spread 150 µL of the cells on to LB agar-Amp plates:

- Lyticase protocol DNA transformation (two plates per group)
- Glass bead protocol DNA transformation (two plates per group)
- Organic protocol DNA transformation (two plates per group)
- Positive control transformation (one plate shared by all groups)
- Negative control transformation (one plate shared by all groups)

**Selection agar plates stored in a 4 °C refrigerator should be warmed up to room temperature before plating the transformation mixture. If you incubate a cold agar plate at 37 °C right after you spread the transformation mixture, many colony-like air bubbles will be produced in the bottom of the agar plate.*

15. Allow the plates to dry in the laminar flow hood until most of the surface liquid disappears.
16. Stack up your plates and tape them together. Put your group name on the bottom of the stacks.
17. Incubate the plates upside down at 37 °C for 14 to 16 h.

Post-lab assignment

1. List at least six factors that influence transformation efficiency of *E. coli* and yeast.
2. Describe the different regions on pRY121 plasmid that confer the functions of replication and propagation in both *E. coli* and *S. cerevisiae*.
3. Compare and contrast the chemical transformation method of *E. coli* and yeast. Exclude incubation temperature and medium.
4. Do you expect that *E. coli* transformants carrying the pRY121 plasmid express β-galactosidase enzyme in a medium containing galactose or glucose? Explain your choice.
5. Why are most yeast plasmids used in biotechnology a shuttle vector that is able to replicate in both yeast and *E. coli*? Why is the plasmid isolated from yeast transformant retransformed into *E. coli*?

Further reading

Inoue, H., Nojima, H., and Okayama, H. (1990). High efficiency transformation of *Escherichia coli* with plasmids. *Gene*, **96** (1): 23–28.

Hengen, P.N. (1996). Preparing ultra-competent *Escherichia coli*. *Trends in Biochemical Sciences*, **21** (2): 75–76.

Sharma, A.D., Singh, J., and Gill, P.K. (2007). Ethanol mediated enhancement in bacterial transformation. *Electronic Journal of Biotechnology*, **10** (10). Available at: http://www.ejbiotechnology.info/content/vol10/issue1/full/10/index.html. ISSN 0717-3458. DOI: 10.2225/vol10-issue1-fulltext-10.

33 AMB 1 experiment 33: X-gal filter lift assay

Purpose: This is up to you to write down.

Introduction

In both *E. coli* and *S. cerevisiae*, glycolysis provides important intermediates for the synthesis of amino acids, nucleotides, and ATP energy, and thus glucose is the most preferred carbon source. When glucose is not available, however, cells turn on other alternative catabolic genes that enable them to utilize other available carbon sources such as lactose and galactose. In yeast, expression of the genes required for the uptake and catabolism of galactose (*GALI, GAL2, GAL7,* and *GAL10*) is induced by galactose and repressed by glucose. This phenomenon is referred to as catabolite repression, which allows cells to adapt to changes in the carbon sources present in their environment. The degree of catabolite repression widely varies depending on strains, showing 80- to 600-fold repression in wild-type yeast strains grown in the presence of a mixture of glucose and galactose.

Unlike *E. coli,* wild type yeast cells are unable to catabolize lactose due to the absence of lactose metabolic genes. Plasmid pRY121 contains an entire *E. coli* β-galactosidase coding sequence that is under the control of yeast *GAL1-10* promoter, which is very sensitive to galactose. In the presence of galactose, the *GAL1* transcription level increases as much 1000 times more than occurred in the presence of glucose.

In this lab exercise, you will use three different YNB dropout plates streaked with YNN281 transformants harboring pRY121 DNA that was reassembled from overlapping linear DNA fragments (Experiment 30). The three plates have different carbon sources: one has glucose, the second has galactose plus raffinose, and the third has galactose plus glucose. Because β-galactosidase expressed in yeast is not secreted very well, a yeast cell should be made more permeable before its substrate is added. To do so, yeast cells that have grown on Petri agar plate are transferred to a piece of filter paper, quickly frozen in liquid nitrogen, and thawed. As the yeast cells thaw, ice crystals break the cells open, releasing the proteins. The replica filters are then laid on to filter paper impregnated with a solution containing X-β-gal chromogenic substrate. If β-galactosidase enzyme is expressed, it will cleave a colorless X-β-gal on the filter to produce a blue color. This assay is relatively simple and quick, allowing for simultaneously screening a large number of yeast colonies. However, this qualitative assay is not so sensitive that slight differences in expression levels cannot be distinguished. Once yeast cells are grown, the filter lift assay takes about 30 min followed by 5 min to several hours for color development, depending on the signal strength.

Materials and equipment

- YNN281 (reassembled pRY121)-streaked YNB –Ura/+ glucose plate (Experiment 30, Part D)
- YNN281 (reassembled pRY121)-streaked YNB –Ura/+ galactose/ + raffinose plate (Experiment 30, Part D)
- YNN281 (reassembled pRY121)-streaked YNB –Ura/+ galactose/ +glucose plate (Experiment 30, Part D)
- Z-buffer (60 mM $Na_2HPO_4.7H_2O$, 40 mM $NaH_2PO_4.H_2O$, 10 mM KCl, 1 mM $MgSO_4.7H_2O$: adjust pH to 7.0; add 50 mM β-mercaptoethanol just before use)
- X-β-Gal: 2% (w/v) dissolved in *N,N*-dimethylformamide; store at −20 °C
- 1 M Na_2CO_3 (stop solution)
- Whatman 3MM paper
- Liquid nitrogen
- Forceps, aluminum foil, Petri dish (100 × 15 mm)
- 30 °C incubator

Procedure

1. Cut six circular pieces of Whatman 3MM paper (or nitrocellulose membrane) that fit into the inside of a Petri dish.
2. Mark the orientation of the three cut filter papers and label plate names using a pencil. Mark the orientation of the three different plates having YNN281 (reassembled pRY121) streaked cells (Experiment 30, Part D) with a Sharpie pen.
3. Using forceps, gently layer a filter paper on to each surface of the three different plates.
4. Gently press the filter paper on to the agar plate so that there are no bubbles or wrinkles. Let it sit for 1 to 2 min to allow the entire filter paper to be completely wet.
5. Gently lift the filter off the plate using forceps to avoid smearing the colonies.

**A large number of the cells should stick to the filter and leave a mirror image of the streak-outs on the filters.*

6. Submerge the filter blotted with colonies into a shallow dish made of aluminum foil that contains just enough liquid nitrogen to cover the filter for 15 s.

Methods in Biotechnology, First Edition. Seung-Beom Hong, M. Bazlur Rashid and Lory Z Santiago-Vázquez.

Companion Website: www.wiley.com\go\hong\Methodsinbiotechnology

**Freezing in this way serves to permeabilize cell membrane and allows the subsequent entry of X-β-gal, the substrate for β-galactosidase.*

7. Remove the filter carefully from the liquid nitrogen, place the blotted filter with the colony side up on a piece of dry paper towel, and allow it to thaw at room temperature.

**The filter gets very brittle when frozen.*

8. Place one circular piece of filter paper into each of three empty Petri plates. Add 2 mL of Z buffer containing 5.4 μL of β-mercaptoethanol and 20 μL of 2% X-β-Gal to each Petri plate, and remove any superfluous fluid in the plate.

**Flooding the filter paper will allow the colonies to be diffused and make screening difficult to interpret.*

9. Place the filter with CELL SIDE FACING UP on to the filter paper that had been wetted with a solution of Z buffer containing X-β-gal.

**Be sure that there is good contact between the filter blotted with the cells and the filter impregnated with the X-gal. Remove any bubbles trapped between the two filters.*

10. Cover the Petri dish, seal with Parafilm, and incubate at 30 °C for a few minutes to hours until a blue color starts to develop. Do not invert the plate.

**Blue color may start to appear with 10 min for a positive control strain.*

11. Remove the filters and briefly place in 1 M Na_2CO_3 stop solution.

Post-lab assignment

1. Describe which portion of X-β-gal structure is responsible for the blue color when it is cleaved by β-galactosidase enzyme. Is this blue color compound soluble in aqueous solution or not?
2. If X-gal lift filters are left too long, all the filters including those containing cells grown on YNB –Ura/+glucose will turn blue. Explain this observation.
3. What is the purpose of adding raffinose in addition to galactose in YNB –Ura/+galactose/+raffinose medium?

Further reading

Becker, J.M, Caldwell, G.A., and Zachgo, E.M. (1990). *Biotechnology: A Laboratory Course*, 2nd edition. Academic Press. ISBN 0-12-084562-8.

Gancedo, J.M. (1992). Carbon catabolite repression in yeast. *European Journal of Biochemistry*, **206**: 297–313.

Johnston, M. (1987). A model of fungal regulatory mechanism: the *GAL1* genes of *Saccharomyces cerevisiae*. *Microbiological Review*, **51**: 458–476.

Reynolds, A., Lunblad, V., Dorris, D., and Keaveney, M. (2001). Unit 13.6. Yeast vectors and assays for expression of cloned gene. In *Current Protocols in Molecular Biology* (Edited by F.M. Ausubel), pp. 13.6.1–13.6.6. John Wiley & Sons, Inc. ISBN 0-471-50338-X.

34 AMB 1 experiment 34: Protein quantitation assay

Purpose: This is up to you to write down.

Introduction

Bradford, Lowry, and bicinchoninic acid (BCA) assays are widely used for the colorimetric detection and quantitation of protein. All these assays can be performed in either test tubes or microplates. The test tube procedure employs a high ratio of working reagent to protein sample in a large volume of each protein assay, which makes handling of many samples difficult. However, a protein sample in the test tube assay becomes further diluted as a result of a much larger volume (milliliter) of working reagent added to a smaller volume (microliter) of sample, and thus the effect of interfering substances on color development can be minimized by dilution. The microassay procedure requires a much smaller volume of working reagent that allows many samples to be analyzed more easily. It is important to note that all of the colorimetric assays are neither entirely specific to protein due to the interference by non-protein components nor respond uniformly to all different proteins owing to the differences in their primary and secondary structures. The developed color intensities may considerably vary when different proteins having the same molecular weight are assayed. Because of this property, specific activity of a protein may vary depending on the type of protein assay and protein standard used.

The amount of protein in a sample is routinely estimated using a standard curve of a selected standard protein solution. Accurate pipetting is essential for the precise, accurate, and reproducible results; pipetting small volumes of protein sample in glycerol tends to increase inaccuracy. If samples are so concentrated that you have to pipet less than 2 μL, it would be better to simply dilute samples 10 times and use a larger volume of the diluted solutions. More importantly, the absorbance value does not change very much at high concentrations. Each sample data point must fall within the standard curve ranges for accurate quantitation. It is advisable to use triplicate samples for each assay so that large pipetting errors can be readily recognized.

Microvolume protein samples can be directly quantified without the use of cuvettes using the NanoDrop 2000c spectrophotometer:

- http://www.jove.com/video/1610/microvolume-protein-concentration-determination-using-nanodrop-2000c/

However, the colorimetric NanoDrop quantification is still prone to protein-to-protein variation. In contrast, infrared (IR) spectrum-based Direct Detect® Spectrometer quantifies microliter volumes of protein samples more accurately without variation and interference by non-protein components because it measures amide bonds, an intrinsic feature of protein chains:

- http://www.emdmillipore.com/US/en/life-science-research/protein-detection-quantification/direct-detect-spectrometer/videos/WHKb.qB.ao8AAAFBI.0RRkxA, nav

Part I. Bradford assay

The Bradford method depends on the change in absorbance of Coomassie Blue G-250 when it binds to protein. The dye binding causes a shift in its absorption maximum from 465 nm to 595 nm. Unknown protein concentrations are determined by reference of their absorptions to a series of standard protein dilutions. Coomassie dye primarily binds to basic and aromatic side chains. The interaction with arginine is very strong and less strong with tryptophan, phenylalanine, tyrosine, histidine, and lysine. Depending on the amino acid composition of protein, about 1.5 to 3 dye molecules bind per positive charge on the protein. The quantification procedure is simple and rapid, since the dye binding process is completed in approximately 5 min with color stability for 1 hour. The detection range is 1 to 2000 μg/mL of protein solution. Unlike many other assays, this assay is not susceptible to interference from a wide variety of chemicals present in samples (e.g. β-mercaptoethanol or DTT) but is affected by more than 0.1% detergents such as SDS, Triton X-100, and Nonidet P-40.

Materials and equipment

- Bradford reagent: Bio-Rad sells a ready-to-use reagent (Cat. No. 500-0006) that can be stored at 4 °C. Before use, dilute the Bradford reagent fivefold in water and filter the diluted reagent through Whatman No. 1 paper. The reagent also can be made by dissolving 100 mg of Coomassie Blue G-250 in 50 mL of 95% ethanol or methanol (stirring for about 10 min), adding 100 mL of 85% (w/v) phosphoric acid (H_3PO_4) to this solution and diluting the mixture to 200 mL with water. Store in a dark bottle at 4 °C (stable indefinitely).
- Bovine serum albumin (BSA) (1 mg/mL): dissolve in the same solution that is used to dissolve the protein to be assayed; store it frozen in small aliquots.

Methods in Biotechnology, First Edition. Seung-Beom Hong, M. Bazlur Rashid and Lory Z Santiago-Vázquez.

Companion Website: www.wiley.com\go\hong\Methodsinbiotechnology

- γ-Globulin, gelatin, lysozyme, *E. coli* cell-free lysate
- 1 × PBS (137 mM NaCl, 10 mM phosphate, 2.7 mM KCl, pH 7.4)
- Spectronic 20, cuvette
- 13 × 100 mm disposable borosilicate test tubes
- MilliQ H_2O

Procedure

A. Direct reading assay

1. Add 1 volume of Bradford dye to 4 volumes of H_2O (1/5 dilution).
2. Filter the diluted dye through Whatman No. 1 paper.
3. Label 18 glass test tubes and prepare duplicate aliquots of 5, 10, 30, 60, 80, 100 μL of BSA (1 mg/mL) standard sample and 20 μL aliquots of test samples in a final volume of 100 μL, as shown below.

Volume (μL) of BSA standard (1 mg/mL)	μg BSA	Volume (μL) of 1 × PBS	Sample tube
0*	0	100	Single
5	5	95	Duplicate
10	10	90	Duplicate
30	30	70	Duplicate
60	60	40	Duplicate
80	80	20	Duplicate
100	100	–	Duplicate
BSA† 20 μL	–	80	Single
γ-Globulin† 20 μL	–	80	Single
Gelatin† 20 μL	–	80	Single
Lysozyme† 20 μL	–	80	Single
E. coli lysate 20 μL	–	80	Single

*The first 0 serves as a blank solution (no duplicate).
†The same concentrations of the proteins are given based on the ratio of mass to volume.

4. Add 3.0 mL of the 1/5 diluted Bradford dye solution to each tube, mix well, and let it stand for ~5 min at room temperature.
5. Zero Spec20 (pre-warm up) at 595 nm using 1× PBS as a blank solution.
6. Place the sample in the cuvette. Read the absorbance at 595 nm in Spectronic 20.

**If the unknown protein concentration is too high, dilute the protein, and assay smaller aliquots of the unknown.*

7. Return all the samples to the original test tubes and save for the Part B experiment.

B. Normalized reading assay

1. Zero Spec20 at 595 nm using *H_2O as a blank.*
2. Dilute each Bradford assay sample (Part A) to 1/2 by adding 3.1 mL of H_2O to each of 18 tubes.
3. Place each sample in the cuvette. Read the absorbance at 595 nm in Spectronic 20.
4. Return the samples to new test tubes and save.
5. Zero Spec20 at 465 nm using *H_2O as a blank.*
6. Read the absorbance of each sample at 465 nm in Spectronic 20.

**Free dye molecules are read at its maximum 465 nm.*

BSA (μg)	Direct assay	Normalized assay		
	A_{595}	A_{595}†	A_{465}†	A_{595}/A_{465}
0*				
5				
10				
30				
60				
80				
100				
BSA				
γ-Globulin				
Gelatin				
Lysozyme				
E. coli lysate				

*The first 0 serves as a blank solution for direct assay but as a zero reference solution for normalized assay.
†Reading values are the result of ½ dilution; if the values are still out of range (i.e., Spec20 Absorbance Readout is blinking, further dilution of the sample is needed.

7. Subtract the 595 and 465 nm absorbance measurement of a zero reference from the 595 and 465 nm absorbance measurements of all other individual standard and test samples.
8. Plot the standard curve of BSA in Microsoft Excel.
- Direct Reading: average A_{595} in the *y* axis versus μg protein in the *x* axis.
- Normalized reading: average A_{595}/A_{465} in the *y* axis versus μg protein in the *x* axis.

9. Determine the concentrations of test samples by extrapolation of test samples' absorbance values into the standard curve. Show your work. Compare the concentrations of the test samples determined by the direct reading assay with those of the normalized reading assay.

C. Microtiter protocol (This procedure will not be performed in the lab)

The assay here is designed for use in microtiter plates. This is an easy assay format for those who can have an access to a microtiter plate spectrophotometer.
1. Pipette 0, 2, 4, 6, 10, 15, and 20 μL of BSA (1 mg/mL) into assigned wells of a 96-well plate.
2. Pipette up to 20 μL of unknown test samples into individual wells of a 96-well plate.
3. Add 40 μL of the undiluted Bradford reagent into all wells containing a standard or sample.
4. Add 140 μL of H_2O to each well to bring the final volume to 200 μL.
5. Read absorbances of samples at 595 and 465 nm after a 5-min incubation using H_2O as blank.
6. Subtract the 595 and 465 nm absorbance values of a zero reference solution from the 595 and 465 nm absorbance values of all other individual standard and test samples.

7. Plot the standard curve of BSA with A_{595}/A_{465} in the *y* axis and μg protein in the *x* axis.
8. Determine the concentrations of test samples using the standard curve.

D. NanoDrop 2000 bradford assay protocol

1. Prepare aliquots of 1, 4, 8, and 10 μL of BSA (10 mg/mL) standard samples and 5 μL of aliquots of unknown samples in a final volume of 10 μL into 1.5-mL tubes, as shown below.

Volume of BSA (1 mg/mL) standard solution (μL)	μg BSA	Volume (μL) of 1 × PBS	Concentration (μg/mL)‡
0*	0	10	0
1	1	9	100
4	4	6	400
8	8	2	800
10	10	0	1000
BSA† 2 μL		8	
γ-Globulin† 2 μL		8	
Gelatin† 2 μL		8	
Lysozyme† 2 μL		8	
E. coli lysate 2 μL		8	

*The first 0 serves as a zero reference solution.
†The same concentrations of the proteins are given based on the ratio of mass to volume (mg/ml).
‡Concentration is based on the total sum 10 μL of protein solution plus buffer. Therefore,The unknown sample concentration shown in the NanoDrop assay is the result of a 1/10 dilution of the original sample. If the reading is out of range, the sample is either diluted or concentrated.

2. Add 500 μL of 1/5 diluted Bradford dye to each tube, briefly mix, and allow to stand for ~5 min at room temperature.
3. Select the "Protein Bradford" application from the main menu.
4. Pipette 2 μL of 1 × PBS on to the bottom pedestal, lower the arm, and click "Blank."

**After measurement, wipe the upper and lower pedestals using Kimwipe and the instrument is ready to measure the next sample.*

5. Click the "Standards Radio" button, type the values for each standard concentration in the right pane table, and click "Measure" to initiate the measurement starting from zero reference.

6. After all the standards have been measured, click on the "Samples Radio" button. Enter a sample name. Load 2 μL of sample. Click "Measure" to initiate the measurement of sample.
7. Compare the concentrations of the samples with those obtained from Spec20-based direct and normalized Bradford assay.

Part II. Lowry assay

The Lowry method consists of the Biuret reaction followed by the reduction of Folin Ciocalteu (FC) reagent (phosphomolybdic/phosphotungstic acid) under alkaline conditions. When cupric ions (Cu^{+2}) bind to the peptide bonds, they are reduced to cuprous ions (Cu^{+1}) to form Cu^{+1}–peptide bond complex, which subsequently facilitates the reduction of FC reagent with tyrosine, tryptophan, and polar amino acids to produce a blue color that can be read at a maximum absorption of 750 nm. Therefore, the color intensity varies with different proteins depending on their content of aromatic amino acids. Unlike the Bradford assay, the Lowry reaction is interfered by sucrose, lipids, phosphate buffers, monosaccharides, and hexoamines to varying degrees. The conventional Lowry assay is a two-step procedure that requires a minimum of a 40-min incubation time. However, an improved Lowry assay reagent, known as DC (detergent compatible) protein assay reagent, developed by Bio-Rad, reaches 90% of its maximum color development within 15 min and the color is stable for 1 h. This reagent is compatible with up to 1% ionic or non-ionic detergents and many common buffer components like EDTA, Tris, DTT, etc.

Materials and equipment

- FC (Folin–Ciocalteu) reagent (related to Bio-Rad DC Reagent B). Dissolve 10 g of sodium tungstate and 2.5 g of sodium molybdate in 70 mL of water. Add 5 mL of 85% phosphoric acid and 10 mL of concentrated HCl. Reflux for 10 h. Add 15 g of lithium sulfate, 5 mL of H_2O, and 1 drop of bromine. Reflux for 15 min. Cool to room temperature and bring to 100 mL with H_2O.
- Copper tartrate/carbonate (CTC) solution (related to Bio-Rad DC Reagent A). Mix by adding slowly while vigorously stirring in the following order: 100 mL of alkaline carbonate (0.4% NaOH + 2% Na_2CO_3) + 1 mL of 2.37% sodium tartrate + 1 mL of 1.56% of $CuSO_4.5H_2O$.
- Bio-Rad sells a ready-to-use, improved Lowry assay reagent known as Bio-Rad DC protein assay (Cat. No. 500-0116).
- 1 × PBS
- Bovine serum albumin (BSA) (1 mg/mL)
- γ-Globulin, gelatin, lysozyme, *E. coli* cell-free lysate
- Spectronic 20, cuvette
- 13 × 100 mm disposable borosilicate test tubes

Procedure

1. Label 18 glass test tubes and prepare duplicate aliquots of 5, 10, 30, 60, 80, 100 μL of BSA (1 mg/mL) standard sample and 20 μL aliquots of test samples in a final volume of 100 μL, as shown below.

Volume (μL) of BSA standard (1 mg/mL)	μg BSA	Volume (μL) of 1 × PBS	Sample tube
0*	0	100	Single
5	5	95	Duplicate
10	10	90	Duplicate
30	30	70	Duplicate
60	60	40	Duplicate
80	80	20	Duplicate
100	100	0	Duplicate
BSA† 20 μL	–	80	Single
γ-Globulin† 20 μL	–	80	Single
Gelatin† 20 μL	–	80	Single
Lysozyme† 20 μL	–	80	Single
E. coli lysate 20 μL	–	80	Single

*The first 0 serves as a blank solution (no duplicate).
†The same concentrations of the proteins are given based on the ratio of mass to volume.

2. Add 375 μL of Reagent A to each tube and mix immediately and briefly.

**Use 100 μL of Reagent A instead of 375 μL if the spectrophotometer sample holder accommodates a smaller cuvette.*

3. Add 3.0 mL (8 × volume of Reagent A) of Reagent B to each test tube, and vortex immediately and briefly.
4. Incubate for 15 min (up to 1 h) at room temperature.
5. Zero Spec20 (pre-warm-up for at least 10 min) at 750 nm using a blank solution containing 100 μL of water or buffer, 375 μL of Reagent A, and 3 mL of Reagent B.
6. Read the absorbance of all the samples at 750 nm.

**If the concentration of test protein is too high, dilute the protein for assay or use smaller aliquots of the test protein.*

7. Plot the standard curve of BSA (A_{750} in the *y* axis versus μg of protein in the *x* axis) in Microsoft Excel.
8. Determine the concentrations of test samples based on the standard curve. Compare with the data obtained with the Bradford assay.

Microassay protocol (This procedure will not be performed in the lab)

1. Pipet 5 μL of standards (0, 0.2, 0.5, 0.6, 1.2, 1.5 mg/mL) and samples into a microtiter plate.
2. Add 25 μL of Reagent A into each well.
3. Add 200 μL of Reagent B into each well.
4. Agitate the plate to mix the reagents.
5. Read absorbance at 750 nm after 15 min.

Part III. BCA assay

The BCA (bicinchoninic acid) protein assay is similar to the Lowry reaction, except that BCA is used instead of Folin–Ciocalteu reagent. The first step is the Biuret reaction in which Cu^{+2} is reduced to Cu^{+1} when it binds to peptide bonds of protein in alkaline solution containing sodium tartrate. This reaction develops to form a pale blue complex. In the second step, two molecules of BCA react with one cuprous ion (Cu^{+1}) to form an intense purple-colored reaction product. This water-soluble BCA/Cu^{+1} complex can be read at a maximum absorption of 562 nm. The developed color intensity considerably differs depending on the incubation time and temperature; the detection sensitivity is 20 to 2000 μg/mL and 5 to 250 μg/mL when the assay is performed at 37 °C for 30 min and at 60 °C for 30 min, respectively. Although the color continues to develop slowly over time, no significant errors are made within a 10-min measurement (~2.5% drift per 10 min).

The BCA color formation is strongly influenced by the number of peptide bonds and the presence of four particular amino acids (cysteine, cystine, tryptophan, and tyrosine). Unlike the Bradford method, variability caused by differences in protein composition is reduced because the universal peptide backbone contributes to color formation. In addition, the BCA assay is more tolerant to a range of ionic and non-ionic detergents and denaturing agents; however, BCA is easily affected by reducing reagents such as DTT and β-mercaptoethanol (BME), metal chelator EDTA, and reducing sugars sucrose and glucose, which are frequently used in protein extraction buffer. Use of a new version of Reducing Agent Compatible BCA (BCA-RAC) assay may overcome the interference by up to 5 mM DTT and 35 mM BME.

Material and equipment

- BCA Protein Assay Kit (Thermo Fisher Scientific, Cat. No. 23225)
  BCA Reagent A (100 mL):
  1 g 4,4′-dicarboxy-2,2′-biquinoline, disodium salt (26 mM final concentration)
  2 g $Na_2CO_3 \cdot H_2O$ (0.16 M final concentration)
  160 mg disodium sodium tartrate (7 mM final concentration)
  0.4 g NaOH (0.1 M final concentration.)
  0.95 g $NaHCO_3$ (0.11 M final concentration) in MilliQ H_2O
  After dissolving the components, adjust the pH to 11.3 ± 0.2, using solid NaOH to increase pH or $NaHCO_3$ to decrease pH. Store at 4 °C.
  BCA Reagent B (100 mL):
  4 g $CuSO_4 \cdot 5H_2O$ (16 mM final concentration) in MilliQ H_2O; store at room temperature
  BCA reagent A/B mix:
  Mix 50 volume of BCA Reagent A with 1 volume of BCA Reagent B; prepare before use.
- Bovine serum albumin (BSA) (1 mg/mL)
- γ-Globulin, gelatin, lysozyme, *E. coli* cell-free lysate
- 1× PBS
- Spectronic 20, cuvette
- 13 × 100 mm disposable borosilicate test tubes

Procedure

1. Label 18 glass test tubes and prepare duplicate aliquots of 5, 10, 30, 60, 80, 100 μL of BSA (1 mg/mL) standard sample and 20 μL of aliquots of test samples in a final volume of 100 μL, as shown below.

Volume (μL) of BSA standard (1 mg/mL)	μg BSA	Volume (μL) of 1× PBS	Sample tube
0*	0	100	Single
5	5	95	Duplicate
10	10	90	Duplicate
30	30	70	Duplicate
60	60	40	Duplicate
80	80	20	Duplicate
100	100	0	Duplicate
BSA† 20 μL	–	80	Single
γ-Globulin† 20 μL	–	80	Single
Gelatin† 20 μL	–	80	Single
Lysozyme† 20 μL	–	80	Single
E. coli lysate 20 μL	–	80	Single

*The first 0 serves as a blank solution (no duplicate).
†The same concentrations of the proteins are given based on the ratio of mass to volume.

2. Add 3.0 mL of light apple-green BCA Reagent A/B mix to each test tube.

3. Incubate the tubes at 37 °C water bath for 30 min, and cool to room temperature.

4. Zero Spec20 (pre-warm-up for at least 10 min) at 562 nm using a blank solution consisting of 100 μL of water or buffer and 3 mL of the BCA Reagent A/B mix.

5. Measure the absorbance of all the samples within 10 min.

6. Plot the standard curve of BSA (A_{562} in the *y* axis versus μg of protein in the *x* axis) in Microsoft Excel.

7. Determine the concentrations of test samples based on the standard curve. Show your work. Compare with the data obtained using the Bradford and Lowry assays.

Post-lab assignment

1. In addition to the Bradford, Lowry, and BCA assays, there are Biuret, UV absorbance, and Quant-iT protein assays. Describe the advantage and disadvantage of each of these six methods. Present succinctly in a table format.

2. Go to the NCBI web site (http://www.ncbi.nlm.nih.gov/gene).

- Type "*E. coli* lacZ" and click "Search."
- Click "*lacZ*" under Name/Gene ID.
- Click "EcoGene: EG10527" in the Primary source of Summary.
- Click "Protein" menu.
- Copy the protein sequence and paste into Microsoft Word as Courier New font (size 9).
- Go to the ProtParam tool web site (http://web.expasy.org/protparam/).
- Paste the protein sequence only in the box, and click "Compute parameters."
- What is a molar extinction coefficient value at 280 nm in water?

3. Once a molar extinction coefficient has been established for a given protein, the concentration of a purified protein sample can be calculated from its absorbance according to the Beer's Law ($A = \varepsilon cl$). The molar extinction coefficient of protein X is 210 000 $M^{-1}cm^{-1}$ at 280 nm in water. You dissolved 0.01 g of protein X in 10 mL of water and measured protein X solution using a 1-cm cuvette at 280 nm. The optical density at A_{280} was 0.14.

(a) Calculate the molar concentration of the protein X solution.

(b) Calculate the molecular weight (Dalton) of protein X.

4. Only the amino acids tryptophan (W) and tyrosine (Y) and cysteine (C) contribute to protein absorbance at 280 nm. Thus, the molar extinction coefficient of a protein can be calculated within ± 5% accuracy by the weighted sum of molar extinction coefficients of the three amino acids using the following equation:

$$\varepsilon(280\ \text{nm}) = (\#\ \text{of tryptophan}) \times 5500 + (\#\ \text{of tyrosine}) \times 1490 + (\#\ \text{of cysteine}) \times 125$$

where ε (280 nm) indicates the ε at 280 nm of a native protein in water and values are the amino acid average ε at 280 nm.

- Visit the NCBI web site (http://www.ncbi.nlm.nih.gov/).
- Choose "Nucleotide" in the drop-down menu, type "V00296" in the Search box window, and click "Search."
- Copy the *lacZ* CDS translation amino acid sequence and paste into a Microsoft Word document.

(a) Highlight W, Y, and C residues with different colors on the amino acid sequence data. Show this work.

(b) Based on the numbers of W, Y, and C, calculate the ε value of β-galactosidase protein.

Further reading

Olson, B.J.C. and Markwell, J. (2007). Unit 3.4. Assays for determination of protein concentration. In *Current Protocols in Protein Science* (Edited by F.M. Ausubel). John Wiley & Sons, Inc. ISSN 1934-3655.

Pace, C.N., Vajdos, F.V., Fee, L, Grimsley, G., and Gray, T. (1995). How to measure and predict the molar absorption coefficient of a protein. *Protein Science*, **4**: 2411–2423.

Thermo Scientific Pierce Protein Assay Technical Handbook (Version 2): https://tools.lifetechnologies.com/content/sfs/brochures/1602063-Protein-Assay-Handbook.pdf.

Zor, T. and Selinger, Z. (1996). Linearization of the Bradford protein assay increases its sensitivity: theoretical and experimental studies. *Analysis of Biochemistry*, **236**: 302–308.

35 AMB 1 experiment 35: Quantitative β-Galactosidase assay in yeast

Purpose: This is up to you to write down.

Introduction

The *E. coli lacZ* gene encoding β-galactosidase enzyme is one of the most popular reporter genes because the assay is so simple and well established. Since many eukaryotes except for plants do not have endogenous β-galactosidase activity, the *lacZ* gene is often placed behind a promoter of a target gene in order to monitor the temporal and spatial expression of the target gene of interest under different conditions. There are two methods for quantitation of β-galactosidase activity. They differ mainly in the cell disruption method to prepare the material for the assay. In the first method, cells are permeabilized to allow the enzyme substrate ONPG (*o*-nitrophenyl-β-D-galactoside) to enter the cells, and enzyme activity is normalized to the number of cells as determined by measuring the A_{600}. Because the cell wall is not broken apart by permeabilization treatment, most proteins remain within the cell. In the second method, cells are physically broken open and centrifuged to obtain a cell-free protein extract, and enzyme activity is normalized to the protein concentration (as determined by protein assay) for calculating specific activity. The latter method is more reliable when comparing cells that are grown under very different conditions or that have different genetic backgrounds. The former method is adapted from the Miller β-galactosidase assay of *E. coli* and is suitable for the analysis of changing levels of activity within a single strain grown under different conditions.

After incubation with the chromogenic substrate ONPG at 28 °C, the reaction is stopped by adding Na_2CO_3 to bring the pH to ~11 to inactivate β-galactosidase. The amount of colored products formed is quantified by the spectrophotometer reading.

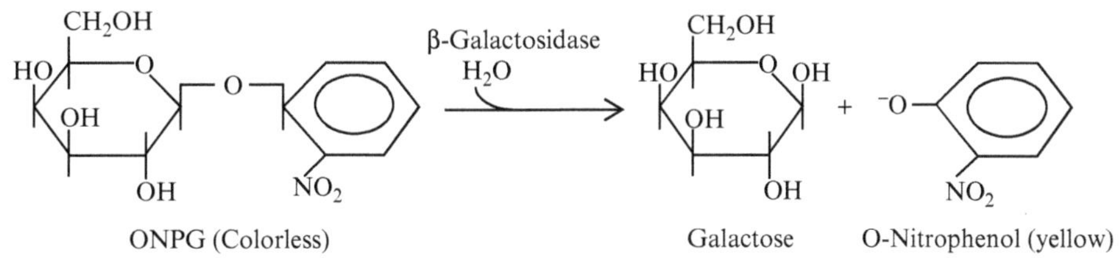

β-Galactosidase activity determined by permeabilized assay is expressed in the Miller unit in honor of Jeffrey Miller (1972). Uninduced *E. coli* cells have typically 1 Miller unit and IPTG-induced *E. coli* cells have 1000 to 1800 Miller units. It is important to note that the Miller activity unit differs from the specific activity unit (μmole/min/mg) obtained from protein extract-based β-galactosidase assay.

In this lab exercise, you will quantitatively measure β-galactosidase activity using both permeabilized cells and cell-free extracts of yeast (*S. cerevisiae*) YNN281 (pRY121) cultures grown under uninduced and induced conditions. To induce the expression of the *lacZ* gene in pRY121 plasmid, galactose is added to the culture medium because *E. coli lacZ* is under the control of the yeast *GAL* gene promoter (p*GAL*). The *E. coli*–yeast shuttle plasmid pRY121 map is shown on the right. The yeast YNN281 strain is able to transport but not to metabolize galactose so that the concentration of galactose remains unchanged for maximum *lacZ* expression.

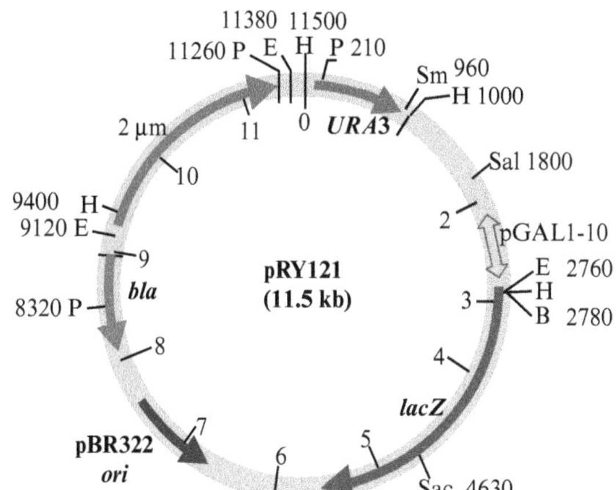

Methods in Biotechnology, First Edition. Seung-Beom Hong, M. Bazlur Rashid and Lory Z Santiago-Vázquez.

Companion Website: www.wiley.com\go\hong\Methodsinbiotechnology

Pre-lab assignment

(Typing and submission must be completed before lab work begins.)

1. What would be the compositions of the inducing and uninducing media for YNN281 (pRY121) to be used in step 1 of Parts I and II experiments based on the information described in Experiment 17 (Part B) and pRY121 plasmid?

2. Draw a flowchart of Part I experiment and Part II, A, B, and C experiments. You do not need to describe it in detail.

3. Calculate the total volumes of ONPG (4 mg/mL), Z buffer, and 1 M Na_2CO_3 solutions to be needed for Parts I and II experiments. Show your work.

4. Calculate the actual pre-harvest cell culture volume (mL) that corresponds to 10 μL cells used in the permeabilized assay reaction (Part I, step 8). Note that this information is required for the Post-Lab Assignment 1.

5. Calculate (a) an approximate total volume (X μL) of the liquid extract obtained from step 9 of Part II, A and (b) an approximate pre-harvest cell culture volume (mL) that corresponds to 1 μL liquid extract (initial harvest cell volume/X), assuming that there is neither a contribution of cells to volume nor a loss of volume during the sample processing. Note that this information is required for Post-Lab Assignment 1.

Materials and equipment

Part I:

- *Saccharomyces cervisae* YNN 281 strain genotype: MATa, *trp1Δ, his3Δ200, ura3-52, lys2-801(amber mutation), ade2-1(ochre mutation), gal, mal, CUP(r)*
- YNN281 (pRY121) cultures grown under inducing and uniducing conditions
- Z buffer (see Experiment 33 protocol), β-mercaptoethanol
- 5% sodium dodecyl sulfate (SDS)
- Triton X-100
- Toluene
- 4 mg/mL ONPG in Z buffer; store at –20 °C
- 1 M Na_2CO_3
- Spectronic 20, 28 °C water bath, vortexer

Part II:

- Breaking buffer (100 mM Tris-HCl (pH 8), 1 mM dithiothreitol (DTT), 20% glycerol, 1 mM EDTA, 10 mM $MgCl_2$)
- Acid-washed glass beads (0.45–0.53 mm)
- PMSF (phenylmethylsulfonyl fluoride; stock 100 mM): 17.4 mg in 1.0 mL of isopropanol; store at –20 °C
- Borosilicate glass tubes (13 × 100 mm)

Procedure

Part I. Permeabilized cell assay

1. Grow yeast cell (5 mL) to mid or late log phase (3 to 5 × 10^7 cells/mL: A_{600} = 0.7 to 1.0) in an appropriate medium under induced and uninduced conditions. (This will be done by TA.)

2. Harvest 5 mL of each culture in a disposable 15-mL culture tube by centrifuging at 2700 rpm for 10 min, discarding all the media, inverting, and blotting the tube to a paper towel.

3. Wash the cell pellet by suspending each pellet in 1.0 mL Z buffer, transferring the suspension to a 2.0-mL tube, spinning at maximum speed for 40 s, and aspirating off the supernatant.

4. Resuspend the cell pellet in a 500-μL Z buffer.

5. Take 100 μL of the cell suspension, add to 2.9 mL of Z buffer, and read at A_{600} in Spectronic 20 using 3.0 mL of Z buffer as a blank.

**Calculate the original sample of absorbance (A_{600} value × 30). If readings are above 1.0, dilute the sample further and read again to obtain an accurate absorbance reading.*

6. Add 1.5 μL of β-mercaptoethanol, 3 μL of toluene, 3 μL of 5% SDS and 2.4 μL of Triton X-100 to the remaining 400-μL cell suspension, and vortex at full speed several times for 1 min.

7. Shake-incubate the tube with its cap opened at 37 °C, at 150 rpm for 30 min.

8. Label and set up 8 reaction tubes for the two uninduced (tubes 1 and 2), four induced (tubes 3, 4, 5, and 6) and two blank (tubes 7 and 8) samples (total volume = 1 mL) as shown in the table:

	Uninduced		Induced				Blank	
Tube No.	1	2	3	4	5	6	7	8
Permeabilized cell (μL)	10	10	5	5	10	10	–	–
Z buffer (μL)	990	990	995	995	990	990	1000	1000

9. Pre-incubate the tubes at 28 °C for 5 min.

10. Add 0.2 mL of ONPG (4 mg/mL) to each tube, vortex briefly, incubate in a 28 °C water bath, and start timing.

11. When a pale medium-yellow color has developed, add 500 μL of 1 M Na_2CO_3 to stop the reaction, and record the time.

**Yellow color will develop within 10 min. If yellow color appears immediately, use less volume of cell suspension or diluted cell suspension in Z buffer (step 5). Ideally the length of reaction should be 30 min to 4 h since reaction times shorter or longer than this will affect the accuracy of the measurement.*

12. Centrifuge the tubes to pellet the cells at full speed for 5 min, transfer the supernatants to new glass tubes, combine the duplicate supernatant together in one tube, and determine A_{420} and A_{550} of the samples using 2 mL of Z buffer plus 1 mL of 1 M Na_2CO_3 as a blank in Spectronic 20.

**If the cell debris is well pelleted, an A_{550} reading of the supernatant is usually zero and thus is not necessary to read.*

13. Calculate the Miller units of each sample using the equation below:

$$\text{Miller units} = \frac{1000 \times [A_{420} - (1.75 \times A_{550})]}{T \times V \times A_{600}}$$

where

T = reaction time (min)
V = volume of cells used in assay (mL)
A_{600} = cell density at the start of assay
A_{550} = light scattering by cell debris
A_{420} = combination of absorbance by o-nitrophenol and light scattering by cell debris 1.75 × A_{550} is the correction for light scattering by cell debris at A_{420}

14. Calculate the ß-galactosidase activity of each sample using the equation below:

$$\text{Activity} = \frac{A_{420} - (1.75 \times A_{550})}{T \times V \times A_{600}} \times \frac{1\ \text{nmole}}{0.0045} \times 1.7$$

where T = reaction time (min)
V = volume of cells used in assay (mL)
A_{600} = cell density at the start of assay
A_{550} = light scattering by cell debris
A_{420} = combination of absorbance by *o*-nitrophenol and light scattering by cell debris
0.0045 = 1 µM (1 nmole/mL) ONP at A_{420}
1.7 = final total reaction volume of Miller assay (mL)
Activity = nmole/min/$A_{600}\cdot$ mL

15. Compare the Miller unit (step 13) with the activity (step 14). How much activity is equivalent to 1 Miller unit (activity/Miller unit)?

Part II. Cell-free protein extract assay

A. Yeast cell extract

1. Grow yeast cell (10 mL) to mid or late log phase (2 to 5 × 10^7 cells/mL: A_{600} = 0.7 to 1.0) in an appropriate medium under inducing and uninducing conditions. (This will be done by TA.)

2. Transfer 10 mL of culture into a disposable 15-mL culture tube.

3. Spin the 10 mL of culture volume in a clinical centrifuge at 2700 rpm for 10 min. Decant the culture supernatant, invert, and blot the tube to a paper towel.

4. Wash the cell pellet by adding 1.0 mL of sterile dH_2O, suspending by pipetting up and down, transferring the suspension to a 2.0-mL microfuge tube, and spinning at 8000 rpm for 40 s.

5. Discard the supernatant. Resuspend the cells in 250 µL of breaking buffer.

**The cells can now be frozen at –20 °C and assayed at a later date.*

**All of the subsequent steps can be performed in a 2.0 mL microfuge tube.*

6. Add 12.5 µL of PMSF stock solution and an equal volume (265 µL) of glass beads to the cell suspension.

**The level of beads is about just below the meniscus of the liquid.*

7. Vortex at maximum speed for 30 s and chill on ice for 30 s. Repeat the vortexing/cooling cycle 6 times more.

8. Add 400 µL of breaking buffer and mix well by inverting the tube 5 times. Let the beads settle. Withdraw the liquid extracts and transfer to new labeled and chilled microfuge tubes.

9. Clarify the extract by centrifuging at 13 000 rpm, at 4 °C for 10 min. Transfer the clarified supernatants to new labeled and chilled microfuge tubes. Measure and record the volume, and keep the clarified crude extracts on ice or store at –80 °C.

B. β-gal assay

10. Label and set up 8 reaction tubes for the induced, uninduced, and blank duplicate sample (total volume = 1 mL), as shown in the table.

	Uninduced				Induced		Blank	
Tube no.	1	2	3	4	5	6	7	8
Extract (µL)	10	10	5	5	10	10	–	–
Z buffer (µL)	990	990	995	995	990	990	990	990
Breaking buffer (µL)	–	–	–	–	–	–	10	10

11. Incubate all the tubes in a 28 °C water bath for 5 min.

12. Begin the reaction by adding 0.2 mL ONPG stock solution (4 mg/mL) and record the time of addition. Incubate at 28 °C until the mixture has developed a pale yellow color (10–30 min).

**Yellow color will develop within 10 min of incubation. You may use a lesser amount of extract. Ideally the length of reaction should be 30 min to 4 h since reaction times shorter or longer than this will affect the accuracy of the reaction.*

13. Stop the reaction by adding 0.5 mL of 1 M Na_2CO_3 stop solution and record the time of termination. Pool the duplicate sample together into one glass tube. Measure the absorbance at 420 nm in Spectronic 20.

C. Bradford assay

14. Add 0, 5, 10, 20, 30, and 50 µL of 1 mg/mL (1 µg/µL) BSA to breaking buffer in each of the duplicate test tubes so that a final volume is 50 µL in breaking buffer.

**Add a required volume of breaking buffer to each test tube first and then mix BSA directly into the buffer.*

15. Add 50 µL of the uninduced and induced cell extract to each test tube labeled (U and I).

16. Add 3.0 mL assay reagent (1/5 diluted), starting with the lowest concentration of BSA and cell extract samples last.

17. Mix well gently by shaking the test tubes. Incubate for 5 min at room temperature.

18. Read the absorbance at 595 nm in Spectronic 20.

**If reading is out of range or above 1.0, dilute protein sample to 1/2 in the breaking buffer for reading.*

19. Plot the standard curve (µg protein on the x axis versus A_{595} on the y axis) and use it to determine the protein concentration (µg of sample A_{595} ÷ 50 µL in this lab exercise = µg/µL = mg/mL).

**Typical extracts prepared contain 0.1 to 0.2 mg/mL of protein.*

20. Calculate the specific activity of each sample in the equation below:

$$\text{SA(nmole/min/mg)} = \frac{A_{420} \times 1.7}{0.0045 \times \text{Protein(mg/mL)} \times \text{Extract volume(mL)} \times \text{Time(min)}}$$

- A_{420} is the optical density of the product, *o*-nitrophenol, at 420 nm.
- The factor 1.7 corrects for the total reaction volume in mL (1.7 mL).
- The factor 0.0045 is the A_{420} of a 1 nmole/mL (1 µM) solution of *o*-nitrophenol.
- Protein concentration is expressed as mg/mL.
- Extract volume is the volume assayed in mL.

- Time is in minutes.
- Specific activity (SA) is expressed as nmole/minute/mg protein.

The above formula can also be expressed as:

$$\text{SA (nmole/min/mg)} = \frac{A_{420} \times 378}{\text{Protein (mg)} \times \text{Time (min)}}$$

To get the most accurate measure of activity, the absorbance at 420 nm (A_{420}) should range from 0.6 to 0.9. Readings as low as 0.1 and as high as 1.2 are acceptable. Tubes that have become as yellow as the tube of fresh LB broth would be sufficiently yellow.

The above formula can be expressed as a unit (μmole/min/mg):

$$\text{SA (μmole/min/mg)} = \frac{A_{420} \times 378 \times 10^{-3}}{\text{Protein (mg)} \times \text{Time (min)}}$$

Post-lab assignment

1. What are the advantages and disadvantages of enzyme assays on permeabilized cells versus cell extracts?
2. Why is galactose the major carbon source in the growth medium for this experiment?
3. Why does 1M Na_2CO_3 stop the activity of β-galactosidase?
4. How much specific activity is equivalent to 1 Miller unit? This question is about comparison of the enzyme activity of β-galactosidase obtained from permeabilized cells (Part I exercise) with that of cell extracts (Part II exercise).

The activity of permeabilized assay should be converted to a specific activity in order to compare:

$$\text{Activity} = \frac{A_{420} - (1.75 \times A_{550})}{T\text{ (min)} \times V\text{ (ml)} \times A_{600}} \times \frac{1\text{ nmole}}{0.0045} \times 1.7 \text{ (Part I, step 14)}$$

To do this, the V (mL) × A_{600} value must be replaced with the protein amount (mg). Perform the following calculations only for the induced samples.

(a) Calculate the actual pre-harvest cell culture volume (x mL) that corresponds to the 10 μL of cell suspension used in the permeabilized assay reaction (Pre-Lab Assignment 4).
(b) Calculate the protein amount (mg) per mL of cell culture from (i) 10 mL of harvested culture volume used for making a cell-free extract (Part II, A, step 3), (ii) volume Y μL of clarified extract (Part II, A, step 9: Pre-Lab Assignment 5), and (iii) Bradford assay of 50 μL cell-free protein extract (Part II, C, step 15).

Protein amount (mg) per mL cell culture

$$= \frac{\text{Protein con. (mg/mL)} \times 50\ \mu\text{L}}{(10\text{ mL}/Y\ \mu\text{L}) \times 50\ \mu\text{L}}$$

(c) Calculate the total amount (mg) of protein used in a permeabilized assay from (a) and (b) values (a × b).
(d) Calculate specific activity (nmole/min/mg) by replacing V (mL) × A_{600} value (Part I, step 14 formula) with the calculated (c) value.
(e) How much SA (nmole/min/mg) is equivalent to 1 Miller unit? Divide the (d) value by the Miller unit at step 13 of Part I.
(f) How much activity (nmole/min/A_{600}·mL) is equivalent to SA (nmole/min/mg)? Divide the (d) value by the activity at step 14 of Part I.

5. What is the purpose of adding PMSF to the Z buffer?
6. Calculate the specific activity unit (μmole/min/mg) of β-galactosidase using the following information and procedure. Show your work.

- The A_{420} value is 1.2 for a total of 3.4 mL of solution containing 0.5 mL of Na_2CO_3 stop solution, 0.2 mL of ONPG substrate, 0.99 mL of Z buffer, and 10 μL of induced sample.
- The reaction time is 30 min.
- The path length of the cuvette you used to measure the absorbance in Spectronic 20 is 1.2 cm.
- The molar extinction coefficient (ε) for *o*-nitrophenol is $2.13 \times 10^4\ M^{-1}\ cm^{-1}$ in the assay buffer condition.
- The concentration of the protein sample is 0.2 mg/mL from the protein assay you performed.

(a) Use Beer's formula ($A = \varepsilon cl$) to convert the absorbance (A_{420}) to a concentration (c) of *o*-nitrophenol in the solution you read for a 10-μL induced pooled sample.
(b) Calculate the rate of *o*-nitrophenol produced per min during the reaction. Use the final reaction volume to convert from *o*-nitrophenol concentration to the number of μmoles generated. Be careful with your unit conversion here.
(c) Calculate the protein amount (mg) used in the enzyme assay reaction.
(d) Calculate the specific activity unit (μmole/min/mg) from (b) and (c).

7. In the process of β-galactosidase protein purification, you wanted to determine the specific activity of your extract. You diluted the protein extract sample to 1/20 in the extraction buffer before you performed the Bradford protein assay. You then transferred 20 μL of the diluted sample to a tube containing 780 μL of H_2O, added 200 μL of dye to the tube, and measured the optical density of the mixture at A_{595}, which was 0.732. A_{595} values of 800 μL of BSA plus 200 μL of dye are shown below.

BSA standard (μg in 800 μL)	A_{595}
0.0	0.000
1.0	0.055
4.0	0.225
8.0	0.442
12.0	0.625
16.0	0.846
20.0	0.993

(a) Prepare the BSA protein standard curve.
(b) What is the concentration (mg/mL) of total protein in the diluted sample?
(c) What is the concentration (mg/mL) of total protein in the original sample?
(d) If the enzyme activity of the undiluted sample is 500 000 units/mL, what is the specific activity (μmole/min/mg) of β-galactosidase? 1 unit (U) of enzymatic activity = 1 μ mole/min.

Reference

Miller, J.H. (1972). *Experiments in Molecular Genetics*, pp. 352–355. Cold Spring Harbor Laboratory, Cold Spring Harbor, New York.

Further reading

Chow, C.-K. and Palecek, S.P. (2004). Enzyme encapusulation in permeabilized *Saccharomyces cerevisae* cells. *Biotechnology Progress*, **20**: 449–456.

Guarente, L. (1983). Yeast promoters and *lacZ* fusions designed to study expression of cloned genes in yeast. *Methods in Enzymology*, **101**: 181–191.

Rose, M. and Botstein, D. (1983). Construction and use of gene fusions *lacZ* (β-galactosidase) which are expressed in yeast. *Methods in Enzymology*, **101**: 167–180.

36 AMB 1 experiment 36: Gel filtration chromatography (GFC)

Purpose: This is up to you to write down.

Introduction

The two most common methods used to determine the molecular weights of proteins are gel filtration chromatography (GFC) and sodium dodecylsulfate (SDS)-polyacrylamide gel electrophoresis (PAGE). Both methods are based on the differential movement of proteins having different sizes through a porous gel matrix. Both will give an estimated or apparent molecular weight of a protein. A more accurate molecular weight (MW) of a protein can be deduced from its primary sequence by adding up the masses of its component amino acids. The estimated and actual molecular weight may considerably differ depending on the shape and physical properties of a protein.

GFC separates molecules according to their sizes, but GFC separation is formally based on hydrodynamic radius rather than size (MW); for similarly shaped molecules the hydrodynamic radius is proportional to the molecular weight. During chromatography, proteins in the mobile phase interact with the stationary phase. The general term to describe the substance that makes up the mobile and stationary phases is called solvent and gel matrix, respectively. In GFC, the matrix is composed of tiny beads (~0.1 mm diameter) made from highly cross-linked polymers. Commonly used gels are made of Sephadex (polymers of dextran), Sepharose or BioGel A (polymers of agarose). and BioGel P (polymers of polyacrylamide). The beads swell in the presence of solvent to form microscopic porous sponges. In our case, the solvent as a mobile phase will be a buffer. The hydrated beads as a stationary phase are packed into a column to form the matrix for chromatography. A porous nature of the hydrated beads, called gel matrix, acts as a molecular sieve and forms the basis of the separation method, also termed the size exclusion chromatography (SEC) or gel permeation chromatography. The gel beads should be chosen based on the fractionation range so that the expected molecular weight of the protein of interest falls approximately in the middle of the working range.

Small molecules can penetrate the pores and enter the beads, whereas larger molecules do not enter the beads and are therefore "excluded." A very large dye blue dextran molecule that cannot enter the gel is usually used to determine the "exclusion volume" or "void (outside the bead) volume (V_0)." The elution volume (V_e) is the volume where a molecule is eluted at the fraction with the highest peak of a molecule. To determine which fraction has the highest and lowest concentration of target protein, column fractions are read either by absorbance at 280 nm or by colorimetric assay. A linear inverse relationship between the size and V_e is obtained when the logarithms of the molecular weights of standard proteins are plotted against their respective V_e value. The molecular weight of the unknown protein can be estimated by extrapolating from the standard graph. Unlike the desalting column for group separation, the columns used to separate one macromolecule from another should have a height to diameter ratio of 20:1 to 100:1. It is recommended that blue dextran or dye mix is run alone, not mixed with the standard proteins or sample proteins, because the fraction of blue dextran or dye mix is broad and may overlap the protein peaks and dye may bind to some proteins.

GFC will give you the apparent molecular weight of the native protein. Proteins do not undergo any harsh denaturing treatments prior to chromatography. Therefore, they maintain their secondary, tertiary, and quaternary structures. In this lab exercise, you will learn how to run a gel filtration column to separate components from a mixture and how to collect fractions as they elute from the column. The column will be eluted by gravity flow and the eluent will be fractionated manually. Since the compounds in the mixture are colored, their migration, separation, and elution from the column should be easily visible.

Pre-lab assignment

(Typing and submission must be completed before lab work begins.)

1. What are the total bed volume (V_t), void volume (V_0), and elution volume (V_e) and how do you experimentally determine them in gel filtration chromatography?

2. Approximately how long does it take to pack the column and elute all the samples if the column is eluted at a flow rate of 0.5 mL/min? Show your work.

3. What are the total amounts of sample mixture and the individual amounts of catalase, cytochrome c, hemoglobin, and vitamin B-12 that will be loaded on to the column? Show your work.

4. What is the expected order of elution of catalase, cytochrome c, and vitamin B-12? Which molecule is expected to elute at V_0 and V_t? Explain why.

Methods in Biotechnology, First Edition. Seung-Beom Hong, M. Bazlur Rashid and Lory Z Santiago-Vázquez.

Companion Website: www.wiley.com\go\hong\Methodsinbiotechnology

Materials and equipment

- Blue dextran (20 mg/mL; non-ionic; MW 2 000 000; ε is 8970 $M^{-1}cm^{-1}$ at 620 nm)
- Loading sample: mixture of catalase (10 mg/mL), cytochrome c (5 mg/mL), hemoglobin (5 mg/mL) and vitamin B-12 (1 mg/mL) in 1× PBS

Molecule	MW (Da)	pI	Other properties
Catalase (bovine liver)	240 000 (tetramer)	5.4	Enzyme that produces water and oxygen from hydrogen peroxide (H_2O_2), weakly brownish; λ_{max} at A_{406} Sigma-Aldrich, Product No. C9322-5G
Cytochrome c (equine heart)	12 400	10.7	Transports electrons across cell membranes, contains a hem group, strong red color; λ_{max} at A_{410} (both oxidized and reduced form) Sigma-Aldrich, Product No. C2506-100MG
Hemoglobin (bovine)	64 500	6.8	Reddish brown; λ_{max} at A_{410} or A_{550} Sigma-Aldrich, Product No. H2500-1G
Vitamin B-12	1355		Pink; λ_{max} at A_{360}, A_{550} Sigma-Aldrich, Product No. V2876-1G

- Sephadex G-100 (Pharmacia): fractionation range of globular proteins MW 4–150 kDa
- Glass Econo-Column (1.0 × 30 cm; Bio-Rad Cat. No. 737-1032) in a ring stand with two clamps
- Equilibration and elution buffer: 1× PBS (137 mM NaCl, 2.7 mM KCl, 4.3 mM Na_2HPO_4, and 1.47 mM KH_2PO_4, pH 7.4): de-gas for 10 min prior to use
- 10 % hydrogen peroxide H_2O_2
- 50-mL graduated cylinder, waste beaker (100 mL), Pasteur pipette, Parafilm, 1.5-mL microcentrifuge tubes
- Spectronic 20 and cuvette, Kimwipe

Procedure

A. Column packing (24 mL) and equilibration (instructor will do the steps 1 to 6)

Good packing is essential to obtain good resolution. Ideally, the column should be packed at the highest pressure possible without deforming the beads. We will pack the column at room temperature under gravity flow. Therefore, make sure that the degassed buffer and hydrated Sephadex resin are at the same temperature before pouring into the column to avoid air bubbles forming in the column. The flow rate for column packing must be higher than the flow rate used for separation.

1. Mount the column vertically and tightly on a ring stand so that the top of the column is about 60 cm above your lab station. Make sure the column is level. Place a beaker under the column to catch the buffer that will flow through.

2. Close the column outlet, add 24 mL of water, mark a 24-mL position on the outside of the column, and empty the water.

3. Close the column outlet, add ~3 mL of buffer with no air bubbles trapped (DO NOT DROP the buffer directly but let the buffer slide down the column).

4. Swirl the settled Sephadex G50 suspension and place the uniform gel slurry on to buffer in the column using a 5-mL pipette.

5. Open the column outlet when there is ~1 cm of packed resin in the bottom of the column. The height of the gel bed will decrease. Continue to allow buffer to drain into a beaker until the height of the gel bed becomes constant. Add more buffer to the top to keep the buffer level above the level of the resin, if necessary.

**The column should be poured in one step if possible, using a homogeneous gel suspension of no more than two times the settled volume to get a 24 mL column volume.*

6. Run at least two bed volumes (48 mL) of degassed buffer through the column.

**This is to ensure equilibration and stabilization of the gel bed; students will run additional 10 mL as a starting point of this lab exercise.*

7. When the buffer level is about 1 cm above the top of the gel bed, close the outlet and adjust the column height so that the top of the column is about 30 cm above your lab station. Make sure the column is level.

**Always use a lower flow rate for separation than the flow rate used for column packing.*

8. Add more buffer up to the top of the column using a Pasteur pipette and open the outlet.

**During this time, determine the flow rate (mL/min). Discard the flow-through.*

**The column flow rate will depend on how well it is packed and the hydrostatic head.*

**The same buffer level should be maintained throughout the rest of the experiment to keep the constant hydrostatic operating pressure so that the flow rate is constant.*

**The hydrostatic operating pressure is created by the difference between the top of the buffer and the column outlet, which are in direct contact with the atmosphere.*

B. Loading blue dextran to determine V_0

9. Let the buffer run out until the buffer meniscus just touches the top of the column matrix, and immediately close the column outlet using a clamp. Watch carefully the surface of the top. Do not let the column dry.

10. Put a test tube or a small flask underneath the column outlet to start collecting the flow-through buffer to measure the void volume (V_0).

11. Use a *Pasteur pipette* to apply blue dextran (125 μL) by placing the tip on the inside of the column about 1 to 2 mm above the surface of the gel bed and letting it *slowly* flow down. Do this by circling the pipette in order to slowly and carefully lay the sample on to the bed surface. As soon as sample loading is finished, open the column outlet. The dye component will very quickly enter the gel.

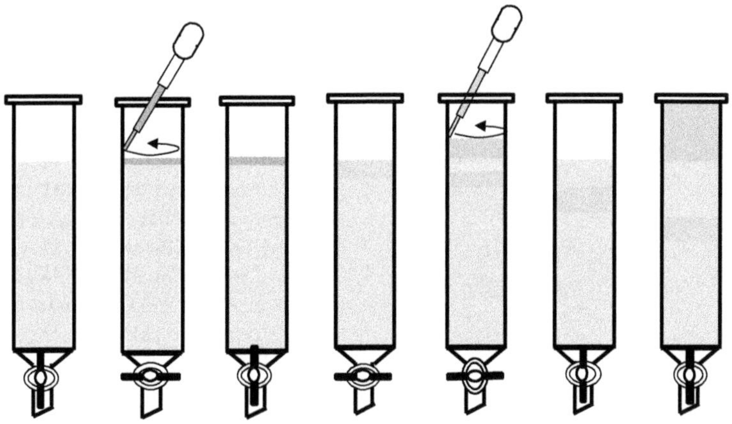

**Do not drop the sample directly on to the bed. Do not use a micropipettor to apply the sample.*

12. Let the sample enter the packed column bed completely until the meniscus of the sample touches the column top. Close the column outlet.

**Do not let the column dry; watch the top surface.*

13. Wash the inside wall of the column by letting *slowly* about two sample volumes (250 µL) of buffer flow down, open the column outlet, and let the buffer enter the column bed completely.

**Do not allow the buffer to fall below the surface of the gel bed or your column will go dry and be useless. Therefore, you must pay close attention to the level of the buffer as it approaches the bed surface.*

14. Once the buffer enters the column bed completely, close the column outlet. Add more buffer to the gel bed surface by circling a Pasteur pipette and allowing it to flow slowly down the side of the column tube, and add up to the top of the column.

**Do not disturb the flat bed surface.*

15. Open the column outlet to let the buffer enter the packed column. Keep the buffer level maintained to the top of the column during fractionation for a constant pressure flow.

**The shape of the blue dextran band as it elutes through the column can also be used to assess the column packing and resolution: you should see a discrete horizontal band travelling through the column.*

**Note that the more the sample penetrates the gel pores, the more it spreads out on the column.*

C. Collecting fractions

16. Collect the eluent until the blue dextran reaches the lower part (~1 cm above the bottom) of the column in a glass test tube. Measure and record the total volume (X mL) of the eluent.

17. Start collecting 0.5-mL fractions in numbered microcentrifuge tubes. (Label the number and mark with a Sharpie pen the location of 0.5 mL before you start collecting; this enables you to collect 0.5-mL fractions with a reasonable accuracy.)

18. Replenish the top of the column reservoir periodically with buffer during fractionation to keep the level of buffer constant. This will ensure that the flow rate through the column is constant.

19. Determine the void volume (V_0) by measuring the volume collected up until the blue dextran *begins to elute* from the column ($V_0 = X +$ pooled 0.5-mL fractions).

**V_0 is usually 30 to 35% of total column bed volume (V_t).*

D. Loading the sample mixture

20. As soon as the blue dextran elutes, close the column outlet and carefully remove the elution buffer from above the resin using a Pasteur pipette. Drain the residual buffer to the bed surface and close the outlet.

21. Using the Pasteur pipette, very slowly layer the sample mixture (125 µL) on top of the resin as described in the steps 11 to 15 of Part B.

**Do not disturb the flat bed surface.*

22. Collect 1.5-mL fractions in 4 microcentrifuge tubes and then start to collect 0.5-mL fractions in numbered microcentrifuge tubes. Once the slowest moving band has eluted, you may stop collecting fractions.

23. Examine each of the column fractions and determine which fractions contain the highest concentrations of the colored substances. This is most easily accomplished by holding the fractions up against a sheet of white paper.

E. Catalase activity test

24. Withdraw 20 µL from each fraction and place the drops on a numbered clean Parafilm sheet on the flat surface.

25. Add 10 μL of 10 % H_2O_2 to each of all the drops to be tested. Observe bubbles (oxygen gas formation) and record the activity, that is, amounts of bubbles produced in each fraction: very high ++++; high +++; medium ++; low +; very low +/–; none –.

F. Absorbance reading

26. Place 0.5 mL of each fraction into a Spectronic 20 cuvette containing 2.5 mL H_2O (1/6 dilution) and read at 410 nm using H_2O as a blank.

**The peak of vitamin B12 will not be shown unless the sample is read at λ_{max} of A_{360} or A_{550}. The actual A_{410} value will be the reading value × 6 (DF).*

27. Determine the elution volume (V_e): the total collection volume starting from sample application to the highest peak (absorbance) fraction of elution.

28. Draw the elution profiles of absorbance (y axis) versus elution volume (x axis).

**Indicate the positions of V_0 (void volume), V_e (elution volume) of blue dextran, catalase, cytochrome c, hemoglobin, and vitamin B-12 in the elution profile.*

**Indicate the location of catalase activity (very high to low) on the graph.*

29. Plot the V_e/V_0 (x axis) versus $\log_{10}$ MW (y axis) for hemoglobin and cytochrome c. (*Note:* use MW if you use semi-log graph paper; use $\log_{10}$MW for the Excel program.)

**Do not include the V_e/V_0 for blue dextran, catalase, and vitamin B-12 because they elute outside of the fractionation range.*

Post-lab assignment

1. Do you expect that the concentrations of the fractions of the highest peaks (blue dextran, catalase, cytochrome c, vitamin B-12) to be lower or higher as compared with those of each protein in the starting loading sample? Explain your choice.

2. List at least 6 factors that influence the resolution of size exclusion chromatography and briefly explain how each factor affects it.

3. List at least 4 applications of gel filtration chromatography other than analysis of mixtures of molecules of different molecular weights.

4. A protein was purified to homogeneity. Determination of the apparent molecular weight by gel filtration chromatography yields 60 kDa. Chromatography in the presence of 6 M urea yields a 30 kDa species. However, the chromatography in the presence of 6 M urea and 10 mM β-mercaptoethanol resulted in a single peak of 15 kDa. Describe the structure of the protein.

5. A calibrated GFC column having a fractionation range of 5000 to 250 000 Da gives a V_0 of 40 mL and V_t of 135 mL. Molecular weight standards are applied to the column and their V_e are measured and shown in the table below. A purified protein has a V_e of 60 mL. Prepare the standard curve ($\log_{10}$ MW versus V_e/V_0) and determine the apparent MW of the purified protein.

Protein	MW (Da)	V_e (mL)
A	12500	82.50
B	15300	81.25
C	22000	77.50
D	25100	75.00
E	28700	70.00
F	33300	68.75
G	45200	67.50
H	74500	55.00
I	96800	52.50
J	120000	50.00
K	244000	41.25

Further reading

Gel Filtration Principles and Methods (2010). GE Healthcare Handbook 18-1022-18AK, 09/2010.

Mallik, B., Chakravarti, B., and Chakravarti, D.N. (2012). Unit 6.1. Overview of chromatography. In *Current Protocols and Essential Laboratory Techniques* (Edited by S.R. Gallagher and E.A. Wiley). John Wiley & Sons, Inc. ISBN 978-0-470-08993-4.

Robert, K.S. (1982). *Protein Purification.* Springer-Verlag New York, Inc. ISBN 978-0-470-94241-3.

37 AMB 1 experiment 37: Ion exchange chromatography (IEC)

Purpose: This is up to you to write down.

Background

Proteins bind to ion exchangers by electrostatic interactions between the surface charges of proteins and the clustered charged groups on the exchanger adsorbent. The strength of binding depends on the buffer pH. If the pH is raised, deprotonation occurs and the protein will become more negative. As the pH is lowered, protein will become more protonated and positive. If the isoelectric point (pI) of the protein of interest is known, buffer pH should be adjusted to at least 1 unit away from the pI to ensure that it is fully charged. There are two types of ion exchanger, an anion exchanger carrying positively charged groups (e.g., diethylaminoethyl, DEAE) that exchanges negative ions and a cation exchanger carrying negatively charged groups (e.g., carboxymethyl, CM) that exchanges positive ions. DEAE and CM groups are covalently linked to cellulose, agarose, or dextran as an inert gel matrix. The choice of ion exchanger depends on the stability of the protein of interest in a pH range; when a protein is stable at a pH above pI, an anion exchanger is used, but if it is stable at a pH below pI, a cation exchanger is required. DEAE (pK_a = 9) and CM (pK_a = 5) cellulose are examples of weak ion exchangers, which are ionized over a limited pH range. DEAE-cellulose begins to lose its positive charge above pH 9, while CM-cellulose begins to lose its negative charge below about pH 5. Thus, the effective starting buffer pH range for DEAE-cellulose or CM-cellulose is only about pH 5 to 9.

The charges imposed by ionic groups attached to a gel matrix are initially neutralized by counter ions (ions bound to an exchanger) that are present in the buffer during the equilibration step. When a sample is added to the column, an exchange with the weakly bound counter ions takes place and the charged molecules bind to the solid support. It is recommended that the charged form of buffer should be the same as that of the ion exchanger to avoid possible changes in the strengths of ionic interaction between proteins and exchanger resins. For instance, Tris and phosphate buffers are used for DEAE-cellulose and phosphocellulose, respectively. In order to ensure adsorption of weakly binding proteins, the buffering power should be maximized at the least possible ionic strength. For this, the pK_a of the buffer should lie within ± 0.5 unit of the buffer pH being used, and one of the buffering species should be uncharged. Tris-Cl buffer (pH 8.0) meets these rules for DEAE-cellulose equilibrated at pH 8.0; pK_a of Tris is 8.1 at 25 °C, the counter ion that is associated with DEAE is Cl^-, and the buffering species are Tris-H^+ (non-interactive) and Tris (neutral).

The protein samples to be applied should be prepared in the same buffer used to pre-equilibrate the ion exchange chromatography (IEC) column. This can be done by either dialysis or desalting gel filtration using an equilibration buffer. Inclusion of EDTA in the buffer for the anion exchanger is not recommended because polyanionic EDTA binds to an anion exchanger (e.g., DEAE) and competes with proteins for binding sites.

Salt gradients of KCl or NaCl are most commonly used to elute bound proteins from the ion exchanger. The increasing concentrations of counter ions of salt displace the proteins by occupying the charged sites and blocking reattachment by proteins. To change the buffer pH also alters the ionization of the protein's charged groups and results in disruption of the interaction with the ion exchangers. As salt concentration increases and/or pH changes, electrostatic forces between the protein and adsorbent are weakened, and the protein begins to move down.

Unlike GFC, a column with a higher ratio of diameter to height yields better resolution. Any sample volume may be applied as long as the sample solution does not exceed the ionic strength of the starting buffer and binding capacity of resin. In this experiment you will pack a simple ion exchange column using a Poly-Prep chromatography column and pre-equilibrated DEAE-cellulose resin, load a protein mix consisting of catalase, hemoglobin, and cytochrome c, and separate by stepwise ascending salt elution. Once the loaded protein sample enters the bed of resin, the column should be washed for two column volumes so that all unbound protein has passed through the column before elution by increasing the ionic strength of the elution buffer.

Pre-lab assignment

(Typing and submission must be completed before lab work begins.)

1. What is the expected order of elution of catalase, hemoglobin, and cytochrome c? Explain why it is.

2. What happens if the equilibration and elution buffers used are TE (pH 8.0) instead of Tris-Cl (pH 8.0)? Is the expected outcome of elution the same or not? Explain your choice.

Methods in Biotechnology, First Edition. Seung-Beom Hong, M. Bazlur Rashid and Lory Z Santiago-Vázquez.

Companion Website: www.wiley.com\go\hong\Methodsinbiotechnology

Materials and equipment

- Buffer A (autoclaved): 20 mM Tris-HCl (pH 8.0)
- Buffer B (autoclaved): 20 mM Tris-HCl (pH 8.0), 2.0 M NaCl
- Loading sample: a mixture of catalase (5 mg/mL) and hemoglobin (2 mg/mL), cytochrome c (1 mg/mL) in buffer A

Molecule	MW (Da)	pI	Other properties
Catalase	240 000	5.4	Enzyme that produces water and oxygen from hydrogen peroxide (H_2O_2), weakly brownish; λ_{max} at A_{406} Sigma-Aldrich, Product No. C9322-5G
Hemoglobin (bovine)	64 500	6.8	Reddish brown; λ_{max} at A_{410} or A_{550} Sigma-Aldrich, Product No. H2500-1G
Cytochrome c	12 400	10.7	Transports electrons across cell membranes, contains a hem group, strong red color; λ_{max} at A_{410} Sigma-Aldrich, Product No. C2506-100MG

- DEAE-Sephacel (GE Healthcare, Product No. 17-05000-01) equilibrated with buffer A
- Pasteur pipette, waste beaker, 2.0-mL and 1.5-mL microcentrifuge tubes, Parafilm
- Poly-Prep Chromatography Columns (Bio-Rad, Cat. No. 731-1550)

Procedure

1. Prepare the following salt buffers in 2.0-mL tubes by mixing buffers A and B, as shown below.

Buffer #	Salt concentration (mM)	Buffer A (μL)	Buffer B (μL)	Total volume (mL)
1	0	1000	–	2
2	50	975 × 2	25 × 2	2
3	100	950 × 2	50 × 2	2
4	150	925 × 2	75 × 2	2
5	200	900 × 2	100 × 2	2
6	300	850 × 2	150 × 2	2
7	400	800 × 2	200 × 2	2
8	500	750 × 2	250 × 2	2

2. Set a Polyprep column vertically on a stand: secure with a clamp.

3. Close the bottom outlet of the column with a cap.

4. Fill the column with 0.1 mL buffer A to remove air pockets.

5. Carefully add ~2 mL of pre-equilibrated ion exchange resin suspension to the column. Avoid air bubbles by allowing it to flow gently along the side of the column.

6. Place a small beaker underneath the column and remove the bottom cap.

7. Wait for the resin to pack. Add more resin if necessary to pack the column to a 1-mL position.

8. Add 2 mL of buffer A and let it enter to stabilize the 1-mL packed resin.

9. Place the column outlet over a rack with 32 of 1.5-mL tubes and apply 125 μL of your sample. Collect 0.5-mL fractions.

10. Once the sample touches the top surface of the column bed, wash the wall of the column by slowly letting 2 mL of buffer #1 flow down the column and continue to collect 0.5-mL fractions.

11. Apply 2 mL each of the next step elution buffers to the column, starting with buffer #2, and continue to collect 0.5-mL fractions.

12. Move to the next tube before applying the next elution buffer until elution with buffer #8. Collect all of the 0.5-mL fractions in 32 separate 1.5-mL tubes.

13. Analyze fractions for color and catalase activity as in Part E of Experiment 36 (very strong ++++; strong +++; medium ++; weak +; very weak +/–; none –).

14. Place 0.5 mL of each fraction into a Spectronic 20 cuvette containing 2.5 mL of H_2O (1/6 dilution) and read at 410 nm using H_2O as a blank.

**The actual A_{410} value will be the reading value × 6 (DF).*

15. Draw the elution profiles of absorbance (*y* axis) versus elution volume (*x* axis). The elution volume must begin with collecting fractions when the samples are loaded on to the column.

**Indicate the location of catalase activity (very high to low) on the graph.*

**Rate each fraction for the presence of color (0 = no color, + = weak color, ++ = medium color, +++ = strong color).*

Post-lab assignment

1. Explain your choice of answers to the following questions.

(a) If you had used an elution buffer at pH 5 for the DEAE ion exchange experiment, how would this influence the elution pattern of the three proteins you used?

(b) If you had used an elution buffer at pH 10 for the DEAE ion exchange experiment, how would this influence the elution pattern of the three proteins you used?

(c) If you had used an elution buffer at pH 7 for the DEAE ion exchange experiment, how would this influence the elution pattern of the three proteins you used?

2. List at least 4 factors that influence the resolution of ion exchange chromatography and briefly explain how each factor affects it.

3. What is band broadening? What factors can be controlled to minimize band broadening?

Further reading

Ion Exchange Chromatography and Chromatofocusing Principles and Methods (2010). GE Healthcare Handbook 11-0004-21AB, 04/2010.

Mallik, B., Chakravarti, B., and Chakravarti, D.N. (2012). Unit 6.1. Overview of chromatography. In *Current Protocols and Essential Laboratory Techniques* (Edited by S.R. Gallagher and E.A. Wiley). John Wiley & Sons, Inc. ISBN 978-0-470-08993-4.

Robert, K.S. (1982). *Protein Purification*. Springer-Verlag New York, Inc. ISBN 978-0-470-94241-3.

Advanced Methods in Biotechnology (AMB) 2 Laboratory Exercises

38 AMB 2 experiment 38: *E. coli* culture media preparation

Purpose: This is up to you to write down.

Introduction

Bacterial culturing is the most commonly used laboratory task in biotechnology. Not all bacteria grow optimally on the same kind of culture medium and condition. However, they are well established for *E. coli*, the most widely used bacterium for gene cloning and protein expression in molecular biology. The powdered mixture of necessary nutrients, either as a rich or minimal medium, can be used for a liquid medium (broth) or a solid agar medium. Agar is a natural gelatinous polysaccharide produced by marine algae seaweed and is the most commonly used solidifying agent added to media to a final concentration of 1.5 to 2.0%.

Growth media are often classified as defined (or synthetic) and complex (or rich) media according to composition, as selective, differential screening, and general media according to usage, and as liquid (or broth), solid, and semi-solid media according to fluidity. Instead of weighing individual components, you may use commercially available ready-to-use dehydrated powder to prepare growth media. Before inoculation with bacterial cells, culture media and all other materials directly coming into contact with cells must be sterile. Sterilization is usually accomplished by the use of pressurized steam for heat-stable compounds or 0.2 or 0.45 μm filtration for heat-labile compounds.

For steam sterilization, an autoclave must be operated according to the manufacturer's instructions. There are two basic autoclaving cycles, "fast exhaust" or "gravity" and "slow exhaust" or "liquid." The gravity cycle is used only when you autoclave empty glassware or dry materials. Under the gravity mode, the steam pressure is dumped all at once without cool-down time as soon as the sterilization cycle is finished. Under the liquid cycle, the steam pressure is released slowly in order to prevent media from boiling over. It is important to remember that when you use a screw-cap bottle to autoclave liquid medium, you must loosen the cap to prevent the bottle from shattering caused by internal pressure buildup during autoclave. You must fill a vessel less than half full with liquid to be autoclaved. When autoclaving is done, open the door carefully while wearing autoclave gloves, remove and place a flask of hot agar solution in the water bath (set at about 50 to 55 °C). This will allow the agar to cool, but not to go less than 45 °C, at which it will rapidly solidify. Let it cool to 50 to 55 °C before you add antibiotic solution and pour into Petri dishes. A flask at this temperature can be held in bare hands. As soon as a medium flask is taken out of the water bath, an appropriate amount of antibiotic should be added and swirled to mix well. Addition of the right amount of antibiotics is important; adding too little will increase background colonies, making it difficult to select transformant colonies, but adding too much will result in no cell growth. For working concentrations of some commonly used antibiotics, refer to http://www.addgene.org/mol_bio_reference/antibiotics/.

Air bubbles on the surface must be removed right after the agar medium is poured into plates, and the lids should be left open in a laminar flow hood to prevent evaporated water from condensing on the inside of the lid. Since the efficacy of many antibiotics does not last long at room temperature, plates are usually stored at 4 °C for long-term storage. If you autoclave broth in a screw-cap bottle, you must cool down to below 40 °C before you tighten the cap as the vacuum resulting from steam cooling in the bottle may cause the cap or bottle to shatter. Also you must not put autoclaved hot glassware on to a cold workbench since this sudden temperature shift will cause the glass bottom to crack, leaking its contents. It is recommended to watch the YouTube video regarding how to make an agar plate (http://www.youtube.com/watch?v=OljTdYH_Wtg).

From this experiment you will gain the basic laboratory skills in preparing bacterial media for selection and culturing. The media you are going to make will be used for the next three weeks in lab exercises, *E. coli* transformation for site-directed mutagenesis (Experiment 39) and for protein expression (Experiment 40) and crude cell lysate preparation for affinity chromatography (Experiment 41). During agar media preparation and any subsequent handling of the bacterial cultures, unwanted microbes must be excluded by employing aseptic techniques. It is important to remember that the successful outcomes of your intended experiments are dependent on your aseptic skill in nearly all lab exercises of this course.

Pre-lab assignment

(Typing and submission must be completed before lab work begins.)

1. Calculate the final percentages of tryptone, yeast extract, and glycerol in TB medium.
2. Calculate the final molarities of monobasic and dibasic potassium phosphates in TB medium.

Methods in Biotechnology, First Edition. Seung-Beom Hong, M. Bazlur Rashid and Lory Z Santiago-Vázquez.

Companion Website: www.wiley.com\go\hong\Methodsinbiotechnology

3. Calculate the final percentage and molarity of NaCl in LB medium.
4. Calculate each microliter (μL) volume of stocks of ampicillin, X-β-Gal, IPTG, chloramphenicol, and L-arabinose needed for the Parts D, E, and F experiments.
5. Describe your aseptic techniques for media preparation in today's experiment.

Materials and equipment

- Tryptone, yeast extract, glycerol, NaCl, KH_2PO_4, K_2HPO_4
- Bacto-agar
- Ampicillin (100 mg/mL in dH_2O; filter-sterilized, then store at −20 °C)
- Chloramphenicol (30 mg/mL in 100% ethanol; no filter sterilization, store at −20 °C)
- X-β-Gal (50 mg/mL in dimethylformamide or DMSO; no filter-sterilization; wrapped in aluminum foil to prevent damage by light and should be stored at −20 °C)
- IPTG (isopropyl-β-D thiogalactopyranoside, 0.1 M: 0.0238 g in 1 mL of water; filter-sterilized, then store at −20 °C)
- L-arabinose (1.5 M in dH_2O; filter-sterilize, then store at 4 °C)
- Petri plates (100 × 15 mm)
- Balance, graduated cylinder
- 50 to 60 °C water bath
- Autoclave
- Laminar flow hood cabinet

Procedure

A. Terrific broth (TB)

Note: you may use commercial TB powder that contains all components except glycerol.

1. Prepare the following in a 250-mL Erlenmeyer flask or screw-cap bottle (only one group). Calculate how much you would need for 200 mL.

Component*	Per liter	Per 200 mL
Tryptone	12 g	
Yeast extract	24 g	
Glycerol	4 mL	
KH_2PO_4 monobasic	2.31 g	
K_2HPO_4 dibasic	12.54 g	

*If ready-to-use dehydrated TB powder is used, follow the manufacturer's instruction.

2. Weigh tryptone, yeast extract, and glycerol per 200 mL, transfer to a flask containing 100 mL dH_2O, dissolve completely using a magnetic stirring bar, and bring up to 180 mL.
3. Autoclave for 15 min at 121 °C, 15 psi on a liquid cycle, and cool to room temperature.
4. Weigh monobasic and dibasic potassium phosphate per 200 mL, dissolve in dH_2O, and bring up to a final volume of 20 mL with dH_2O.
5. Filter-sterilize the potassium salt solution.
6. Add the filtered potassium salt solutions (20 mL) to the autoclaved medium (180 mL) in the laminar flow hood.
7. Store TB broth at room temperature.

B. Luria bertani (LB) broth

1. Prepare 200 mL of LB Broth in a 500-mL Erlenmeyer flask or screw-cap bottle (only one group). Below is the recipe for 1 L of media. Calculate how much you would need for 100 mL (the recipe below is for 1 L).

Chemical*	Per liter	Per 200 mL
Tryptone	10 g	
Yeast extract	5 g	
NaCl	10 g	

*If ready-to-use dehydrated LB powder is used, follow the manufacturer's instruction.

2. Add components to the flask containing ~180 mL of dH_2O.
3. Place the foil and autoclave tape over the mouth of the flask and label.

**Do not write on flask but use the tape for labeling.*

4. Autoclave for 15 min at 121 °C, 15 psi liquid cycle (slow exhaust mode).
5. Store the LB broth at room temperature.

**If the broth is in a screw-cap bottle, cool down to below 40 °C before tightening the cap.*

C. LB/Amp (ampicillin)/X-β-gal agar plates

1. Prepare 100 mL of LB agar by adding the components to a 250-mL flask containing ~96 mL of dH_2O (each group).

Chemical	Per liter	Per 100 mL
Bacto-tryptone	10 g	
Bacto-yeast extract	5 g	
NaCl	10 g	
Bacto-agar	15 g	

2. Autoclave for 15 min at 15 psi liquid cycle (slow exhaust mode).
3. During autoclave, label 4 Petri plates on the edge of the bottom as LB/Amp/X-β-Gal.
4. After autoclaving, place the agar media flasks in a 56 to 60 °C water bath for 20 min.
5. Add ampicillin and X-β-Gal to a final concentration of 100 μg/mL and 50 μg/mL, respectively. *Calculation is needed in the pre-lab notebook.*
6. Swirl the flask gently.

**A dense agar solution settles down in the bottom of the flask. However, do not shake vigorously.*

**Remove your flask from the water bath only when you are ready to pour your plates.*

**Wipe out the drenched water outside the flask with a clean paper towel before pouring.*

7. Wrap a clean paper towel around the neck of the flask and hold it to pour into Petri plates in the lamina flow hood; remove any bubbles on the agar surface by aspirating off with a sterile Pasteur pipette or by pushing air bubbles to the edge of the plate.

**As soon as the liquid medium touches the circled edge of the plate, stop pouring.*

**The paper towel around the flask neck is to catch dribbles that flow down the outside of the flask.*

**Immediately after pouring, rinse the empty flask with tap water.*

8. Allow the plates to solidify with the lids open.

**The agar surface should be smooth and moist, but without excessive moisture.*

9. Put your stack of four plates into a Petri plate plastic bag.
10. Close the cover lid, place the agar plate upside down into a vinyl bag, and store at 4 °C.

**You may incubate all the plates upside down at 37 °C overnight and then store at 4 °C to check for any contamination.*

D. LB/Amp (ampicillin)/X-β-gal/IPTG agar plates

1. Prepare 100 mL of LB agar in a 250-mL flask as before (each group).
2. Autoclave for 15 min at 15 psi liquid cycle (slow exhaust mode).
3. During autoclave, label four Petri plates on the edge of the bottom as LB/Amp/X-β-Gal/IPTG.
4. After autoclaving, place the agar media flasks in a 56 to 60 °C water bath for 20 min.
5. Add ampicillin, X-β-Gal, and IPTG to a final concentration of 100 µg/mL, 50 µg/mL, and 0.1 mM, respectively. *Calculation is needed in the pre-lab notebook.*
6. Follow the same procedures of steps 6 to 10 of Part C.

E. LB/Amp (ampicillin)/Cm (chloramphenicol)/IPTG agar plates

1. Prepare 200 mL of LB agar in a 500-mL flask as before (each group).
2. Autoclave for 15 min at 15 psi liquid cycle (slow exhaust mode).
3. During autoclave, label eight Petri plates on the edge of the bottom as LB/Amp/Cm/IPTG.
4. After autoclaving, place the agar media flasks in a 56 to 60 °C water bath for 20 min.
5. Add ampicillin, chloramphenicol, and IPTG to a final concentration of 100 µg/mL, 30 µg/mL, 50 µg/mL, and 0.1 mM, respectively. *Calculation is needed in the pre-lab notebook.*
6. Follow the same procedures of steps 6 to 10 of Part C.

F. LB/Amp (ampicillin)/Cm (Chloramphenicol)/arabinose agar plates

1. Prepare 50 mL of LB agar in a 100-mL flask as before (each group).
2. Autoclave for 15 min at 15 psi liquid cycle (slow exhaust mode).
3. During autoclave, label two Petri plates on the edge of the bottom as LB/Amp/Cm/Arabinose.
4. After autoclaving, place the agar media flasks in a 56 to 60 °C water bath for 20 min.
5. Add ampicillin, chloramphenicol, and L-arabinose to a final concentration of 100 µg/mL, 30 µg/mL, and 2×10^{-3} M, respectively. *Calculation is needed in the pre-lab notebook.*
6. Follow the same procedures of steps 6 to 10 of Part C.

Post-lab assignment

1. What is the purpose of phosphate and glycerol components in the TB broth recipe?
2. Why is TB used instead of LB broth to grow cultures?
3. Why is it necessary to sterilize liquid media under a steam pressurized condition as opposed to a non-steam hot oven?
4. What is the standard operating practice (SOP) for (a) preparing media for autoclave and (b) pouring media into plates in the lab protocol used?
5. What is the most common way to contaminate the stock liquid broth media besides the use of a non-sterile pipette tip and pipettor?

Further reading

Brent, R. and Lech, K. (2000). Unit 1.1. Media preparation and bacteriological tools. In *Current Protocols in Molecular Biology* (Edited by F.M. Ausubel). John Wiley & Sons, Inc. ISBN 0-471-50338-X.

Tartoff, K.D. and Hobbs, C.A. (1987). Improved media for growing plasmid and cosmid clones. *Bethesda Research Laboratory Focus*, **9**: 12.

39 AMB 2 experiment 39: Site-directed mutagenesis

Purpose: This is up to you to write down.

Introduction

In vitro site-specific mutagenesis is an invaluable technique for studying the protein structure–function relationship, gene expression, protein engineering, and vector modification. Because any amino acid in a protein can be selectively replaced with another amino acid, the resulting mutant protein will provide valuable information on the role of the amino acid that had been replaced in terms of stability, catalytic activity, substrate specificity, and affinity. There are several protocols and commercial kits for altering the nucleotide sequence of cloned DNA fragments. One of the most popular methods is a PCR-based site-directed technique that allows site-specific base substitutions, deletion, and insertion to be made in virtually any plasmid DNA. In this procedure, oligonucleotides containing the desired changes are used as amplification primers. Depending on the primer design, the resulting products are either a defined segment of linear DNA carrying a mutation or an entire plasmid DNA containing a mutated cloned gene. The major advantages of this PCR-based mutagenesis system are technical simplicity and high frequency of incorporating specific mutations. The main disadvantage is the introduction of unwanted mutations due to the infidelity of thermostable polymerases. However, the chance of a polymerase-induced error can be minimized by using a thermostable enzyme with high fidelity and processivity and reducing the number of PCR cycles.

Site-directed mutagenesis can be performed using overlapping or non-overlapping mutagenic primers. A major advantage of non-overlapping primer-based mutagenesis over overlapping primer-based mutagenesis is that any size of insertion and deletion can be introduced into a desired site. However, unlike overlapping primer-based mutagenesis, the 5′ end of forward and reverse primers must be phosphorylated and then non-overlapped PCR products must be circularized by blunt-end ligation before transformation. After PCR, *Dpn*I enzyme, which cleaves methylated GATC, is often added to remove the plasmid DNA templates that are methylated when they are replicated in *E. coli*. The removal of template DNA is important for reducing background transformant colonies, especially when large amounts of template DNA are used. Although several site-directed mutagenesis kits are commercially available, most of the approaches are designed to introduce one mutation in a one-step reaction. However, the Gibson assembly method enables several mutations to be introduced simultaneously into a very large DNA in a one-step reaction. For more detailed information on this multimutagenesis, refer to http://www.dddmag.com/articles/2013/05/site-directed-mutagenesis-gibson-assembly.

In this lab exercise, you will use a non-overlapping divergent primer-based mutagenesis protocol in order to introduce a single base change into pUC19WHITE plasmid (2686 bp) that contains a stop codon (TAA) at the 8th amino acid position in the *lacZα* gene encoding alpha (α) fragment of β-galactosidase enzyme. The successful mutagenesis followed by *E. coli* transformation leads to the formation of blue colonies on LB-ampicillin agar plates containing X-β-Gal. The "blue and white screening" of pUC plasmid transformants is based on the alpha complementation of β-galactosidase fragments. DNA sequencing should be carried out for the final confirmation of mutagenesis, though this step is not performed in this exercise. Plasmids up to 10 kb in length can be successfully mutagenized using Phusion polymerase enzyme that has a high rate of processivity (15 to 30 s/kb extension rate) and fidelity (5×10^{-7} mutation/bp/doubling error rate).

Pre-lab assignment

(Typing and submission must be completed before lab work begins.)

1. What is the Gibson assembly method? Describe how it works.
2. Calculate the mole of pUCWHITE plasmid template used in Part A. pUC19 is 2686 bp and a molecular weight of 1 bp dsDNA is 650 g/mole.
3. Why must the 5′ end of the primers used for this experiment be phosphorylated?
4. What is the composition of the Quick Ligation buffer? What is the main difference between the normal and Quick Ligation buffer?

Materials and equipment

- Phusion® high-fidelity DNA polymerase (NEB, Cat. No. M0530S)
- Quick Ligation™ kit (NEB, Cat. No. M2200S)
- 5′-Phosphorylated primer 1: 5′ GTC GAC TCT AGA GGA TCC CCG GGT 3′
- 5′-Phosphorylated primer 2: 5′ CTG CAG GCA TGT AAG CTT GGC GTA 3′

Methods in Biotechnology, First Edition. Seung-Beom Hong, M. Bazlur Rashid and Lory Z Santiago-Vázquez.

Companion Website: www.wiley.com\go\hong\Methodsinbiotechnology

- pUC19WHITE and pUC19 plasmid (~70 ng/μL)
- Competent *E. coli* DH5 α (*F- endA1 glnV44 thi-1 recA1 relA1 gyrA96 deoR nupG Φ80dlacZΔM15 Δ(lacZYA-argF)U169, hsdR17(rK- mK+), λ–*)
- SOC or LB medium
- LB agar plates with ampicillin (100 μg/mL), X-Gal (40 μg/mL) and IPTG (0.1 mM)
- DNA size marker (Thermo Scientific GeneRuler 1 kb DNA ladder; *λ/Hin*dIII/*Eco*RI)
- Agarose, 10× lithium borate (LB) buffer (100 mM, pH 8.2), ethidium bromide (10 mg/mL), 10× DNA loading dye
- Ice box
- Sterile 1.5-mL microcentrifuge tubes, thin-walled PCR tubes
- Thermocycler
- 42 °C water bath, 37 °C shake incubator

Procedure

A. PCR mutagenesis

1. Add the following components to the bottom of your labeled thin-walled PCR tube *on ice* in this order.

Addition order	Component	Volume (μL)	Final
1	Nuclease-free sterile dH_2O	35.5	
2	5x Phusion HF buffer	10.0	1×
3	10 mM dNTPs	1.0	200 μM each
4	25 μM primer mix	1.0	0.5 μM each
5	pUC19WHITE plasmid (10 pg/μL)	2.0	20 pg
6	Phusion® DNA polymerase (2U/μL)	0.5*	0.02 U/μL

*Pipette carefully and slowly because the high content (50%) of glycerol in the storage buffer may lead to pipetting errors. Keep the addition order and set up in ice because 3′→5′ exo activity degrades primers and template in the absence of dNTP.

2. Program the thermal cycler using the following parameters.

Cycle	Steps	Temperature (°C)	Time
1	Initial denaturation	98	30 s
	Denaturation	98	10 s
25	Annealing and extension	72	50 s
1	Final extension	72	5 min
1	Hold	4	∞

3. Begin the program cycle. Once the initial denaturation temperature reaches 90 °C, quickly spin your sample tube and load in the thermal cycler.

B. Gel electrophoresis

4. During PCR cycle, prepare 0.5% agarose gel in 1× LB buffer and 0.5 μg/mL ethidium bromide (one 50 mL mini-gel for whole group).

**Calculate the mass of agarose and the volumes of 10× LB and 10 mg/mL of ethidium bromide in the pre-lab notebook.*

5. After the PCR cycle, take 5 μL of PCR sample and add to a tube containing 7 μL of TE buffer (pH 8.0) and 2 μL of 6× DNA loading dye.

6. Load the sample along with the DNA size marker (Lane 1) and electrophoresis at 300 V for 20 min.

7. View the gel under a UV transilluminator and take a photograph.

8. Estimate the amount of the PCR product by comparing the band intensity of the PCR sample to that of a known amount of DNA standards.

**See an example of the estimation below.*

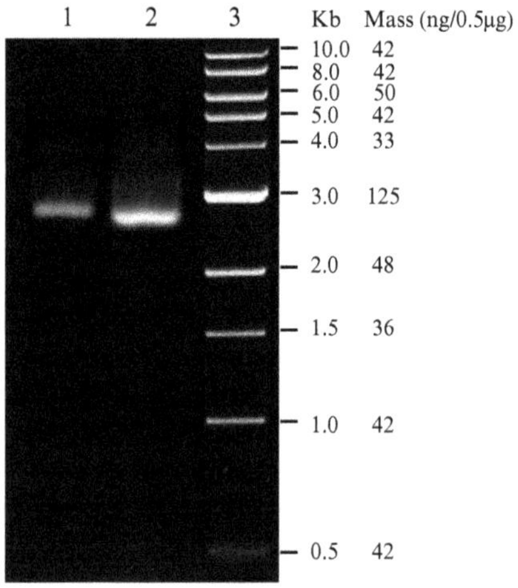

Lane 1: 2 μL DNA sample

Lane 2: 2 μL DNA sample

Lane 3: 5 μL (0.1 μg/μL) 1-kb DNA ladder (Copyright NEB; reproduced with permission)

The intensity of the DNA band in Lane 1 is approximately the same as that of a 6.0-kb marker band that contains 50 ng DNA. Thus, 2 μL of DNA ≈ 50 ng; concentration of Lane 1 DNA = 25 ng/μL.

The intensity of the DNA band in Lane 2 is approximately the same as that of a 3.0-kb marker band that contains 125 ng of DNA. Thus, 2 μL of DNA ≈ 125 ng; concentration of Lane 1 DNA = 62.5 ng/μL.

C. Ligation

9. Set up the ligation reaction based on the estimated DNA amount, as shown below.

Component	Volume (μL): total 10 μL
Nuclease-free sterile dH_2O	–
2x Quick ligation buffer (NEB)*	5
PCR mutagenesis product	– (25 ng)
Quick T4 DNA ligase (NEB)	0.5†

*Avoid repeated freeze/thaw cycles that will degrade ATP. Make aliquots once it has thawed.

†Pipette carefully and slowly because the high content (50%) of glycerol in the storage buffer may lead to pipetting errors.

10. Mix gently by pipetting up and down.
11. Incubate at room temperature (25 °C) for 5 min and chill on ice (or store at −20 °C).

**Neither extend the incubation nor heat inactivate the Quick Ligation reaction sample.*

D. Transformation

12. Label three sterile 1.5-mL tubes (+ Control; − Control; Expt) and put into ice for ~10 min.
13. Obtain competent *E. coli* DH5α, thaw on ice, and dispense 100-μL cells into the three pre-chilled sterile microcentrifuge tubes.

**Cells can also be thawed by hand, but warming to above 0 °C will decrease the transformation efficiency.*

14. Add 10 μL of the ligation mixture to the "Expt" tube, 0.5 μL of pUC19 DNA to the "+ Control (19)" tube, 0.5 μL of pUCWHITE DNA to the "+ Control (W)" tube, and 10 μL of TE buffer to the "− Control" tube.

**Be very gentle when working with competent cells. Competent cells are very sensitive to changes in temperature or mechanical lysis caused by pipetting. Transformation should be started immediately as soon as the last bit of ice disappears. Mix by swirling or tapping the tube gently, not by pipetting.*

15. Incubate the tubes on ice for 30 minutes. Do not disturb.
16. Place each transformation reaction into a 42 °C water bath for 1 min.
17. Immediately chill the tubes on ice for 1 min.
18. Add 500 μL of SOC medium to each transformation reaction.
19. Cap the tubes tightly and shake horizontally at 37 °C, 200 rpm for 1 hour.
20. While the tubes are being shaken in the incubator, label your agar plates on the edge of the bottom (not on the lid) as follows:

(a) Label one LB/Amp/X-β-Gal plate:	− Control; Group #
(b) Label two LB/Amp/X-β-Gal/IPTG plates:	Expt; Group #
(c) Label two LB/Amp/X-β-Gal plates:	Expt; Group #
(d) Label one LB/Amp/X-β-Gal/IPTG plates:	+ Control (19); Group #
(e) Label one LB/Amp/X-β-Gal/IPTG plates:	+ Control (W); Group #
(f) Label one LB/Amp/X-β-Gal plate:	+ Control (19); Group #

21. Plate the transformation reaction on to agar plates as follows:
(a) 100 μL of − Control on to + LB/Amp/X-β-Gal plate
(b) 50 and 100 μL of Expt on to each volume-labeled LB/Amp/X-Gal/IPTG plate
(c) 50 and 100 μL of Expt on to each volume-labeled LB/Amp/X-Gal plate
(d) 100 μL of + Control (pUC19) on to + LB/Amp/X-β-Gal/IPTG plate
(e) 100 μL of + Control (pUCWHITE) on to + LB/Amp/X-β-Gal/IPTG plate
(f) 100 μL of + Control (pUC19) on to + LB/Amp/X-β-Gal plate

**You do not need to re-flame the glass hockey spreader between (b) and (c) plating.*

22. Incubate the plates upside down at 37 °C overnight (16 to 18 h).
23. Examine the color and number of colonies on the plates.

**Maximum blue color forms after a few hours at 4 °C.*

24. Estimate the efficiency of mutagenesis by the number of blue (mutated) colonies divided by the total number of blue and white (unmutated) colonies.

**The white colonies are due to the co-transformation of PCR template pUCWHITE plasmid.*

25. Calculate the fold increase in amplification (mole of PCR products ÷ mole of starting template) from the step 1 and 8 result.

**The molar mass of ds-DNA is 660 g/mole/bp; mole of DNA = g DNA/(DNA size × 660).*

26. Determine the transformation efficiency using the formula:

$$\frac{\#\text{Transformants}}{\mu\text{g Plasmid DNA}} = \frac{\#\text{Transformants}}{\text{Cell plate volume } (\mu\text{L})} \times \frac{\text{Total cell volume } (\mu\text{L})}{\text{Total DNA added } (\mu\text{g})}$$

where

- Total cell volume = total transformation reaction volume after addition of SOC medium
- Cell plate volume = fraction of total cell volume used for plating
- Total DNA added = total DNA amount added to total cell volume
- Show all works and all units for your calculations.

Post-lab assignment

1. Locate and underline the primer #1 and #2 sequences in the pUC19c sequence. The pUC19c GenBank sequence accession number is L09137.

- Visit the NCBI GenBank database (http://www.ncbi.nlm.nih.gov/genbank/). On the main entry page, you will see a search menu bar at the very top. Select (*v*) "Nucleotide" from the drop-down search bar and type in "L09137" in the blank box, and then click "Search."
- Copy and paste the first 540-bp numbered sequence as an HTML Format into a Microsoft Word document. Locate and underline the start codon of *lacZα*; its translation starts at nucleotide position 469 of the pUC19c sequence (L09137).
- Visit NCBI Align Sequences Nucleotide BLAST (blastn):

http://blast.ncbi.nlm.nih.gov/Blast.cgi?PAGE_TYPE=BlastSearch&PROG_DEF=blastn&BLAST_PROG_DEF=megaBlast&BLAST_SPEC=blast2seq

- Copy the primer 1 sequence and paste into the "Enter Query Sequence" box.
- Type L09137 in the "Enter Subject Sequence" box and click "BLAST."

Note: For this process, you may use pUC19 sequence (GenBank Acc. No. M77789) where *lacZα* translation starts at nucleotide position 216 of pUC19 sequence.

(a) Copy the highest match in the alignments and paste as an HTML format into a Microsoft Word document. Indicate the primer 1 sequence on the 540-bp pUC19c sequence by underlining.

Note: the Query sequence is always a "Plus" strand. The "Minus" strand is a reverse complementary sequence.

(b) Repeat the procedure using the primer 2 sequence and L09137. Indicate the primer 2 sequence on the 540-bp pUC19c sequence by underlining italic letters.

Which one is a mutagenic primer based on BLAST alignment? In which direction does polymerization from the two primers proceed?

2. The pUC19WHITE sequence contains a stop codon (TAA) at the 8th codon position in the gene coding for *lacZα*. Based on your answer to question 1, show all of your workings by comparing the mutagenic primer sequence with its cognate pUC19 and pUC19WHITE sequences in order to answer the following questions.

(a) What are the original codon and amino acid sequence of pUC19 that was converted to a stop codon?

(b) What bases are different between the pUC19 and pUC19 WHITE plasmid?

(c) What are the codon sequence and amino acid after mutagenesis of the pUC19 WHITE plasmid using the mutagenic primer?

(d) What restriction enzyme site can be used to distinguish between the pUC19 and the mutagenized pUC19WHITE?

Hint: refer to the pUC19 map to read codon and amino acid sequences and restriction sites (https://www.neb.com/products/n3041-puc19-vector). *Information on the stop codon in pUCWHITE is described in the Introduction.*

3. What is alpha (α) complementation that is briefly described in the Introduction?

4. Describe the molecular basis of blue-white screening of colonies using X-β-Gal (5-bromo-4-chloro-3-indoyl-β-D-galactopyranoside) and IPTG (isopropyl-β-thiogalactopyranoside). Why is IPTG used instead of lactose?

5. Are there any significant differences in the color and number of the experimental blue colonies between the LB/amp/X-β-Gal plates with and without IPTG? Why are all of the transformant DH5α colonies of positive control in the LB/amp/X-β-Gal plate blue even in the absence of IPTG?

Further reading

Improved methods for site-directed mutagenesis using Gibson Assembly Master Mix. Application Note, New England BioLabs (https://www.neb.com/~/media/Catalog/All-Products/0AA961B294E444AFBEDD5C4A904C76E6/Application%20Notes/appNoteE2611.pdf).

Schildkraut, E. (2013). Site-Directed Mutagenesis Via Gibson Assembly (http://www.dddmag.com/articles/2013/05/site-directed-mutagenesis-gibson-assembly).

Zheng, L., Baumann, U., and Reymond, J.L. (2004). An efficient one-step site-directed and site-saturation mutagenesis protocol. *Nucleic Acids Research*, **32**: e115.

40 AMB 2 experiment 40: Protein expression in *E. coli*

Purpose: This is up to you to write down.

Introduction

The central act of gene cloning and protein expression is the introduction of recombinant DNA molecules into bacterial host cells by the process called transformation. If a cloned gene fragment coding for a specific protein is placed behind a promoter in a plasmid vector containing an *E. coli* replication origin, many copies of that translated protein would be produced in the transformed host cell. Expression relies on the positioning of the insert in the correct orientation and the correct reading frame with the vector's expression elements. This process is important for the large-scale production of proteins that are of great commercial value in biotechnology industries. Transformation also refers to the insertion of a new genetic material into non-bacterial cells including animal and plant cells. However, the term is avoided in animal cells because it often indicates a progression to a cancerous state, and thus introduction of foreign DNA into animal cells is usually referred to as transfection. Besides bacterial transformation, transfection of insect and mammalian cells and transformation of yeast and plant cells can be utilized for expression and production of heterologous proteins.

Protein expression requires both upstream and downstream *cis*-acting regulatory sequences. A clustered region carrying promoter, ribosomal binding, polylinker, and transcription terminator sequences in the expression vector is referred to as an expression cassette. In a prokaryotic expression vector, the ribosome binding site (RBS), termed the Shine–Dalgarno sequence (AGGAGG), is located 4 to 8 bases upstream of the start codon (ATG). The eukaryotic sequence for efficient translation, termed the Kozak sequence (A/CCAATGG), is often present in a eukaryotic expression vector. Unlike the Shine–Dalgarno sequence, however, the Kozak sequence is not required for ribosome binding since eukaryotic ribosome binds to 5′-cap sequence of mRNA and scans the downstream sequence for the first start codon for translation.

There are two choices of protein expression cassettes, transcriptional fusion and translational fusion. The transcriptional fusion vector does not have the RBS and start codon but a promoter and polylinker site followed by a transcription termination site, so the target gene has to carry its own RBS, and start and stop codons. Some transcription vectors contain an entire reporter gene lacking only a promoter, and are useful for studying the temporal and spatial expression activities of the target gene promoter. On the other hand, all translational fusion vectors have a promoter and RBS immediately upstream of the start codon; RBS is distantly separated from the downstream polylinker site, where either a short peptide coding sequence such as a protease recognition residue or a polypeptide coding sequence of a structural reporter or tagging gene separates the RBS from the polylinker site. Thus, it is critical that when a translational fusion vector is used, cDNA or an open reading frame (ORF) of the target gene is integrated into the polylinker site in such a way that its insertion results in a continuous open reading frame between the transgene protein and the structural protein to be fused. Any target protein coding sequence can be in-frame fused to either the amino (N)-terminal or the carboxyl (C)-terminal end of a fusion tag sequence depending on the site of insertion into the gene sequence to be fused. The reason for N-terminal or C-terminal fusion is that it can minimize the inactivation of fusion protein activity as a result of joining two distinct functional entities. The translational fusion vector has been routinely used to purify a target protein by affinity chromatography or to study the subcellular location of a variety of target proteins. To enhance the expression and purification as well as to avoid problems of a heterologous protein in bacteria, many *E. coli* strains and other bacterial hosts have been engineered. For more detailed information on protein expression system, refer to the articles given in Further Reading.

Once a protein expression gene cassette is carefully constructed, its introduction into the host cell and selection of transformants must be effectively made. While transformation is a relatively rare event under natural conditions, the efficiency of transformation can be improved by carefully manipulating conditions in the laboratory. *E. coli* is a common host for transformation because of advanced knowledge of its genetics and protein expression mechanisms, a rapid growth at a rate of 20 to 25 min of doubling time under typical growth conditions, a well-established fermentation system, and a greater ability to host foreign DNA than other microorganisms. Bacterial cells capable of taking up DNA are called "competent" cells. This competency becomes considerably higher when actively growing cells are used and are pre-treated, especially for *E. coli* cells, with cold $CaCl_2$ solution followed by heat shock to 42 °C for a short time. The metal ions and a temperature shift work together to affect the structure and permeability of the cell wall and membranes so that DNA molecules can pass through by still unknown mechanisms. This method works very well for circular

Methods in Biotechnology, First Edition. Seung-Beom Hong, M. Bazlur Rashid and Lory Z Santiago-Vázquez.

Companion Website: www.wiley.com\go\hong\Methodsinbiotechnology

plasmid DNA but does not work well for linear DNA because the cell's native double-stranded exonuclease (RecBCD) rapidly degrades linear DNA. The transformation efficiency significantly varies with the size of the plasmid DNA; the larger the size, the lower is the efficiency. Electroporation results in significantly higher transformation efficiency than chemical transformation by a factor of 10 to 100 and is generally recommended for large constructs (>10 kb). Putative transformants must be checked for the presence of plasmid by min-prep DNA analysis or colony PCR, though this will not be carried out in this lab due to time limitations.

In this lab exercise, you will use two different approaches to make an in-frame fusion of a GFP coding sequence to the *E. coli* expression vector pCAL-kc. In the first method of restriction sequence/ligation-dependent cloning, you will amplify the GFP_{uv} from pGLO plasmid using the primers containing restriction enzyme sites in the 5′ end and directionally clone into pCAL-kc. To do this, you will cut both the amplified GFP_{uv} DNA fragment and the pCAL-kc plasmid with *Nhe*I and *Kpn*I enzymes, purify the DNA fragments, and ligate to each other. This allows not only the carboxyl-terminal end of GFP_{uv} protein to be translationally fused to a 26-amino acid CBP (calmodulin binding peptide) but also the expression of the fusion protein to be under the control of a strong IPTG-inducible T7 promoter. Finally, the ligated hybrid DNA molecule is transformed into protein expression *E. coli* strain Lemo 21(DE3) cells.

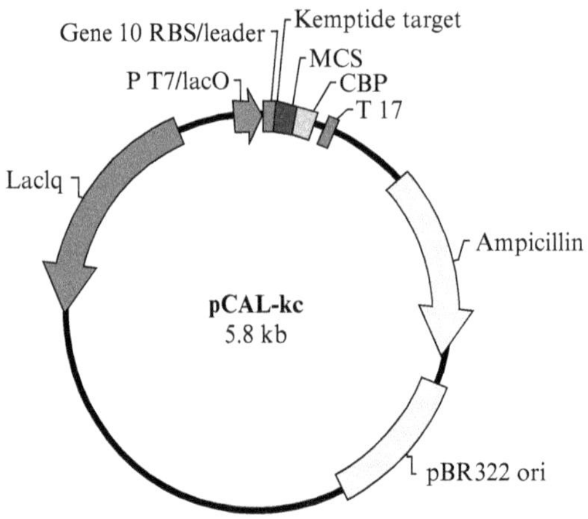

Reverse Complementary Sequence of MCS Region (322-510) in Expression Vector pCALkc (GenBank Acc. No. U36453)

```
                          NcoI    NheI                    BamHI
TAAGAAGGAGATATACC ATG GCT AGC ATG ACT GGT GGA CAG CAA ATG GGT C GGA TCC
    RBS                M   A   S   M   T   G   G   Q   Q   M   G   *  G   S
                     start         T7 gene 10 leader peptide           KpnI
CTT AGA CGC GCA TCA CTT GGT AGA TCC ATG TAT CCA CGT↓GGG AAT GGT ACC AAG
 L   R   R   A   S   L   G   R   S   M   Y   P   R   G   N   G   T   K
       Kemptide Target                    Thrombin Target
CGA CGA TGG AAA AAG AAT TTC ATA GCC GTC TCA GCA GCC AAC CGC TTT AAG AAA
 R   R   W   K   K   N   F   I   A   V   S   A   A   N   R   F   K   K
                                  Calmodulin Binding Peptide (CBP: 26 aa)
ATC TCA TCC TCC GGG GCA CTT TGA
 I   S   S   S   G   A   L  stop
```

* ATG is not in frame with the C-termninal fusion tags

↓ Thrombin Cleavage Site

pCAL-kc Map: ©Agilent Technologies, Inc, (2003). (Reproduced with permission. Courtesy of Agilent Technologies, Inc.)

In the second method of restriction sequence/ligation-independent cloning, you will amplify the GFP_{uv} from pGLO plasmid using the primers containing flanking overlapping sequences between the linearized vector (pCAL-kc/*Nhe*I-*Kpn*I) ends and GFP coding sequence ends. Then, you will mix the purified amplified DNA with the linearized vector and treat with T4 DNA polymerase. This will produce complementary ssDNA 5′ overhang ends because T4 DNA polymerase has 3′→ 5′ exonuclease activity in the absence of dNTPs, facilitating complementary base pairing of flanking ends between the vector and GFP insert DNA. At last, the assembled fragments are directly transformed into *E. coli*, where ssDNA gaps between the fragments are repaired *in vivo*.

Pre-lab assignment

(Typing and submission must be completed before lab work begins.)

1. What is the recovery percentage of DNA obtained from a spin column protocol? What is the smallest size of DNA fragment to be recovered? Refer to the following websites:

- http://www.epochlifescience.com/Product/
- http://www.qiagen.com/literature/default.aspx (QIAquick® Spin Handbook)

2. Calculate the quantities (ng) that correspond to 0.06 pmole of insert (PCR-GFP/*Nhe*I + *Kpn*I) and 0.02 pmole of vector DNA (pCAL-kc/*Nhe*I + *Kpn*I) for ligation (Part F). MW of 1 bp dsDNA is 650 g/mole.

3. Determine the approximate concentrations of the PCR DNA/*Nhe*I + *Kpn*I and vector pCAL kc/*Nhe*I + *Kpn*I (step 18 of Part E) if you have the following result after gel electrophoresis:

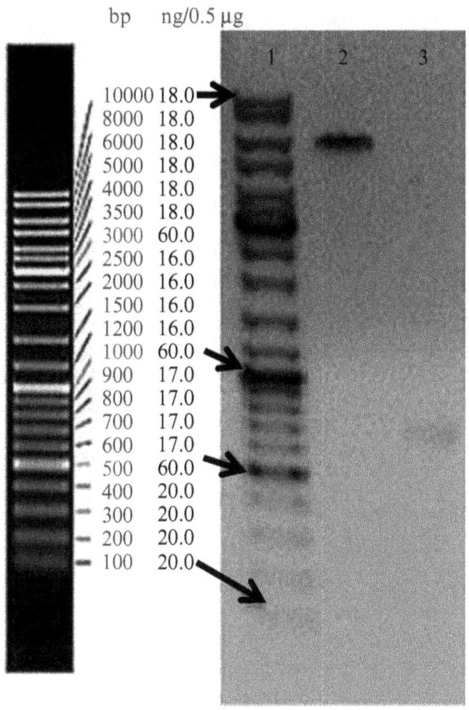

Lane 1: 0.5 µg GeneRuler 1 kb plus DNA ladder
Lane 2: pCAL-kc/*Nhe*I+*Kpn*I (2 µL)
Lane 3: PCR-GFP/*Nhe*I+*Kpn*I (2 µL)
Mass values of the individual bands are per 0.5 µg of GeneRuler 1 kb plus DNA ladder. (Used with permission from Thermo Fisher Scientific, copyright 2015.)

4. **(a)** What is the purpose of the positive and negative control transformation experiment?
(b) What does it mean that a positive control experiment results in very few or no transformant colonies?
(c) What does it mean that a negative control experiment gives rise to a few or many colonies?

5. **(a)** What things would you change if a negative control experiment shows colonies?
(b) What things would you change if a positive control experiment shows no colonies?

Materials and equipment

- FNhe I primer: GGAGATATACATATGGCTAGCAAAGGAGAAG
- RKpn I primer: AGTCGAATGGTACCTTTGTAGAGCTCATCCAT GCCATG
- SLIC 1_GFP-pCAL primer: AAGAAGGAGATATACCATGGC-TAGCAAAGGAGAAGAAC
- SLIC 2_GFP-pCAL primer: CTTTTTCCATCGTCGCTTGTATTTG-TAGAGCTCATCCAT
- Phusion® high-fidelity DNA polymerase (NEB, Cat. No. M0530S; 2 units/µL)
- pGLO plasmid (100 ng/µL)
- pCAL-kc plasmid (0.1 µg/µL; Agilent Technologies, Cat. No. 214301; Stratagene, Cat. No. 204300)
- *Nhe*I-HF™ (NEB Cat. No. R3131S), *Kpn*I-HF™ (NEB Cat. No. R3142S), *Bam*HI-HF™ (NEB, Cat. No. R3136S), *Dpn*I (NEB, Cat. No. R0176S)
- T4 DNA polymerase (NEB, Cat. No. M0203S, 3 units/µL)
- 10× NEB buffers 2 and 4, 10× BSA
- Quick T4 DNA ligase and 2× Quick Ligation buffer (NEB, Cat. No. M2200S)
- EconoSpin™ mini -spin column (Epoch Life Science, Cat. No. 1920-050)
- PX DNA binding buffer (5.5 M guanidine-HCl, 20 mM Tris-HCl, pH 6.6)
- PB wash buffer (5 M guanidine-HCl, 20 mM Tris-HCl, pH 6.6, 30% isopropanol)
- PE wash buffer (10 mM Tris-HCl pH 7.5, 80% ethanol)
- QG buffer (6 M guanidine thiocynate, 50 mM Tris-HCl, pH 7.5, 20 mM EDTA, pH 8.0, 0.006% cresol red)
- EB ($T_{10}E_{0.1}$) buffer (10 mM Tris-HCl pH 8.5, 0.1 mM EDTA; TE:EB, 0.1:0.9, v:v)
- 10× DNA loading dye (no xylene cynol but only BPB dye)
- 10× LB (lithium borate) and 10× TAE (Tris-acetate-EDTA) electrophoresis buffers
- Agarose, ethidium bromide (10 mg/mL)
- GeneRuler 1 kb plus DNA ladder (Life Technologies, Cat. No. SM1331)
- *E. coli* Lemo21 (DE3) competent cells
- LB + Amp (100 µg/mL) + Cm (30 µg/mL) + IPTG (0.1 mM) agar plates
- LB + Amp (100 µg/mL) + Cm (30 µg/mL) + arabinose (2×10^{-3} M) agar plates
- SOC medium
- 37 °C and 42 °C and 60 °C water baths, ice box
- Agarose gel electrophoresis apparatus
- Two thermal cyclers
- NanoDrop 2000 spectrophotometer

Procedure

Part I. Restriction sequence/ligation-dependent cloning

A. PCR

1. PCR amplify the GFP using FNheI and RKpnI primers and Phusion high-fidelity enzyme.

- Add the components listed in the table to your labeled thin-walled PCR tube *kept on ice*.
- Program the thermal cycle parameters shown in the table.
- Begin the program cycle. Once the initial denaturation temperature reaches to 90 °C, quickly spin your sample tube and load into the thermal cycler.
- During PCR cycling, prepare a 60 mL of 0.5% agarose gel (+0.5 µg/mL ethidium bromide) in 1× LB buffer using a 12-well comb. Solidify the gel at 4 °C. This gel will be used in Part I, E and Part II, D).
- During PCR cycling, set up restriction digestion of pCAL-kc DNA.

Addition order	Component	Volume (μL)	Final concentration	Master mix (4.5×)
1	Nuclease-free sterile dH_2O	35.5		164.25
2	5× Phusion HF buffer	10.0	1×	45
3	10 mM dNTPs	1.0	200 μM	4.5
4	25 μM primer mix (FNheI + RKpnI)	1.0	0.5 μM	4.5
5	pGLO plasmid (1 ng/μL)	2.0	2 ng	
6	Phusion® high-fidelity DNA polymerase (2 U/μL)	0.5*	0.02 /μL	2.25
	Total volume	50		216

*Pipette carefully and slowly because the high content (50%) of glycerol in the storage
buffer may lead to pipetting errors.

Cycle	Steps	Temperature (°C)	Time
1	Initial denaturation	98	30 s
	Denaturation	98	10 s
25	Annealing	69	15 s
	Extension	72	15 s
1	Final extension	72	5 min
1	Hold	4	∞

B. Restriction digestion of vector pCAL-kc

2. Set up the reaction for vector pCAL-kc digestion as follows (by two groups).

Addition order	Component	Volume (μL)
1	Sterile dH_2O	x
2	10× NEB buffer 4	5
3	pCAL-kc DNA (1 μg)	y
4	*Nhe*I-HF™	1.5*
4	*Kpn*I-HF™	1.5*
4	Bam HI- HF™	1.0*
	Total volume	50

*The combined volume of all added enzymes should not exceed 1/10 of the total reaction volume. Add each component directly to the bottom of a sterile 1.5-mL microcentrifuge tube and mix well by gently pipetting up and down.

3. Mix the reaction mixture gently by pipetting up and down, and incubate at 37 °C until the PCR is completed.

C. Silica-based purification of restriction DNA and PCR DNA (spin column protocol)

4. Add 5× volume of PX buffer to one volume of the restriction digestion sample and mix. Do the same for the PCR DNA sample.

5. Place a spin column in a provided 2 mL collection tube.

6. To bind DNA, apply the sample to the spin column and centrifuge for 30 to 60 s.

7. Discard flow-through. Place the spin column back into the same tube.

**Collection tubes are reused for recycling.*

8. To wash, add 750 μL of PB buffer to the spin column and centrifuge for 30 to 60 s.

9. Discard flow-through and place the spin column back in the same tube.

10. Repeat steps 8 and 9 using 750 μL of PE buffer. Centrifuge the washed column for an additional 1 min at maximum speed.

**Residual ethanol from PE buffer will not be completely removed unless the flow-through is discarded before this additional centrifugation. Any residual PE buffer may interfere with subsequent enzymatic reactions.*

11. Place the spin column in a sterile labeled 1.5-mL microcentrifuge tube.

12. To elute DNA, add 32 μL of EB (warmed to ~55 °C) to the center of the spin column membrane, let the column stand for 1 min, and centrifuge the column for 1 min.

**Ensure that the elution buffer is dispensed directly on to the center of the QIAquick membrane for complete elution of bound DNA. Typically, recovery of the yield is 60 to 80%.*

D. Restriction digestion and purification of PCR-GFP DNA fragment

13. Set up the double digestion reactions as follows.

Addition order	Component	Volume (μL)
1	Sterile dH_2O	x
2	10× NEB buffer 4	5
3	Purified PCR-GFP DNA	y
4	*Nhe*I-HF™	1.5*
4	*Kpn*I-HF™	1.5*
4	*Bam*HI- HF™	1.0*
	Total volume	50

*The combined volume of all added enzymes should not exceed 1/10 of the total reaction volume.

14. Mix the reaction mixture gently and incubate for 15 min at 37 °C.

15. Follow steps 4 to 11 of Part C to purify the restriction-digested PCR fragment, and elute the DNA with 25 μL of EB buffer (warmed to ~55 °C).

E. DNA quantification (NanoDrop 2000 and gel electrophoresis)

16. Spot 2 μL each of purified PCR-GFP/*Nhe*I + *Kpn*I + *Dpn*I and pCAL-kc/*Nhe*I + *Kpn*I sample on to NanoDrop 2000 for DNA quantification using EB as a blank.

17. Add 2 μL each of the purified PCR-GFP/*Nhe*I + *Kpn*I and pCAL-kc/*Nhe*I + *Kpn*I samples to 8 μL of TE buffer and 1.5 μL of 10× DNA loading dye.

18. Run the samples, along with a 1-kb DNA ladder (0.5 μg/5 μL), side by side on to a 0.5% agarose gel in 1× LB buffer at 300 V for 20 min.
19. Estimate the approximate concentrations of the PCR-GFP/*NheI* + *KpnI* + *DpnI* and vector pCAL-kc/*NheI* + *KpnI* by the band brightness method and compare them with those determined by NanoDrop 2000.

F. Ligation

20. Set up the ligation reaction based on the NanoDrop quantification in this order.

Addition order	Component	Total volume 10 μL
		Experimental (pCAL-GFP)
1	Nuclease-free sterile dH_2O	–
2	2× Quick Ligation Buffer (NEB)	5
3	pCAL-kc/*NheI-KpnI* (5723 bp)*	
4	PCR-GFP/*NheI-KpnI* (714 bp)*	
5	Quick T4 DNA Ligase (NEB)	0.5†

*The overall concentration of vector + insert should be between 1 and 10 ng/μL for efficient ligation. The optimal single insertions are insert/vector molar ratios between 2 and 6.
†Pipette carefully and slowly because the high content (50%) of glycerol in the storage buffer may lead to pipetting errors.

21. Mix gently by pipetting up and down.
22. Incubate at 25 °C for 5 min and chill on ice (or store at −20 °C).

**Do not rely on the ambient air temperature but place the tube in a 25 °C water bath.*

**Neither extend the incubation nor heat-inactivate the sample.*

G. Transformation

23. Obtain competent *E. coli* Lemo21 (DE3), thaw on ice, and dispense 100 μL of cells into the *pre-chilled* microcentrifuge tubes labeled "Expt," "+ Control," and "−Control."
24. Add 10 μL of experimental reaction mixture to the 'Expt" tube, 1 μL of positive control (pGLO plasmid) to the "+ Control" tube, and 1 μL of negative control (TE buffer) to the "−Control" tube.

**Be very gentle when working with competent cells. Competent cells are very sensitive to changes in temperature or mechanical lysis caused by pipetting. Transformation should be started immediately as soon as the last bit of ice disappears. Mix by swirling or tapping the tube gently, not by pipetting.*

**The volume of the reaction mixture should not exceed 10% of the competent cell volume in the transformation process.*

25. Incubate tubes on ice for 30 min. Do not disturb.
26. Heat the tubes for exactly 60 s in a 42 °C water bath; do not shake.
27. Immediately chill tubes on ice for 1 min.
28. Add 300 μL of SOC medium to each transformation reaction.
29. Cap the tubes tightly and shake horizontally at 37 °C, at 200 rpm for 60 min.
30. Spread 120 μL of each transformation mixture on to two LB + Amp + Cm + IPTG agar plates. Spread 120 μL of positive control and negative control mixtures on to each LB + Amp + Cm + Arabinose agar plate.
31. Incubate the plates upside down at 37 °C overnight (16 to 18 h).

H. Mini-prep plasmid DNA analysis (extra lab work)

32. Follow steps 24 to 26 of Part II.

Part II. Restriction sequence/ligation-independent cloning)

Procedure

A. PCR

1. PCR-amplify the GFP using SLIC1 and SLIC2 primers and Phusion high-fidelity enzyme.

- Add the following components to your labeled thin-walled PCR tube *kept on ice*.

Addition order	Component	Volume (μL)	Final concentration	Master mix (4.5×)
1	Nuclease-free sterile dH_2O	35.5		164.25
2	5× Phusion HF buffer	10.0	1×	45
3	10 mM dNTPs	1.0	200 μM	4.5
4	25 μM primer mix (SLIC1 + SLIC2)	1.0	0.5 μM	4.5
5	pGLO plasmid (1 ng/μL)	2.0	2 ng	
6	Phusion® high-fidelity DNA polymerase (2 U/μL)	0.5*	0.02/μL	2.25
	Total volume	50		216

*Pipette carefully and slowly because the high content (50%) of glycerol in the storage buffer may lead to pipetting errors.

- Program the thermal cycler using the following parameters.

Cycle	Steps	Temperature (°C)	Time
1	Initial denaturation	98	30 s
	Denaturation	98	10 s
25	Annealing	54	15 s
	Extension	72	15 s
1	Final extension	72	5 min
1	Hold	4	∞

- Begin the program cycle. Once the initial denaturation temperature reaches 90 °C, quickly spin your sample tube and load in the thermal cycler.

B. Agarose gel electrophoresis

2. During PCR cycling, prepare 60 mL of 0.7% agarose gel (+0.5 µg/mL ethidium bromide) in 1× TAE buffer using a large well comb. Solidify the gel at 4 °C.

3. Add 1/10 volume of 10× loading dye to the PCR sample, load the sample on to gel, and run at 100 V for 20 min.

C. Silica-based gel purification of PCR-GFP DNA (Qiagen spin column protocol)

Note: this step is to remove the template vector DNA from the PCR reaction.

4. Weigh an empty sterile 1.5-mL microcentrifuge tube.

5. Place the electrophoresced gel on to a UV transilluminator covered with plastic wrap.

6. Cut out the gel band with a clean razor blade as quickly as possible and place into the empty microcentrifuge tube.

**Take special care to avoid sources of nuclease contamination.*

**Cut out the bands as quickly as possible to minimize the UV exposure time because DNA is damaged by UV light, causing sequence errors.*

**Minimize the size of the gel slice by removing extra agarose.*

7. Weigh the gel-containing microcentrifuge tube and calculate the gram quantity of gel slice.

**Put gel slice up to 0.25 g in a 1.5-mL microcentrifuge tube.*

8. Add 3× volume of QG buffer to 1 volume of gel (1 g equals about 1 mL).

9. Incubate at 50 to 57 °C for 5 min (or until the gel slice has completely dissolved). Mix by tapping the tube every 2 min during the incubation.

**After the gel slice has dissolved completely, check that the color of the mixture is yellow. If the color of the mixture is orange or violet, add 10 µL of 3 M sodium acetate (pH 5.2) and mix to turn the color of the mixture to yellow.*

10. Place a spin column in a provided 2-mL collection tube, apply the sample to the spin column, and centrifuge for 30 to 60 s.

11. Discard flow-through. Place the spin column back into the same tube.

**Collection tubes are reused for recycling.*

12. To wash, add 750 µL of PE buffer to the spin column and centrifuge for 30 to 60 s.

13. Discard flow-through and place the spin column back in the same tube. Centrifuge the column for an additional 1 min at 13 000 rpm.

**Residual ethanol from PE buffer will not be completely removed unless the flow-through is discarded before this additional centrifugation. Any residual PE buffer may interfere with subsequent enzymatic reactions.*

14. Place the spin column in a sterile 1.5-mL microcentrifuge tube.

15. To elute DNA, add 32 µL of EB buffer (warmed to 55 °C) to the center of the spin column membrane, let the column stand for 1 min, and centrifuge the column for 1 min.

**Ensure that the elution buffer is dispensed directly on to the QIAquick membrane for complete elution of bound DNA. Typically, recovery of the yield is 60 to 80%.*

D. DNA quantification by NanoDrop2000 and gel electrophoresis

16. Spot 2 µL each of the purified PCR insert (SLIC_PCR GFP) on to NanoDrop 2000 using EB buffer as a blank.

17. Add 2 µL each of the purified PCR insert to 8 µL of TE buffer and 1.5 µL of 10× DNA loading dye.

18. Run the samples, along with a 1-kb DNA ladder side by side on to a 0.5% agarose gel in 1× LB buffer with 0.5 µg/mL EtBr at 300 V for 20 min.

19. Estimate the approximate concentrations of the SLIC_PCR GFP by the band brightness method and compare them with those determined by NanoDrop 2000.

E. SLIC reaction

20. Based on the NanoDrop quantification, mix the linearized vector and insert at a molar ratio of 1: 2.0 to 2.5 in a final total volume of 10 µL in a sterile 0.5-mL microcentrifuge tube as shown in the table.

21. Add 0.2 to 0.5 µL of T4 DNA polymerase (3 U/µL) to the mixture and incubate at a 25 °C water bath for 2.5 min.

22. Put the reaction mixture on ice immediately to stop the reaction and incubate on ice for 10 min. Add 5 or 10 µL of the reaction mixture to 50 or 100 µL competent *E. coli* cells for transformation.

F. Transformation

23. Follow steps 23 to 31 of the transformation protocol in Part I, G, except for + and – controls.

24. Next day, remove the plates from the incubator, check the transformant colonies for green fluorescence under UV light, and calculate the transformation efficiencies of pCAL-GFP and pGLO and compare them.

Addition order	Components	Stock concentration	Volume (µL)	Final concentration	Total amount
1	Nuclease-free H_2O				–
2	10× BSA	10×	1 µL	1×	10×
3	10× NEB buffer 2	10×	1 µL	1×	10×
4	pCAL-kc/*NheI*-*KpnI* (5723 bp)			10 ng/µL	100 ng
5	SLIC_GFP PCR (754 bp)			2.6–3.3 ng/µL	26–33 ng
	Total volume		10 µL		

G. Mini-prep plasmid DNA analysis (optional extra lab work)

25. Transfer the transformed green fluorescent colonies to a new LB + Amp + Cm + IPTG plate with sterile toothpicks and incubate at 37 °C overnight for the master grid plate; insert five independently inoculated toothpicks into five labeled test tubes, each containing 2 mL of LB + Amp (100 µg/mL) medium (no Cm), and grow at 37 °C, 200 rpm overnight for mini-prep plasmid DNA.

26. Prepare mini-prep plasmid DNA, set up double digestion of 5 µL of mini-prep DNA with *Kpn*I and *Nhe*I, and run on a 0.5% agarose gel at 300 V for 25 min in 1× lithium borate buffer.

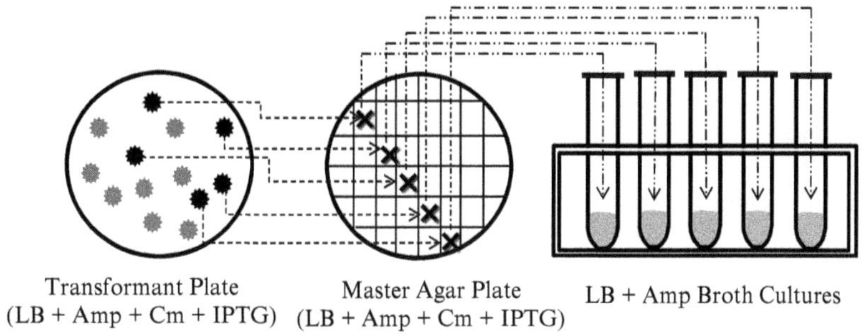

Post-lab assignment

1. How would you set up restriction enzyme digestion at step 26 of Part II, G, in a total reaction volume of 20 µL? Details are needed as to how to add each component and incubation time and temperature.

2. Visit NCBI Align Sequences Nucleotide BLAST (blastn): http://blast.ncbi.nlm.nih.gov/Blast.cgi?PAGE_TYPE=BlastSearch&PROG_DEF=blastn&BLAST_PROG_DEF=megaBlast&BLAST_SPEC=blast2seq

The GenBank accession number of pGLO plasmid is U62637. Give answers for (a) to (f):

(a) Copy the FNheI primer sequence and paste on to the "Enter Query Sequence" box.
- Type "U62637" in the "Enter Subject Sequence" box.
- Click "BLAST."
- Copy the alignment and paste in HTML format in a Microsoft Word document.

(b) Repeat the above procedure using the RKpnI primer sequence.

(c) Visit http://www.ncbi.nlm.nih.gov/nuccore/U62637. Copy the entire of pBAD-GFPuv (pGLO plasmid) and paste in HTML Format in a Microsoft Word document. Note the CDS position of GFP_{uv} from the annotated U62637 sequence and mark the positions of the start codon and stop codon of GFPuv on the pGLO DNA sequence.

(d) Using the position numbers of a subject sequence in the BLAST alignments, locate and underline the matched sequence of the pGLO plasmid.

(e) Present the GFP sequence between the start and stop codons and underline the F and R primer sequences.

(f) Indicate the matched and unmatched primer sequences in each primer. What are the unmatched primer sequences for?

3. Visit NCBI Align Sequences Nucleotide BLAST (blastn): http://blast.ncbi.nlm.nih.gov/Blast.cgi?PAGE_TYPE=BlastSearch&PROG_DEF=blastn&BLAST_PROG_DEF=megaBlast&BLAST_SPEC=blast2seq

The GenBank accession of pCALkc plasmid is U36453.

(a)
- Copy the SLIC1 primer sequence and paste on to the "Enter Query Sequence" box.
- Type "U62637" in the "Enter Subject Sequence" box.
- Click "BLAST."
- Copy the alignment and paste in HTML Format in a Microsoft Word document. Indicate the overlap matching sequence.

(b) Repeat the above procedure using the SLIC 1 primer sequence and U36453.

(c) Repeat the above procedure using the SLIC 2 primer and U62637.

(d) Repeat the above procedure using the SLIC 2 primer and U36453.

(e) Based on the alignments, distinguish between the pCALkc vector and GFP_{uv} sequence in SLIC primers 1 and 2.

4. Visit the NEB Tm calculator (http://tmcalculator.neb.com/#!/).

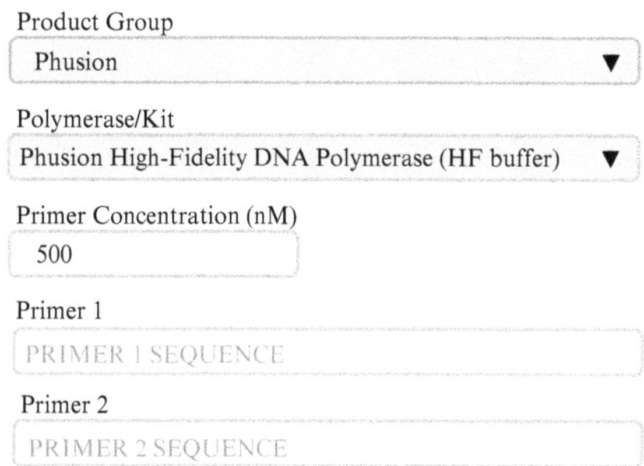

- Select "Phusion" in Product Group and "Phusion High-Fidelity DNA Polymerase (HF Buffer)" in Polymerase/Kit.
- Copy only the matched FNheI primer sequence and paste on the "Primer 1 sequence" box.

- Copy only the matched RKpnI primer sequence and paste on the "Primer 2 sequence" box.
- Write down "Anneal Temp (Ta)," "Primer 1 Tm," and Primer 2 Tm."
- Do the same using SLIC1 and SLIC2 primer sequences that match the pGLO sequence.

5. **(a)** What is in-frame fusion and why is it important for cloning into the expression vector?

(b) Show your work as to why insertion of the *Nhe*I-*Kpn*I fragment of pGLO DNA into the *Nhe*I and *Kpn*I site of the pCAL-kc vector DNA result in an inframe fusion to the CBP tag. To do this, perform the following steps.

- Based on the information of the primer sequence and GFP, the pCAL-kc multiple cloning site sequence (see Introduction), and restriction enzyme digestion ligation, write the newly assembled sequence in the FASTA format (i.e., assembled sequences under >pCAL-GFP name of the sequence):

  >pCAL-GFP
  Your assembled nucleotide sequences

Note: the entire pCAL-kc vector sequence is not needed to assemble this.

(c) Visit the online Translate Tools software (http://web.expasy.org/translate/).

- Enter the entire copy of the assembled sequence in FASTA format.
- Choose "Verbose ("Met," "Stop," spaces between residues)" in the Output Format.
- Click "TRANSLATE SEQUENCE."
- Copy the longest protein sequence that begins with Met (methionine) and ends with Stop.
- Distinguish between GFP and CBP amino acids by underlining CBP tag residues.

6. In step 32 of the Part I, H experiment, you selected five Amp^R non-fluorescent colonies, cultured in LB+Amp media, and isolated plasmids. You then cut the plasmids with *Nhe*I + *Kpn*I, and analyzed them, along with vector pCAL-kc/*Nhe*I + *Kpn*I by agarose gel electrophoresis. The cleavage patterns of several mini-prep plasmids show the presence of only 5.8 kb of the pCAL-kc vector DNA fragment. How would you explain this plasmid? How can you verify that your reasoning is correct? Assume that plasmid DNA is cut completely by both enzymes.

7. **(a)** List genotype and explain every genotypic characteristics of Lemo21 (DE3) strain.

(b) What are advantages of the use of this strain?

8. At step 32 of Part I, H, you confirmed the presence of the pCAL-GFP plasmid. Now you have transformed this pCAL-GFP plasmid into *E. coli* DH5α (*F⁻ endA1 glnV44 thi-1 recA1 relA1 gyrA96deoR nupG Φ80dlacZΔM15 Δ (lacZYA-argF) U169, hsdR17* (r_K^- m_K^+), *λ–) and E. coli TOPO10 (F⁻ mcrA Δ (mrr-hsdRMS-mcrBC) φ80lacZΔM15 ΔlacX74 nupG recA1 araD139 Δ (ara leu) 7697 galE15 galK16 rpsL* (Str^R) *endA1 λ*). You analyzed DH5α and TOP10 transformants for the presence of pCAL-GFP by *Kpn*I-*Nhe*I double digestion and sequencing, and confirmed the presence of the plasmid. You then streaked the individual transformants on to LB + Amp + IPTG plate to look for green fluorescence of GFP. Surprisingly, none of the confirmed transformants showed the green fluorescence. Explain the reason.

9. **(a)** Why are the transformed cells allowed to grow in a non-selective LB growth media in the course of transformation?

(b) What are the satellite colonies? What should you do to prevent them from your selection plates?

10. How are the animal (mammalian and insect) cell transfection and plant cell transformation performed for protein expression? Briefly describe the most widely used method, one in each case, as to how transfection or plant transformation is performed.

Further reading

Affinity® Protein Expression and Purification System and Affinity Protein Expression Vector. Instruction Manual (2003, Revision #073007d). Stratagene, Inc.

Jeong, J.-Y., Yim, H.S., Ryu, J.Y., Lee, H.S., Lee, J.-H., Seen, D.-S., and Kang, S.G. (2012). One-step sequence- and ligation-independent cloning as a rapid and versatile cloning method for functional genomics studies. *Applied and Environmental Microbiology*, **78**: 5440–5443.

Kozak, M. (1987). An analysis of 5′-noncoding sequences from 699 vertebrate messenger RNAs. *Nucleic Acids Research*, **15**: 8125–8148.

Shine, J. and Dalgarno, L. (1975). Determinant of cistron specificity in bacterial ribosomes. *Nature*, **254**: 34–38.

Terepe, K. (2006). Overview of bacterial expression system for heterologous protein production: from molecular and biochemical fundamentals to commercial systems. *Applied Microbiology and Biotechnology*, **72**: 211–222.

41 AMB 2 experiment 41: Protein purification by affinity column chromatography

Purpose: This is up to you to write down.

Introduction

Inducible protein expression system is designed to produce large amounts of a desired protein within a host cell only under a certain changing condition. One of the most popular bacterial expression approaches is the so-called pET expression system. In this scheme, target genes in the pET series of vectors are under the control of the T7 promoter, which is specifically bound by a bacteriophage T7 RNA polymerase to drive transcription. The entire LacI repressor gene (*lacI*) and T7 polymerase gene are integrated into the protease-defective *E. coli* BL21 (DE3) host chromosome, and the T7 gene expression is under the control of *lacUV5* promoter and *lac* operator (*lacO*) regulatory sequences. The mutant *lacUV5* promoter is less subject to catabolite repression than the wild-type *lac* promoter. High levels of Lac repressors are constitutively expressed from both chromosome-borne *lacI* (two copies) and pET vector-borne *lacI^q^* genes to block T7 polymerase

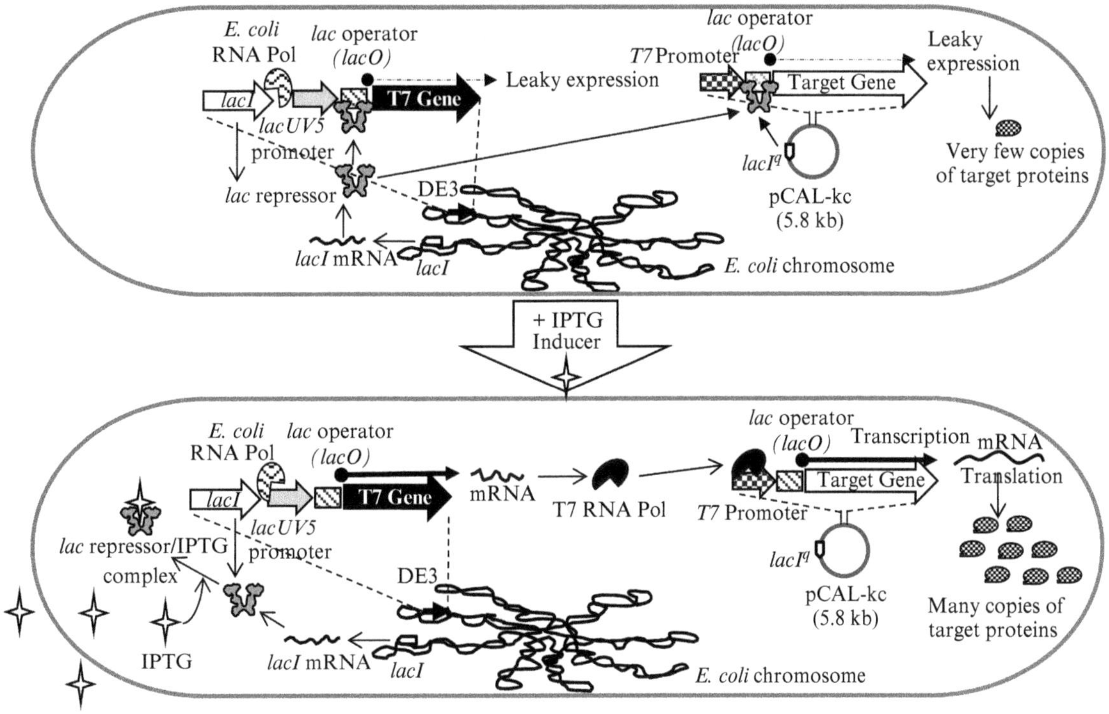

TAATACGACT CACTATAGGG GAATTGTGAG CGGATAACAA TTCCCCTCTA GAAATAATTT
T7 promoter — Lac operator (*lacO*)
TGTTTAACTT TAAGAAGGAGATATACCATG – – Target Gene In-frame fused to CBP Tag-
RBS — start
TGATCCGGCT GCTAACAAAG CCCGAAAGGA AGCTGAGTTG GCTGCTGCCA CCGCTGAGCA
stop
ATAACTAGCA TAACCCCTTG GGGCCTCTAA ACGGGTCTTG AGGGGTTTTT TGCTGAAAGGA
T7 terminator

Methods in Biotechnology, First Edition. Seung-Beom Hong, M. Bazlur Rashid and Lory Z Santiago-Vázquez.

Companion Website: www.wiley.com\go\hong\Methodsinbiotechnology

expression and T7 promoter-driven transcription. A mutation in the *lacI* promoter (*lacI^q*) leads to a tenfold increase in LacI repressor production, thereby considerably reducing the basal level of a leaky expression. When non-metabolizable lactose analog, isopropyl-β-D-thiogalactoside (IPTG), is added, Lac repressors can no longer occupy *lacO* sites due to their conformational change, leading to expression of T7 RNA polymerase followed by T7 promoter transcription. T7 polymerase transcribes eight times faster than *E. coli* RNA polymerase, allowing high protein expression. Thus, only in the presence of IPTG can the target protein be highly produced. Also the amount of target protein produced can be controlled by changing induction time and inducer dosage to reduce the toxic or growth inhibition effect caused by overexpression. However, this BL21 (DE3)-pET expression system does not completely abolish the leaky expression despite the overexpression of the LacI repressor. The diagram illustrates how BL21 (DE3) cell harboring pET expression plasmid is induced to express a target gene.

When foreign proteins, especially membrane-associated proteins, are expressed at high levels by induction, they frequently form insoluble aggregates known as inclusion bodies. Sometimes the induction system does not always bring about overproduction but rather little or no production when codon bias occurs. To circumvent these problems, *E. coli* BL21 (DE3) has been engineered to produce SoluBL21® strain (Genlantis Inc.), Rosetta-gami™ 2 host strain (Novagen Inc.), or ArcticExpress (DE3) strain (Agilent Technologies Inc.), which helps to express insoluble proteins in more soluble forms, or Lemo21 (DE3) strain, which is able to optimize the protein expression level for solubilztion by fine tuning T7 expression system. Alternatively, either cell lysate is prepared using BugBuster® Protein Extraction Reagent (Novagen Inc.) or a coupled wheat germ *in vitro* transcription–translation reaction may be employed. When target protein requires post-translational modifications, there is no choice but to use the eukaryotic expression system. For more detailed information on protein expression, refer to the articles in Further Reading.

In pCAL-GFP plasmid that was constructed in the previous lab exercise, GFP_{uv} coding sequence is in-frame fused to the coding sequence of a 26-amino acid calmodulin binding peptide (CBP) fragment. In this lab exercise, you will purify the CBP-tagged GFP_{uv} fusion protein from Lemo21 (DE3) transformant culture using calmodulin affinity chromatography. You will also prepare soluble and insoluble proteins after cell lysis to determine what fractions contain GFP_{uv}-CBP fusion protein by SDS-PAGE (Experiment 42) and Western blot analysis (Experiment 43). Because insoluble proteins are not amenable to any column chromatography purification, the pilot experiments to optimize expression should be conducted prior to large-scale purification by column chromatography. Calmodulin binds proteins through hydrophobic interactions and hydrophobic sites are exposed on its surface in the presence of a low (1 to 2 mM) Ca^{+2} at neutral pH. The binding capacity was reported to be approximately 1 to 3 mg of CBP fusion protein per milliliter of calmodulin resin (http://www.gbiosciences.com/CalmodulinResin-desc.aspx). This system purifies CBP-tagged proteins in *E. coli* at high specificity with 80 to 90% recovery because there are no endogenous proteins able to interact with calmodulin in bacteria.

Pre-lab assignment

(Typing and submission must be completed before lab work begins.)

1. How many and what types of tubes should be labeled for Part B, C, and D experiments? Briefly describe the purpose for each tube.
2. Where should you keep those tubes during the experiment and why?
3. What volumes of lysis buffer, column binding buffer, column wash buffer, column elution buffer, and column regeneration buffer are needed to conduct today's experiment?
4. How many and what samples are collected in aliquots and stored at –20 °C or –80 °C for the subsequent SDS-PAGE analysis in today's experiment? What are they for?
5. Draw the flow diagram of today's experiment.

Materials and equipment

- Transformed *E. coli* Lemo21 DE3 (pCAL-GFP) obtained from Experiment 40
- L-rhamnose (0.5 M)
- Ampicillin (100 mg/mL in dH_2O; filter-sterilize, store at –20 °C)
- Chloramphenicol (30 mg/mL in 100% ethanol, no filter-sterilization, store at –20 °C)
- 0.1 M IPTG (dissolve 0.0238 g IPTG in 10 mL of dH_2O, filter-sterilize, store at –20 °C)
- 1 M imidazole (1.702 g imidazole in 20 mL of dH_2O; adjust to pH 7.4 with HCl and bring to a final volume of 25 mL)
- 0.1 M PMSF (0.0174 g/mL isopropanol; store at –20 °C)
- Lysozyme (10 mg/mL in lysis buffer; prepare just before use)
- Cell pellet washing buffer (1 M urea, 1% Triton-100, 50 mM Tris-Cl, pH 7.5)
- Resolubilization buffer (8 M urea, 50 mM Tris-Cl, pH 7.5)
- Lysis/binding base buffer (50 mM Tris-HCl, pH 8.0, 150 mM NaCl, 1 mM Mg $(CH_3COO)_2 \cdot 4H_2O$, 2 mM $CaCl_2$; autoclave)
- Washing buffer (50 mM Tris-HCl, pH 7.5, 300 mM NaCl, 1 mM $Mg(CH_3COO)_2 \cdot 4H_2O$, 0.2 mM $CaCl_2$; autoclave)
- Elution buffer I (50 mM Tris-HCl, pH 7.5, 150 mM NaCl, 2 mM EGTA, 10 mM β-mercaptoethanol)
- Elution buffer II (50 mM Tris-HCl, pH 7.5, 1 M NaCl, 2 mM EGTA, 10 mM β-mercaptoethanol)
- Regeneration buffer 1 (0.1 M ammonium carbonate buffer, pH 8.6, 2 mM EGTA)
- Regeneration buffer 2 (1 M NaCl, 2 mM $CaCl_2$)
- Regeneration buffer 3 (50 mM Tris-HCl, pH 7.5, 2 mM EGTA, 1.0 M NaCl)
- 4× SDS sample buffer (250 mM M Tris-HCl, pH 6.8, 40% glycerol, 8% SDS, 0.02% bromophenol blue)
- 20% ethanol
- Ice box
- Sonicator
- Calmodulin affinity resin: Calmodulin Sepharose 4B (GE Healthcare Life Sciences, Product Code: 17-0529-01)
- 1.5- or 2.0-mL microcentrifuge tubes and 15-mL tubes
- Polyprep (Bio-Rad) column, column stand, and clamp

Procedure

Part I. Protein expression and cell harvest (TA will do this part)

1. Pick a single colony of Lemo21 DE3 (pCAL-GFP$_{uv}$) from LB + Amp + Cm plate and inoculate into 2 mL of LB media containing 100 µg/mL ampicillin and 30 µg/mL chroamphenicol.
2. Grow overnight at 37 °C in a shaking incubator at 200 rpm.
3. Add 0.4 mL of overnight culture each into two 20 mL of LB + Amp + Cm broth in two separate sterile 150-mL flasks (labeled as "Uninduced" and "Induced").
4. Add L-rhamnose to the induced flask to a final concentration of 0 µM of L-rhamnose (Group 1), 100 µM of L-rhamnose (Group 2), 200 µM of L-rhamnose (Group 3), and 300 µM of L-rhamnose (Group 4).

**Various levels (0 to 2000 µM) of L-rhamnose can be tested to determine the optimizing expression of soluble protein.*

5. Shake at 37 °C, 200–250 rpm until OD_{600} = 0.4 to 0.6.
6. Add IPTG to a final 0.4 mM concentration to the induced flask. *Calculation is needed in the pre-lab notebook.* There is no IPTG addition to the uninduced flask.
7. Shake both flasks at 30 °C, 200 rpm for 8 h to overnight (~16 h).

**Various temperatures (20 to 37 °C) and incubation times can be tested to determine the optimum expression of soluble protein.*

8. Centrifuge 12 mL each of induced and uninduced cells in 15-mL conical tubes at 3500 rpm for 10 min, and aspirate off the supernatant completely.

**Check the cell pellet under UV light for green fluorescence. Fluorescence may be either weak or not be detectable in the IPTG-induced culture containing high rhamnose levels.*

9. Store the induced and uninduced cell pellets at –80 °C if not used on the same day.

Part II. Cell lysate preparation and affinity column chromatography

A. Preparation of buffers

Obtain the following materials and prepare freshly made buffers.

1. Uninduced and induced *E. coli* Lemo21 DE3 (pCAL-GFP) cell pellets.
2. Three 2.0-mL sterile microcentrifuge tubes and three sterile 15-mL tubes.
3. Cell pellet washing buffer.

Component (final concentration)	MW	Per 10 mL
1 M urea	60	g of urea
1% (v/v) TritonX-100	647	mL of TritonX-100
50 mM Tris-Cl (pH 7.5)	121.14	mL of 1 M Tris-Cl (pH 7.5)

4. Resolubilization buffer.

Component (final concentration)	MW	Per 10 mL
8 M urea	60	g of urea
50 mM Tris-Cl (pH 7.5)	121.14	mL of 1 M Tris-Cl (pH 7.5)

5. Lysis/binding base buffer.

Component (final concentration)	MW	g per 200 mL	mL per 200 mL
50 mM Tris-HCl, pH 7.5	121.14		of 1 M Tris-Cl (pH7.5)
150 mM NaCl	58.44		of 4 M NaCl
1 mM magnesium acetate (tetrahydrate)	214.46		of 0.1 M MgAc
2 mM $CaCl_2.2H_2O$	147.00		of 0.1 M $CaCl_2$

*For the lysis buffer (5 mL), add the following components to the base buffer just before use.

- Lysozyme (10 mg/mL) to a final concentration of 200 µg/mL
- PMSF (0.1 M) to a final concentration of 2 mM
- After addition, keep the lysis buffer in ice.

*For binding buffer (10 mL), add the following components to the base buffer just before use.

- Imidazole to a final concentration of 1 mM
- β-Mercaptoethanol (14.3 M) to a final concentration of 10 mM
- PMSF to a final concentration of 2 mM
- After addition, keep the binding buffer at room temperature.

6. Washing buffer.

Component (final concentration)	MW	g per 100 mL	mL per 100 mL
50 mM Tris-HCl, pH 7.5	121.14		of 1 M Tris-Cl (pH7.5)
300 mM NaCl	58.44		of 4 M NaCl
1 mM magnesium acetate (tetrahydrate)	214.47		of 0.1 M MgAc
0.2 mM $CaCl_2.2H_2O$	147.00		of 0.1 M $CaCl_2$

*For washing buffer (10 mL), add the following components to the base buffer just before use.

- Imidazole to a final concentration of 1 mM
- β-Mercaptoethanol (14.3 M) to a final concentration of 10 mM
- PMSF to a final concentration of 2 mM
- After addition, keep the binding buffer at room temperature.

7. Elution buffer I.

Component (final concentration)*	MW	g per 50 mL	mL per 50 mL
50 mM Tris-HCl, pH 7.5	121.14		of 1M Tris-Cl (pH7.5)
150 mM NaCl	58.44		of 4 M NaCl
2 mM EGTA	380.4		of 0.2 M EGTA

*β-Mercaptoethanol (final concentration 10 mM) and PMSF (final concentration 2 mM) must be added just before use.

8. Elution buffer II.

Component (final concentration)*	MW	g per 50 mL	mL per 50 mL
50 mM Tris-HCl, pH 7.5	121.14		of 1M Tris-Cl (pH7.5)
1 M NaCl	58.44		of 4 M NaCl
2 mM EGTA	380.40		of 0.2 M EGTA

*β-Mercaptoethanol (final concentration 10 mM) and PMSF (final concentration 2 mM) must be added just before use.

B. Preparation of soluble proteins (all steps are carried out at 4 °C)

9. Suspend each uninduced and induced cell pellet in 1.5 mL of pre-chilled lysis buffer, transfer to a 2.0-mL microcentrifuge tube, and incubate on ice for 5 min.
10. Place the tubes in liquid nitrogen for ~ 1 min so that the solution completely freezes and thaw at 37 °C in a water bath. Repeat this and keep in ice.
11. Sonicate the cell suspension with a micro-tip at 24–25% amplitude for 20 s ON and 25 s OFF cycles over 3 min.

**Keep the sample tube in ice during sonication.*

12. Centrifuge at 13 000 rpm, at 4 °C for 10 min.
13. During centrifugation, label two 1.5-mL tubes with "induced supernatant (**IS**)" and label one 1.5-mL tube with "uninduced supernatant (**US**)" and keep all the tubes on ice.
14. Transfer the clarified induced lysate to a pre-chilled 1.5-mL tube on ice.

**This sample will be used for Part D on column chromatography).*

15. Take 30 μL of induced supernatant (**IS**) and transfer to a labeled pre-chilled 1.5-mL tube, add 8 μL of 4× SDS sample dye, mix, and store at −80 °C.

**This sample will be used for SDS-PAGE analysis of Experiment 42 next week.*

16. Take 30 μL of uninduced supernatant (**US**) and transfer to a labeled pre-chilled 1.5-mL tube, add 8 μL of 4× SDS sample dye, mix, and store at −80 °C. Discard the remaining uninduced supernatant.

**This sample will be used for SDS-PAGE analysis of Experiment 42 next week.*

C. Preparation of insoluble proteins

17. Add 1 mL of cell pellet washing buffer to each of the uninduced and induced pellets, suspend the pellet completely, spin at 13 000 rpm, at 4 °C for 10 min, and aspirate off the supernatant.
18. Resuspend each pellet in 1 mL of resolubilization buffer (8 M urea, 50 mM Tris-Cl, pH 7.5) and stand at room temperature for 1 h.
19. Take 30 μL each of the induced pellet (**IP**) and uninduced pellet (**UP**), transfer to a labeled pre-chilled 1.5-mL tube, add 8 μL of 4× SDS sample dye to each tube, mix, and store at −80 °C. (*This sample will be used for SDS-PAGE analysis of Experiment 42 next week.*) Save the remaining IP and UP samples on ice for protein quantification.

D. Column chromatography of induced lysate supernatant

**Before you begin, label all the tubes needed to complete the protocol.*

**Make sure that all the material components of the column including the resin are at room temperature to avoid air bubble formation in the packed column.*

**Do not use a micropipettor but use a plastic Pasteur pipette to load the sample and buffer.*

Column equilibration

20. Set a Polyprep (Bio-Rad) column vertically on a stand and secure with a clamp.
21. Close the bottom outlet of the column with a cap.
22. Fill the column with ~0.1 mL of binding buffer to remove air pockets.
23. Slowly pour 0.4 mL of calmodulin affinity resin suspension (approximately 0.2 mL packed resin) into the column.
24. Place a glass test tube underneath the column and remove the bottom cap.
25. Wait for the resin to pack and drain.
26. Add 4 mL of binding buffer and let it flow through. Check the flow rate (mL/min) during this step. Empty out the flow-through buffer and immediately proceed to do the next step.

Sample loading

27. Carefully load the pooled induced soluble protein sample (step 14 of Part B) on to the column bed surface by slowly releasing the sample against the wall near the bed surface).

**Do not disturb the resin surface.*

28. Collect all the flow-through in a test tube and transfer 1 mL to a 1.5-mL microcentrifuge tube labeled "**FT**." Discard any remaining flow-through.

Column washing

29. Carefully apply 4 mL of binding buffer to remove any unbound proteins and collect all the flow-through. Do not discard the flow-through and combine with the next wash buffer flow-through.

30. Carefully apply 4 mL of washing buffer to remove non-specifically bound proteins and collect all the flow-through fractions.

31. Transfer 1 mL of wash fraction to a 1.5-mL microcentrifuge tube labeled "**W**." Discard any remaining wash eluent.

Protein elution

32. Carefully apply 1.5 mL of elution buffer I. Collect a 0.5-mL fraction (7 to 8 drops) per tube on ice. You will need 3 tubes labeled "**EFI-1**," "**EFI-2**," and "**EFI-3**."

33. Carefully apply 1.5 mL of elution buffer II. Collect a 0.5-mL fraction (7 to 8 drops) per tube on ice; you will need 3 tubes labeled as "**EFII-1**," "**EFII-2**," and "**EFII-3**."

34. Check all the fractions under UV light for green fluorescence.

35. Store all the samples at −80 °C for SDS-PAGE/Western blot next week.

E. Regeneration of calmodulin affinity resin (all steps are carried out at room temperature)

36. Wash with 3 column bed volumes (3 mL) of regeneration buffer 1.

37. Wash with 3 column bed volumes of regeneration buffer 2.

38. Wash with 3 column bed volumes of binding buffer.

39. Wash with 3 column bed volumes of regeneration buffer 3.

40. Re-equilibrate with 3 column bed volumes of binding buffer.

41. Remove the equilibrated resin from the column using 20% ethanol and store at 4 °C in a tube marked with the number of times used.

**Do not regenerate more than 3 times but discard.*

Part III. Small-scale quick batch method (optional)

Small-scale purification of CBP fusion proteins (50 to 150 μg) can be done in a microcentrifuge tube. This method is useful for optimizing conditions to determine the amount of resin per unit volume of extract prior to a larger-scale purification.

1. Place aliquots ~50 μL of calmodulin affinity resin slurry into a few 1.5-mL microcentrifuge tubes, spin at 1000 rpm at 4 °C for 2 min, and remove the supernatant fluid.

2. Add 200 μL of binding buffer, resuspend the resin using a P200 micropipettor, spin as above, and remove the supernatant. Repeat this washing step three more times to equilibrate the resin.

**Do not make foams during resuspension.*

3. Resuspend the equilibrated resin in clarified *E. coli* lysate and bring the slurry to a total volume of 300 μL with the binding buffer.

**Test different lysate volumes for the constant 50 μL resin.*

4. Rotate the tube containing the lysate–resin mixture on a shaker at 4 °C for 2 h.

5. Spin at 1000 rpm at 4 °C for 2 min and collect and save the unbound material. This will be used for SDS-PAGE analysis.

6. Add 300 μL f binding buffer to the bead, resuspend, spin at 1000 rpm at 4 °C for 2 min, and collect the wash supernatant. Repeat this step four more times (a total of 5 washes).

**The final wash fraction should contain no detectable protein by SDS-PAGE or a negligible amount of protein by spectrophotometric protein determination assays.*

7. Elute the fusion protein by two sequential washes, each with 200 μL of elution buffer I, and pool the elution fractions.

**While most proteins elute efficiently with buffers containing 2 mM EGTA and low salt, some proteins require an additional elution with 50 mM Tris-HCl (pH 8.0), 2 mM EGTA, and 1 M NaCl to recover immobilized fusion proteins.*

Post-lab assignment

1. What are special about Rosetta-gami™ 2 host and ArcticExpress (DE3) strains described in the Introduction?

2. IPTG and L-rhamnose were used to express GFP-CBP fusion protein in a Lemon 21 (DE3) culture.

- **(a)** Why is IPTG a more potent inducer of the *lac* promoter than lactose?
- **(b)** Why is L-rhamnose added to the culture and how does it work?

3. What are the functional roles of the following components in the lysis, binding, wash, elution, and regeneration buffers in this experiment?

- **(a)** Lysozyme
- **(b)** Magnesium acetate
- **(c)** PMSF
- **(d)** ß-Mercaptoethanol
- **(e)** Imidazole
- **(f)** EGTA
- **(g)** NaCl (0.15 M, 0.3 M, and 1 M)

4. The difference between elution buffers I and II is the concentration of NaCl. What roles does NaCl play in affinity column chromatography? Why is elution buffer I followed by elution buffer II during the elution step?

5. Can EDTA be used in the elution steps in this experiment? Explain why EGTA was used for elution instead of EDTA.

6. On the basis of your answer to Post-Lab Assignment 4(c) of Experiment 40, what is the approximate molecular weight (dalton) of the target protein eluted from calmodulin affinity chromatography? Show the fusion protein sequence that clearly distinguishes between GFP and CBP sequences by underlining the CBP sequence. Calculate this by assuming that an average amino acid has 120 dalton.

Further reading

Affinity® Protein Expression and Purification System and Affinity Protein Expression Vectors. Instruction Manual (Revision # 073007d), Stratagene, Inc. (http://www.chem-agilent.com/pdf/strata/204300.pdf).

Hartley, J.L., Salehi-Ashtiani, K., and Hill, D.E. (2008). Proteome expression moves *in vitro*: resources and tools for harnessing the human proteome. *Nature Methods*, **5** (12): 1001–1002.

pET System Manual (2006). TB055 11th Edition, Novagen, Inc.

Studier, F.W. and Moffatt, B.A. (1986). Use of bacteriophage T7 RNA polymerase to direct selective high-level expression of cloned genes. *Journal of Molecular Biology*, **189**: 113–130.

Terpe, K. (2003). Overview of tag protein fusions: from molecular and biochemical fundamentals to commercial systems. *Applied Microbiology and Biotechnology*, **60**: 523–533.

Wagner, S., Klepsch, M.M., Schlegel, S., Appel, A., Draheim, R., Tarry, M., Hogbom, M., Wijk, K.J., Slotboom, D.J., Persson, J.O., and Gier, J.-W. (2008). Tuning *Escherichia coli* for membrane protein overexpression. *PNAS*, **105** (38): 14371–14376.

42 AMB 2 experiment 42: SDS-PAGE analysis of affinity column fractions

Purpose: This is up to you to write down.

Introduction

Polyacrylamide gel electrophoresis (PAGE) in the presence of sodium dodecyl sulfate (SDS) is the most popular procedure for separating proteins because it is relatively simple to use and highly reproducible. SDS binds to proteins so that only 0.1% SDS is sufficient to saturate the polypeptide chains with about one detergent molecule per two amino acid residues. For complete denaturation, a reducing agent such as β-mercaptoethanol and dithiothreitol is added to the protein sample to break disulfide bonds. As negatively charged SDS molecules bind to the unfolded proteins, all proteins are assumed to have a virtually identical charge-to-mass ratio with uniform shape. Thus, electric mobility of proteins in such a denaturing buffer-filled porous gel medium is mainly influenced by size; the smaller they are, the faster they move. Using a mixture of standard proteins of known molecular weights, the whole gel can be calibrated in terms of their migrated distances against logarithms of molecular weights. However, incomplete denaturation, unusual amino acid sequences (highly acidic or basic), and/or the presence of non-protein residues can affect mobility, resulting in a considerable error in estimating molecular mass. Hydrophilic proteins and glycoproteins bind less SDS, while hydrophobic proteins bind more SDS because SDS has a 12-carbon tail of hydrophobic molecule that binds roughly to two amino acid residues per molecule.

Two of the most commonly used SDS-PAGE techniques are glycine–SDS-PAGE (also known as Laemmli SDS-PAGE) and Tricine–SDS-PAGE, which employ Tris-glycine and Tris-tricine buffer systems, respectively. The choice depends on the optimal separation range of the buffer system. Laemmli SDS-PAGE is used for separating proteins of >30 kDa, whereas Tricine–SDS-PAGE is preferentially employed for proteins of <30 kDa. These two different separation characteristics are directly related to the very different pK_a values of the functional groups of glycine and tricine. For the detailed protocol of Tricine-SDS-PAGE, refer to Herman Schägger (2006).

Laemmli SDS-PAGE consists of two distinct gel regions referred to as stacking gel and separating (or resolving) gel for a vertical electrophoresis setup. Hence, it is also called a discontinuous SDS-PAGE gel. The stacking gel in the upper portion has a lower acrylamide concentration (~4%), a lower pH 6.8 and a lower ionic strength (0.125 M Tris) than the separating gel (6–20%; pH 8.8; 0.375 M Tris) in the lower portion. The lower ionic strength of the stacking gel creates a higher local electric field (V/cm) as compared to the separating gel. The higher electric field combined with the lower acrylamide percentage makes proteins of very different sizes move similarly with higher mobility through the large pore gel matrix. In addition, glycine molecules in the electrode buffer move behind while chloride ions move ahead of proteins in the stacking gel. When proteins reach the separating gel, their mobilities are reduced because of the increased acrylamide concentration and decreased field strength, whereas pH increase in the resolving gel results in glycine having a higher mobility than proteins due to its increased net negative charge. All these conditions together cause all of the proteins to become more compressed at the interface between the two parts of gel matrix, thereby sharpening the starting sample bandwidth for separation. The advantage of a discontinuous system is the use of a stacking gel to concentrate dilute protein samples. However, some proteins may be aggregated due to the stacking effect.

A continuous buffer system will be required if the discontinuous system cannot be used due to the stacking-induced protein aggregation, as indicated by a heavily stained band at the gel origin. In a continuous buffer system, constant pH and concentration of the same buffer ions are used for the gel made of one continuous % polyacrylamide, and the sample is loaded directly into the separating gel. Because of the absence of the stacking gel, the protein band width is determined in part by the height of the sample loaded in the well, so the sample should be concentrated to above 1 mg/mL to minimize the loading volume for a good resolution. It is recommended to watch the following YouTube videos as to the principle and how to assemble the gel apparatus, make a discontinuous SDS-PAGE gel, load sample, and run the gel:

- Overview: http://www.youtube.com/watch?v=orKtTwHLwIc
- Apparatus assembly: http://www.youtube.com/watch?v=IAfUEe4Db2U
- Making gel: http://www.youtube.com/watch?v=EDi_n_0NiF4
- Sample loading: http://www.youtube.com/watch?v=1KLADhc4cog
- Gel running: http://www.youtube.com/watch?v=koe0Xv7a2MM

For a reliable SDS-PAGE analysis, both the gel and protein sample must be of a high quality. Pre-cast SDS-PAGE gels are available from several manufacturers. Protein samples should be prepared in a buffer free of interfering substances such as chaotropic agents,

Methods in Biotechnology, First Edition. Seung-Beom Hong, M. Bazlur Rashid and Lory Z Santiago-Vázquez.

Companion Website: www.wiley.com\go\hong\Methodsinbiotechnology

detergents, lipids, and salts (especially potassium) and at protein concentrations adequate for analysis. Once proteins are separated by electrophoresis, they can be visualized in a gel by various stains in order to monitor the purification process by examining the intensity/thickness of bands (amount) and the number of bands (purity) in fractions obtained at each purification step. The separated proteins can also be blotted (transferred) on to a membrane for Western blot analysis or excised and extracted for mass spectrometry analysis. SDS-PAGE is, therefore, a common step in proteomics analyses. In this lab exercise, you will prepare and run two Laemmli discontinuous SDS-PAGE gels; one is used for staining protein bands and the other is used for blotting the membrane.

Pre-lab assignment

(Typing and submission must be completed before lab work begins.)

1. What is the final NaCl concentration after protein is diluted in 50 mM Tris-HCl (pH 8.0) at step 3 of the Part I experiment?
2. What toxic substance is used in SDS-PAGE and what should you do when working with the substance?
3. List four things that can be learned about proteins by conducting SDS-PAGE.
4. What are the two properties of SDS that make it important in this experiment?
5. What are the compositions of 4× SDS sample buffers? Besides SDS, what does each component in the SDS protein sample buffer do?

Materials and equipment

- Protein samples (Experiment 41)
- Amicon Centricon 10 (MWCO 10 000; Millipore, Cat. No. 4205)
- Mini-protean 3 electrophoresis system (Bio-Rad)
- 30% acrylamide/0.8% bisacrylamide
- 1.5 M Tris-Cl (pH 8.8)
- 0.5 M Tris-Cl (pH 6.8)
- 10% SDS
- Ammonium persulfate (APS)
- TEMED
- 4× SDS sample buffer (250 mM M Tris-HCl, pH 6.8, 40% glycerol, 8% SDS, 0.02% bromophenol blue)
- 10 × SDS electrophoresis buffer (0.25 M Tris base, 1.92 M glycine, 1% SDS)
- Water-saturated butanol
- 50 mM Tris-HCl (pH 8.0; autoclaved)
- Untagged GFP protein (ProSpec, Cat. No. PRO-687)
- Purified *E. coli* β-galactosidase (tetramer molecular weight 465 kDa; each subunit 116.3 kDa; Sigma G3153-5MG)
- Molecular weight markers: PageRuler plus pre-stained protein ladder (Thermo Fisher Scientific, Cat. No. 26619)
- GelCode™ blue staining solution (Thermo Fisher Scientific, Cat. No. 24590)
- Round floating 1.5 mL microcentrifuge tube racks, microcentrifuge safety cap locks
- Boiling water bath

Procedure

Part I. Protein concentration and buffer exchange

Refer to the instruction manual for Amicon® Ultra-2 Centrifugal Filter Devices:

http://www.emdmillipore.com/US/en/product/Amicon-Ultra-2-mL-Centrifugal-Filters-for-DNA-and-Protein-Purification-and-Concentration,MM_NF-C86533#documentation.

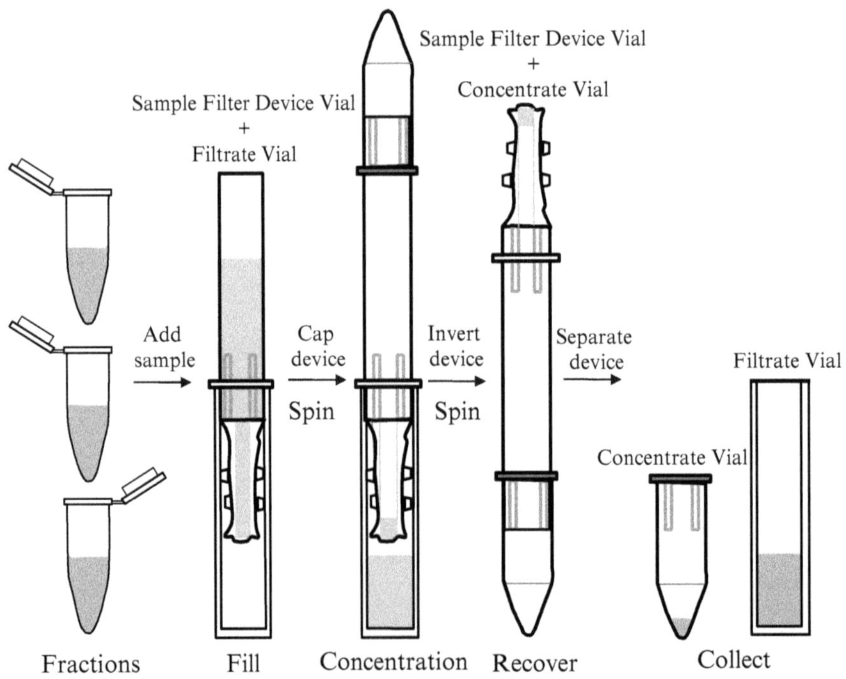

1. Thaw EB I and II fractions (obtained from Experiment 41) and place on ice.
2. Label two concentrate vials and sample filter device vials of a Amicon Centricon 10 unit.
3. Load 0.4 mL each of three EB I fractions and 0.8 mL of 50 mM Tris-HCl (pH 8.0) into each labeled sample filter device vial. Do the same for the three EB II fractions. Combine and save the remaining fractions on ice.
4. Place concentrate vial and filtrate vial on to the top and bottom of the sample filter device vial, respectively.
5. Centrifuge at 4000× g at 4 °C for 60 min (during this time, proceed to do Part II).
6. Invert the sample filter device vial and place over the concentrate vial.
7. Centrifuge at 1000× g for 2 min to transfer the concentrated sample to the concentrate vial.

**The volume of concentrate should be between 30 and 50 μL.*

8. Keep the sample on ice: 15 μL (or half of the concentrate) will be used for gel staining and the remaining 15 μL will be used for Western blot analysis.

**Check the concentrate for green fluorescence under UV light.*

Part II. Sodium dodecyl sulfate/poly acrylamide gel electrophoresis

A. Gel casting apparatus assembly

Refer to the instruction manual for a description of the Mini-Protean 3 system:

http://www.bio-rad.com/webroot/web/pdf/lsr/literature/4006157B.pdf.

1. Clean small and large glass plates with 70% ethanol.

**Each group needs two sets of small and large glass plates for making two gels.*

2. Place a short plate on top of a long plate that has spacers at the edge.

**Make sure that the spacer is in between the two plates to create a seal.*

3. Slide the two plates into the casting frame, keeping the short plate facing the front.

**Make sure that the bottom of the plates are flushed to avoid leakage.*

4. Set up the gel cassette assembly by locking the two pressure cams of the casting frame outwards to secure the glass plates.
5. Secure the gel cassette assembly in the casting stand by engaging the spring loaded lever.
6. Insert a P1000 pipette tip into the spring-loaded lever to make an additional pressure to seal the bottom of the gel cast assembly to the gasket of the casting stand.
7. Insert the comb and mark ~1 cm below the comb. This is the level to which the resolving gel solution is poured.

B. Pouring the separating gel

8. Prepare a 12% resolving gel solution in a 25-mL Erlenmeyer side-arm vacuum flask. Degas the acrylamide gel solution by stirring under a vacuum for ~3 to 5 min. Add APS and TEMED just before pouring and stir gently to mix.

**Degassing can be omitted with the decreased rate of polymerization.*

	12% Resolving gel (pH 8.8)*	
1	ddH_2O	1.9 mL
2	1.5 M Tris-Cl (pH 8.8)	4.0 mL
3	30% acrylamide†/0.8% bisacrylamide	3.9 mL
4	10% SDS	100 μL
5	Mix together and degas under vacuum for 2 min	
6	10% ammonium persulfate (APS): freshly prepared	55 μL
7	TEMED	15 μL
	Total volume	10 mL

*Use of Bio-Rad spacer 1.0 mm and 0.75 mm spacers in one glass plate sandwich requires 4.0 mL and 3.3 mL of resolving gel solution, respectively.
†Unpolymerized acrylamide is a neurotoxin, so always handle with gloves.

9. Using a 5-mL disposable pipette, apply the gel solution to the edge of the gel cassette assembly sandwich in a slanted position up to the marked line, and then stand it vertically.
10. Using a Pasteur pipette, *immediately* layer either dH_2O or water-saturated butanol slowly against the edge of one of the spacers until it reaches the other edge. Fill up to the top of the gel cassette sandwich.

**Overlaying provides a barrier to oxygen and allows a flat interface to form during polymerization so that all samples would pass through the resolving gel at the same position and time. Skip this step if a continuous gel is made, and instead pour the solution up to the top and insert the comb.*

11. Leave it to polymerize for 10 to 15 min at room temperature.

**A sharp line between the water layer and gel indicates completion of polymerization. Keep the remaining gel solution in a flask to check for gel polymerization. The amounts of APS and TEMED affect polymerization time.*

C. Pouring the stacking gel

12. Pour off the layered water and remove the residual water by inverting the casting sandwich into a paper towel. Use a piece of filter paper to dry the top of the resolving gel.
13. Prepare a 4% stacking gel solution in a 25-mL Erlenmeyer side-arm vacuum flask. Degas the acrylamide gel solution by stirring under a vacuum for ~3 to 5 min. Add APS and TEMED just before pouring and stir gently to mix.

	4% Stacking gel (pH 6.8)	
1	ddH_2O	3.00 mL
2	0.5 M Tris-Cl (pH 6.8)	1.25 mL
3	30% acrylamide/0.8% bisacrylamide	660 μL
4	10% SDS	50 μL
5	Mix together and degas under vacuum	
6	10% ammonium persulfate (APS)	40 μL
7	TEMED	10 μL
	Total volume	5 mL

14. Pour the stacking gel solution until it reaches the top of the glass plate and insert a comb immediately.

**Be careful not to trap air bubbles in the tooth edges of the comb.*

15. Let it polymerize until the gel becomes firm at room temperature (~10 min).
16. Remove the gel cassette sandwich, carefully remove the comb, and rinse the wells with dH_2O to remove any unpolymerized polymers.

**Do not remove the comb before the gel cassette sandwich is removed from the casting frame in the casting stand. Otherwise, distorted or broken wells will be formed.*

**Measure the dimension (length × width) of the resolving gel for Part III (see A and B procedures in Day 1 of Experiment 43).*

D. Assembly of gel electrophoresis unit

17. Prepare 450 mL of 1× SDS running buffer from a 10× buffer.
18. Place the two sets of gel sandwiches into the electrophoresis inner chamber (electrode assembly + clamping frame in the Bio-Rad Mini-PROTEAN 3 system) with the short plates facing inward. Put this assembled inner chamber into the mini tank (= lower chamber).
19. Fill the assembled inner chamber (= upper chamber: gel sandwich + electrode assembly + clamping frame in the Bio-Rad Mini-PROTEAN 3 system: ~125 mL) with a 1 × SDS running buffer so that the buffer level reaches the top of the large glass plate.

**Wait a few minutes to see if there is any leaking. If there is leaking, take apart the inner chamber and reassemble it. Pour buffer in the inner chamber again and look for leaks again. Repeat as necessary.*

**The level of buffer in the upper buffer chamber must be above the top of the short plate.*

**Flush out the wells thoroughly to remove any unpolymerized polymers with a stream of buffer using a Pasteur pipette.*

20. Pour 1 × SDS running buffer into the lower chamber (= outer chamber: mini tank in the Bio-Rad Mini-PROTEAN 3 system: ~200 mL) until the level is above the bottom of the glass plate. Avoid air bubbles trapped underneath the bottom of the gel.

E. Denaturing and loading protein samples

21. Keep the protein samples listed below *on ice* and transfer 30 μL to a pre-chilled and labeled (1, 2, 3, etc.) safety cap microcentrifuge tube that contains 10 μL of 4× SDS sample buffer.

**Do not leave the samples in the SDS sample buffer at room temperature until heating.*

Lane #	Sample	Sample (μL)	4× SDS sample buffer (μL)
1	Protein ladder*	N/A	N/A
2	Uninduced insoluble protein (UP)	30	10
3	Unduced soluble protein (US)	30	10
4	Induced insoluble protein (IP)	30	10
5	Induced soluble protein (IS)	30	10
6	Flow-through fraction (FT)	30	10
7	Wash fraction (W)	30	10
8	Affinity fraction I (concentrated)	30	10
9	Affinity fraction II (concentrated)	30	10
10	Untagged GFP (0.1 μg) + β-galactosidase (0.1 μg)	30	10

*5 μL PageRuler plus pre-stained protein ladder (Life Technologies).

22. Heat the samples at 90 to 100 °C for 4 min, spin all the samples at the top speed for 2 min before loading (to remove any aggregates), and cool to room temperature.
23. Load the same sample twice into the two mini-gels; each well of this gel holds ~15 μL of sample.

*Take a sample using P20, insert the pipet tip between the two glass plates, and eject the entire sample by steadily pressing down the plunger of the micropipettor.

**Do not release the plunger until the entire sample in the tip is expelled.*

**Be careful not to spill each sample over to the other sample well during ejection.*

24. Change the pipette tip and continue with the next different sample.

F. Running and removal of the protein gel

25. Place the lid on the mini tank; make sure to align the color-coded banana plugs and jack (red: + electrode; black: – electrode). Insert the electrical leads into the power supply.
26. Run at 100 V for 10 min and then run at 200 V for ~50 min until the bromophenol blue (BPB) dye has moved to the bottom of the gel.

**Check the gel after running for 10 min or so. If the buffer level in the upper buffer chamber has dropped below the top of the shorter plate, the gel will stop running.*

**During the running time, prepare one membrane and 6 filter papers whose size is identical to the size of the resolving gel without the stacking gel (see steps 5 to 8 of Experiment 43, Part I, A); if you correctly prepared the mini-gel, the size will be about 9.5 × 7.0 cm.*

27. Turn off the power supply, disconnect the electrical leads, remove the tank lid, and lift out the inner chamber assembly.
28. Open the cams of the clamping frame, remove the gel cassette, and pry the gel cassette carefully using a razor blade.

**Do not trim off the upper stacking gel portion for staining, but trim off the upper stacking gel portion for blot transfer to the membrane.*

G. Staining protein gel

29. Put the gel attached to the glass plate into a small tray containing ~50 mL dH_2O, detach and wash the gel on a platform shaker for 10 min.
30. Discard the water, and wash the gel twice more each for 10 min.

**This step is important to remove SDS because the SDS can interfere with staining.*

31. Add enough GelCode blue stain solution to immerse the gel, cover the staining tray with plastic wrap, and stain for 1 h to overnight on a platform shaker.
32. Wash the gel with ~50 mL of dH_2O twice each for 10 min on a platform shaker.
33. Take a picture of the gel with a digital camera and study the band patterns.

**The lower level of detection is 10 ng of protein per band.*

Part III. Protein blotting on to protran® nitrocellulose membrane

Refer to Parts A and B in the Experiment 43 protocol.

Post-lab assignment

1. **(a)** Glycine has two ionizable groups at pK_{a1} of the α-carboxyl group (=2.45) and pK_{a2} of the α-amino group (=9.60). Describe and calculate the percentage of the anionic form of a glycine molecule in stacking gel (pH 6.8) and resolving gel (pH 8.8). *Hint: use the Henderson–Hasselbach equation to calculate percentage.*
   (b) Predict the order of mobilities of glycine, Cl^-, and proteins in the stacking gel. Explain your choice.
2. Polyacrylamide gels are described by reference to two characteristics: (i) the total monomer concentration (%T: acrylamide + bis) and (ii) the cross-linking monomer concentration (%C: bis). "T" stands for the total percentage of acrylamide [acrylamide + bisacrylamide], and "C" stands for the percentage of cross-linker [bisacrylamide] in the total. What are %T and %C of the 30% acrylamide/0.8% bisacrylamide solution we used? Show your work.
3. You have been using a 30% acrylamide stock solution (29:1, acrylamide/bis ratio) to prepare a 10% SDS-PAGE gel for your routine analysis of a 55-kD target protein. In this gel, this protein is found to be in a mid-position of your gel after electrophoresis. Now the 30% stock solution you have used runs out, but in the lab you found two different 30% stock solutions with acrylamide/bis ratios of 19:1 and 37.5:1. What will happen to the target protein if you use 30% (19:1, acrylamide/bis) and 30% (37.5:1, acrylamide/bis) solutions? Describe the expected mobility of the target protein as compared to the mid-position of the target protein in the 30% (29:1) SDS-PAGE gel. Explain why it is.
4. How can you experimentally determine your answers to Pre-Lab Assignment 3 (four things that can be learned about proteins by conducting SDS-PAGE)?
5. Single-step affinity chromatography fractions often do not yield a single protein band but several bands when analyzed by SDS-PAGE. What are the reasons for this even with proper SDS-PAGE and chromatographic equilibration, binding, washing, and elution conditions? List at least three possible reasons.
6. A series of protein molecular weight standards were run on an acrylamide gel with the following results:

Protein	Molecular mass (kDa)	Migration (cm)
Phosphorylase B	92	1.11
Bovine serum albumin	68	1.72
Carbonic anhydrase	31	3.34
Soybean trypsin inhibitor	22	4.24
α-Lactalbumin	14	5.05

An unknown protein was run on the same gel and migrated 2.58 cm. What is the molecular mass of the unknown protein? Use a graph to show your work.

Reference

Schägger, H. (2006). Tricine-SDS-PAGE. *Nature Protocols*, **1**: 16–22.

Further reading

A Guide to Polyacrylamide Gel Electrophoresis and Detection. Bio-Rad Bulletin 6040 Rev. B. http://www.bio-rad.com/webroot/web/pdf/lsr/literature/Bulletin_6040.pdf.

Bollag, D.M., Rozycki, M.D., and Edelstein, S.J. (1996). *Protein Methods*, 2nd edition. John Wiley & Sons, Inc. ISBN 0-471-118370-0.

Gallagher, S.R. (2000). One-dimensional SDS gel electrophoresis of proteins. In *Current Protocols in Molecular Biology*, pp 10.2A.1–10A.34. John Wiley & Sons, Inc. ISBN 0-471-50338-X.

43 AMB 2 experiment 43: Western blot analysis of affinity column fractions

Purpose: This is up to you to write down.

Introduction

Western blot, also called immunoblot or protein blot, is commonly used to detect a specific protein in a complex protein mixture. Protein bands separated by SDS-PAGE are transferred to a membrane electrophoretically and are probed with the antibody directed against a target protein of interest. Stained gel can also be directly blotted on to a membrane. Ponceau S dye, a reversible red stain, is typically used to visualize protein bands on nitrocellulose membrane because it is compatible with all types of immuno-detection labels and substrates. This staining protocol is simple to use and yields results in about 10 minutes. When unstained molecular standards are used, their individual positions should be marked with a pencil directly on the blot membrane when they become visible after staining.

Before blotting, SDS-PAGE gel must be equilibrated in transfer buffer to remove the electrophoresis buffer components. Transferring protein bands from the gel to membrane, termed as protein blotting, can be achieved using either a buffer-filled tank transfer apparatus or a semi-dry transfer apparatus. In a tank transfer blotting system, a cassette assembly that holds the gel and membrane (nitrocellulose, nylon membrane, or PVDF) between stacks of filter papers and buffer pads is placed vertically into a tank filled with a large volume of cold buffer. In a semi-dry transfer blotting system, the gel and membrane are sandwiched between stacks of buffer-saturated filter papers and placed between two closely spaced cathode and anode plates. As compared to the tank transfer system, the semi-dry system requires a much shorter transfer time, typically less than 30 min at room temperature, and much less volume of transfer buffer with the ability to blot several gels simultaneously. Protein binding occurs via hydrophobic and ionic interactions between the membrane and proteins when an electric current is applied across the system. It is recommended to watch the following animation and video as to the principle and how to set up protein blotting using the semi-dry electrophoretic transfer system:

- http://www.youtube.com/watch?v=pRIn3p7x3V8 (Principle of immunoblot)
- http://www.benchfly.com/video/95/how-to-use-a-semi-dry-transfer-apparatus/(Blot transfer)

Western blot analysis is not limited to one-dimensional SDS-PAGE but can be applied to two-dimensional electrophoresis gel. After blot transfer, the membrane surface that has no transferred proteins needs to be blocked before antibody addition so that antibody protein cannot bind the unoccupied membrane sites non-specifically. To detect a protein antigen blotted on to membrane, a primary antibody in the form of either monoclonal or polyclonal (serum) antibody is added at an appropriate dilution concentration and incubated with the membrane regardless of the detection methods used. Any excess of primary antibody that does not bind should be washed away after incubation. For colorimetric detection of the bound antibody, anti-immunoglobulin (Ig) antibody conjugated to a reporter enzyme such as alkaline phosphatase (AP) and horseradish peroxidase (HRP) are added. Such an anti-Ig-enzyme conjugate is commonly called a secondary antibody or detection antibody. When a substrate is added after an unbound secondary antibody is washed out, its reaction with the conjugate enzyme produces a visible precipitated band where the antigen is located. Unlike colorimetric detection, the fluorescent detection method utilizes secondary antibodies linked to fluorescent dyes, rather than enzymes, and thus does not need substrates. Although detection sensitivity depends on the affinity of the primary antibody and the type of detection methods in use, Western blot can detect as little as 100 pg of protein in the best case.

To accurately interpret the result of a Western blot experiment requires an internal reference loading control, which is helpful in determining whether the same amount of protein sample is loaded into each lane. Housekeeping protein is typically used for the reference control since its expression level in cells remains unchanged under different conditions. If the molecular weight of the loading control differs from that of a target protein of interest, probing with primary antibodies of loading control and target protein can be carried out simultaneously. If both loading control and target protein are of the same size, the blot should be stripped of previously bound antibody before probing with a second primary antibody. However, if secondary antibodies labeled with different fluorophores are used, two proteins of the same size can be detected on a single blot by changing excitation and emission wavelength channels. The signal intensity of the loading control is typically used to normalize the signal intensity of target protein in order to equalize the differences in the gel loading and membrane transfer between samples. Different sample preparations need to use different loading controls.

Methods in Biotechnology, First Edition. Seung-Beom Hong, M. Bazlur Rashid and Lory Z Santiago-Vázquez.

Companion Website: www.wiley.com\go\hong\Methodsinbiotechnology

For whole cell/cytoplasmic proteins, the most commonly used controls are α- or β-actin (43 kDa), GAPDH (37 kDa), and α- or β-tubulin (55 kDa). GADPH can conveniently be used for the loading control of both prokaryotic and eukaryotic cells. For a serum sample, transferrin (77 kDa) is commonly used for the loading control.

In this lab exercise, mouse anti-GFP primary antibody and anti-mouse IgG (H+L)-AP conjugate secondary antibody are used in order to probe GFP protein immobilized on nitrocellulose membrane. The signal intensity of the band probed with mouse anti-β-galactosidase primary antibody will be used to estimate the amount of GFP protein by the ImageJ processing analysis program. However, it is important to understand that the signal data obtained from Western blot does not give an absolute quantity but a semi-quantitative measure. This is because the rates of blot transfer among samples may vary, a certain amount of protein may move through the membrane and on to filter papers, and the signal intensity may not be linear across a wide range of sample concentrations.

Pre-lab assignment

(Typing and submission must be completed before lab work begins.)

1. What volume of TBS solution is needed to conduct Parts C, D, and E experiments if you use the TBS solution to make both blocking and washing solutions? Show your work.
2. How would you make 100 mL of blocking solution from TBS solution?
3. What volumes of blocking and washing solutions are needed to conduct Parts C, D, and E experiments? Show your work.
4. List at least four things that you can learn from Western blot analysis.

Materials and equipment

- Trans-Blot® semi-dry transfer cell (Bio-Rad, Cat. No. 170-3940)
- Protran® nitrocellulose membranes BA83 (0.2 or 0.45 μm pore size)
- Whatman 3MM filter paper, overhead projector transparency film
- Electroblot transfer buffer (25 mM Tris, 192 mM glycine, 20% methanol)
- Living Colors® A.v. monoclonal antibody (JL-8) (mouse IgG2a isotype; 1 mg/mL: Clontech, Cat. No. 632380 or 632381)
- Mouse anti beta-galactosidase monoclonal Ab (100 μg, 2.0–2.5 mg/mL: Promega, Cat. No. Z3781)
- Anti-mouse IgG (H+L) AP conjugate (Promega, Cat. No. S3721)
- NBT (nitro blue tetrazolium chloride)/BCIP (5-bromo-4-chloro -3-indolyl phosphate): Fisher Scientific, Cat. No. PI-34042
- Electro-blot transfer buffer (25 mM Tris base, 192 mM glycine, 20% methanol)
- TBS (Tris-buffered saline) buffer (20 mM Tris-HCl, pH 7.5, 300 mM NaCl)
- Washing solution: TTBS (0.05% (v/v) Tween 20 freshly made in TBS)
- Blocking solution: TTBS/milk (0.05% Tween 20 and 5% non-fat dry milk freshly made in TBS)
- MilliQ H_2O
- Image J software (download from: http://rsb.info.nih.gov/ij/download.html and install according to the instructions; see AMB 1 Appendix 4 'Band Quantification by ImageJ Program')

Procedure

Part I. Blot transfer (day 1)

A. Gel equilibration and membrane/filter paper preparation (15–30 min)

1. Equilibrate the SDS-PAGE gel in electro-blot transfer buffer by adding enough transfer buffer to cover the gel and gentle shaking on a rotatory platform for 5 min.

**Low percentage gels (<12% acrylamide) will shrink in methanol-containing buffers.*

**The length of time required for equilibration is dependent on the gel thickness. For example, 15 minutes for a 0.75 mm SDS-PAGE gel.*

2. Change the transfer buffer and repeat twice (total three buffer changes in 15 minutes).While you wait for gel equilibration, prepare the membrane and filter papers following steps 3 to 7. You may do this during SDS-PAGE running time.

3. Cut one membrane and six filter papers to exactly the same size as the resolving gel.

**Handle the membrane with forceps and touch only the edges using a gloved hand so as not to damage the surface. If you neglect this rule, your blot will produce artifacts including fuzzy and smearing bands.*

4. Using a blunt pencil, write the group # on the edge of the membrane.

**Use of a sharp pencil will puncture the membrane. Never use a pen.*

**Neither touch nor press down the membrane surface with your gloved hand when you write.*

5. Cut a small piece on the upper corner of the membrane as an orientation guide.

6. Wet the membrane by slowly sliding into the transfer buffer and soaking for 10 min.

**If the PVDF membrane is used, immerse the membrane in methanol for 1 min before soaking in the transfer buffer.*

7. Wet the six filter papers by soaking in transfer buffer.

B. Semi-dry electro blot transfer (25 min)

Refer to the instruction manual for a description of the Trans Blot system:

- http://www.bio-rad.com/webroot/web/pdf/lsr/literature/4006066B.pdf
- http://www.bio-rad.com/webroot/web/pdf/lsr/literature/M1703940.pdf

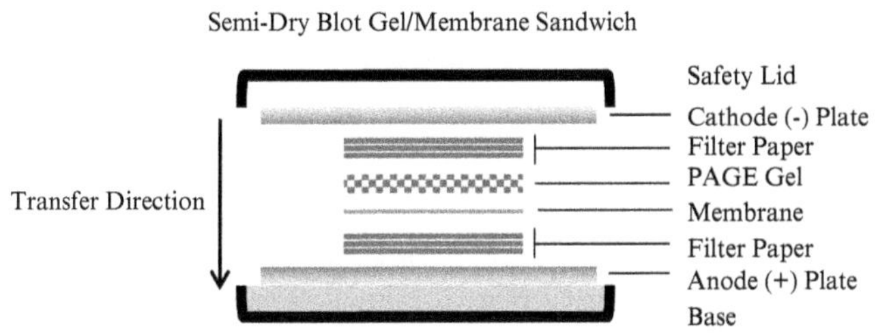

8. Remove the safety cover and stainless-steel cathode assembly, and place three pre-soaked filter papers on to the platinum anode.
9. Roll a clean glass test tube over the surface of the filter paper to push out any air bubbles.
10. Put the pre-soaked membrane on top of the filter paper and roll out any air bubbles trapped.
11. Place the pre-equilibrated gel on a pre-wet overhead projector transparency film, slide the gel on to the membrane surface by slowly pulling out the film while aligning the gel correctly on the membrane, and roll out any air bubbles trapped between the gel and membrane.

**To avoid ghost prints and other artifacts, do not move around the gel once it is positioned.*

**Do not use a hand to transfer the thin gel (<1 mm) as this will cause breakage. Alternatively, place the pre-soaked membrane on to the gel still attached to the glass plate, invert the membrane/gel sandwich, carefully put on to the pre-wetted filter papers, and then remove the glass plate.*

12. Placed three pre-soaked filter papers on top of the gel and remove bubbles.

**Make sure the upper filter paper does not contact with the lower filter paper. Otherwise it will inhibit or decrease transfer of proteins from the gel.*

13. Wipe up any excess buffer around the gel/membrane sandwich to prevent current leaks.
14. Place the cathode assembly on to the sandwich stack and the safety cover back on to the unit.
15. Connect the cables to the power supply (Bio-Rad PowerPack 200) and turn on the power at 10 V for 10 min for a 0.75 mm thick 12% mini-gel.

**The transfer time depends on the gel thickness and percentage and target protein MW.*

16. Remove the cathode assembly and disassemble the gel/membrane sandwich.
17. Rinse the membrane briefly in MilliQ H_2O to ensure that no residual gel pieces or smeared sample adhere to the membrane.
18. Put the membrane on to a clean paper towel, dry briefly in the hood, and store in a plastic bag at 4 °C until use.

Part II. Antibody detection (day 2)

C. Membrane blocking (30 min)

19. Prepare a fresh 12 mL of blocking solution TTBS/milk just before use and pour into a clean small tray slightly larger than the membrane.
20. Incubate the blotted membrane with gentle shaking for 30 min.

D. Primary antibody binding (45 min)

21. Prepare 10 mL of 1/5000 diluted primary antibody (mouse anti-GFP mAb) plus 1/4000 diluted primary antibody (mouse anti-β-galactosidase Ab) in a freshly made blocking solution of TTBS/milk.

**Optimization of the dilution (1:100 to 1:5000) may be needed according to the results. Too much primary antibody will result in non-specific bands.*

22. Pour out the blocking solution, replace with the diluted primary antibody solution, and incubate for 30 to 60 min on a rotary platform shaker.
23. Decant the solution and wash the membrane three times each for 5 min using 12 mL washing solution at a time.

E. Secondary antibody binding (50 min)

24. Prepare 10 mL of 1/7500 diluted secondary antibody (anti-mouse IgG-AP conjugate) in TTBS/milk.

**Optimization of the dilution (1:1000–1:20,000) may be needed according to the results.*

25. Add the diluted secondary antibody to the washed membrane and incubate for 30 to 60 min.
26. Decant the solution and wash four times each for 5 min using 12 mL of TTBS solution at a time.
27. Rinse briefly twice in MilliQ H_2O.

**Residual Tween 20 can affect substrate precipitation and cause band smearing.*

F. Color reaction (5 min to a few hours)

28. Incubate the membrane in BCIP/NBT substrate solution until the bands of interest have acquired the desired intensity.

**The substrate solution is light-sensitive; cover the lid with aluminum foil to protect it from light.*

29. Wash the membrane in MilliQ H_2O for several minutes and change the water at least once.
30. Take a photograph with a digital camera.

**The membrane can be air-dried and kept in a notebook.*

G. Densitometric quantification by digital image processing analysis program

31. Open the ImageJ program and follow the instruction of AMB 1 Appendix 4 'Band Quantification by ImageJ Program'.
32. Estimate the amounts of the protein bands by comparing the band intensities (band areas or percent data from ImageJ analysis).

**Remember that the E. coli Lemo21 (DE3) strain contains an endogenous intact lacZ encoding β-galactosidase protein.*

33. Estimate the concentration and total amount of GFP based on the β-galactosidase (~120 kDa) or untagged GFP band (27 kDa) signal intensity, gel loading volume of sample, and collected fraction volume.

Post-lab assignment

1. What role does each component of TTBS/milk play in the antibody detection step?
2. When your SDS-PAGE gel blotted membrane was probed with monoclonal and polyclonal primary antibodies, you consistently observed different band patterns. Probing with monoclonal antibody yielded a single 90-kDa band, whereas probing with polyclonal antibody gave two bands of 30 kDa and 90 kDa. Explain this observation. What would you do to verify your rationale?
3. What happens if blocking and/or washing steps are omitted or improperly performed? Describe the reasons and results in each case.
4. The detection protocol calls for the use of diluted antibodies in the primary and secondary antibody binding step. Why are they diluted? What happens if very high and low concentrations of antibodies are used, respectively?
5. Three types of membranes, nitrocellulose, polyvinyldifluoriden (PVDF), and nylon can be used for Western blotting. What are the advantage and disadvantage of the use of each membrane?
6. **(a)** What molecular forces are involved in antigen and antibody interaction?
(b) Can the used Western blot membrane be re-probed with different antibodies? If no, explain the reason. If yes, explain how you can perform this.
7. List at least (a) three ways in which ELISA is different from Western blotting and (b) three ways in which ELISA is similar to Western blotting. (c) Which test do you think would be more specific?

Further reading

Bolt, M.W. and Mahoney, P.A. (1997). High-efficiency blotting of proteins of diverse sizes following sodium dodecyl sulfate polyacrylamide gel electrophoresis. *Analytical Biochemistry*, **247**: 185–192.

Dionisi, H.M., Checa, S.K., and Viale, A.M. (1995). Protein immunoblotting of stained gels. *BioTechniques*, **19**: 348–350.

Gallagher, S.R., Winston, S.E., Fuller, S.A., and Hurrell, J.G.R. (2000). Immunoblotting and immunodetection. *Current Protocols in Molecular Biology*, pp. 10.8.1–10.8.21. John Wiley & Sons, Inc. ISBN 0-471-50338-X.

Protein Blotting Handbook: Tips and Tricks, 6th edition, EMD Millipore (http://static.fishersci.com/cmsassets/downloads/segment/Scientific/pdf/Millipore/protein_blotting_hndbk.pdf).

Trans-Blot® SD Semi-Dry Electrophoretic Cell Instruction Manual. Bio-Rad Cat. No. 170-3940.

Western Blotting – A Beginner's Guide (http://www.abcam.com/ps/pdf/protocols/WB-beginner.pdf).

44 AMB 2 experiment 44: Yeast media preparation and phenotypic analysis of yeast strains

Purpose: This is up to you to write down.

Introduction

The yeast *Saccharomyces cerevisiae* is a non-pathogenic single-cell fungus that is important in the baking and brewing industries and basic scientific studies. It is by far the best understood eukaryote because a wealth of information on molecular genetics and entire genome sequence is available. Moreover, it can be accessible to high-throughput mutant collections and easily manipulated for the genome-wide gene expression studies exploiting rapid growth, switching between diploid and haploid cell status, easy transformation, and mating. Importantly, it has a very high efficiency of the homologous recombination system so that a 50-bp DNA homology is enough for specific gene targeting in the genome to create a knockout, delete a gene, tag a gene at its endogenous locus with a reporter gene, exchange promoters, and integrate a transgene into a specific locus.

To work on yeast requires use of different types of growth media (also called culture media) as a rich complex medium, a synthetic minimal medium lacking an amino acid supplement, a synthetic defined medium containing a mixture of amino acids, or a selective medium. Nutritional markers are most commonly used for selection of yeast transformants and mating diploids. Most widely used selection markers are biosynthetic genes *URA3*, *HIS3*, *TRP1*, *LEU2*, and *ADE2* for auxotrophic complementation, though antibiotics geneticin (G418 sulfate), hygromcin B (*hph*), and aureobasidin A (*AUR1-C*) resistance genes are often used in yeast. Canavanine, a toxic analog of arginine, is often used as a negative selection of a wild-type yeast cell producing arginine permease (*CAN1*) as well as a positive selection of arginine permease-deficient mutant (*can1*). Combined with a defined mixture of amino acids and nucleotide precursors, YNB (yeast nitrogen base), SD (synthetic defined) base, SD minimal medium, and SC (synthetic complete) medium provide an excellent choice for cultivating and selecting auxotrophic strains of yeast. To prepare a minimal or a dropout medium, careful attention to the components of YNB and SD base and SD medium should be made with respect to the carbon and nitrogen source and supplement mixture because their commercial products may or may not contain a carbon and/or nitrogen source and have different types of supplement mixture depending on the manufacturers. SD or SC dropout medium lacks a single or more than one nutrient that allows selection for plasmids with specific nutritional markers or induction or repression of specific genes.

If you autoclaved the defined agar medium containing amino acids, it is important to pour the medium into plates as soon as it cools to 55 °C rather than letting it solidify and then re-melting because amino acid mixture makes the medium more acidic and hydrolyzes the agar during its re-melting. Unlike YPD media, re-melting SD or SC agar media presents a problem setting hard again. Thus, if you want to make SD or SC agar for re-melting, amino acids should not be included. Instead, a separate 50× filter-sterilized amino acid stock solution is added after re-melting and cooling to 55 °C. Use of high-quality MilliQ H_2O is unnecessary to prepare a culture medium, single-deionized water (dH_2O) being enough to prepare the culture medium. It is advised not to autoclave yeast agar medium longer than 18 minutes at 15 lb/in^2 for less than 1 liter volume since this will tend to make 2% agar plate soft. Use of the filter-sterilized SD or SC broth mixed with sterile (autoclaved and cooled to ~55 °C) 4% water agar solution produces a solid agar plate with less carmelization of sugar, which supports a healthier and faster growth of yeast than the entire autoclaved counterpart.

In this lab exercise, you will prepare both rich and SD dropout media that will be used for yeast transformation and mating in the next two lab exercises. You will also streak agar plates to isolate single yeast colonies. Then you will use them to verify the phenotypes of yeast strains before working on transformation and mating. Because different media are utilized for nutritional selection and phenotypic verification, it is important for you to implement aseptic techniques. Contamination is the most frequent event in student laboratory exercises, wasting time and resources. Never pour medium into Petri plates while you hold an autoclaved flask or bottle with an autoclave gloved hand. If pre-poured plates have been refrigerated for long-term storage, warm them to room temperature prior to plating or streaking.

It is recommended to watch the following YouTube video for the streak plate technique, http://www.youtube.com/watch?v=AaG3Pt3nwLQ, before coming to the lab. It is advisable to familiarize yourself with the genetic nomenclature of yeast, as well as the difference between the bacterial and yeast genetic nomenclature in AMB 2 Appendix 5.

Methods in Biotechnology, First Edition. Seung-Beom Hong, M. Bazlur Rashid and Lory Z Santiago-Vázquez.

Companion Website: www.wiley.com\go\hong\Methodsinbiotechnology

Pre-lab assignment

(Typing and submission must be completed before lab work begins.)

1. How many Petri plates are needed for four groups in today's experiment? Show your work.
2. How many and what size of flasks or bottles are needed for four groups in today's experiment? Show your work.
3. List all the components of YNB and briefly categorize them in general terms.
4. Which of the following strains can and cannot grow on YNB –Leu + galactose plate? Explain the reason for each case.

(a) *gal, LEU;*	(b) *GAL, leu;*	(c) *LEU, trp;*	(d) *leu, TRP.*

5. A strain with the yeast genotype *arg2-9, PDR3-11* has:
(a) a recessive mutation in the *ARG2* gene and a dominant mutation in the *PDR3* gene;
(b) a dominant mutation in the *ARG2* gene and a recessive mutation in the *PDR3* gene;
(c) a recessive mutation in the *ARG2* gene and no mutation in the *PDR3* gene;
(d) a dominant mutation in the *ARG2* gene and no mutation in the *PDR3* gene
(e) a recessive mutation in the ARG2 gene and a wild type *PDR3-11* gene.

Materials and equipment

- *S. cerevisiae* Y2HGold
  MATa, trp1-901, leu2-3, 112, ura3-52, his3-200, gal4Δ, gal80Δ, LYS2::GAL1$_{UAS}$*-GAL1*$_{TATA}$*-HIS3, GAL2*$_{UAS}$*-GAL2*$_{TATA}$*-ADE2, URA3:: MEL1*$_{UAS}$*-MEL1*$_{TATA}$*-AUR1-C, MEL1*
- *S. cerevisiae* Y187
  MATα, ura3-52, his3-200, ade2-101, trp1-901, leu2-3,112, met–, gal4Δ, gal80Δ, URA3::GAL1$_{UAS}$ *-GAL1*$_{TATA}$*-lacZ, MEL1*
- YPD medium (Clontech, Cat. No. 630409): a pre-mixed yeast extract (for vitamins and other nutrients), peptone (broken-down proteins), and dextrose (carbon source for yeast)
- Minimal SD base medium (Clontech, Cat. No. 630411): a pre-mixed YNB, ammonium sulfate, and glucose
- Adenine hemisulfate (0.4% in dH_2O; filter-sterilized)
- Dropout (DO) supplements (Clontech): –Trp, –Leu, –Leu/Trp, –Trp/–Ura, –Ura
- Arginine DO supplement (CSM-ARG, MP Biomedicals, Cat. No. 114510-112)
- Uracil, 0.4% adenine hemisulfate
- L-Canavanine (60 mg/mL in dH_2O; filter-sterilized)
- X-α-Gal (4%: 0.04 g in 1 mL of dimethylformamide; kept in the dark at –20 °C for ~6 months)
- Agar (Difco)
- Petri plates (100 × 15 mm)
- Autoclave
- Laminar flow hood cabinet

Procedure

Part I. Media preparation

A. YPDA broth (100 mL/group)

1. Weigh 5 g of YPD powder and add to 100 mL of dH_2O in a 250-mL flask or bottle.
2. Write "YPDA broth, Group #, Date" on the labeling tape attached to the flask.

**Do not directly label on the flask.*

3. Cover the opening top with aluminum foil and autoclave for 15 min at 121 °C and 15 psi.
4. After autoclaving, allow the medium to cool to ~60 to 55 °C.
5. Add 1.5 mL of sterile 0.4% adenine hemisulfate to the medium in the laminar flow hood.
6. Store at room temperature.

B. YPDA agar and YPD agar plates (250 mL/group 1 only)

1. Write "YPD agar" and "YPDA agar" on the labeling tapes of two 500-mL flasks.
2. Weigh 12.5 g of YPD base and 5 g agar and add to 233 mL of dH_2O in the YPDA agar-labeled flask. Do the same in the YPD agar-labeled flask.
3. Cover the opening top with aluminum foil and autoclave for 15 min at 121 °C and 15 psi.
4. Mark "YPD" and "YPDA" on the bottom edge of the Petri plate, each labeling10 plates.
5. After autoclaving, allow the agar media to cool to ~60 to 55 °C in a water bath.
6. Add 4 mL of sterile 0.4% adenine hemisulfate to the medium in the YPDA agar flask.
7. Swirl the flask gently.
8. Wrap a clean paper towel around the neck of the vessel and hold it to pour into 12 labeled Petri plates in the laminar flow hood. Push any surface air bubbles to the edge of the plate using a sterile pipette tip.

**As soon as the liquid medium touches the circled edge of the plate, stop pouring to dispense about 20 mL in each Petri plate.*

**Immediately after pouring, rinse the empty flask with tap water.*

9. Follow steps 7 and 8 using the YPD agar flask to pour into 12 labeled Petri plates.
10. Once the agar media are firm, carry out streaking with two yeast strains (Part II).

C. SD –Trp agar plates (500 mL/group 1 only: prepare for 4 groups)

1. Weigh the following components. *Calculation is needed in the pre-lab notebook.*

Chemical	For 1 L	For 500 mL
Minimal SD base medium	26.7 g	
−Trp dropout supplement	0.74 g	
Agar	20.0 g	

2. Add to 476 mL dH_2O in a 1-liter bottle labeled "SD –Trp Agar" on its tape.
3. Cover the opening top with aluminum foil and autoclave for 15 min at 121 °C and 15 psi.
4. Label 24 Petri plates (each plate ~20 mL) on the bottom edge: "SD –Trp, Date."
5. After autoclaving, allow the medium to cool to ~60 to 55 °C.
6. Swirl the flask gently.
7. Wrap a clean paper towel around the neck of the vessel and hold it to pour into Petri plates in the laminar flow hood. Push any air bubbles to the edge of plate using a sterile pipette tip.

**As soon as the liquid medium touches the circled edge of the plate, stop pouring and pour the next.*

**Immediately after pouring, rinse the empty flask with tap water.*

8. Once the agar media are firm, invert the plates in a stack, put into a vinyl bag or a plastic wrap, tape, label, and store at 4 °C (long-term storage) or at room temperature.

**You may stack up all the plates upside down in a vinyl bag and incubate at room temperature until next week to check for contamination before long-term storage at 4 °C.*

D. SD –leu agar plates (500 mL/group 2 only: prepare for 4 groups)

1. Weigh the following components. *Calculation is needed in the pre-lab notebook.*

Chemical	For 1 L	For 500 mL
Minimal SD base medium	26.7 g	
−Leu dropout supplement	0.69 g	
Agar	20.0 g	

2. Add to 476 mL of dH_2O in a 1-liter bottle labeled "SD –Leu Agar" on its tape.
3. Cover the opening top with aluminum foil and autoclave for 15 min at 121 °C and 15 psi.
4. Label 24 Petri plates (each plate ~20 mL) on the bottom edge: "SD –Leu, Date."
5. Do the same next steps as in the Part C experiment.

E. SD –leu/–trp agar plates (400 mL/group 2 only: prepare for 4 groups)

1. Weigh the following components. *Calculation is needed in the pre-lab notebook.*

Chemical	For 1 L	For 400 mL
Minimal SD base medium	26.7 g	
−Leu/−Trp dropout supplement	0.64 g	
Agar	20.0 g	

2. Add to 381 mL of dH_2O in a 1-liter bottle labeled "SD –Leu/–Trp Agar" on its tape.
3. Cover the opening top with aluminum foil and autoclave for 15 min at 121 °C and 15 psi.
4. Label 20 Petri plates (each plate ~20 mL) on the bottom edge: "SD –Leu/–Trp, Date."
5. Do the same next steps as in the Part C experiment.

F. SD –ura agar plates (100 mL/group 3 only: prepare for 4 groups)

1. Weigh the following components. *Calculation is needed in the pre-lab notebook.*

Chemical	For 1 L	For 100 mL
Minimal SD base medium	26.7 g	
−Ura dropout supplement	0.77 g	
Agar	20.0 g	

2. Add to 96 mL of dH_2O in a 250-mL flask labeled "SD –Ura Agar" on its tape.
3. Cover the opening top with aluminum foil and autoclave for 15 min at 121 °C and 15 psi.
4. Label 5 Petri plates on the bottom edge: "SD –Ura, Date."
5. Do the same next steps as in the Part C experiment.

G. SD +ura agar plates (100 mL/group 3 only: prepare for 4 groups)

1. Weigh the following components. *Calculation is needed in the pre-lab notebook.*

Chemical	For 1 L	For 100 mL
Minimal SD base medium	26.7 g	
−Ura dropout supplement	0.77 g	
Uracil	2.00 g	
Agar	20.0 g	

2. Add to 96 mL of dH_2O in a 250-mL flask labeled "SD +Ura Agar" on its tape.
3. Cover the opening top with aluminum foil and autoclave for 15 min at 121 °C and 15 psi.
4. Label 5 Petri plates on the bottom edge: "SD +Ura, Date."
5. Do the same next steps as in the Part C experiment.

H. SD –trp/–ura agar plates (600 mL/group 4 only: prepare for 4 groups)

1. Weigh the following components. *Calculation is needed in the pre-lab notebook.*

Chemical	For 1 L	For 600 mL
Minimal SD base medium	26.7 g	
–Trp/–Ura dropout supplement	0.72 g	
Agar	20.0 g	

2. Add to 572 mL of dH_2O in a 1-liter bottle labeled "SD –Trp/–Ura Agar."
3. Cover the opening top with aluminum foil and autoclave for 15 min at 121 °C and 15 psi.
4. Label 30 Petri plates on the bottom edge: "SD –Trp/–Ura, Date."
5. Do the same next steps as in the Part C experiment.

I. SD –arg/+can agar plates (100 mL/group 4 only: prepare for 4 groups)

1. Weigh the following components. *Calculation is needed in the pre-lab notebook.*

Chemical	For 1 L	For 100 mL
Minimal SD base medium	26.7 g	
–Arg dropout supplement	0.74 g	
Agar	20.0 g	

2. Add to 98 mL of dH_2O in a 250-mLflask labeled "SD –Arg/+Can Agar" on its tape.
3. Cover the opening top with aluminum foil and autoclave for 15 min at 121 °C and 15 psi.
4. Label 5 Petri plates on the bottom edge: "SD –Arg/+Can, Date."
5. Do the same next steps as in the Part C experiment.

J. SD –L/–W/–H/–A +X-α-gal (QDO) agar plates (250 mL/group 4 only: prepare for 4 groups)

1. Weigh the following components. *Calculation is needed in the pre-lab notebook.*

Chemical	For 1 L	For 100 mL
Minimal SD base medium	26.7 g	
–L/–W/–A/–H dropout supplement	0.60 g	
Agar	20.0 g	
4% (40 mg/ml) X-α-Gal*	1.0 mL	

*Add X-α-Gal to a final concentration of 40 μg/mL after autoclave.

2. Add to 240 mL of dH_2O in a 500-mLflask labeled "QDO" on its tape.
3. Cover the opening top with aluminum foil and autoclave for 15 min at 121 °C and 15 psi.
4. Label 12 Petri plates on the bottom edge: "QDO."
5. Do the same next steps as in the Part C experiment.

Part II. Single colony isolation (quadrant streaking)

1. Label Y187 on the bottom edge of one YPD agar plate and one YPDA agar plate.
2. Flame an inoculating loop, cool down the loop, and touch a well-isolated single colony (or a loopful of broth culture) of Y187 yeast strain.
3. Lift one edge of the agar plate lid while the plate rests on the lab bench.
4. Place the inoculating loop at the edge of agar surface and make parallel zigzag strokes over 1/4th of the plate (quadrant section 1).
5. Close the plate, re-flame and cool the loop, rotate the plate to the 1/4 turning left, open the plate lid, make 2 or 3 cross-overs from section 2 into section 1, and do a continuous zigzag streak in section 2 without crossing over section 1. Perform the same procedure for the next two sections 3 and 4.

**If you use a tiny amount of a single colony or liquid culture for inoculum, perform the above quadrant streaking without reflaming the loop (→ continuous streaking).*

**Ensure the re-flamed loop is cooled by briefly touching on the agar surface edge.*

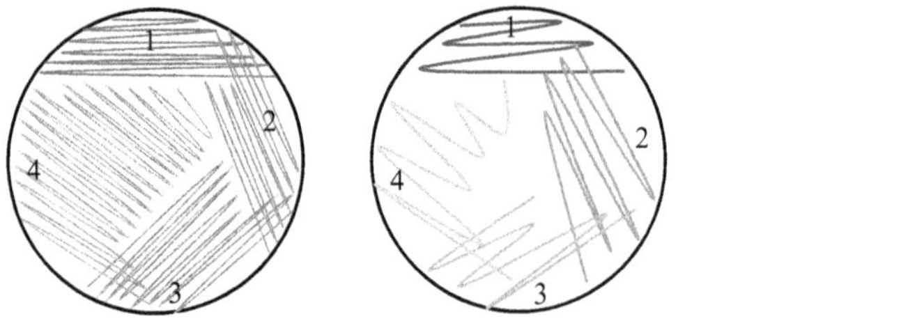

Good Streaking — Bad Streaking

Note: the more streaking, the more isolated are the single colonies. Therefore, make the maximum use of the space.

6. Repeat steps 1 to 5 using Y2HGold yeast strain.
7. Turn all the plates upside down and incubate at 30 °C until the yeast colonies reach ~2 mm in diameter. Observe the plates 2, 3, and 4 days after incubation. Fill in the table data.

Number (N) and color (C) of isolated colonies on the streaked plates

YPDA PLATE	QUAD 1		QUAD 2		QUAD 3		QUAD 4	
	N	C*	N	C*	N	C*	N	C*
	2-day	4-day	2-day	4-day	2-day	4-day	2-day	4-day
Y187								
Y2HGold								
YPD PLATE	**QUAD 1**		**QUAD 2**		**QUAD 3**		**QUAD 4**	
	N	C*	N	C*	N	C*	N	C*
	2-day	4-day	2-day	4-day	2-day	4-day	2-day	4-day
Y187								
Y2HGold								

*Record the data 2 and 4 days after incubation at 30 °C.

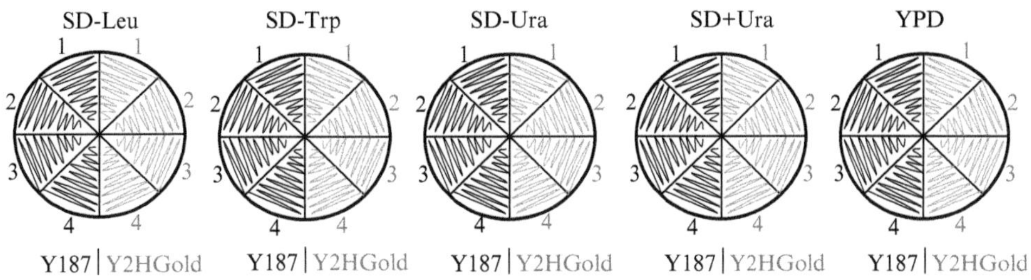

8. Take photographs, seal the plates with Parafilm, label, and store at 4 °C.

Part III. Phenotype Verification

1. As soon as you have obtained streaked colonies of Y187 and Y2HGold, circle 4 well-isolated single colonies on the back of the plate, *touch lightly* a single colony on the YPDA plate (NOT from the YPD plate) with a sterile inoculating loop and streak on to the labeled SD –Leu, SD –Trp, SD –Ura, SD +Ura, and YPD agar plates, as shown below.

**Do not use a heavy inoculum for streaking. The same number of streaks come from the same colony of each strain.*

2. Incubate at 30 °C for 3 to 4 days. Inspect your streaking results and check the phenotypes of growth on five plates.

Discussion

1. Did you successfully isolate more than 50 colonies in the 3rd and 4th quadrant sectors? If not, what should you have done more to have many well-isolated colonies?
2. Why is it important to isolate many single colonies rather than a few single colonies?
3. Did your streaked YPD plate have either white and pink-reddish colonies or only white colonies on every quadrant sector? What are they? What is your expected colony color as indicated by the yeast strain genotype?
4. Explain the observations of the Part II and III experiments. Are the growth patterns of yeast strains consistent with your expectation in Part III? If not, what is the problem?
5. What happened to Y187 and Y2HGold colonies on the streaked YPD and YPDA agar plates in terms of their size and color? Explain why this is.

Post-lab assignment

1. How can you tell if streaked colonies on a plate come from a pure culture?
2. Both *E. coli* and *S. cerevisiae* are most commonly used in the laboratory, and thus their cultures are often contaminated by each other. Both colonies may look similar at first. How can you simply distinguish between the yeast and *E. coli*? List at least three approaches.
3. List and explain *ALL* of the phenotypic characteristics or functionalities of Y187 and Y2HGold strains based on their genotypes listed in the Materials and Equipment section.
4. Why are Y2HGold and Y187 yeast strains URA positive when the genotype indicates that they have a *ura3-52* mutation, which normally confers Ura auxotrophy (the inability to grow on medium lacking uracil)?
5. In the Part III procedure you are instructed: *"Do not use a heavy inoculum for streaking."* Why do you think auxotrophic yeast strains show a growth in the minimal selection medium deficient in the auxotrophic nutrients in a relatively short incubation time if the cells are heavily inoculated?
6. Distinguish among the SD base (DOB), SD minimal medium, SD medium, SD dropout medium, SC medium, and SC dropout medium. Refer to the information in the following websites:

- http://www.formedium.com/us/products/saccharomyces-cerevisiae.html
- http://www.mpbio.com/index.php?cPath=1_7_45

Further reading

Bergman, L.W. (2001). Growth and maintenance of yeast. In *Methods in Molecular Biology* (Edited by P.N. MacDonald), Vol. **177**, pp. 9–14. Humana Press Inc., NJ.

Sherman, F. (1997). An introduction to the genetics and molecular biology of the yeast *Saccharomyces cerevisiae*. In *The Encyclopedia of Molecular Biology and Molecular Medicine* (Edited by R.A. Meyers), Vol. **6**, pp. 302–325. VCH Publisher, Weinheim, Germany.

Sherman, F. (2002). Getting started with yeast. *Methods in Enzymology*, **350**: 3–41.

45 AMB 2 experiment 45: Yeast transformation for yeast two-hybrid (Y2H) assay and genome editing by CRISPR-Cas system

Purpose: This is up to you to write down.

Introduction

Yeast can be transformed using spheroplast, lithium acetate, and electroporation techniques. The spheroplast method requires enzymatic removal of the cell wall with lyticase or zymolase prior to addition of transforming DNA. The lithium acetate method involves pre-treatment of cells with lithium acetate along with polyethylene glycol and dimethyl sulfoxide in order to render the cell wall permeable to nucleic acids. The electroporation method forces DNA to directly enter yeast cells without removing or pre-treating the cell wall through a high voltage electric pulse. The spheroplast method is very efficient and thus used for introducing several hundred kb of DNA fragments into yeast. The lithium acetate protocol has been routinely practiced for strain manipulation and gene cloning experiments mainly due to the simple, rapid, and inexpensive procedure with good transformation efficiency. However, the electroporation system is more efficient so that yeast cells can be directly transformed without any preparation of cell cultures with defined growth phase, just by scraping the cells off the agar surface and using them for electroporation. High transformation efficiency is important especially when transformation is performed to construct a cDNA library or to screen the library using a conventional library-scale yeast transformation. Unlike the spheroplast method, lithium acetate and electroporation techniques are not amenable to the uptake of large DNA segments (>30 kb).

Although *E. coli* is the first choice of heterologous protein expression, the yeast *S. cerevisiae* has some attractive features. Unlike *E. coli*, yeast has several post-translational modification mechanisms, enabling the expression of several eukaryotic proteins with appropriate functional modifications. In addition, some heterologous proteins secreted from specially engineered strains are correctly cleaved and folded and are easily harvested from yeast culture media. The use of signal peptides has allowed authentic maturation of protein products secreted by yeast. The importance of yeast transformation is illustrated by the fact that human vaccine hepatitis B virus core antigen, cheese-making enzyme, rennin, and synthesis of morphine from glucose were accomplished in the yeast for the first time. Above all, yeast is able to recombine large DNA segments that have overlapping sequences. A 583-kb synthetic *Mycoplasma genitalium* chromosome was assembled and cloned in yeast. The yeast *S. cerevisiae* can take up and assemble at least 38 overlapping single-stranded oligonucleotides and a linear double-stranded vector correctly in one transformation event, though the efficiency of correct assembly is markedly reduced as the number of DNA fragments increases. The oligonucleotides can overlap by as few as 20 nucleotides and can be as long as 200 nucleotides in length to produce kilobase-sized synthetic DNA molecules.

In the first part of our lab exercise, you will transform two yeast mating strains Y187 and Y2HGold with plasmids pGBKT7-53, pGADT7-T, and pCL1 for a yeast two-hybrid assay. The bait plasmid pGBKT7-53 encodes a DNA binding domain (BD) of yeast *GAL4* transcriptional factor gene that is in-frame fused to a 975-bp fragment of murine p53 bait gene. The prey plasmid pGADT7-T encodes an activation domain (AD) of yeast *GAL4* gene in-frame fused to a 2.0-kb fragment of SV40 large T-antigen prey gene. pCL1 is a control plasmid that contains a full-length wild-type *GAL4* gene. Maps of these expression plasmids and vectors are described in AMB 2 Appendix 6.

In the second part of the experiment, you will co-transform BY4733 strain with Cas9 expression and gRNA CAN1.Y expression cassette plasmids for RNA-mediated *CAN1* gene-targeted genome editing. Expression of Cas9 endonuclease and single guide RNA (sgRNA) of *CAN1* sequence makes a specific double-strand break in the genomic *CAN1* locus. Once a double-strand break is made, the cell's natural repair processes of homologous recombination (HR) and non-homologous end joining (NHEJ) are activated. In the absence of homologous *CAN1* DNA segment, NHEJ creates nucleotides deletion, insertion, or modification, resulting in a site-specific genome mutagenesis. For the detailed information on the genome editing plasmids used in this lab exercise, refer to DiCarlo *et al.* (2013) and www.addgene.org (Plasmid 43802, 43803). This system requires PAM (protospacer-adjacent motif) sequence NGG following the 20-bp crRNA (CRISPR RNA) in the form of the genomic target $GN_{20}GG$ sequence. Cas9 will cut 3 or 4 bases upstream of the PAM sequence. Scientists have further engineered to create mutant forms of wild-type Cas9; the so-called Cas9 nickase introduces only a single-strand break (nick), while Cas9 null nuclease does not make any DNA break but still retains its ability to bind DNA that matches the sgRNA sequence. It is recommended to watch the following videos regarding the bacterial CRISPR (Clustered

Methods in Biotechnology, First Edition. Seung-Beom Hong, M. Bazlur Rashid and Lory Z Santiago-Vázquez.

Companion Website: www.wiley.com\go\hong\Methodsinbiotechnology

Regularly Interspaced Short Palindromic Repeats)-Cas9 system to understand the principle of the CRISPR-Cas system mediated genome engineering:

- https://www.youtube.com/watch?v=SuAxDVBt7kQ (Genome Engineering with CRISPR-Cas9)
- https://www.youtube.com/watch?v=1aJxXWkE3Ek (CRISPR Cas9 – A Brief Introduction)

Because of the ease of expression cassette construction, genome-scale knockout screening of mammalian cells has been achieved recently using lentiviral vectors expressing both Cas9 and a library of sgRNA. Although the CRISPR-Cas9 mediated genome engineering is highly efficient, Cas9 has significant off-target mutations in its wild-type format. However, replacing three positively charged amino acids of wild-type *Streptococcus pyogenes* Cas9 with neutral ones has been shown to dramatically minimize off-target mutations. This newly engineered enzyme, termed as "enhanced" *S. pyogenes* Cas9, or eSpCas9(1.1), was shown to confer a high level of specificity with non-toxicity to human cells (Slaymaker *et al.*, 2016).

Pre-lab assignment

(Typing and submission must be completed before lab work begins.)

1. What are the bait and prey plasmids? Briefly explain how the yeast two-hybrid (Y2H) system works.

2. **(a)** What are the Cas9 expression cassette and sgRNA CAN1.Y expression cassette plasmids? Briefly explain how the CRISPR-Cas system works.

(b) What is the genomic target sequence in the sgRNA CAN1.Y expression cassette? (Refer to DiCarlo *et al.*, 2013.)

3. Calculate the volume of 1.1× TE/LiAc at step 9 (Part I) if you harvest 25 mL of cell culture that reads 0.8 at A_{600}.

4. What is the expected growth of transformants in step 19 of the Part I,C experiment? Indicate + (growth) or – (no growth) in the data table and explain your choice in each case.

5. **(a)** What is a canavanine and why is this included in the selection medium?

(b) What is the expected growth of BY4733 transformants, BY4733, and DPB276 on SD –Trp/–Ura and SD –Arg/+Can plate in the Part II experiment? Explain your choice in each case. To answer this, pay attention to the genotypes of BY4733 and DPB276 strains as well as the selection markers and a genome editing gene sequence carried by plasmids p414-TEF1p-Cas9-CYC1.t and p426-SNR5gRNA.CAN1 .y-Sup4t.

Materials and equipment

Part I (Y2H assay):

- SD –Leu, SD –Trp, and SD –Leu/–Trp agar plates
- pGBKT7-53 (0.2 μg/μL): 8.3-kb bait insert plasmid (*TRP1*, Km^R)
- pGADT7-T (0.2 μg/μL): 10.0-kb prey insert plasmid (*LEU2*, Amp^R)
- pCL1 (0.2 μg/μL): 15.3-kb control plasmid (*LEU2*; Amp^R)
- *S. cervisiae* Y187 strain: *MATα, ura3-52, his3-200, ade2-101, trp1-901, leu2-3,112, met–, gal4Δ, gal80Δ, URA3::GAL1*$_{UAS}$*-GAL1*$_{TATA}$*-lacZ, MEL1*
- *S. cervisiae* Y2HGold strain: *MATa, trp1-901, leu2-3, 112, ura3-52, his3-200, gal4Δ, gal80Δ, LYS2::GAL1*$_{UAS}$*-GAL1*$_{TATA}$*-HIS3, GAL2*$_{UAS}$*-GAL2*$_{TATA}$*-ADE2, URA3::MEL1*$_{UAS}$*-MEL1*$_{TATA}$*-AUR1-C, MEL1*

Part II (CRISPR/Cas-mediated genome editing):

- SD-Ura/-Trp and SD-Arg/+Canavanine agar plates
- p414-TEF1p-Cas9-CYC1t: 9.5-kb Cas9 expression plasmid (*TRP1*, Amp^R)
- p426-SNR5p-gRNA.CAN1.y-Sup4t: 6.3-kb gRNA.CAN1.y expression plasmid (*URA3*, Amp^R)
- *S. cervisiae* BY4733 strain: MATa *ade2::hisG, his3200, leu20, lys2delta0, met150, trp163, ura30*
- *S. cervisiae* BY4733 (p414-TEF1p-Cas9-CYC1.t)
- *S. cervisiae* DPB276 strain: *LEU2::pTEF-Dcr1, TRP1::pTEF-Ago1, can1-100, ade2-1, HIS3::pGAL1-hpSC_URA3*

Media and solutions for transformation:

- YPDA (1% yeast extract, 2% peptone, 2% dextrose, 0.003% adenine hemisulfate)
- Carrier DNA (10 mg/mL): sonicated, salmon sperm carrier DNA; prior to use, denature the carrier DNA by placing in a boiling water bath for 10 min and immediately cooling on ice; use only high-quality carrier DNA
- 1.1× TE/LiAc solution: prepare fresh just before use by mixing 1.1 mL of 10× TE with 1.1 mL of 10× LiAC and 7.8 mL of sterile dH_2O
- Sterile PEG/LiAc solution (polyethylene glycol/lithium acetate): prepare fresh aseptically just prior to use:

Components	Final concentration	To prepare 10 mL of solution
PEG 4000	40%	8 mL of 50% PEG
TE buffer	1× (10 mM Tris, 1 mM EDTA)	1 mL of 10× TE
LiAc	1× (0.1 M)	1 mL of 10× LiAc

- 50% PEG 4000 (polyethylene glycol, average MW = 3350): prepare with dH_2O; if necessary, warm solution to 50 °C to help the PEG go into solution; autoclave
- 10× TE buffer: 0.1 M Tris-HCl, 10 mM EDTA, pH 7.5; autoclave
- 10× LiAc: 1 M lithium acetate; adjust to pH 7.5 with dilute acetic acid; autoclave
- Sterile 0.9% (w/v) NaCl and 100% dimethyl sulfoxide (DMSO)

Procedure

Part I. LiAc-mediated transformation for Y2H assay

A. Preparation of yeast competent cells (steps 1 to 6 will be done by TA)

1. Inoculate 0.4 mL of YPDA with a single colony (2 to 3 mm in diameter) of Y187 strain; repeat the inoculation for Y2HGold strain.

2. Vortex vigorously to disperse any clumps.
3. Transfer 0.4 mL cell suspension into a 100-mL flask containing 20 mL of YPDA.
4. Incubate at 30 °C at 250 rpm overnight (~16 to 18 h) to stationary phase (A_{600} >1.5).
5. Next morning, transfer 10 mL of the overnight culture to a flask containing 100 mL of YPDA (A_{600} = 0.1 to 0.2).
6. Shake-incubate as above until A_{600} reaches 0.8 to 2.0 (for 4 to 6 h).

**Check A_{600} using the YPDA medium as a blank.*

**Cells must be freshly prepared. Do not store the cells in a refrigerator.*

7. Obtain 25-mL each of Y187 and Y2HGold cultures; label two 50-mL sterile conical tubes, place cell cultures into the tubes, and centrifuge at 3000 rpm for 5 min at room temperature.
8. Discard supernatant and suspend each pellet in a sterile 10 mL dH_2O and spin as above.
9. Re-suspend the cell pellet in a freshly prepared, sterile 1.1× TE/LiAc solution to a final cell density of 10^9 cells/mL, assuming that optical density at A_{600} of 1 is equivalent to 3×10^7 cell/mL.

B. Transformation of yeast competent cells

10. Add salmon sperm carrier DNA (50 μg) and plasmid DNA (~0.4 μg) to sterile 1.5 mL tubes.

TUBE 1: 5 μL (50 μg) carrier DNA + 1.5 μL pGBKT7-53 (bait) + Y2HGold
TUBE 2: 5 μL (50 μg) carrier DNA + 1.7 μL pGADT7-T (prey) + Y187
TUBE 3: 5 μL (50 μg) carrier DNA + 1.5 μL pGBKT7-53 (bait) + 1.7 μL pGADT7-T (prey) + Y2HGold (co-transformation)
TUBE 4: 5 μL (50 μg) carrier DNA + 2.5 μL pCL1 + Y187 (control)
TUBE 5: 5 μL (50 μg) carrier DNA + 2.5 μL pCL1 + Y2HGold (control)

**The order of addition matters; add carrier DNA plus plasmid first and cells last.*

11. Add 50 μL of yeast competent cells to each tube and mix well.
12. Add 500 μL of sterile PEG/LiAc solution to each tube and mix well by tapping.
13. Incubate at 30 °C for 30 min.
14. Add 25 μL of DMSO to each tube and mix well.
15. Place in a 42 °C water bath for 15 min.
16. Centrifuge cells at 8000 rpm for *15 s*, remove the supernatant, gently (no vortex) re-suspend the cells in 0.5 mL of sterile H_2O.

**Yeast cells at this point are fragile.*

C. Plating transformation mixtures

17. Spread 100 μL on each SD agar plate as shown below.

**Make sure your plates are well-labeled and that you are able to identify them from each other and from other groups' plates.*

**When spreading on to the plate, gently distribute the fluid completely while minimizing strokes with a sterile glass rod. Let the plates dry for a few minutes prior to incubation.*

(a) Y2HGold/pGBKT7-53 (bait): every group
- Leu:1 plate
- Trp: 2 plates

(b) Y187/pGAD7-T (prey): every group
- Leu: 2 plates
- Trp: 1 plate

(c) Y2HGold/pGBKT7-53 + pGAD7-T: every group
- Leu/-Trp: 2 plates

(d) Y187/pCL1 (+/– control): one setup for the whole group
- Leu: 2 plates
- Trp: 1 plate

(e) Y2HGold/pCL1 (+/– control): one setup for the whole group
- Leu: 2 plates
- Trp: 1 plate

18. Incubate all plates upside down at 30 °C until colonies appear (generally, 2 to 4 days).
19. Record the data.

Transformation mixture	Average number of colonies		
	SD –Leu	SD –Trp	SD –Leu/–Trp
Y2HGold + pGBKT7-53			N/A
Y187 + pGADT7-T			N/A
Y187 + pCL1			N/A
Y2HGold + pCL1			N/A
Y2HGold + pGBKT7-53 + pGADT7-T	N/A	N/A	

20. To calculate the transformation efficiency, count the colonies (cfu) growing on the dilution plate which has 30 to 300 cfu:

$$\frac{\text{cfu} \times \text{Total suspension volume (mL)}}{\text{Volume plated (mL)} \times \text{DNA amount used } (\mu\text{g})} \times \text{Dilution factor} = \text{cfu}/\mu\text{g DNA}$$

Sample calculation:
- 100 colonies grew on the 1/100 dilution plate (dilution factor = 100)
- Plating volume: 0.1 mL
- Total suspension volume (step 16 of Part I. B) = 0.5 mL
- Amount of plasmid = 0.1 μg

$$[(100 \text{ cfu} \times 0.5 \text{ mL}) \div (0.1 \text{ mL} \times 0.1 \ \mu\text{g})] \times 100 = 5 \times 10^5 \text{cfu}/\mu\text{g DNA}$$

21. Pick large colonies using sterile tooth picks or pipette tips, make cross marks on the selection media, and incubate at 30 °C for 2 to 3 days.

**Each transformant will be used for yeast mating and re-streaking on to a QDO plate (step 23 of Experiment 46).*

**Each number indicates the same colony from the selection plates.*

22. Seal plates with Parafilm and store at 4 °C.

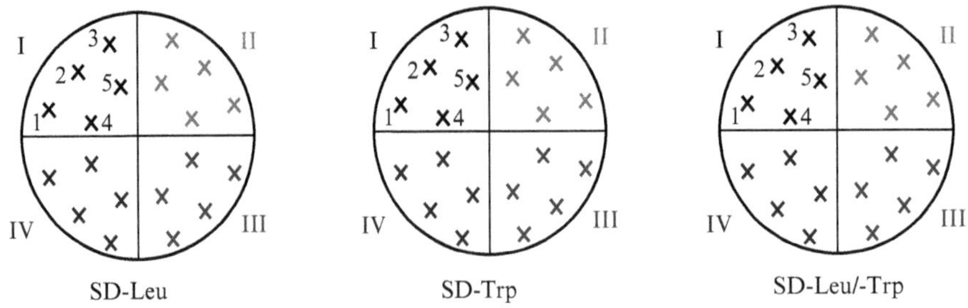

I: Y2HGold (pGBKT-7-53 + pGADT7-T): ✕
II: Y2HGold (pGBKT-7-53): ✕
III: Y187 (pGADT-7T): ✕
IV: Y2HGold (pCL1) or Y187 (pCL1): ✕

Part II. LiAc-mediated transformation for genome editing

1. Follow the same procedure as Part I, A (steps 1 to 9) using BY4733 and BY4733 (p414-TEF1pCas9-CYC1.t) strains.
2. Follow the same procedure as Part I, B (steps 10 to 16) using the BY4733 strain and plasmids p414-TEF1pCas9-CYC1.t and p426-SNR5p-gRNA.CAN1.y-Sup4t.
   TUBE 1: 5 µL (50 µg) carrier DNA + 2.5 µL p414-TEF1p-Cas9-CYC1.t + 2.5 µL p426-SNR5p-gRNA.CAN1.y-Sup4t + 50 µL BY4733 (set up duplicate co-transformation)
   TUBE 2: 5 µL (50 µg) carrier DNA + 2.5 µL p426-SNR5p-gRNA.CAN1.y-Sup4t + 50 µL BY4733 (p414-TEF1p-Cas9-CYC1.t)
   TUBE 3: 5 µL (50 µg) carrier DNA + 5 µL sterile H_2O + 50 µL BY4733 (– control)
3. After centrifugation at 8000 rpm for *15 s*, remove the supernatant and suspend the cell pellet in sterile water (150 µL for each Tube 1 and 300 µL for each of Tubes 2 and 3), and combine with the duplicate Tube 1 cell suspension.
4. Spread 150 µL each of TUBE 1, TUBE 2, and TUBE 3 on SD –Trp/–Ura plates (two plates each for TUBE 1 and TUBE 2, one plate for TUBE 3).

**Make sure your plates are well labeled and that you are able to identify them from each other and from other group plates.*

**When spreading on to the plate, gently distribute the fluid completely while minimizing strokes with a sterile glass rod. Let the plates dry for a few minute prior to incubation.*

5. Incubate and record data as in Part I, C.
6. Pick all of the individual transformant colonies of BY4733 (p414-Cas9, p426 gRNA.CAN1), streak them on SD –Trp/–Ura and SD –Arg/+canavanine plates, along with BY4733 (negative control) and DPB276 (positive control) cells, and incubate at 30 °C for 3 to 4 days.
7. Calculate the percentage of genome editing (Number of *can1* mutant/total # of Trp^+Ura^+ colonies × 100) from the above plates streaked with randomly selected transformants.

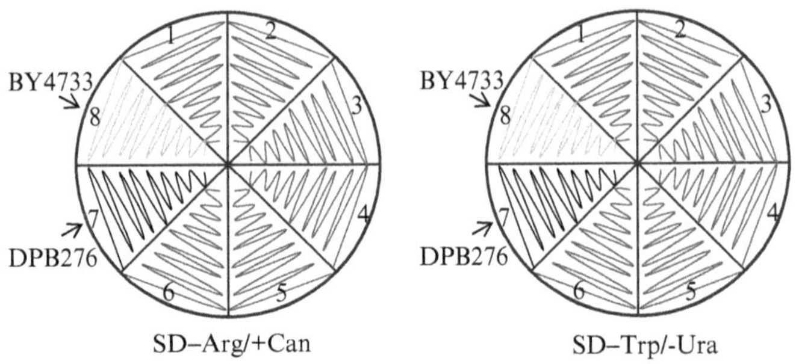

**Each number indicates the same colony from the selection plates.*

Post-lab assignment

1. **(a)** Calculate the number of each plasmid DNA molecule (pGBKT7-53, pGADT7-T, and pCL1) used in Part I, B. transformation. The MW of 1 bp is 660 g/mole. Show your work.
   (b) Which of the three plasmids used in the Part I lab exercise would you expect to have the highest and lowest transformation efficiency when you use the same moles of plasmid DNAs? Explain your choice.
   (c) If you want to determine the effect of plasmid size on transformation efficiency, what data in the lab protocol should you use in the Part I experiment? Provide the two independent lab protocols and explain your choice.
   (d) If you want to determine if transformation efficiency is influenced by yeast strains, what data in the lab protocol should you use in the Part I experiment? Explain your choice.
2. What is the purpose of adding and the function of (a) a single-stranded carrier DNA, (b) PEG, and (c) DMSO in the LiAc transformation protocol? How do they work for transformation?
3. Plasmid incompatibility is important for the success of transformation. What is the plasmid incompatibility and why does this happen?
4. Should plasmid incompatibility also be taken into consideration in yeast transformation? Explain your choice.
5. In the Part II genome editing experiment, all six transformants that carry both p414-TEF1-pCas9-CYC1.t and p426-SNR5p-gRNA.CAN1.y-Sup4t plasmids grew on SD –Trp/–Ura medium. However, five did not grow but only one grows on the SD-Arg + canavanine plate. What is not expected in the experiment? Explain your choice of the unexpected thing and why it may happen.

References

DiCarlo, J.E., Norville, J.E., Mali, P., Rios, X., Aach, J., and Church, J.M. (2013). Genome engineering in *Saccharomyces cerevisiae* using CRISPR-Cas systems. *Nucleic Acids Research,* **41** (7): 4336–4343. DOI: 10.1093/nar/gkt135. http://nar.oxfordjournals.org/content/early/2013/03/03/nar.gkt135.full.pdf+html.

Slaymaker, I.M., Gao, L., Zetsche, B., Scott, D.A., Yan, W.X., and Zhang, F. (2016). Rationally engineered Cas9 nucleases with improved specificity. *Science,* **351**: 84–88.

Further reading

Becker, D.M. and Lundblad, V. (2000). Introduction of DNA into Yeast Cells, Current Protocols in Molecular Biology. Unit 13.7.1-13.7.10, John Wiley & Sons, Inc. ISBN 0-471-50338-X.

Fossati, E., Narcross, L, Ekins, A., Falgueyret, J.-P., and Martin, V.J. (2015). Synthesis of morphinan alkaloids in *Saccharomyces cerevisiae. PLoS ONE,* **10** (4): e0124459. DOI: 10.1371/journal.pone.0124459.

Fu, Y., Foden, J.A., Khayter, C., Maeder, M.L., Reyon, D., Joung, J.K., and Sander, J.D. (2013). High-frequency off-target mutagenesis induced by CRISPR-Cas nucleases in human cells. *Nature Biotechnology,* **31**: 822–826.

Fu, Y., Sander, J.D., Reton, D., Cascio, V.M., and Joung, J.K. (2014). Improving CRISPR-Cas nuclease specificity using truncated guide RNAs. *Nature Biotechnology,* **32**: 279–284.

Gibson, D.G. (2009). Synthesis of DNA fragments in yeast by one-step assembly of overlapping oligonucleotides. *Nucleic Acids Research,* **37**: 6984–6990.

Gibson, D.G, Benders, G.A., Andew-Pfannkoch, A., Dennisova, E.A., Baden-Tillson, H., Zaveri, J., Stockwell, T.B., Brownley, A., Thomas, D.W., Algire, M.A., Merryman, C., Young, L., Noskov, V.N., Glass, J.I., Venter, J.G., Hutchison, C.A., and Smith, H.O. (2008). Complete chemical synthesis, assembly, and cloning of a *Mycoplasma genitalium* genome. *Science,* **319**: 1215–1220.

Mali, P., Aach, J., Stranges, P.B., Esvelt, K.M., Moosburner, M., Kosuri, S., Ynag, L., and Church, G.M. (2013). Cas9 transcriptional activators for target specificity screening and paired nickases for cooperative genome engineering. *Nature Biotechnology,* **31**: 833–888.

Matchmaker® Gold Yeast Two-Hybrid System User Manual (2013). Clontech Laboratories, Inc. Protocol No. PT4084-1, Version No. 092413.

46 AMB 2 experiment 46: Yeast mating-mediated Y2H assay and genomic PCR

Purpose: This is up to you to write down.

Introduction

Yeast (*Saccharomyces cerevisiae*) is a unicellular eukaryote that can live and reproduce stably as either a diploid or a haploid. It typically grows and divides by budding as a diploid in its natural environment, but upon facing stressful conditions such as nutrient depletion, heat, osmotic shock, and oxidative stress, it undergoes meiosis to produce an ascus containing four haploid spores (ascospores) that consist of two "*a*" spores and two "α" spores. The ejected spores are easily spread and resistant to harsh living conditions. When growth conditions become favorable, the ascospores germinate and establish the haploid phase of the life cycle. The ascospores can be isolated and examined individually in the laboratory, allowing genetic analysis of the inheritance patterns of specific genes.

The mating of yeast only occurs between "*a*" and "α" haploids of the opposite mating type. This is determined by a single locus, *MAT*, which exists in two alternative states of *MATa* or *MAT*α. Through a gene conversion mechanism, haploid yeast can freely switch the mating type as often as every cell cycle. Each mating type cell secretes its own pheromone, a small diffusible peptide hormone, to communicate as a mating signal. In response, yeast cells reprogram the cell cycle to stop dividing, change their shape to an elongated pear-shaped structure called "shmoo," contact each other, and fuse together to make a single cell of diploid zygote. Thus, production of mating factor pheromone brings about the growth inhibition of the opposite mating type, giving rise to a distinct halo around cells that produce the pheromone in the agar plate assay.

Yeast mating is a convenient method of introducing two different plasmids into the same host cells as an alternative to yeast co-transformation. Yeast mating can be performed in a microcentrifuge tube or a microtiter plate. If you have many prey plasmids of interest to introduce into several strains carrying bait plasmids for the yeast two-hybrid (Y2H) interaction assay, it is more convenient to use a sterile microtiter plate. The principle of yeast mating is that when the experimental haploid strain having one or more auxotrophic markers (e.g., *ura3* or *leu2*) is mated to the tester haploid strain having other auxotrophic markers (e.g., *trp1* or *his3*), any resultant diploids will be able to grow on dropout medium lacking auxotrophic nutrients but neither haploid will be able to grow.

In yeast cells some foreign bait proteins fused to the GAL4 DNA binding domain (BD) may have transcriptional activating properties. Some prey proteins fused to the GAL4 activation domain (AD) may have DNA binding properties or interact with TATA box promoter-binding protein. Some yeast protein may bridge the bait and prey proteins. Therefore, it is important to check the phenotypes of both bait and prey transformants before a Y2H assay. To avoid false positives, stringent selection criteria should be adopted to obtain useful information. The Y2H assay is useful for screening drug substances that disrupt or stimulate a specific protein–protein interaction. If bait and prey proteins interact with each other and the interaction is affected by a drug compound, the expression level of reporter genes will be changed.

In the first part of the lab exercise, you will use two opposite mating type cells carrying auxotrophic markers, one of which harbors bait plasmid (pGBKT7-53) and the other prey plasmid (pGADT7-T). The time of incubation in dropout selection media may vary and depend on the strength of the interacting proteins. For instance, weaker protein–protein interactions may lead to growth that is visible after 5 to 8 days. The selection criteria is the size of the yeast colony; it should be 2 mm or larger. Often colonies do not grow to be larger than 2 mm after 2 to 3 days. A feature to which attention must be paid is the size and the color of yeast in the restrictive dropout media since they are signal indicators of the strength of protein–protein interaction.

In the second part of the lab exercise, you will isolate genomic DNA from Trp^+Ura^+ yeast cells (Experiment 45), amplify short DNA fragments of wild type *CAN1* and putative genome-edited *can1* genes, and gel-purify the PCR fragments. These fragments will be used for DNA sequencing in Experiment 47.

Pre-lab assignment

(Typing and submission must be completed before lab work begins.)

1. Why do you spread cells on different types of SD media (step 13 of Part I)? What will the growth on each agar plate tell you?
2. Is it all right to suspend cells and make your dilutions in YPD instead of H_2O when you are going to plate the cells on dropout media (steps 14 and 15 of Part I)? Explain your choice.

Methods in Biotechnology, First Edition. Seung-Beom Hong, M. Bazlur Rashid and Lory Z Santiago-Vázquez.

Companion Website: www.wiley.com\go\hong\Methodsinbiotechnology

3. **(a)** What is the purpose of using SD –L/–W/–H/–A +X-α-gal (quadruple dropout QDO) plate (step 10 of Part I)?
(b) Which ones are transformation markers or reporter markers?
(c) What will the growth and color on this QDO plate tell you?
4. What are the expected colors of all the individually streaked colonies on QDO and YPD plates after the agar plate assay (steps 10 and 23 of Part I)? Explain why they are.
5. Perform the NCBI (http://blast.ncbi.nlm.nih.gov/Blast.cgi) nucleotide blast search against the database of "Others (nr etc.)" under Choose Search Set using *CAN1*-F and *CAN1*–R primer nucleotides. What is the expected size of the genomic PCR fragment? Show the sequences of amplified PCR fragment within the CAN1 mRNA coding sequence.

Materials and equipment

Part I (Yeast mating):
- YPD medium (Clontech, Cat. No. 630409), YPDA medium (Clontech, Cat. No. 630464)
- Kanamycin sulfate (50 mg/mL)
- 50% PEG 6000 (autoclaved)
- YPD and YPDA agar plates
- SD agar plates: Two SD – Leu plates
  Two SD – Trp plates
  Three SD – Leu/–Trp (double dropout DDO) plates
  Two SD – L/–W/–H/–A +X-α-gal (qudraple dropout QDO) plates
- Y2HGold (pGBKT7-53) culture: bait plasmid transformant (Experiment 45)
- Y187 (pGADT7-T) culture: prey plasmid transformant (Experiment 45)
- Sterile cotton swabs, inoculating loops, sterile glass test tubes (13 × 100 mm)
- 0.5 McFarland standard (Fischer Scientific, NC9016286)

Part II (Genomic PCR):
- *CAN1* F-primer: CACAAACACACCACAGACGT
- *CAN1* R-primer: CCACCAAGGGCAATCATACC
- YPDA culture or colonies of canavanine-resistant (*can1*) $Trp^{+}Ura^{+}$ (Experiment 45), canavanine sensitive BY4733 (control), and canavanine-resistant DPB276 (control)
- 200 mM LiAc + 1% SDS solution
- 96% and 70% etanol, TE (pH 8.0) buffer
- GoTaq® DNA polymerase (Promega, Cat. No. M3001; 5 U/μL)
- DNA size marker (1 kb DNA ladder: 0.5 μg/5 μL)
- 10× LB buffer (lithium borate 200 mM, pH 8.2)
- EconoSpin™ mini spin column (Epoch Life Science, Cat. No. 1920-050)
- PB buffer (5 M guanidine-HCl, 20 mM Tris-HCl pH 6.6, 30% isopropanol)
- PE buffer (10 mM Tris-HCl pH 7.5, 80% ethanol)
- QG buffer (6 M guanidine thiocynate, 50 mM Tris-HCl, pH 6.6, 20 mM EDTA, pH 8.0, 0.006% cresol red)
- EB buffer (10 mM Tris-HCl pH 8.5)
- Thermal cycler, NanoDrop 2000 spectrophotometer

Procedure

Part I. Y2H assay

A. Agar plate mating assay I (indirect selection) (day 1)

1. Pick one transformed colony of each mating type streaked plates of Y2HGold (pGBKT7-53) and Y187 (pGADT7-T), inoculate into 10 mL YPDA medium, and grow overnight (A_{600} = 1.0 to 2.0; instructor will do this part).
2. Read A_{600} (A_{600} = 1 corresponds to ~3 × 10^7 cells/mL). Spin at 5000 rpm for 3 min to pellet the cells, discard the supernatant, and suspend the cells to a density of 2 × 10^8 cells/mL in YPDA medium containing 50 μg/mL kanamycin (Km).

**Km is added to YPDA to prevent bacterial contamination. However, Km does neither stop contaminating fungal growth nor select for any plasmids in yeast.*

3. Mix 0.1 mL of bait strain culture with 0.1 mL prey strain culture in a sterile 1.5-mL microcentrifuge tube.
4. Add a sterile 50 μL of 50% PEG 6000 solution (optional) and mix well.
5. Spread a 150-μL cell suspension on to one YPDA plate.
6. Incubate the plate upright at 30 °C overnight (~16 h).

B. Agar plate mating assay II (direct selection) (day 1)

7. Immerse a sterile cotton swab or inoculating loop into a cell suspension of Y2HGold (pGBKT7-53) and make two parallel streaks on to the center and right side of the SD -Leu/–Trp plate.

**Immerse cotton swab or inoculating loop into the cell suspension for each streak line.*

**Streak a few times gently on the middle center line.*

8. Immerse a sterile cotton swab or inoculating loop into the cell suspension of Y187 (pGADT7-T) and make two parallel streaks, one at the left side and the other at the center. Mix Y2HGold (pGBKT7-53) streak at the center with Y187 (pGADT7-T) by dabbing up and down a few times, as shown below.

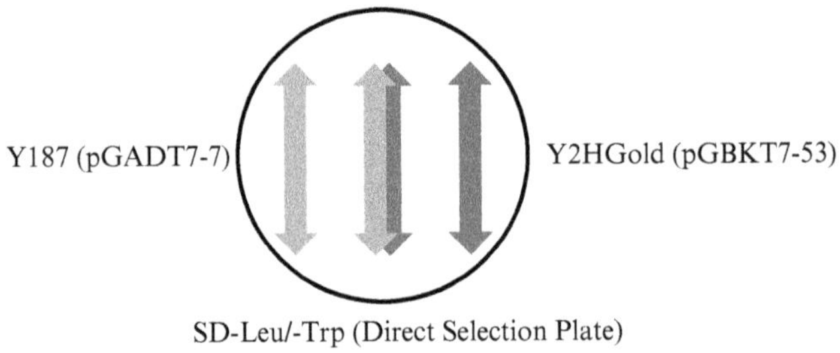

9. Allow the liquid streaks to dry, invert the plate, and incubate at 30 °C for 2 to 3 days.

**This allows mating to take place only at the center and checking if there are any reversion mutants or contaminants in the cultures tested and if bait protein affects yeast mating. Do not incubate too long as reversion mutants will grow larger and merge to diploid cells.*

**Check with all other groups' results; compare the number of diploid colonies. What do you think causes the difference in mating colonies if they are significantly different?*

10. Use a sterile inoculating loop to touch eight large colonies grown on the mating line and make streaks on to the QDO-X-α-Gal plate starting from the outer edge to the center, as shown below. Incubate the plate at 30 °C for 3 to 5 days to confirm the phenotype.

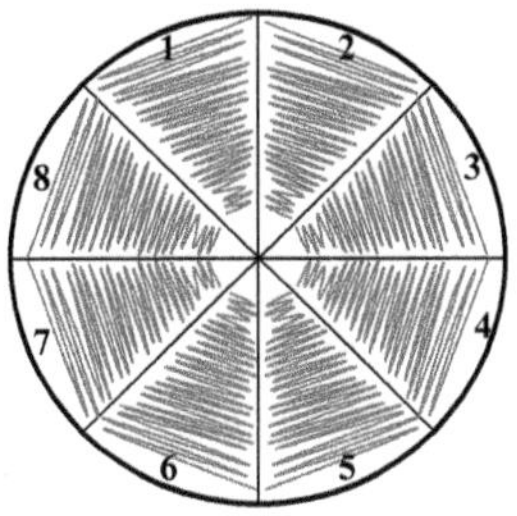

QDO X-α-Gal Plate (Direct Selection)

C. Selection (day 2)

11. Add 3 mL of sterile dH_2O to the surface lawn of mated yeast on the plate (step 6). Make the cell suspension using a sterile glass rod or 5-mL pipette (up and down motion).
12. Transfer 1 mL of homogenous cell suspension into a sterile glass test tube.

**Make sure that cells are completely suspended. If not, vortex the tube.*

13. Adjust yeast suspension to the turbidity of a 0.5 McFarland standard by adding sterile dH_2O.

**Yeast cells adjusted to a 0.5 McFarland standard contain approximately 1 to 5 × 10^6 yeast cells/mL.*

14. Make 1 mL each of 1/10, $1/10^2$, and $1/10^3$ dilutions in sterile dH_2O by mixing 100 µL cell suspension with 900 µL sterile dH_2O in a consecutive series in 1.5-mL sterile tubes.
15. Spread 100 µL of the above dilutions on each labeled plate:
- $1/10^2$ and $1/10^3$ dilutions on to SD –Leu
- $1/10^2$ and $1/10^3$ dilutions on to SD –Trp
- 1/10 and $1/10^2$ dilutions on to SD –Leu/–Trp

**Make sure that the glass spreader is just briefly alcohol (not 70% ethanol but 96–99% denatured ethanol)-flamed and cooled down between the platings. If you use 70% ethanol, you still see 70% ethanol liquid running down on the glass spreader rod after brief flaming. This does not happen when undiluted ethanol is used.*

16. Incubate all the plates upside down at 30 °C for 3 to 4 days.

D. Calculating mating efficiency and phenotype verification (day 3)

17. Count colonies (cfu) growing on SD –Leu, SD –Trp, and SD –Leu/–Trp dilution plates.
18. Calculate the viable cfu/mL on each type of SD medium:
- (cfu ÷ plated volume mL) × dilution factor = number of viable cfu/mL
- Number of cfu/mL on SD –Leu = viability of Y187 partner
- Number of cfu/mL on SD –Trp = viability of Y2HGold partner
- Number of cfu/mL on SD –Leu/–Trp = viability of diploid

19. Compare the viable cfu/mL of the two mating partners. The strain with the lower viability is the limiting partner.
20. Calculate the mating efficiency (% diploid):
- (Number of diploid cfu/mL ÷ Number of limiting partner cfu/mL) × 100 = % diploid

21. Build a table with your results and interpret.
22. Choose six large well-isolated positive colonies of the SD –Leu/–Trp plate and mark them by numbering.
23. Using a sterile inoculation loop, streak (from the edge to center) to generate single colonies on to the QDO X-α-Gal plate and make cross marks on the YPD plate to confirm the reporter phenotype as shown below. Invert all the plates and incubate at 30 °C for 3 to 4 days.

**Each number on the figure indicates the same colony from the SD –Leu/–Trp plate.*

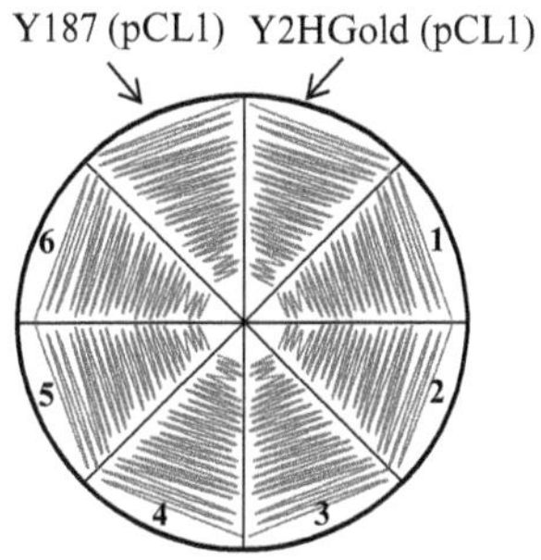

QDO X-α-Gal Plate

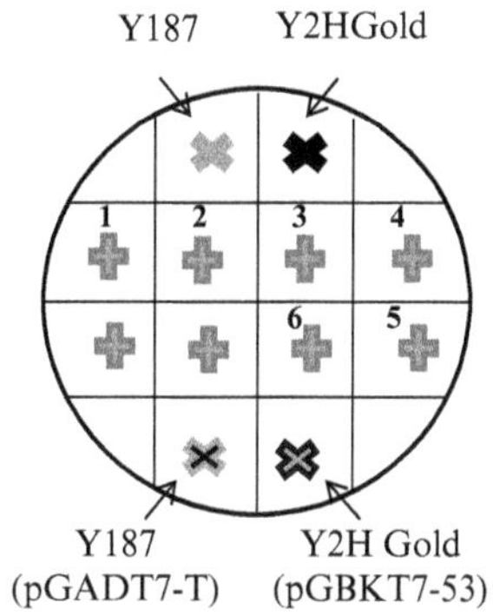

YPD Plate

24. Check the inheritance of a color trait in YPD and QDO X-α-Gal plates. Store the plates at 4 °C.

Part II. Yeast genomic PCR

A. Genomic DNA isolation

1. Pick two yeast colonies from the plate or spin down 120 µL overnight YPD cultures of canavanine-resistant (*can1*) Trp^+/Ura^+ BY4733, BY4733, and DPB276 in 1.5-mL labeled tubes.
2. Suspend the cell pellet in 100 µL of 200 mM LiAc + 1% SDS solution.

3. Incubate at 70 °C for 5 to 10 min.
4. Add 300 μL of 96 to 100% ethanol, mix by brief vortexing, and centrifuge at 15 000× *g* for 3 min.
5. Remove the supernatant, wash the pellet with 500 μL of 70% ethanol, spin for 10 s, and aspirate off the residual fluid in the bottom of tube.
6. Dissolve the precipitated DNA in 100 μL of TE buffer (pH 8.0).
7. Pellet cell debris by centrifuging at 15 000× *g* for 1 min.
8. Add 1 μL of supernatant containing DNA per reaction to the bottom of the labeled thin-walled PCR tube (*CAN1*, *can1*) and keep on ice.

**The isolated genomic DNA is impure with lots of contaminating RNA and proteins, and so cannot be used directly for the DNA sequencing template. However, it is useful for simultaneous PCR-based genotyping analysis of a large number of samples.*

B. Genomic PCR

9. Prepare the PCR master mix in a 1.5-mL tube kept in ice as follows.

Component	Volume (μL)	Master mix (× 4.5)	Master mix (× 17)
Nuclease-free H_2O	17	76.5	289
5× Green GoTaq® buffer*	5.0	22.5	85
CAN1 F primer (10 μM)	1.3	5.85	22.1
CAN1 R primer (10 μM)	1.3	5.85	22.1
10 mM dNTPs	0.5	2.25	8.5
DNA	1.0	–	–
GoTaq® enzyme (5 U/μL)	0.2	0.9	3.4
Total volume (μL)	25	108	430.1

*Vortex thoroughly after thawing and prior to use.

10. Add 24 μL of PCR master mix to the labeled thin-walled tube containing 1 μL of DNA, mix well by gentle pipetting up and down once, and keep on ice until ready to load on the cycler.
11. Program the PCR cycler with the following parameters.

Steps		Temperature (°C)	Time (min)
Initial denaturation		95	5
	Denaturation	94	0.5
30 Cycles	Annealing	52	0.5
	Extension	72	0.25
Final extension		72	5
Hold		8	∞

12. Begin the program cycle. Once the initial denaturation temperature reaches 90 °C, quickly spin your sample tube and load in the thermal cycler.

C. Gel purification of PCR fragment and quantification

13. During the PCR cycle, prepare 60 mL of a 0.5% (w/v) agarose solution containing 0.5 μg/mL of ethidium bromide in 1× lithium borate buffer (two mini-gels with large combs/4 groups).
14. Load directly PCR samples; run the gel at 300 V for 20 min.
15. Weigh empty sterile 1.5-mL tubes (labeled as *can1_PCR* and *CAN1_PCR*).
16. Place the ethidium bromide-stained gel on to a UV transilluminator covered with plastic wrap, and take a gel picture.
17. Cut out the gel bands to the correct size with a clean razor blade as quickly as possible and place into a microcentrifuge tube.

**Minimize the time of UV exposure as much as possible.*

**Minimize the size of gel slice by removing extra agarose that has no DNA band.*

18. Weigh the gel-containing microcentrifuge tube and calculate the gram quantity of gel slice.

**Put the gel slice up to 0.25 g in a 1.5-mL microcentrifuge tube.*

19. Add 3× volume of QG buffer to one gel volume (i.e., 0.3 mL buffer to 0.1 g gel slice).
20. Incubate at 50 to 55 °C for ~10 min or until the gel slice has completely dissolved. Mix by vortexing the tube every 2 min during the incubation.

**If the color of the mixture is orange or violet after the gel slice has dissolved, add 10 μL of 3 M sodium acetate (pH 5.2) and mix to turn the color of the mixture to yellow.*

21. Add one gel volume of isopropanol to the sample and mix.

**0.1 mL of isopropanol is added for 0.1 g of gel slice.*

22. Place a spin column in a provided 2-mL collection tube.
23. Apply the sample to the spin column and spin at maximum speed for 1 min.

**For sample volumes of more than 800 μL, simply load and spin again.*

24. Discard flow-through and place the spin column back in the same collection tube.
25. Add 500 μL of PE buffer and spin at maximum speed for 1 min.
26. Discard flow-through and spin at maximum speed for 1 min again.
27. Place the spin column into a labeled fresh 1.5-mL microtube.
28. Add 25 μL of EB buffer (warmed at 50 to 55 °C) to the *center* of the spin column membrane, let the column stand for 1 min, and centrifuge for 1 min.

**Elution efficiency depends on pH and temperature, and typical yields are >80%.*

29. Determine the concentration and purity by pipetting 2 μL of the sample on to the bottom pedestal of the NanoDrop 2000 spectrophotometer using water as a blank.

**The A_{260}/A_{280} ratio should be between 1.7 and 2.0.*

30. Store the gel-purified sample at –20 °C.

Post-lab assignment

1. The negative results obtained from the Y2H assay does not necessarily mean no interaction between the bait and prey proteins. Conversely, the positive Y2H assay results need to be validated by other techniques. List at least two other techniques that can be used to determine protein–protein interactions independently and briefly explain the principles.

2. What is yeast one-hybrid assay? Describe the principle.

3. **(a)** What experimental factors may affect the mating efficiency of a yeast culture? List at least four things and briefly explain them.

(b) What is the % efficiency of the mating in a normal strain of yeast? Include reference(s) in your answer.

4. What type of genetic analyses can be performed using yeast mating besides the Y2H assay? List at least three things and briefly explain what each case is for.

5. You performed direct and indirect selection of yeast mating samples. What do you think are the advantage and disadvantage of the direct selection method over the indirect selection method?

6. What is the advantage of using X-α-Gal over X-β-Gal in the QDO (–L/–W/–A/–H) plate in the Y2H assay? That is, why is X-α-Gal used instead of X-β-Gal?

Further reading

Albers, M., Kogl, M., Kober, I., and Loser, E. (2002). Method of stimulating the mating of microorganisms in liquid medium. May 23, US Patent No. US2002/0061582 A1.

Hampsey, M. (1997). A review of phenotypes in *Saccharomyces cerevisae*. *Yeast*, **13**: 1099–1133.

Lõoke, M., Kristjuhan, K., and Kristjuhan, A. (2011). Extraction of genomic DNA from yeasts for PCR-based applications. *BioTechniques*, **50** (5): 325–328.

National Committee for Clinical Laboratory Standards (1992). *Reference Method for Broth Dilution Antifungal Susceptibility Testing of Yeasts; Approved Standard, 2nd edition*. M27-A2. NCCLS, Wayne, PA.

Sena, E.P., Radin, D.N., and Gogel, S. (1973). Synchronous mating in yeast. *Proceedings of the National Academy of Science USA*, **70** (5): 1373–1377.

Yeast Protocols Handbook (July 2009). Clontech Laboratories, Inc. Protocol No. PT3024-1, Version No. PR973283.

47 AMB 2 experiment 47: Yeast colony PCR screening and cycle DNA sequencing

Purpose: This is up to you to write down.

Introduction

A direct use of crude samples as templates in PCR eliminates the need for laborious DNA isolation procedures and makes high-throughput genotyping and diagnosis feasible. Direct PCR of intact bacterial or yeast colonies is a simple, rapid, and efficient way to screen a large number of cells for the presence of desired DNA sequences. Furthermore, direct PCR without DNA purification from animal and plant tissue cells is possible with Phusion® and Phire® DNA polymerases. These two engineered DNA polymerases are exceptionally tolerant of many PCR inhibitors present in various sample materials. This allows for robust amplification under the conditions in which conventional thermostable DNA polymerases, such as *Taq* and *Pfu*, are completely inhibited. Clontech company also provides a comparable Terra™ PCR direct polymerase.

It is recommended to watch the following YouTube videos as to how to perform colony PCR:

- http://www.youtube.com/watch?v=h0yRrDWtdA4

(How to perform colony PCR)

- http://www.youtube.com/watch?v=q6MHnuznzpk

(PCR cloning – colony selection)

For amplicons longer than 3 kb, it is recommended to use Phusion or Phire enzyme. Either insert-specific or vector-specific primers can be used for the initial screening. The orientation of the insert in a recombinant plasmid can be determined by the combined use of insert-specific and vector-specific primers. Due to the low copy number of plasmid and tough cell wall, yeast colony PCR often yields unsatisfactory results when a bacterial colony PCR procedure is adopted.

PCR reactions rarely produce only single bands. Hence, the quality of the PCR product must be critically assessed by agarose gel electrophoresis and, if necessary, either PCR conditions should be optimized or a desired fragment should be gel-purified. Although many individual PCR products may have mutated nucleotides, these are scattered randomly and are different in each individual PCR fragment. As a result, the DNA sequence obtained from direct sequencing of the PCR product is the consensus base at each position. This also happens during PCR-based cycle sequencing reactions. Consequently, at any one nucleotide, most of the sequences will be correct, resulting in the original sequence with no mutations. Direct sequencing of the PCR product runs less chance of carrying mutated bases than sequencing of plasmid DNA carrying a cloned PCR DNA.

The BigDye® direct cycle sequencing procedure involves the following four steps:

1. PCR amplification of a target DNA using M13-tailed gene-specific primers, as shown below.

2. PCR sample clean-up and PCR cycle sequencing using an M13-tailed PCR target dsDNA template, chemically enhanced M13 primer (forward or reverse) resistant to exonuclease I, deoxynucleotide triphosphates (dNTPs), four different fluorescent dye-labeled chain terminating nucleotides (e.g., dideoxynucleotide triphosphates ddNTPs dyes) that are used to identify the A, C, G, and T extension products, thermo-stable AmpliTaq DNA polymerase, and exonuclease I. This reaction mix removes the excess PCR amplification single-strand DNA primers during a 15-min incubation at 37 °C, followed by linear amplification of extension products during the PCR cycle. The principle of cycle sequencing is a dideoxy chain terminator (Sanger) method performed in a thermal cycler; it is recommended to watch the following YouTube videos as to how it works:

- https://www.youtube.com/watch?v=SRWvn1mUNMA

(Sanger dideoxy DNA sequencing)

- https://www.youtube.com/watch?v=cqFTGQ_JO1c

(BigDye direct cycle sequencing kit)

3. A sequencing reaction clean-up that purifies the dye-labeled extension products from unincorporated dye terminators, primers, enzymes, and free salts.

4. Capillary electrophoresis that separates extension products by size. An optical detection device and data collection software capture and convert fluorescence signals to digital data. Because each dye emits light at a different wavelength when excited by a laser beam, all four colors from all four bases can be detected and distinguished in a single gel lane.

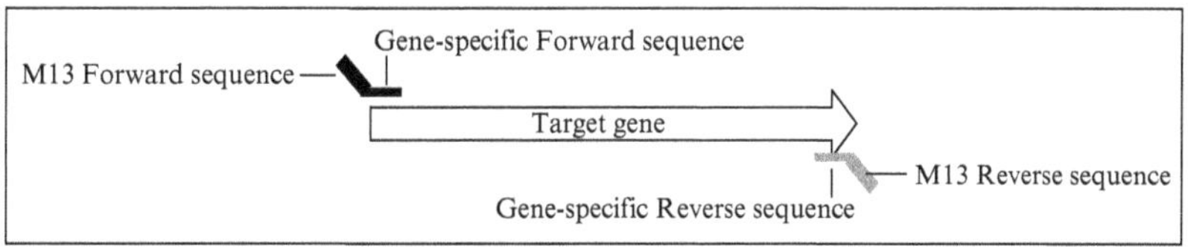

Methods in Biotechnology, First Edition. Seung-Beom Hong, M. Bazlur Rashid and Lory Z Santiago-Vázquez.

Companion Website: www.wiley.com\go\hong\Methodsinbiotechnology

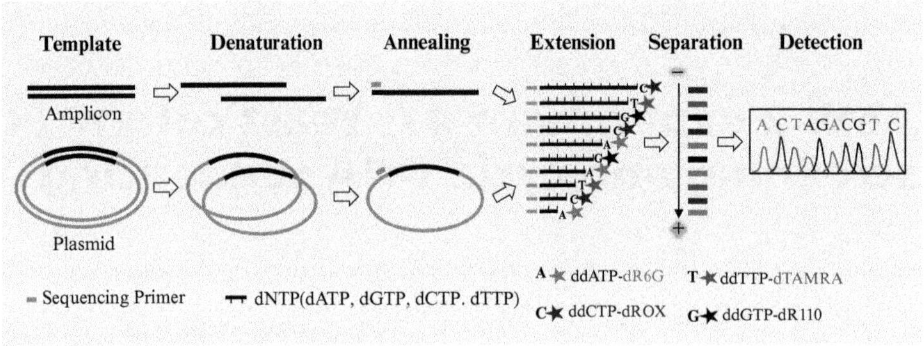

In the first part of the colony PCR experiment, you will treat diploid yeast cells obtained from yeast mating with glass beads to mechanically remove the cell wall before the colony PCR is performed to amplify cDNA inserts. The amplicons are then analyzed by agarose gel electrophoresis to confirm that the diploid colonies truly harbor both bait pGBKT7-53 and prey pGADT7-T plasmid DNA.

In the second part of the DNA sequencing experiment, you will re-amplify PCR fragments of yeast *CAN1* and *can1* locus using M13-tailed gene-specific primers and purified genomic PCR products (obtained from Experiment 46). You will then perform the BigDye® direct cycle sequencing reaction. It is important to sequence both DNA strands to independently confirm the sequence data. For more information, refer to the BigDye® Direct Cycle Sequencing Kit Protocol and the DNA Sequencing by Capillary Electrophoresis Chemistry Guide.

Pre-lab assignment

(Typing and submission must be completed before lab work begins.)

1. What are the expected sizes (bp) of PCR products from the Part I experiment? Show your work using the plasmid maps and MCS site sequences described in AMB 2 Appendix 7.
2. What things would you change if you do not have any PCR products from colony samples but have a PCR product of the expected size from a positive control sample and no PCR product from a negative control sample?
3. What would you change if you have expected PCR products from all the samples including the negative control sample?
4. What would you change if you do not have any PCR product from all the samples including the positive control and negative control samples?
5. What would you change if you have several additional PCR bands along with the PCR band of the expected size?

Materials and equipment

Part I:

- Colonies from QDO plates (Experiment 46)
- GADT7-T (10 ng/μL) and GBKT7-53 (10 ng/μL) plasmid DNA
- GADT7-T primer:
  Forward 5′-GATGAAGATACCCCACCAAACC-3′
  Reverse 5′-CTTGCGGGGTTTTTCAGTATCTACG-3′
- GBKT7-53 primer:
  Forward 5′-ATACGACTCACTATAGGGCGAGC-3′
  Reverse 5′-GGAATTAGCTTGGCTGCAAGC-3′
- GoTaq® DNA polymerase (Promega, Cat. No. M3001; 20 μL, 5 units/μL)
- DNA size marker (NEB 1 kb DNA ladder: 0.5 μg/5 μL)
- Acid-washed glass beads (Sigma-Aldrich; 0.45 to 0.53 mm)
- TE (10 mM Tris-Cl, pH 8.0, 1 mM EDTA) containing 2% Triton X-100
- Thermal cycler (program name: Yeast Colony PCR)
- 10× lithium borate (LB) buffer (100 mM, pH 8.2)
- Agarose gel electrophoresis apparatus

Part II:

- Gel-purified genomic PCR DNA of *can1* and *CAN1* (Experiment 46)
- Control CEPH human DNA (50 ng/μL; Thermo Fisher Scientific Cat. No. 403062)
- M13F/CAN1 F-primer: TGTAAAACGACGGCCAGTCACAAACA CACCACAGACGT
- M13R/CAN1 R-primer: CAGGAAACAGCTATGACCCCACCAA GGGCAATCATACC
- M13F/CEPH F-primer: TGTAAAACGACGGCCAGTGCCCTGG AAGGAGAACAAAGGC
- M13F/CEPH R-primer: CAGGAAACAGCTATGACCTGACCC GAGATGGTGCTTGA
- BigDye® Direct M13 forward primer (1 μM): NCM-TGTAAAAC GACGGCCAG*T; NCM, 5′negatively charged moiety; *, 3′phosphorothioate group
- BigDye® Direct M13 reverse primer (1 μM): NCM-CAGGAAA CAGCTATGAC*C
- BigDye® direct cycle sequencing kit (Thermo Fisher Scientific, Cat. No. 4458689)
- Nuclease-free sterile H_2O
- 96-Well unskirted PCR plate (Bio-Rad, Cat. No. MLP-9601)
- Microseal® "B" adhesive seals (Bio-Rad, Cat. No. MSB-1001)
- 96-Well plate table centrifuge
- Thermal cycler (program name: Cycle DNA Sequencing)

Procedure

Part I. Yeast colony PCR

A. Preparation of master mix and running PCR

1. With a sterile P20 yellow pipette tip, pick three independent colonies in the QDO plate (about this size •) and suspend the cells in three separately labeled tubes, each of which contains 10 μL of TE containing 2% Triton X-100.

2. Add 10 μL volume of glass beads, vortex vigorously for 1 min, spin at maximum speed for 1 min, and transfer 1 μL liquid to the bottom of a labeled PCR tube.

3. One group prepares two separate PCR master mixes using the primer mix for pGADT7-T prey DNA and pGBKT7-53 bait DNA *on ice* as follows.

Reagent	– Control (no DNA and cells)	+ Control (plasmid DNA)	Per sample (μL)	Master mix (× 5.5)
Nuclease-free H_2O	18	17.5	17.5	96.25
5x Green GoTaq® buffer*	5.0	5.0	5.0	27.5
Primer mix (10 μM each)†	1.3	1.3	1.25	7.15
10 mM dNTPs	0.5	0.5	0.5	2.75
DNA or colony	–	1.0‡	Colony 1.0	–
GoTaq® enzyme (5 U/μL)	0.2	0.25	0.25	1.1§
Total volume (μL)	25	25	25	134.75

*Vortex thoroughly after thawing and prior to use.
†F and R primer sets: GAD-F/GAD-R for pGADT7-T prey vector; GBK-F /GBK- R for pGBKT7-53 bait vector.
‡+ Control DNAs are either pGADT7-T or pGBKT7-53.
§Use the P10 micropipettor to take the volumes.

All other groups prepare two separate PCR master mixes using the primer mix for pGADT7-T prey DNA and pGBKT7-53 bait DNA *on ice* as follows.

Reagent	Per sample (μL)	Master mix (× 3.5)
Nuclease-free H_2O	17	59.5
5x Green GoTaq® buffer*	5.0	17.5
Primer mix (10 μM each)†	1.3	4.55
10 mM dNTPs	0.5	1.75
DNA or colony	Colony 1.0	–
GoTaq® enzyme (5 U/μL)	0.2	0.7‡
Total volume (μL)	25	84

*Vortex thoroughly after thawing and prior to use.
†F and R primer sets: GAD-F/GAD-R for pGADT7-T prey vector; GBK-F/GBK-R for pGBKT7-53 bait vector.
‡Use the P10 micropipettor to take the volumes.

4. One group labels 10 thin-walled tubes: five are pGADT7-T and the other five are pGBKT7-53 (+ Control, – Control, Expt 1, –2, –3). All other groups label six thin-walled tubes (Expt –1, –2, –3) for pGADT7-T and for pGBKT7-53.

5. Add 1 μL of H_2O to – Control tubes and add 0.5 μL H_2O plus 0.5 μL control DNA (pGADT7 or pGBKT7-53) to + Control tubes.

6. Add 1 μL colony extract to each of Expt 1, 2, and 3 tubes.

7. Dispense 24 μL of PCR master mix to each PCR tube.

8. Place samples on ice until ready to load in the thermal cycler.

9. Program the thermal cycler with the following parameters.

Steps		Temperature (°C)	Time (min)
Initial denaturation		95	5 min
30 Cycles	Denaturation	94	0.5
	Annealing	52	0.5
	Extension	72	2.5
Final extension		72	2
Hold		8	∞

10. Begin the program cycle. Once the initial denaturation temperature reaches 90 °C, quickly spin your sample tube and load in the thermal cycler.

B. Agarose gel electrophoresis

11. During the PCR cycle, prepare 60 mL of 0.5% (w/v) agarose solution containing 0.5 μg/mL of ethidium bromide in 1× lithium borate buffer.

12. Prepare loading samples on a sheet of Parafilm as shown below.

	Addition order	3	1	2
Lane	Sample	Sample (μL)	TE buffer (μL)	10× DNA loading dye (μL)
1	1 kb DNA ladder	5	–	–
2	Positive control/pGADT7	6	14	2
3	Negative control/pGADT7	6	14	2
4	Colony 1/pGADT7	20	–	2
5	Colony 2/pGADT7	20	–	2
6	Colony 3/pGADT7	20	–	2
7	Positive control/GBKT7	6	14	2
8	Negative control/pGBKT7	6	14	2
9	Colony 1/pGBKT7	10	10	2
10	Colony 2/pGBKT7	10	10	2
11	Colony 3/pGBKT7	10	10	2

13. Load samples and run the gel at 25 V overnight (~14 h).

14. Capture the image of your gel using the gel documentation system.

15. Clean up.

- Rinse the UV transilluminator with dH_2O and wipe dry with a paper towel.
- Dispose of your gel in the designated waste container.

Part II. Cycle sequencing

1. Dilute the gel-purified PCR DNA samples (*CAN1*, *can1*) and control DNA (CEPH: 50 ng/μL) to 4 ng/μL in a nuclease-free sterile H_2O.

2. Add 1 μL of each sample into the bottom of a 96-well plate according to the plate layout specifying reactions: *can1*_PCR DNA in A1/A2 to H1/H2; *CAN1*_PCR DNA in A3 to F3; CEPH control DNA in G3 and H3.

96-Well plate layout*			
	1	2	3
A	can1.1-F (Group 1)	can1.1-R (Group 1)	CAN1-F (Group 1)
B	can1.2-F (Group 1)	can1.2-R (Group 1)	CAN1-R (Group 1)
C	can1.1-F (Group 2)	can1.1-R (Group 2)	CAN1-F (Group 2)
D	can1.2-F (Group 2)	can1.2-R (Group 2)	CAN1-R (Group 2)
E	can1.1-F (Group 3)	can1.1-R (Group 3)	CAN1-F (Group 3)
F	can1.2-F (Group 3)	can1.2-R (Group 3)	CAN1-R (Group 3)
G	can1.1-F (Group 4)	can1.1-R (Group 4)	Control (CEPH)-F
H	can1.2-F (Group 4)	can1.2-R (Group 4)	Control (CEPH)-R

*F and R samples contain the same DNA but sequencing reactions differ at step 6. Samples are injected into capillaries in a consecutive order starting from A1 to H1, A2 to H2, and A3 to H3 by the ABI 3500xL DNA sequencer. Well 1 indicates the first injection into 24 capillaries.

3. Prepare two separate PCR master mixes in 1.5-mL tubes on ice.

Experimental component	Volume per reaction	Master mix (× 22)
M13F/CAN1 F + M13R/CAN1 R primer mix (0.8 μM each)	1.5 μL	33 μL
BigDye® Direct PCR master mix	5.0 μL	110 μL
Nuclease-free sterile H_2O	2.5 μL	55 μL
Total volume	9.0 μL	198 μL

Control component	Volume per reaction	Master mix (× 2)
Control CEPH F + R primer mix (0.8 μM each)	1.5 μL	3 μL
BigDye® Direct PCR master mix	5.0 μL	10 μL
Nuclease-free sterile H_2O	2.5 μL	5 μL
Total volume	9.0 μL	18 μL

4. Pipette 9 μL of master mix to the bottom of each well, mix by gentle pipetting up and down once (avoid generating bubbles), and seal the plate with adhesive film.

5. Run the 96-well reaction plate with the following parameters.

Steps		Temperature (°C)	Time
Initial denaturation		96	5 min
30–35 Cycles	Denaturation	96	3 s
	Annealing	62	15 s
	Extension	68	30 s
Final extension		72	2 min
Hold		4	∞

6. During PCR cycling, prepare forward and reverse sequencing reaction mixes in a separate F and R tube on ice.

**Do not use M13 forward or reverse primers but BigDye® Direct M13 forward or reverse primers.*

Component	Volume per reaction	Master mix (× 12)
BigDye® direct sequencing master mix	2.0 μL	24.0 μL
*BigDye® Direct M13 F primer or *BigDye® Direct M13 R primer	1.0 μL	12.0 μL
Total volume	3.0 μL	36.0 μL

7. After PCR is complete, spin the plate at 3500 rpm for 1 min in a swinging bucket centrifuge.

8. Add 3 μL of forward sequencing reaction mix to the bottom of A1 to H1 and A3, C3, E3, G3 wells; 3 μL of reverse sequencing reaction mix to A2 to H2 and D3, F3, H3 wells; pipette up and down twice to mix. Seal the reaction plate with adhesive film.

9. Run the 96-well reaction plate with the following parameters.

Steps*		Temperature (°C)	Time
Exo I Digestion		37	15 min
Exo I Inactivation		80	2 min
Initial Denaturation		96	1 min
25 Cycles†	Denaturation	96	10 s
	Annealing	50	5 s
	Extension	60	75 s
Hold		4	∞

*Begin PCR cycling without HOT START.
†Set thermal ramp rate to 1 °C/s at each temperature step.

10. Spin the plate briefly at 3000 rpm for 1 min. Store the plate at −20 °C if not used immediately (up to 4 weeks).

Post-lab assignment

1. The performance of PCR is often assessed in terms of sensitivity, specificity, and fidelity. Describe each of the following (a), (b), and (c) cases separately.

(a) What is the difference between "sensitivity" and "specificity" and "fidelity" of PCR?

(b) What PCR conditions affect the sensitivity, specificity, and fidelity? List at least four things in each case separately and briefly explain what happens if each factor is either increased or decreased.

(c) How can you experimentally determine the sensitivity, specificity, and fidelity of PCR?

2. What is the difference between qualitative and quantitative PCR testing? What types of analytical methods are used in each case?

3. PCR has revolutionized molecular biology. List and explain the three major advantages and disadvantages of PCR techniques.

4. List the parameters that affect PCR amplification efficiency when the optimized master reaction mix containing enzyme, dNTP, and buffer as well as the optimized PCR cycling conditions are used.

5. Answer (a), (b), and (c) below by performing the following steps:

- Visit the NCBI homepage (http://www.ncbi.nlm.nih.gov/), select "Nucleotide" in the dropdown menu, type 171557, and click "Search."
- Note the CDS position, then copy the entire numbered sequence (3694 bp), paste as a HTML format into a Microsoft Word document and save. Mark the start and stop codons on the sequence.
- Now visit the Primer3Plus web site (http://biotools.umassmed.edu/cgi-bin/primer3plus/primer3plus.cgi).
- Paste the sequence (3694 bp) into "Paste template sequence below" window (you do not need to worry about the sequence position number; when you paste, numbers of the input sequence will be automatically deleted).
- Select "pick_cloning_primers" in the "Task:" dropdown menu, type the position numbers of CDS with a comma (,) between the numbers in the "Included Region:" click "General Settings" submenu and type 2650-2750 in the "Product Size Ranges" box and click "Load Settings."
- Click the "Pick Primers" menu. This will display the list of Left and Right primers with a start position, length, T_m, and GC% for PCR amplification of CDS.
- Choose, copy, and paste the sequence of primer pair (Left Primer 1 and Right Primer 2) with the highest T_m.

(a) Show the positions of primers and start and stop codons on the downloaded sequence.

- To determine the annealing temperature (T_a), visit the NEB T_m calculator web site (http://tmcalculator.neb.com/#!/).
- Select "Q5" in the "Input Product Group" dropdown menu.
- Copy and paste Primer 1 and 2 sequences. This will automatically display T_a, T_m, and GC%.

(b) Write the detailed PCR reaction setup on the basis of the selected primer set in order to amplify the CDS, assuming that you use primer 1 + 2 mix containing 10 μM each of the primer set, 50 ng of yeast genomic DNA (0.1 μg/μL), and Q5® high-fidelity DNA polymerase (0.02 U/μL) in a total reaction volume of 50 μL.

(c) Write the cycling conditions to amplify the PCR product.

*Refer to the NEB PCR protocol using Q5® high-fidelity DNA polymerase (M0491):

https://www.neb.com/protocols/2013/12/13/pcr-using-q5-high-fidelity-dna-polymerase-m0491

Show your work, along with tables of both reaction (b) and cycling setup (c).

6. To perform overlap extension PCR cloning (BioTechniques, Vol. 48, No. 6: 463–465, 2010), a reaction recipe calls for the use of a 250:1 molar ratio of the insert: plasmid vector. You are given a purified 989-bp gfp insert (1 mg/mL) and 3461-bp pQE30 plasmid vector (1 mg/mL), along with 10× PCR master mix containing Phusion DNA polymerase, dNTPs, and buffer. You are asked to use 3 ng of pQE30 in a total 10-μL reaction volume. How would you set up the reaction using the 10× PCR master mix, gfp insert, and pQE30 plasmid DNA using a P10 micropipettor? The MW of dsDNA is 660 g/mole per bp. Show your work.

Further reading

André, C. (2009). New development of PCR. *BioTechniques Special Issue*, **46** (5): 375–376.

Applied Biosystems 3500/3500xL Genetic Analyzer User Guide, 06/2010, Part Number 4401661.

Arezi, B., Xing, W., Sorge, J.A., and Hogrefe, H.H. (2003). Amplification efficiency of thermostable DNA polymerases. *Analysis of Biochemistry*, **321**: 226–235.

BigDye® Direct Cycle Sequencing Kit Protocol. Life Technologies. 02/2011, Part Number 4458040.

Chemically-Enhanced Primer Compositions, Methods and Kits. Life Technologies Corporation. October 25, 2012. US Patent No. US2A1.

DNA Sequencing by Capillary Electrophoresis: Applied Biosystems Chemistry Guide, Second Edition, 05/2009, Part Number 4305080.

48 AMB 2 experiment 48: DNA sequencing electrophoresis

Purpose: This is up to you to write down.

Introduction

The capillary electrophoresis (CE)-based automated cycle sequencing chain termination (Sanger) method was used for the first completion of the human genome sequencing project. However, whole-genome sequencing using the Sanger system requires tremendous amounts of time and resources, and it is too expensive and slow. For example, whole-genome sequencing of *E. coli* (4.6 Mb) using the CE-based Sanger approach would involve roughly 100 000 sequencing reactions (assuming 10× coverage and 500 bp read length), each conducted in separate tubes or wells. It is recommended to visit the DOE JGI website in order to review step-by-step procedures for automated Sanger DNA sequencing:

- http://jgi.doe.gov/archived-educator-resources/sanger-sequencing-archive/how-sanger-sequencing-is-done/

Using next generation sequencing(NGS) technologies, the same sequencing project including library preparation can be completed within or less than a single day in one sequencing run at 30× coverage or greater. NGS is much more time-saving and cost-effective than CE-based Sanger sequencing, and is thus now routinely used for whole genome sequencing, targeted re-sequencing (mutational profiling) of many individual DNA samples, microbial metagenome sequencing, and RNA-Seq projects. It is recommended to watch all the following YouTube videos about high-throughput NGS platforms.

- http://www.youtube.com/watch?v=4XMO5VfLIKs (Necessity of NGS)
- http://www.youtube.com/watch?v=_ApDinCBt8g (Overview of current NGS)
- http://www.youtube.com/watch?v=womKfikWlxM (Illumina MiSeq/Hiseq)
- http://www.youtube.com/watch?v=JHzkYDyMzOg (Ion torrent proton sequencer)

Despite the comparably low sequence output data, Sanger sequencing is commonly used for fragment analysis, small-scale sequencing projects, and filling in some sequence gaps or ambiguities due to the high error rate of insertions and deletions (indels) in homopolymeric regions by NGS technology. All PCR clones are routinely verified by automated Sanger sequencing.

In this lab exercise, as the last part of the cycle DNA sequencing experiment, you will clean up the DNA sequencing reaction samples (Experiment 47) by mixing with BigDye XTerminator reagents. During mixing, the BigDye XTerminator reagents capture and immobilize unwanted components including salt ions, unincorporated dye terminators, and dNTPs. After brief centrifugation, the insoluble fraction of the captured reaction components is pelleted to the bottom of the reaction well. The purified dye-labeled extension products that remain in the supernatant are electrokinetically injected; a high-voltage charge applied to the sample forces the negatively charged DNA fragments into the capillaries. The extension products are separated by size, based on their total charge by a 24-capillary electrophoresis platform in the ABI 3500xL DNA sequencer.

The following YouTube videos are available regarding the ABI 3500xL DNA sequencer:

- http://www.youtube.com/watch?v=RY2Rn2ggjdA (Functionality overview)
- http://www.youtube.com/watch?v=43-OQTLtrwQ (DNA fragment analysis)

Separation of the fluorescence dye-labeled fragments, laser detection of the individual fragments, and data capture by collection software to display an electropherogram are illustrated in the figure below.

Pre-lab assignment

(Typing and submission must be completed before lab work begins.)

1. If you want to load the DNA sequencing sample into a 3500xL DNA sequencer immediately after the Part I procedure, what should you do with the instrument? Do not copy the protocol and write succinctly in your own words.

2. What is the difference between "Single-End" and "Paired-End" reads? What type of read is performed in this experiment?

Methods in Biotechnology, First Edition. Seung-Beom Hong, M. Bazlur Rashid and Lory Z Santiago-Vázquez.

Companion Website: www.wiley.com\go\hong\Methodsinbiotechnology

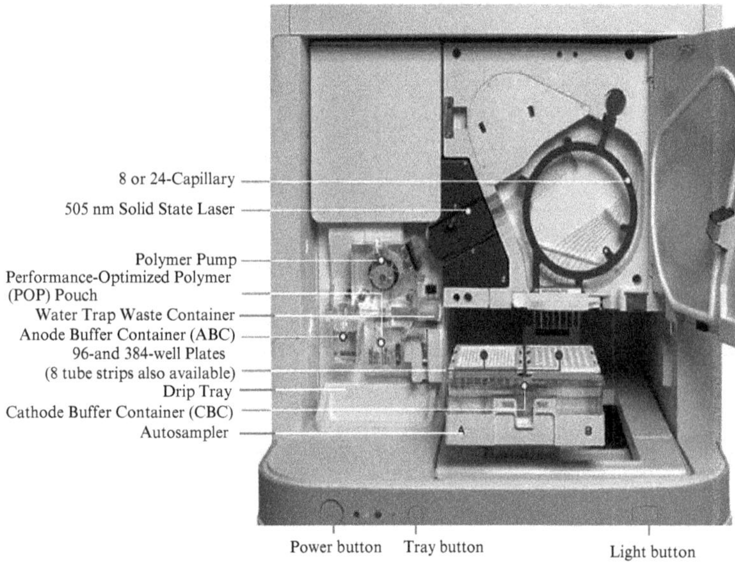

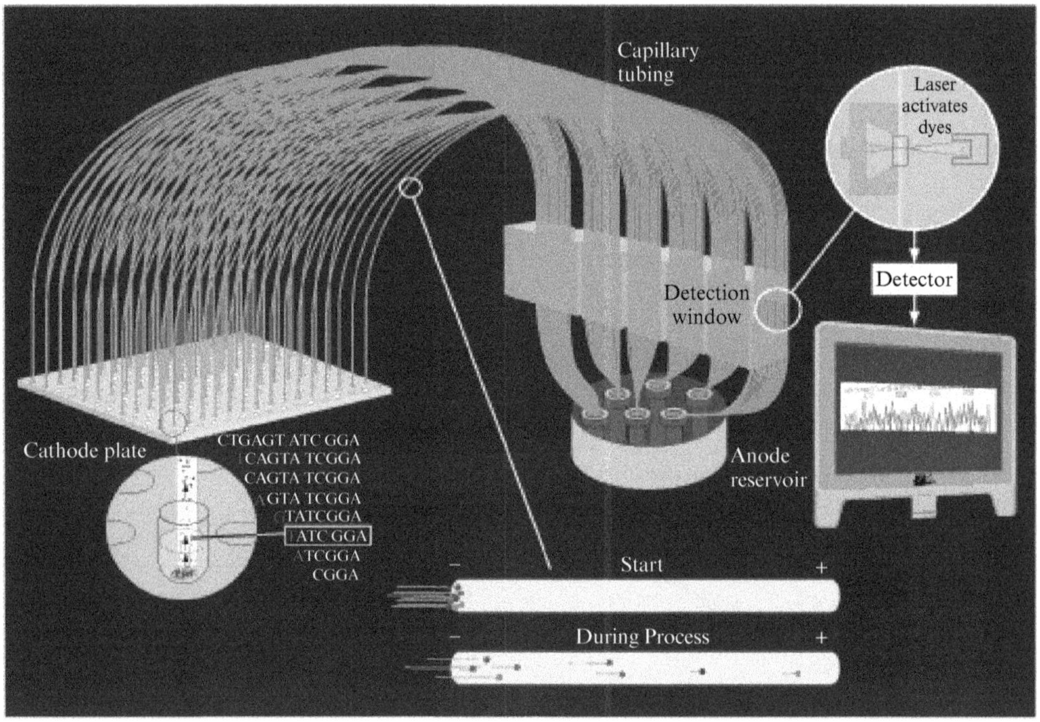

Materials and equipment

- Cycle DNA sequencing reaction samples (Experiment 47)
- BigDye XTerminator purification kit (Thermo Fisher Scientific, Cat. No. 4376486)
- MicroAmp® clear adhesive film (Thermo Fisher Scientific, Cat. No. 4306311)
- MicroAmp® optical 96-well reaction plate (Thermo Fisher Scientific, Cat. No. N8010560)
- 96-Well plate centrifuge
- Vortexer
- ABI 3500xL genetic analyzer (Thermo Fisher Scientific) and user guide
- POP-7™ polymer separation matrix (Thermo Fisher Scientific, Cat. No. 4393708)
- Anode buffer (Thermo Fisher Scientific, Cat. No. 4393925)
- Cathode buffer (Thermo Fisher Scientific, Cat. No. 4408256)
- 3500xL genetic analyzer capillary array, 50 cm (Thermo Fisher Scientific, Cat. No. 4404689; 24 capillaries per array)
- 96-Well (standard) retainer and base set (Thermo Fisher Scientific, Cat. No. 4410228)
- Septa for 3500/3500xL genetic analyzers (Thermo Fisher Scientific, Cat. No. 4410701)

- Conditioning reagent (Thermo Fisher Scientific, Cat. No. 4393718)
- MicroAmp® EnduraPlate™ optical 96-well clear reaction plates (Thermo Fisher Scientific, Cat. No. 4483354)
- Semi-skirted 96-well PCR plates (corner cut: A12) (Life Science Products, Inc., Cat. No. LS-9994-50)

Procedure

Part I. Clean-up of DNA sequencing samples

1. Prepare a premix with SAM™ solution and XTerminator® solution.

Component	Volume per well	Volume for 25 wells
SAM™ solution*	45 µL	1125 µL
XTerminator® solution†	10 µL	250 µL
Total volume‡	55 µL	1375 µL

*Make sure there are no particulates before pipetting. If there are particulates, heat to 37 °C and mix to resuspend. Cool to room temperature before using.
†Vortex at maximum speed until the solution is homogeneous. Using a wide-bore pipette tip, aspirate the XTerminator® solution.
‡Mix the reagents until homogeneous.

2. Add 55 µL of the pre-mix to each well.
3. Seal the plate with MicroAmp® clear adhesive films; cover with aluminum foil.
4. Vortex the reaction plate at 1800 to 2000 rpm for 20 to 30 min.
5. Spin the plate at 1000× g for 2 min in a swinging-bucket centrifuge.
6. Store the plate at room temperature for 48 h or at 4 °C for up to 10 days.

**Protect the plate from light.*

Part II. Capillary electrophoresis using ABI 3500xL DNA analyzer

For detailed information on the ABI 3500xL operation, refer to the Applied Biosystems 3500/3500xL Genetic Analyzer User Guide:
- https://www3.appliedbiosystems.com/cms/groups/mcb_support/documents/generaldocuments/cms_069856.pdf
- http://tools.thermofisher.com/content/sfs/manuals/100025036_3500_3500xL_UG.pdf

A. Start the instrument

- Close the instrument door, press the power on/off button on the front of the instrument, and wait for the green status light to turn on.
- Press the tray button on the front left side of the instrument and wait until the autosampler stops at the forward position.
- Ensure the green status of the light button is on and not flashing before proceeding.

B. Start the computer and launch the application

- Turn on the computer and monitor.
- Log on to the Windows dialog box: enter username and password and click OK.
- The Integration Server Monitor status icon, on the right lower corner of your screen must show up in 2 min. If not, select Start ▶Programs ▶Applied Biosystems ▶3500 ▶Server Monitor.

- The application will launch automatically after the green check mark appears.

– Enter username and password in the Login dialog box→ Click OK→ Dashboard appears.

C. Check system status in the dashboard

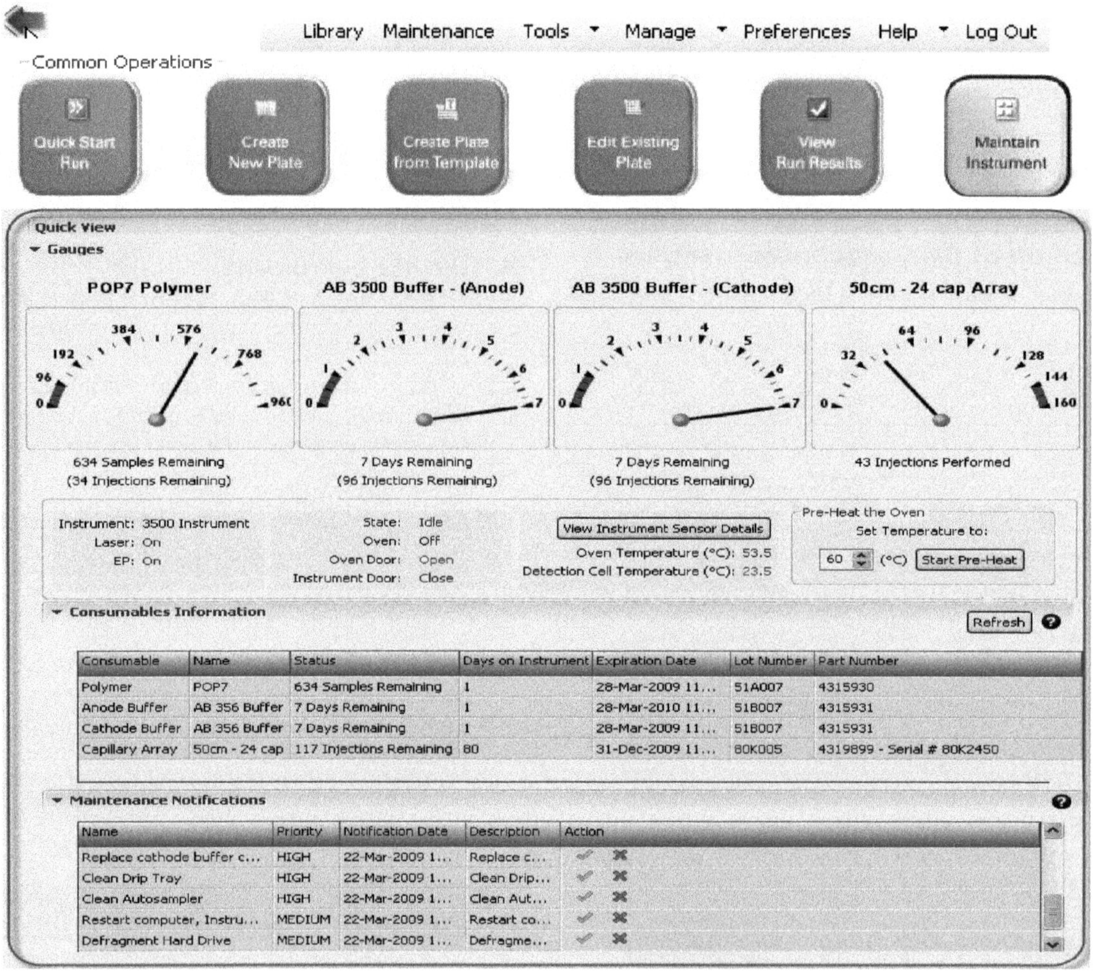

– Click "Refresh" in the Dashboard.
– Check the Maintenance Notifications and Consumable Information status.
– Check the consumables gauges; if consumables have expired or the buffer fill level is too low, replace with a new consumable(s). If you replace consumables, Click "Refresh" to read Radio Frequency Identification (RFID) tags on the product labels and to update the usage parameters on screen.

D. Prepare the instrument

– Set the oven temperature to: 60 °C, then click "Start Pre-heat."

**Pre-heat the oven at least 30 min before starting a run.*

**The pre-heat function automatically turns off after 2 hours of instrument inactivity.*

– Check polymer pump lines for bubbles; if necessary run Remove Bubble wizard by clicking "Remove Bubbles" (it takes 5–15 min to complete).

E. Define plate properties

– Click "Create Plate from Template" in the Dashboard to display "Open Plate Template from Library":

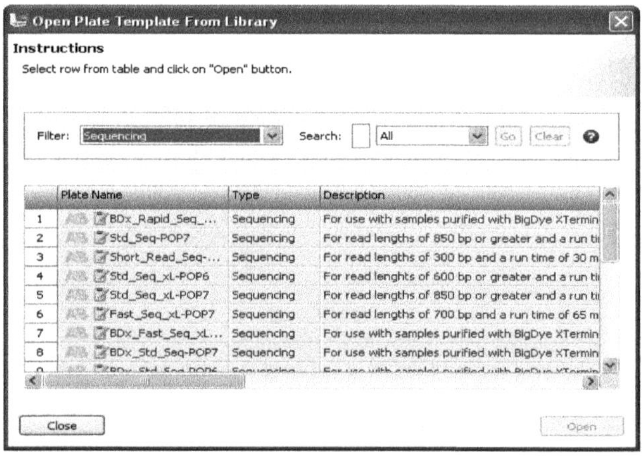

– Select the appropriate template type (sequencing in our experiment) and description.
– Click "Go" to find templates.

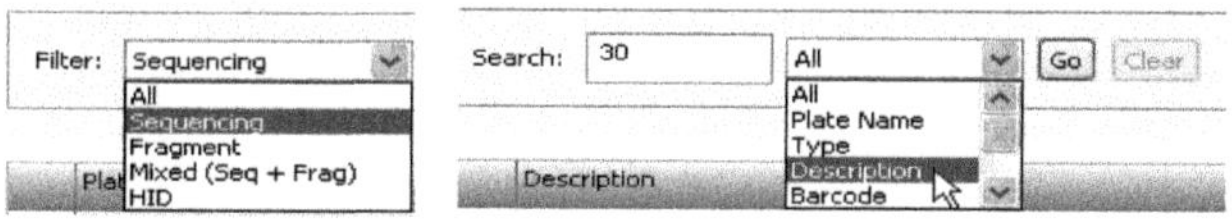

– Select RapidSeq_BDX_50_POP7 (Run module) with KB_3500_POP7_BDTv3direct.mob (Mobility file) in the "Description."
– Click "Open" to display the "Define Plate Properties" screen.

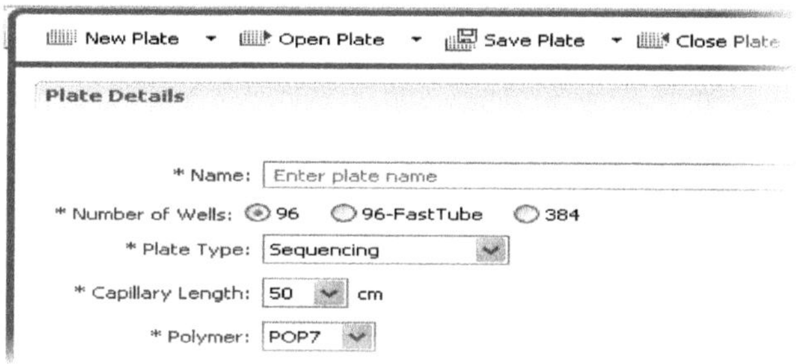

– Enter plate name in the "Name" field (*AMBII in our lab*); select plate details properties (# wells, type of analysis, capillary length, polymer); then select "Save Plate."
– Select the "Assign Plate Contents" button in the navigation pane by clicking "Main workflow arrow (←) in the Dashboard.

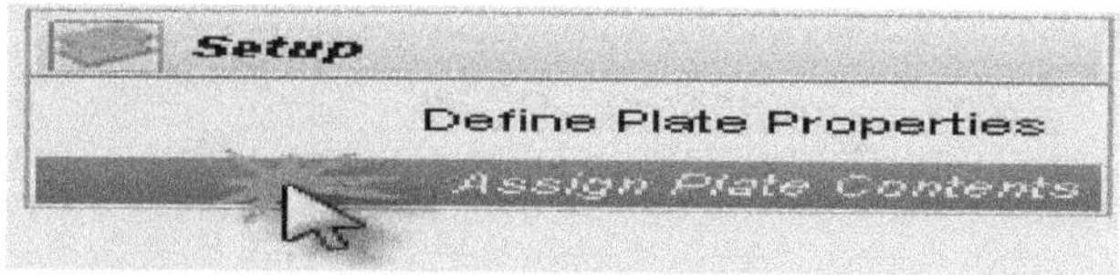

The "Assign Plate Contents Screen" will show up as shown below.

• Click the "Table View" tab and then click a well in the "Plate View."
• Type a sample name directly into the field and then press "Enter."

**For the name, refer to the plate layout in step 2 of Part II in Experiment 47.*

**You can click-drag multiple wells, right click and select the "Fill" option to populate all sample names on the plate.*

• At the bottom right of the screen, expand the "Customize Sample Info" pane.
• In "Plate View," click-drag to select the wells of interest.

**In our lab exercise, A1-H1; A2-H2 and A3-H3.*

• Specify the "Sample Type" for the selected wells and then press "Enter."
• Repeat to assign the 'Sample Type' for all of the named wells.

**In our lab exercise, A1-H1; A2-H2 and A3-F3 are samples; G3 and H3 are positive control.*

• Select the wells for which to specify an assay and enable the checkbox next to the assay name to assign it to the selected wells.

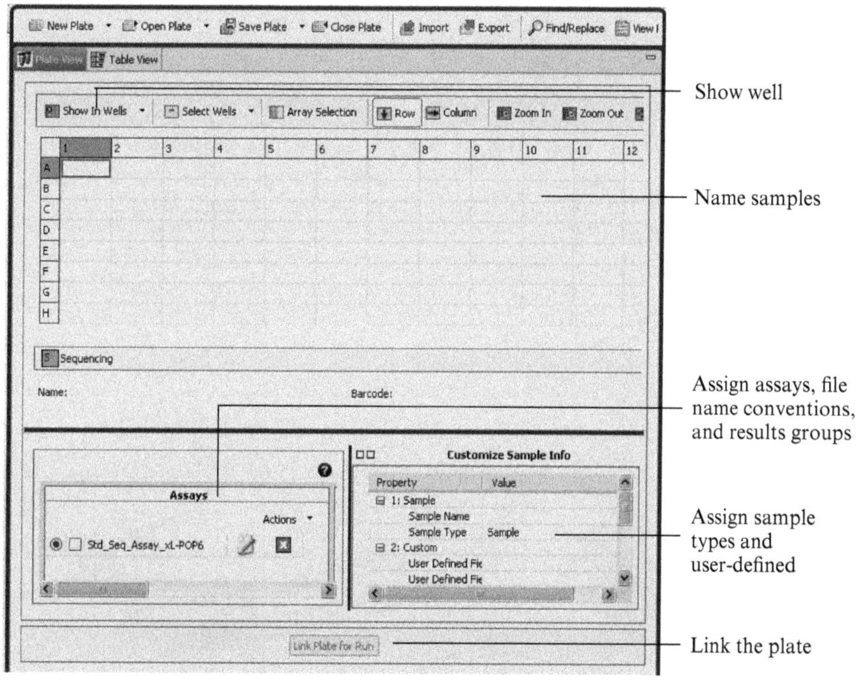

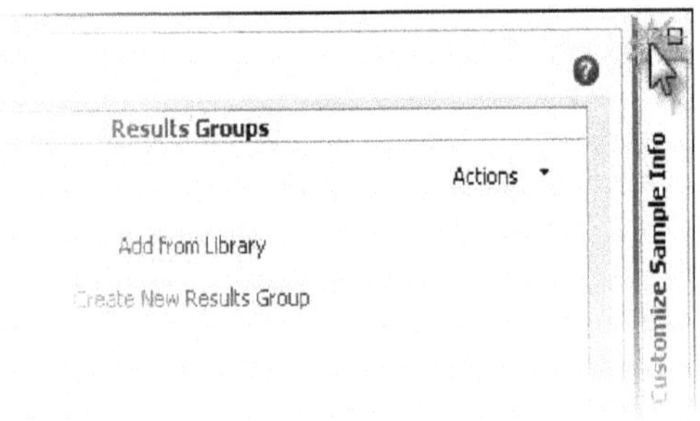

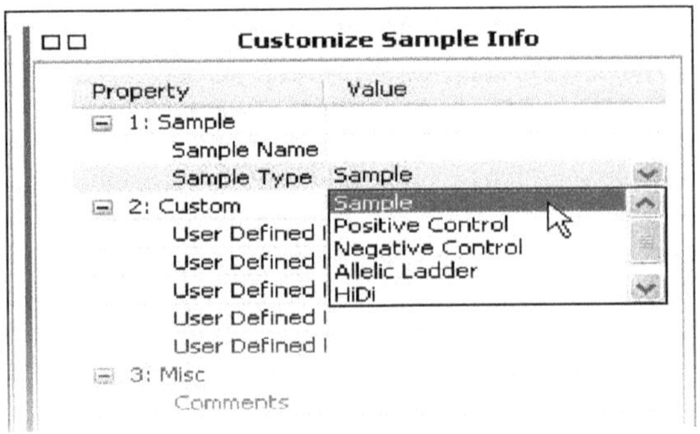

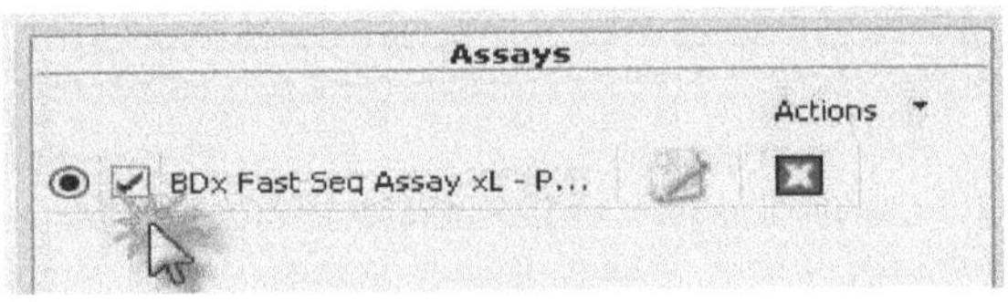

- Repeat for file name conventions and results group.

**If no assay is listed at the bottom of the screen, add at least one assay, file name conventions, and results group from Add from Library.*

- Select "Save Plate" and Click "View Plate Grid Report."

- Select "Print" and close the report by clicking X.

F. Prepare and load sample plate

– Briefly spin the sample plate and verify that each sample is positioned correctly in the bottom of its well.
– On a clean, level surface, align the holes in the septa strip with the wells of the sample plate and firmly press downward on to the sample plate.
– Place the sample plate into the plate carrier. Make sure to use the correct plate base.
– Snap the plate retainer (cover) on to the plate and plate base.

**Verify that the holes of the plate retainer and the septa strip are aligned. If holes are not aligned, reassemble and then assemble the plate assembly. This is important to avoid damaging capillary tips.*

– Place the plate in the autosampler with the labels of the plate base facing you (or the instrument door) and the notched corner of the plate in the notched corner of the autosampler.
– Close the instrument door.
– Check the instrument status in the Dashboard. All indicators display in green.
– In the Assign Plate Contents screen, click "Link Plate for Run." (see the dashboard in the next page.)

– Click the "Main workflow arrow"  in the dashboard and select "Load Plates for Run" in the navigation pane.
"Run Information" screen will show up (see in the next page).

- Enter a Run Name (e.g., AMBII_2013-07-11) in Run Information.
- Check that Plate A (default selection) is indicated by the active 'Unlink' button.

**If the plate is not linked as indicated by the active "Link" button, click "Link Plate" to display "Select Plate" from the library box, select a plate, and then click on "Link Plate."*

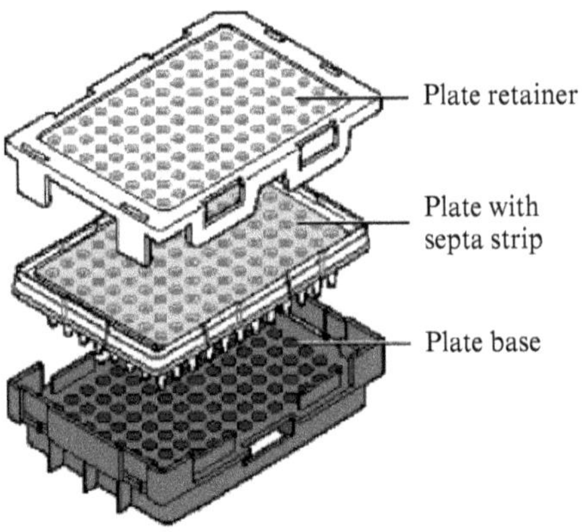

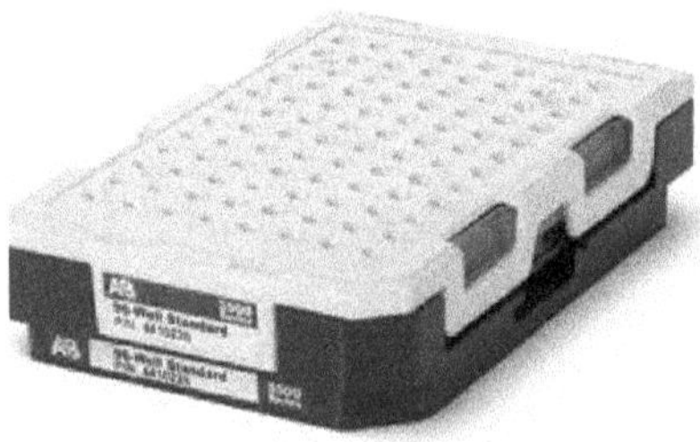

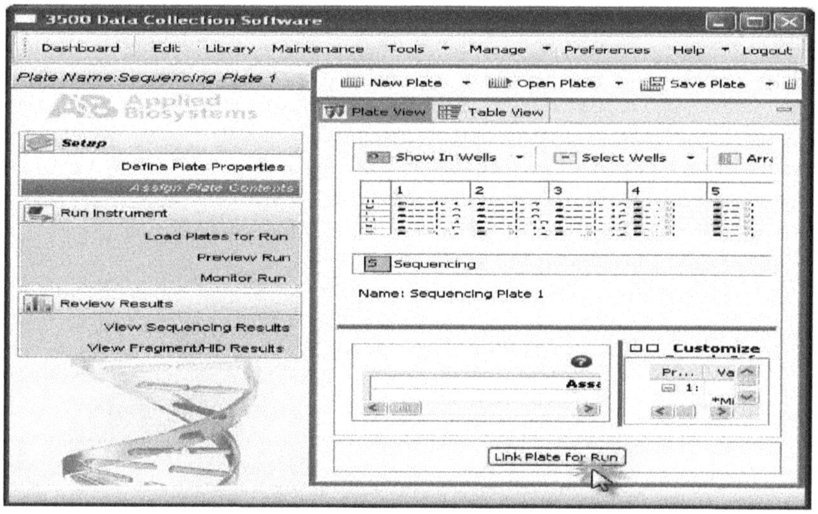

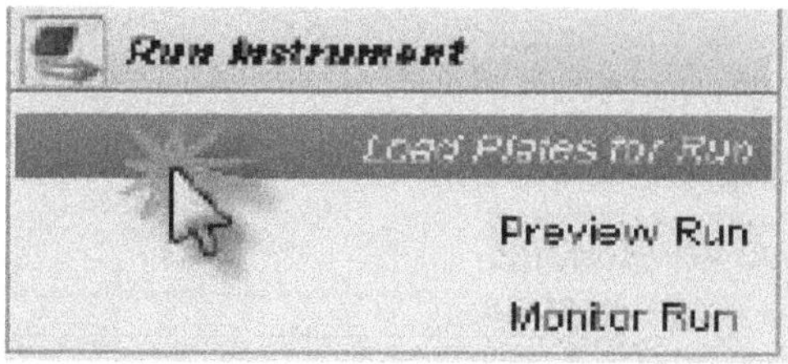

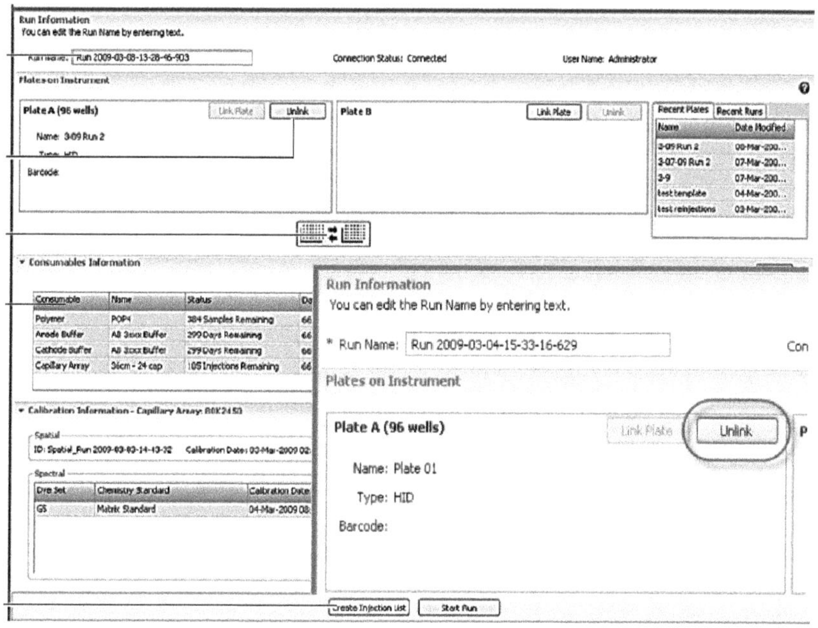

- Make sure that the instrument status light is green.
- Click "Start Run."

**The monitor run screen is automatically displayed and the current injection is highlighted in green in the plate view.*

Part III. Review sequencing results

1. Click "View Run Results" in the Dashboard or "Review Results" in the Monitor Run screen. Two files are generated: the "filename.seq" file is a text file output of the sequence and the "filename.ab1" file is a chromatogram file containing the original sequence peak data. Save all the.ab1 files on your computer.

**Do not use the unedited text file data (.seq) for your exercise.*

To view and edit the.ab1 file on your computer, you need to download a freeware:

- Go to: http://technelysium.com.au/?page_id=13 and download **Chromas Lite2.1.1** for PC Windows.
- Go to: http://nucleobytes.com/index.php/4peaks and download **4Peaks** for Apple Macintosh.

- Once the freeware is activated on your computer, click "Open" to locate the.ab1 file.

2. Review Sequence Quality (ABI 3500xL User Guide, page 81)

The DNA sequence of high quality is characterized by sharp peaks and little to no background. From the peak height and width, a Phred score is assigned to each individual base. A high Phred score indicates a high certainty as to the identity of that particular base. However, you need very often to perform cumbersome manual sequence editing by looking at the resolution of peaks.

Take a careful look at the individual peaks from the beginning to the end. Be critical of the raw data in the chromatogram.ab1 file. Note that the first and last 10 to 20 nucleotides are not well resolved.

- Are peaks spaced very well?
- Is there any missing base or extra base despite the presence or absence of a major conspicuous peak, respectively?
- Is there any extraneous base due to the presence of a minor background peak located between the two major peaks or in a broad merged peak? Is there any base deletion caused by the presence of a broad merged peak? Interpret in terms of the height and width of the peak.
- Are there any ambiguities and noise? If so, what are they?
- What would you do to correct the problems? (Refer to the DNA Sequencing by Capillary Electrophoresis Chemistry Guide).

Part IV. Editing chromatogram sequence and exporting sequence data

Using **Chromas Lite2.1.1** for the PC window, select "Open" from the File menu, click the relevant.ab1 file, and proceed to do the following steps.

1. To delete a nucleotide base in the chromatogram, click the letter to highlight and delete. To insert a base, click the base peak to find the base insert position, click the base position, and type the base. To change a base, click the letter to highlight and type a base. After editing is done, click File menu → select "Save As" and save the edited chromatogram as a file name.edit.

2. To export the sequence data, click the "Edit" menu in the chromatogram → select "Copy Sequence" → FASTA Format (>name, followed by sequence data in the next line). In this format the sequence ends when another line starting with a ">" symbol appears; this indicates the start of another sequence. Next, open a new Microsoft Word document file and paste as a New Courier font (size 9; single space).

3. To produce a reverse complementary sequence, click the "Edit" menu → select "Reverse +Complement" (this will make a reverse complementary sequence chromatogram) → select "Copy Sequence" → FASTA Format. Open a Microsoft Word document file and paste as a New Courier font (size 9; single space). Make sure to type (reverse complement) in the FASTA Format sequence name. Also, click the "Edit" menu in the chromatogram → select "Undo" to return to the original chromatogram.

**Note: for data analysis for sequence alignment, you must convert all reverse primer sequencing data into a reverse complementary sequence.*

Using **4Peaks** for Apple Macintosh, select "Open" from the File menu, click the relevant.ab1 file, and proceed to do following steps:

1. To delete or change a nucleotide base in the chromatogram, double click the base to show a letter encircled by a regular tetragon and delete or type. To insert a base, click the space between the letters, select "Insert Base" from the "Edit" menu, and type. To move between the bases, use arrow keys (▶, ◀). After editing is done, click the "File" menu →select "Save As" and save the edited chromatogram as a file name.edit.

2. To export the sequence data, click a gear wheel (☼▾) in the bottom left corner of the chromatogram → select "Copy Sequence"→ FASTA Format (>name, followed by sequence data in the next line); in this format the sequence ends if another line starting with a ">" symbol appears; this indicates the start of another sequence. Next, open a new Microsoft Word document file, type >name of sequence, and paste in the next line as a New Courier font (size 9).

3. To produce a reverse complementary sequence, click ☼▾ → select "Flip" (this will make a reverse complementary sequence chromatogram) → select "Copy Sequence." Next, open a Microsoft Word document file, type >name of sequence and paste in the next line as a New Courier font (size 9; single space). Make sure to type (reverse complement) in the FASTA Format sequence name. Also click ☼▾ in chromatogram → select "Flip" to return to the original chromatogram.

**Note: for data analysis for sequence alignment, you must convert all reverse primer sequencing data into a reverse complementary sequence.*

Part V. Data analysis using multiple sequence alignment program clustal omega

A. Independent sequence confirmation by pairwise alignment

1. Go to http://www.ebi.ac.uk/Tools/msa/clustalo/.
2. Select "DNA" in a dropdown menu of "STEP 1 - Enter your input sequences" window.

Good resolution | Loss of resolution | Complete loss of resolution

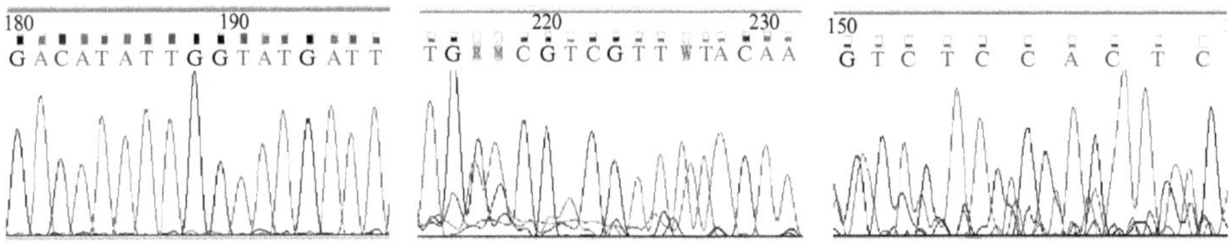

Accept the seq. | Need manual editing and confirm with reverse complementary seq. | Reject the seq.

3. Copy the FASTA format of a pair of forward primer sequence (f) and the reverse complementary sequence of reverse primer sequence (rr) and paste into the input box. An example is as follows:

>can1-1f
CGTGGGTCAATACAATACCATTGAAAG- - - - - - - - - - - - -
>can1-1rr
CCAGTCACAAACCAATACCATTGAAAG- - - - - - - - - - - - -

4. Click "Clustal w/numbers" in a dropdown menu of STEP 2 - OUTPUT FORMAT.

5. Click "Submit."

6. Find any sequence mismatch from the sequence alignment. Check the mismatched bases in the original chromatograms and edit if necessary. If either one has a clear major peak, use this base for editing (deletion, insertion, or change); if both are still not clear, put "N" in the sequence data.

7. Repeat the pairwise alignment again (steps 3 to 6) using the edited sequence so that most of the sequences show a perfect match (*).

8. Do the same (steps 3 to 7) for all other groups' chromatogram.ab1 files.

9. Show your work of all the pairwise alignments and editing in your lab report.

B. Detection of base changes by multiple alignments

10. Remove all of the unmatched tail sequences at both ends of each edited *can1*.seq and *CAN1*.seq on the basis of the step 9 results of the pairwise alignments.

11. Gather all four groups' edited FORWARD *can1* and *CAN1* sequences as a FASTA format in a Microsoft Word document (font size 8; single space). Assuming that each group has one can1.ab1 file, an example is as follows:

>can1-1f (Group 1)
NNNN- - - - - - - - - - - - - - - -
>can1-2f (Group 2)
NNNN- - - - - - - - - - - - - - - -
>can1-3f (Group 3)
NNNN- - - - - - - - - - - - - - - -
>can1-4f (Group 4)
NNNN- - - - - - - - - - - - - - - -

12. Copy all the FASTA format sequences and paste into the input box of the online Clustal Omega program → Click "Clustal w/numbers" and "Submit."

13. Copy and paste into the Microsoft Word document as New Courier font (size 8).

14. Find any sequence mismatch in the *can1* sequence relative to the *CAN1* Sequence. Referring to the answer to the Experiment 45 Pre-Lab Assignment 2 (b), find the PAM (protospacer-adjacent motif) sequence NGG in the form of the genomic target $GN_{20}GG$ sequence. Mutation should occur within the N_{19} region near the PAM. Show your work in your lab report.

Post-lab assignment

1. Shotgun sequencing is the basis of NGS methodologies. What is shotgun sequencing?

2. **(a)** Briefly describe the principles of at least three different NGS technology platforms:
- 454 FLX (Roche)
- Ion Torrent (Thermo Fisher Scientific)
- Hi-Seq (Illumina)
- SOLiD (Thermo Fisher Scientific)

(b) What is the homopolymer? Why is it difficult to sequence accurately?

3. What is bisulfite sequencing? Describe the principle.

4. What is metagenome sequencing? Describe the principle.

5. What is ChIP sequencing? Describe the principle.

6. What is the difference between the DNA fragment analysis and DNA sequencing? What can you do with fragment analysis?

Further reading

Applied Biosystems 3500/3500xL Genetic Analyzer User Guide, 06/2010, Part Number 4401661.

BigDye® Direct Cycle Sequencing Kit Protocol (Life Technologies).

DNA Sequencing by Capillary Electrophoresis, Applied Biosystems Chemistry Guide, Second Edition. Part Number 4305080 Rev. C. 05/2009.

49 AMB 2 experiment 49: RNA interference

Purpose: This is up to you to write down.

Introduction

RNA interference (RNAi) is a gene silencing mechanism that is induced by small double-stranded (ds) RNAs in a sequence-specific manner. RNAi occurs at either post-transcriptional or transcriptional level and is triggered by short interfering RNAs (siRNAs) or microRNAs (miRNAs) that are 21–25 nucleotides long double-stranded RNAs with a characteristic 3′-end UU overhang. The mature siRNAs are produced from endogenous long dsRNAs by the RNase III enzyme Dicer or introduced into a cell exogenously. The mature miRNAs are originated from endogenous non-coding genes as hairpin stem loop precursors that are subsequently processed by the nuclear RNase II enzyme Drosha and cytoplasmic RNase III enzyme Dicer. After being assembled into the so-called RNA-induced silencing complex (RISC) in the cytoplasm, siRNAs cause sequence-specific degradation of its homologous mRNA sequences, whereas miRNAs trigger either sequence-specific degradation or translational inhibition of the transcript depending on the perfect or imperfect sequence match to a target mRNA. RISC is a large ribonucleoprotein silencing complex composed of three proteins, Argonaute (AGO), Dicer (DCR), and transactivating response RNA-binding protein (TRBP). In the RISC, an antisense strand of siRNA or miRNA is used as a template by the AGO slicer to seek out and cleave or inhibit mRNA based on sequence homology. Because of the sequence specificity of the RISC, a synonymous codon change that does not alter amino acid may still bring about an abnormal developmental phenotype. Unlike miRNAs, siRNAs are either sampled randomly from exogenous dsRNAs or produced from endogenous dsRNAs such as bidirectionally transcribed mRNAs that anneal to form dsRNA, transcripts of inverted repeat DNAs, or transcripts of RNA-dependent RNA polymerase (RdRP) acting on a suitable RNA template. In plant and invertebrate, only a few dsRNA molecules per cell are found to be required for the effective interference that may spread systemically in cells and tissues distant from the site of initiation. It is important to note that RNAi gene silencing may not be complete or long-lasting. Incomplete and transient silencing effects lead to a range of phenotypes of differing severity, and so it is a mechanism of "Gene Knockdown" instead of "Gene Knockout." The RNAi gene silencing effect can be diminished with time. In addition, RNAi may also have off-target effects on other mRNAs.

RNAi has been observed in a wide variety of eukaryotes including yeast. However, baker's yeast *Saccharomyces cerevisiae* is known to lack RNAi, though RNAi is present in other yeast species, *Saccharomyces castellii* and *Candida albicans*. It was reported that RNAi works in *S. cerevisiae* when Argonaute *Ago1* and Dicer *Dcr1* genes of *S. castellii* are introduced in the genome of *S. cerevisiae*. Expression of human Ago2, Dicer, and TRBP were found to functionally reconstitute the RNAi system in *S. cerevisiae*.

In this lab exercise, you will transform two yeast strains DPB260 and BY4741 with plasmid pYES2.1-GFP_{uv} and test the transformants for GFP silencing in the presence of galactose. Transcription of GFP_{uv} mRNA is driven by a galactose-inducible promoter in pYES2.1-GFP_{uv}. The GFP_{uv} coding sequence cloned into pYES2.1 is derived from pGLO plasmid and is excited with any long wavelength UV light source, standardly considered to be around 365 nm. GFP_{uv} is known to be expressed well in yeast. You will also test the DPB276 strain for *URA3* silencing in YNB –Ura/+glucose and YNB –Ura/+galactose media. The strains DPB260 and DPB276 used for RNAi are described in Drinnenberg *et al.* (2009). It is recommended to watch the following YouTube videos:

- https://www.youtube.com/watch?v=cK-OGB1_ELE (RNAi)
- https://www.youtube.com/watch?v=dupzE66J8u4 (Introduction to MicroRNAs)

Pre-lab assignment

(Typing and submission must be completed before lab work begins.)

1. What is the genotype of the DPB260 strain? List and explain each genotypic and phenotypic characteristics of this strain.

2. What makes *S. cerevisiae* to be functional in the RNAi system in the DPB260 strain?

3. What is the genotype of the DPB276 strain? List and explain each genotypic and phenotypic characteristics of this strain.

4. What are the expected results under UV light (step 22 of Part I, C) if the transformation of yeast strains BY4741 and DPB260 with pYES2.1-GFP_{uv} plasmid is successful? Explain your choice.

5. What type of two media would give the same growth pattern in Part II that will serve as a reference positive growth control for comparison? Explain your choice.

Methods in Biotechnology, First Edition. Seung-Beom Hong, M. Bazlur Rashid and Lory Z Santiago-Vázquez.

Companion Website: www.wiley.com\go\hong\Methodsinbiotechnology

Materials and equipment

- Sterile 1× TE/1× LiAc (mix equal volumes; prepare prior to use)
- Sterile 1.5-mL microcentrifuge tubes
- YNB –Ura/+glucose, YNB –Ura/+5-FOA/+glucose, YNB –Ura/+ galactose/+raffinose, YNB +Ura/+galactose/raffinose agar plates
- Plasmid pYES2.1-GFP_{uv1} (~0.3 µg/µL)
- *Saccharomyces cerevisiae* BY4741, DPB260 and DPB276 strains of BY4741 genotype: *MATa, his3Δ1, leuΔ0, met15Δ0, ura3Δ0*
- Sonicated denatured salmon sperm carrier DNA (10 mg/mL)
- Sterile PEG/LiAc solution (polyethylene glycol/lithium acetate). Prepare just prior to use.

Components	Final concentration	To prepare 10 mL of solution
PEG 4000	40%	8 mL of 50% PEG
TE buffer	1×	1 mL of 10× TE
LiAc	1×	1 mL of 10× LiAc

- 1.1× TE/LiAc solution: *prepare just before use* by mixing 1.1 mL of 10× TE with 1.1 mL of 10× LiAC and 7.8 mL of sterile H_2O
- Stock solutions:
  - 50% PEG 4000 (polyethylene glycol, average MW = 3350): prepare with dH_2O; if necessary, warm solution to 50 °C to help the PEG go into solution; autoclave
  - 100% DMSO (dimethyl sulfoxide)
  - 10× TE buffer: 0.1 M Tris-HCl, 10 mM EDTA, pH 7.5; autoclave
  - 10× LiAc: 1 M lithium acetate. Adjust to pH 7.5 with dilute acetic acid; autoclave
  - Sterile H_2O

Procedure (day 1)

Part I. LiAc-mediated transformation

A. Preparation of yeast competent cells (steps 1 to 7 will be done by TA)

1. Inoculate 1 mL of YPDA with a single colony (2–3 mm in diameter) of the BY4741 strain. Do the same for the DPB 260 strain.

2. Vortex vigorously to disperse any clumps.

3. Transfer a 400-µL cell suspension into a sterile 100-mL flask containing 20 mL of YPDA.

4. Incubate at 30 °C at 250 rpm overnight (~16 to 18 h) to the stationary phase (A_{600} >1.5).

5. Next morning, transfer 13 mL overnight culture to a sterile 500-mL flask containing 100 mL of YPDA.

6. Shake-incultivate as above until A_{600} reaches 0.8 to 2.0 (for 3 to 6 h).

**Cells must be freshly prepared. Do not store the cells in a refrigerator.*

7. Read A_{600} using YPDA as a blank and calculate the cell density (A_{600} of 1 = ~3 × 10^7 cells/mL).

8. Obtain 25-mL yeast cultures of BY4741 and DPB260; label two 50-mL sterile conical tubes, place cells into the tubes, and centrifuge at 3000 rpm for 5 min at room temperature.

9. Aspirate off supernatant, suspend the cell pellet in sterile 10 mL of dH_2O, and centrifuge as above in step 8; discard the supernatant.

10. Resuspend the cell pellet in freshly prepared sterile 1.1× TE/LiAc solution to a final cell density of 1 × 10^9 cells/mL.

B. Transformation of yeast competent cells

11. Add salmon sperm carrier DNA (50 µg) and plasmid DNA (0.5 to 2.0 µg) to each sterile 1.5 mL tube.

TUBE 1 (DPB260): 5 µL (50 µg) carrier DNA + 3 µL pYES2.1-GFP_{uv1}

TUBE 2 (BY4741): 5 µL (50 µg) carrier DNA + 3 µL pYES2.1-GFP_{uv1}

TUBE 3 (DPB260): 5 µL (50 µg) carrier DNA + No plasmid DNA

TUBE 4 (BY4741): 5 µL (50 µg) carrier DNA + No plasmid DNA

12. Add 50 µL of yeast competent cells to each tube and mix well.

13. Add 280 µL of sterile PEG/LiAc solution to each tube and mix well by tapping.

14. Incubate at 30 °C for 30 min.

15. Add 35 µL of DMSO to each tube (a final 10%) and mix well.

16. Place in a 42 °C water bath for 15 min.

17. Centrifuge cells at 8000 rpm for 15 s, remove the supernatant, and resuspend cells in 1 mL of sterile 1× TE (10 mM EDTA, 1 mM EDTA, pH 7.5) buffer.

**At this time cells are fragile. Do not increase the speed and time of centrifugation.*

C. Plating transformation mixtures

18. Spread 100 µL directly on each YNB –Ura/+glucose plate that will select for the transformants (two plates for each BY4741 and DPB 260 transformations).

**Make sure your plates are well labeled and that you are able to identify them from each other and from other groups' plates.*

**Minimize the strokes of glass rod plating; cells are fragile.*

19. Turn the plates upside down and incubate at 30 °C until colonies appear (generally 2 to 4 days).

Procedure (day 2)

20. Record the data.

Transformation mixture	No. of colonies			DNA (µg)	Transformation efficiency
	Plate 1	Plate 2	Average		
BY4741 + pYES2.1-GFP_{uv1}					
BY4741 + no DNA					
DPB260 + pYES2.1-GFP_{uv1}					
DPB260 + no DNA					

$$\frac{\text{cfu} \times \text{Total suspension volume (mL)}}{\text{volume plated (mL)} \times \text{DNA amount used } (\mu g)} \times \text{Dilution factor}$$

$$= \text{cfu}/\mu g\text{DNA}$$

Sample calculation:

- Average 100 colonies grew on no dilution plate (dilution factor = 1)
- Plating volume: 0.1 mL
- Total suspension volume (step 17) = 1 mL
- Amount of plasmid = 0.1 μg

$$[(100 \text{ cfu} \times 1 \text{ mL}) \div (0.1 \text{ mL} \times 0.1\ \mu g)] \times 1$$

$$= 10^4 \text{cfu}/\mu g \text{ DNA}$$

21. Pick transformant colonies with sterile toothpicks, cross-streak them on the YNB –Ura/+galactose/+raffinose plate, and incubate the plates upside down at 30 °C.
22. Streak BY4741 colonies on to YNB +Ura/+galactose/+raffinose plate and incubate the plates upside down at 30 °C (one plate shared by all groups).

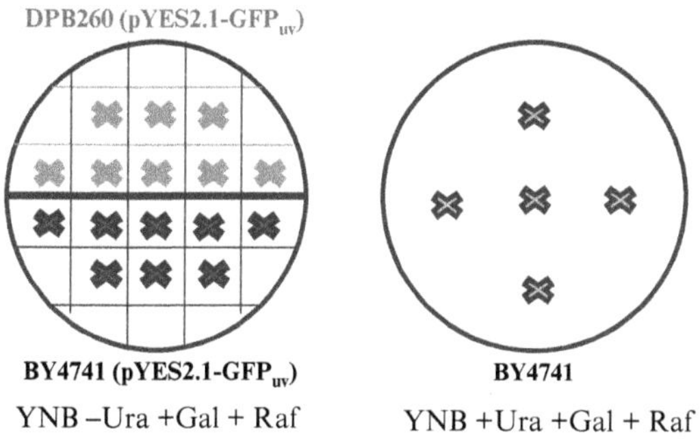

23. Uncover the lids and examine all plates under a UV light transilluminator. Capture the image with a digital camera.

Part II. Growth test

1. Inoculate DPB276 cells into 10 mL of YNB +Ura/+galactose/raffinose and incubate at 30 °C, at 250 rpm for 1 day.
2. Read A_{600} of the cell culture using YNB +Ura/+galactose/raffinose medium as a blank and dilute the cell to A_{600} of 0.6.
3. Spin a 0.5-mL culture of DPB276 at 8000 rpm for 30 s, remove the supernatant, and resuspend the cell pellet in 0.5 mL of sterile water.
4. Perform the above washing step once more.
5. Transfer a 100-μL culture of DPB276 to a 1.5-mL tube containing 900 μL of sterile water (1/10 dilution) and make a series of 1/10 dilutions, $1/10^2$, $1/10^3$, $1/10^4$, $1/10^5$, $1/10^6$, in sterile water.
6. Spot 5-μL aliquots of each dilution ($1/10^2$ to $1/10^6$) on to the following plates marked with each dilution:
  (i) YNB –Ura +glucose
  (ii) YNB –Ura +5-FOA +glucose
  (iii) YNB –Ura +galactose +raffinose
  (iv) YNB +Ura +galactose +raffinose
  (v) YNB –Ura +5-FOA +galactose +raffinose
7. Air-dry the spot in the laminar flow hood.
8. Incubate all plates upside down at 30 °C.

**Do not leave yeast colonies on 5-FOA plates for longer than necessary.*

9. Examine all plates and compare the growth patterns between the plates. Pay attention to the highest and lowest dilution spots on the plates when examined for comparison.

**Do not incubate for too long. As soon as low dilution ($1/10^4$ or $1/10^5$) spots show a sign of growth, begin to examine the plates.*

**Comparison should be made among the same dilution spot that is penultimate to the dilution spot of no growth of the reference control.*

10. Tabulate the growth data on five different media, along with the incubation time.

Medium* \ Dilution	$1/10^2$		$1/10^3$		$1/10^4$		$1/10^5$		$1/10^6$	
	35 h	72 h	35 h	72 h	35 h	72 h	35 h	72 h	35 h	72 h
YNB –Ura+Glc										
YNB –Ura+5-FOA+Glc										
YNB –Ura+Gal+Raf										
YNB +Ura+Gal+Raf										
YNB –Ura+5-FOA+Gal+Raf										

*Growth +; no growth –; incubation can be more or less than the suggested times.

11. Compare the observed growth with your expected growth.

Type of medium	Expected growth of DPB276*				Observed growth	
	Assume complete silencing		Assume incomplete silencing		35 h	72 h
	Silencing†	Growth*	Silencing†	Growth*		
YNB –Ura+Glc						
YNB –Ura+5-FOA+Glc						
YNB –Ura+Gal+Raf						
YNB +Ura+Gal+Raf						
YNB –Ura+5-FOA+Gal+Raf						

*Growth +; poor growth +/–; no growth –.
†Silencing occurs: Yes; no silencing occurs: NO.

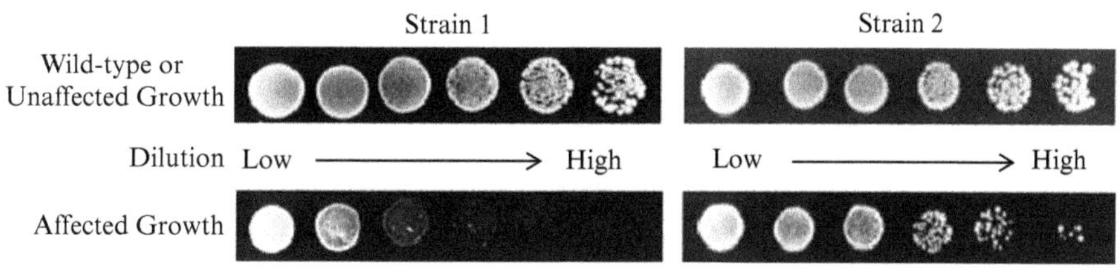

12. Determine whether incomplete silencing has occurred by comparing your expected outcome of growth patterns with the observed growth patterns of the same dilution(s).

**For example, use $1/10^2$ dilution data if there is no growth on both A and B plates of $1/10^3$ dilution but there is + growth on the A plate and – growth on the B plate of $1/10^2$ dilution.*

Post-lab assignment

1. List and describe the practical applications of RNAi.
2. What is difference between siRNA and miRNA in terms of (a) the mode of RNAi action and (b) their origins?
3. What are (a) the advantages and (b) the disadvantages of an RNAi-based screen in functional genomics as opposed to classical genetic screens such as chemical and insertional mutagenesis?
4. Name at least two problems preventing RNAi from being used in therapeutics.
5. Why is 5-flouro-orotic acid (5-FOA) added to YNB –Ura/+glucose medium for the growth test of DPB276? What is the expected growth of DPB276 on this medium? Explain why this is.
6. **(a)** Why is raffinose added to the minimal agar medium in addition to galactose?
(b) What is the chemical structure of raffinose?
(c) How is raffinose metabolized by *S. cervisiae*?
7. On the basis of your study of Pre-Lab Assignment 3, what is the expected growth (Good or Poor or No) of DPB276 on the following media: (i) when *URA3* silencing is complete and (ii) when *URA3* silencing is incomplete? Explain why this is.
(a) YNB +Ura/+FOA/+galactose/raffinose
(b) YNB –Ura/galactose/raffinose
8. Prolonged incubation of the DPB276 cells on YNB –Ura/ +galactose/raffinose and YNB +Ura/+galactose/raffinose plates in the Part II growth test will result in the indistinguishable result. Why do you think this happens?
9. On the basis of your study of Pre-Lab Assignment 1, what is the expected GFP expression of the DBP260 strain on the following media? Explain your choice.
(a) YNB +Ura/+Glucose
(b) YNB +Ura/galactose/raffinose
(c) If you want to transform plasmid carrying the *ADE* marker into this strain, how do you select and screen transformants? Provide two independent medium compositions to be used for selection and screening. Explain your choice.
(d) If you want to transform plasmid carrying the *URA3* marker into this strain, how do you select and screen transformants? Provide two independent medium compositions to be used for selection and screening. Explain your choice.

Reference

Drinnenberg, I.A., Weinberg, D.E., Xie, K.T., Mower, J.P., Wolfe, K.H., Fink, G.R., and Bartel, D.P. (2009). RNAi in budding yeast. *Science*, **326**: 544–550.

Further reading

Gietz, R.D. and Schiestl, R.H. (2007). High-efficiency yeast transformation using the LiAc/SS carrier DNA/PEG method. *Nature Protocols*, **2**: 31–34.

Kaelin, W.J. (2012). Use and abuse of RNAi to study mammalian gene function. *Science*, **337**: 421–422.

Suk, K., Choi, J., Suzki, Y., Ozturk, S.B., Mellor, J.C., Wong, K.H., Mackay, J.L., Gregory, R.I., and Roth, F.P. (2011). Reconstitution of human RNA interference in budding yeast. *Nucleic Acids Research*, **39** (7): e43. DOI:10.1093/nar/gkq1321.

50 AMB 2 experiment 50: Protein preparation for 2D gel electrophoresis

Purpose: This is up to you to write down.

Introduction

BugBuster® Protein Extraction Reagent is formulated to directly release soluble proteins from *E. coli* without mechanical disruption such as sonication and French press. The proprietary reagent contains a mild non-ionic detergent mix that is capable of cell wall perforation without denaturing soluble protein. Following a brief incubation of cells suspended in BugBuster reagent, insoluble cell wall and membranes are removed by centrifugation. However, it is important to note that the proteins in the lysate represent only those that are soluble in the BugBuster® reagent buffer. Because of very different biochemical and physical characteristics among proteins, any single extraction procedure cannot display all the proteins. Accordingly, the protein profiles may vary considerably depending on the extraction methods and type of buffers employed. Proteins obtained from differential extraction, subcellular extraction, chromatography fractions, and immunoprecipitated fractions can also be used for more focused analyses of target proteome. For example, glycoproteins can be enriched with concanavalin A or lectin affinity chromatography prior to two-dimensional (2D) gel analysis for glycome profiling.

Soluble proteins in crude cell lysate are not generally suitable for isoelectric focusing (IEF) because they contain many interfering substances. Small soluble non-protein substances must be removed by dialysis, desalting gel filtration, or organic solvent precipitation methods. Precipitation has an advantage over the other procedures since it permits concentration of proteins as well as elimination of undesirable substances including salts, carbohydrates, nucleotides, and phospholipids. One major disadvantage of precipitation is that proteins may be denatured, making the pellet difficult to re-solubilize. However, it is a simple clean-up technique for the samples that do not require protein activity prior to analyses (e.g., IEF, SDS-PAGE). Besides removal of interfering substances, proteins must be protected from degradation and modification by adding a protease inhibitor cocktail to cell lysis buffer and keeping samples ice-cold during extraction.

In this lab exercise, you will prepare soluble *E. coli* protein samples using BugBuster® reagent buffer from *E. coli* cultures that are untreated or treated with H_2O_2 and use TCA/acetone precipitation protocol to clean up and concentrate soluble protein samples. The precipitated proteins can be stored in the freezer and must be completely solubilized in rehydration buffer prior to use for IEF. Because the TCA/acetone precipitation procedure gives significantly differing recovery yields depending on the type of organisms and tissues, it is recommended to quantify proteins after solubilization of the precipitated protein pellet in a minimum volume of rehydration buffer. However, you will quantify the extracted protein before the precipitation due to the time limitation. You will be using Bio-Rad RC-DC protein assay reagent for quantitation. It is recommended to check if the protein is degraded or intact by running an aliquot of protein sample on SDS-PAGE gel prior to use in 2-D gel analysis. However, this also will not be performed due to the time limitation. It is recommended to watch the following YouTube videos regarding IEF sample preparation:

- http://www.youtube.com/watch?v=PncWHRbhP5k (Protein extraction)
- http://www.youtube.com/watch?v=6d8TtjDTtcE (IEF protein sample)
- http://www.youtube.com/watch?v=O6m0QKPz7Sw (Protein sample cleanup)
- http://www.youtube.com/watch?v=iBV2jAEjZko (protein sample cleanup)

Pre-lab assignment

(Typing and submission must be completed before lab work begins.)

1. What is the final molar concentration of H_2O_2 ($\rho = 1.46\,g/mL$) at step 3 of Part A?
2. What do RC and DC stand for and why is this reagent used in the protein assay?

Materials and equipment

- *E. coli* TOPO10 or HB101 or Lemo21 (DE3)
- LB medium
- Hydrogen peroxide (H_2O_2) 0.3% (v/v)
- BugBuster^{TM} protein extraction reagent (EMD Millipore Chemicals, Cat. No. 70584-3)
- Nuclease mix (GE Healthcare Life Science, Cat. No. 80-6501-42)
- Protease inhibitor cocktail (Life Technologies, Cat. No. 87786; 100×)

Methods in Biotechnology, First Edition. Seung-Beom Hong, M. Bazlur Rashid and Lory Z Santiago-Vázquez.

Companion Website: www.wiley.com\go\hong\Methodsinbiotechnology

- TCA (100% trichloroacetic acid: 1 g TCA in 0.7 mL of ultrapure H_2O; prepare just before use)
- Acetone (HPLC grade, 100% cold –20 °C)
- RC DC protein assay reagent kit (Bio-Rad, Cat. No. 500-0121)
- BSA (1 mg/mL: 0.01 g in 10 mL of BugBuster™ protein extraction reagent)
- 1.5-mL and 2.0-mL microcentrifuges

Procedure

A. Growth and perturbation of cells (TA will do this part)

1. Inoculate 1 mL of an overnight culture of *E. coli* into 50 mL of LB medium in a sterile 250-mL flask, shake-incubate at 150 rpm, at 37 °C to an A_{600} of.
2. Aseptically transfer 25 mL of cell culture into two sterile 125-mL flasks.
3. Label one flask "treated" and add 20 µL of 0.3% (v/v) H_2O_2. Label the other flask "control" and add 20 µL of sterile dH_2O.
4. Shake the H_2O_2-treated culture tubes at 150 rpm, room temperature, for 30 min. Stand the untreated culture tube at room temperature without shaking.

B. Protein extraction

5. Weigh the two empty microcentrifuge tubes labeled "treated" and "control."
6. Fill a microcentrifuge tube with each culture, spin at full speed for 1 min, and aspirate off the supernatant completely without disturbing the pellet.
7. Repeat step 6 in the same microcentrifuge tube until you have harvested a total 5 mL of culture.
8. Determine the wet weight of the cell pellet.

**Obtain at least 0.04 g wet weight of cell pellet; if necessary, harvest more cell culture.*

9. Freeze the cell pellet in dry-ice/ethanol bath or in liquid nitrogen for 1 min.

**This helps lysis. At this point you may store the cells in the freezer until ready for use.*

10. Add BugBuster™ protein extraction reagent (0.5 mL per 0.1 g of wet cell pellet).
11. Add 1 µL of protease inhibitor cocktail (100×) and 1 µL of nuclease mix (100×) per 100 µL of BugBuster™ protein extraction reagent used.
12. Suspend the cell pellet completely by pipetting up and down, and shake-incubate the cell suspension slowly on a rotatory platform shaker for 10 to 20 min at room temperature.
13. Spin at the full speed for 15 min at 4 °C.
14. Transfer each supernatant to the labeled fresh 1.5-mL microfuge tube kept on ice.
15. Transfer 10 µL of each supernatant to the labeled 1.5-mL microfuge tube. This will be used for protein assay (step 28 of Part D).

C. Clean up by precipitation and solubilization

16. Mix the following solutions in a 1:8:1 (v/v/v) ratio by inverting the tube after adding each component (protein sample in 10% TCA in acetone). For example:
0.1 mL of protein extract in a tube (step 14)
0.8 mL of 100% pre-chilled (–20 °C) acetone
0.1 mL of 100% TCA

**Use all your protein extract and precipitate in 1.5-mL microcentrifuge tube(s).*

**The mixed sample should turn milky white if proteins are present.*

17. Precipitate protein by incubating at –20 °C for 30 to 60 min.

**While you are waiting, begin the protein assay (Part D).*

18. Spin at maximum speed for 10 min at 4 °C.
19. Decant the supernatant by pouring and blot the tube on to a paper tissue; be careful not to dislodge the protein pellet.
20. Add 500 µL of cold acetone to the tube and vortex for 5 s.

**The proteins will not dissolve in the acetone. This ensures removal of the TCA.*

21. Spin at maximum speed for 2 min at 4 °C and aspirate off the supernatant.

**Place the cap-hinge of the tube outward in the centrifuge.*

22. Repeat steps 20 and 21 twice more for a total of three acetone washes.

**It is important to wash thoroughly to remove residual TCA, which is very acidic.*

23. Spin the tube briefly for 10 to 20 s.

**Place the tube in the centrifuge in the same orientation as the previous centrifugation so that protein pellet remains attached to the same side of the tube, thus minimizing the protein loss.*

24. Remove the remaining supernatant from the pellet using a P100 micropipettor.
25. Leave the uncapped tube in the laminar flow hood for 10 min to evaporate acetone.

**Do not over-dry the pellet or it may not dissolve properly and consequently results in the loss of bulk proteins and/or incomplete IEF, which causes horizontal streaking on a 2D gel.*

26. Store the tubes (labeled protein ID, group #) in a –20 °C freezer until ready for use.

D. Protein assay

27. Add duplicate volume aliquots of BSA (1 mg/mL) to the labeled 1.5-mL microfuge tubes, as shown in the table.

Volume of BSA (1 mg/mL) standard solution (μL)	Volume of BugBuster™ protein extraction reagent (μL)
0*	100
10	90
20	80
40	60
60	40
80	20
100	–
Volume of your samples (μL)	
H_2O_2 treated: 10	90
Untreated: 10	90

*The first 0 serves as a blank solution (no duplicate).

28. Add BugBuster™ protein extraction reagent to each tube to bring to a final volume of 100 μL.
29. Add 500 μL of RC Reagent I to each tube, vortex, and incubate all tubes for 2 min at room temperature.
30. Add 500 μL of RC Reagent II into each tube and vortex. Centrifuge the tubes at 15 000× *g* for at least 5 min.
31. Discard the supernatant by inverting the tubes on a clean paper towel. Allow the liquid to drain completely from the tubes. If needed, a pipette can be used to carefully remove excess liquid.

**The protein pellet may be difficult to see as it may form a thin layer on the walls of the tube.*

32. Add 510 μL Reagent A′ to each tube, vortex, and incubate tubes at room temperature for 5 min, or until the precipitate is completely dissolved.
33. Transfer the entire sample to each glass test tube and add 4 mL of DC Reagent B to each tube and vortex immediately. Incubate all tubes at room temperature for 15 min.
34. Read the absorbance at 750 nm within 1 h.
35. Plot a BSA standard curve (*x* axis: μg protein amount; *y* axis: A_{750} value).
36. Determine the concentrations of protein samples using the standard curve.
37. Determine the total amounts of treated and untreated proteins precipitated at step 25 of Part C, assuming a 60% recovery yield.

**This information is needed for rehydration of proteins in the Day 2 procedure of Experiment 51.*

Post-lab assignment

1. You will use protein samples prepared from H_2O_2 treated and untreated *E. coli* cultures for 2D-Gel analysis.

(a) What known proteins are expected to be more or newly expressed when *E. coli* cells are exposed to oxidative stress damage caused by hydrogen peroxide (H_2O_2)?

(b) • Visit the ExPASY SWISS-2DPAGE web site (http://world-2dpage.expasy.org/swiss-2dpage/).
• Click "[description, ID or gene]" under the "SWISS-2DPAGE Search by" menu.
• Based on the (a) answer, type the name of one protein of your interest into the box of "Enter search keywords" and click "Execute query."
• Select and click "_ECOLI ID."
• Select and click MAP LOCATIONS: ID under "2D PAGE maps for identified proteins."
• Copy the image of the 2D gel pattern and paste into a Microsoft Word document. Mark the target protein spot with an arrow.

2. The standard curve of the RC-DC protein assay was established using BSA in a total volume of 100 μL, as shown below. Your protein assay sample consisting of 10 μL of protein extract and 90 μL of extraction buffer gave an absorbance value of 0.25. What volume of IEF rehydration buffer should be used to dissolve a protein pellet obtained from 100 μL of protein extract precipitaed by TCA/acetone in order to bring a final protein concentration of 3 mg/mL? Assume that the recovery yield of precipitation is 60%.

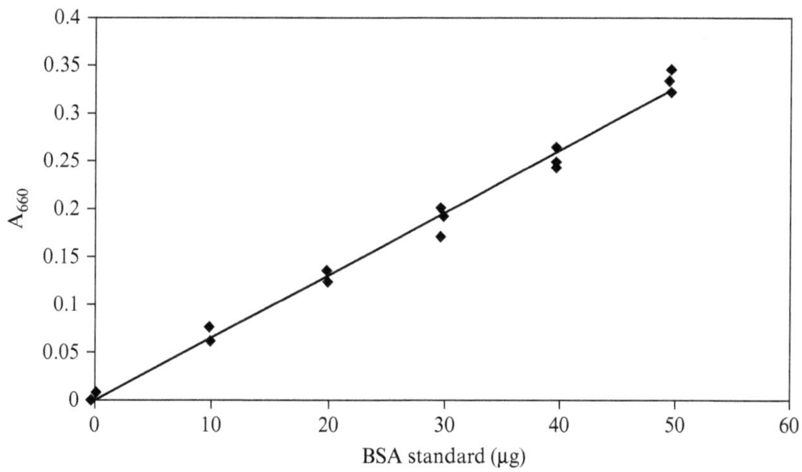

Further reading

Bio-Rad RC DC™ Protein Assay Instruction Manual.

Lypons, K. (2003). TCA/Acetone Precipitation (Large Scale). AfCS Procedure Protocol PP00000148 (http://www.signaling-gateway.org/data/cgi-bin/ProtocolFile.cgi/afcs_PP00000148.pdf?pid=PP00000148).

Novagen BugBuster™ Protein Extraction Reagent Instrucion (TB245 09/00).

Xiaolin, W., Xiong, E., Wang, W., Scali, M., and Cresti, M. (2014). Universal sample preparation method integrating trichloroacetic acid/acetone precipitation with phenol extraction for crop proteomic analysis. *Nature Protocols*, **9**: 362–374.

51 AMB 2 experiment 51: Two-dimensional gel electrophoresis

Purpose: This is up to you to write down.

Introduction

Two-dimensional polyacrylamide gel electrophoresis (2D-PAGE) is a technique for separating a complex mixture of proteins according to an isoelectric point (pI) by isoelectric focusing (IEF) in the first dimension and according to a size (MW) by SDS-PAGE in the second dimension. A good result usually yields a very large number of stained spots scattered across the entire surface of the gel. Thus, thousands of proteins can be resolved in a single experiment, allowing us to detect the major proteins in a sample and to compare protein expression patterns in related samples under different conditions. Detection sensitivity depends on the choice of staining reagent, ranging from ~50 ng/spot with Coomassie blue dye to ~1 ng/spot with SYPRO® ruby dye. The comparative analysis of resolved protein spots can be facilitated by using 2D gel image analysis software. To ensure the reproducibility of gel images, which is a key issue of 2D-PAGE in the bioscience community, the area and intensity of spots as well as the presence or absence of spots are taken into consideration. Currently one major drawback of 2D-PAGE is the difficulty in separation of hydrophobic proteins, especially for those associated with membrane. The protein spots of interest can be extracted and further analyzed by mass spectrometry for identification and quantification. Accordingly, 2D-PAGE is the core technology for expression proteomics.

Since IEF is very sensitive to charge differences (pI around 0.01 pH units), its reproducibility requires that protein must be handled with great care to prevent any protein modifications in chemical composition and primary sequence. During sample preparation, salts content and protease activity must be minimized. In addition, interactions of proteins with lipids, carbohydrates, nucleic acids, and other proteins must be avoided because they make charge modifications that will give rise to the shifted isoelectric mobilities or streaking in the gel. IEF involves using ampholytes (*ampho*teric electro*lytes*), a mixture of small

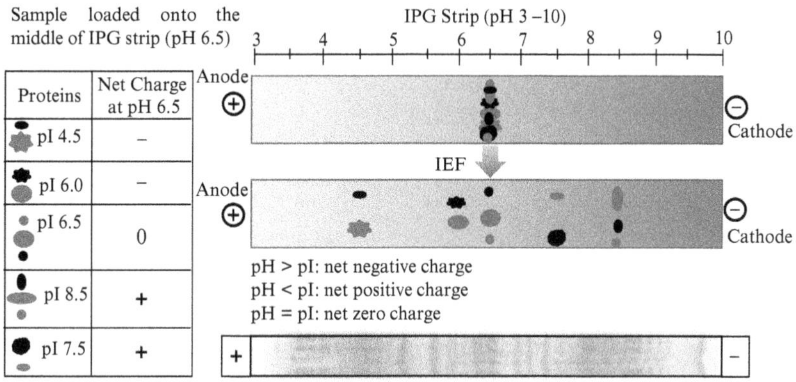

oligomolecules (300 to 1000 Da) that contain both acidic and basic buffering groups with closely spaced pIs, which creates a pH gradient during electrophoresis. When a protein is placed in a pH gradient medium and subjected to an electric field, it will initially move toward the opposite charge. In the course of migration, protein will either gain or lose protons and eventually become a net charge of zero when the positive and negative charges balance at pH equal to its pI, as illustrated in the figure above.

In this lab exercise, a ready-made immobilized pH gradient (IPG) gel strip, in which ampholytes are covalently bonded to the acrylamide gel matrix on a plastic strip, is used to separate protein samples obtained from the previous lab exercise along with IEF marker proteins in the first dimension (IPG-IEF). There are two types of pre-cast IPG gel strips having the same pH range, linear gradient, and non-linear (NL) gradient. The linear gradient means that the pH increases at a constant rate along

Methods in Biotechnology, First Edition. Seung-Beom Hong, M. Bazlur Rashid and Lory Z Santiago-Vázquez.

Companion Website: www.wiley.com\go\hong\Methodsinbiotechnology

the whole IPG strip. The non-linear gradient has a sigmodal pH gradient along the whole IPG strip. As an example of a pH 3–10 NL strip, the pH slope between 5 and 7 is linear, while the pH slopes between 3 and 5 and between 7 and 10 are rather flat. As a result, the pH 5–7 slope of a NL IPG is steeper than that of a linear IPG, enabling the NL IPG to give better resolution between pH 5 and 7. Commercially available IPG strips are also supplied as broad, medium, and various narrow pH ranges. IPG strips with narrow pH ranges are helpful to resolve the crowded spots that were not well separated in an IPG strip having a broad pH range.

Prior to use, immobiline DryStrip gels are rehydrated in rehydration solution containing a matching IPG buffer. It is a standard practice to carry out IEF in a denaturing IPG gel containing urea and non-ionic and/or zwitterionic detergents. To perform IEF, the protein pellet must be suspended in the same rehydration solution, and this protein solution is used to rehydrate the IPG strip (rehydration loading), to load on to the rehydrated gel strip (sample cup loading), or to load on to the paper pad placed between the end of the IPG and the electrode strip (paper bridge loading).

It is recommended to watch the following YouTube videos regarding IEF:

- http://www.youtube.com/watch?v=V3ArPwoRK5k (Introduction)
- http://www.youtube.com/watch?v=P1T4eyOfZNY (IEF sample)
- http://www.youtube.com/watch?v=M9pQo0SgdBs (IPG)
- http://www.youtube.com/watch?NR=1&v=IOVfPezX9q4 (Sample loading)
- http://www.youtube.com/watch?v=nVpZkfC0ezk (Applications)

Three types of IEF equipment, Multiphor™ II and IPGphor (GE Healthcare Life Sciences) and Protean IEF (Bio-Rad) systems are widely used. Whatever system is utilized, isoelectric-focused IPG strips must be equilibrated in the presence of SDS, DTT, urea, glycerol, and iodoacetamide (IAA) before they are subjected to SDS-PAGE in the second dimension.

Pre-lab assignment

(Typing and submission must be completed before lab work begins.)

1. What properties of proteins are used to separate complex protein mixtures in 2D-PAGE?
2. Can you use SDS for protein preparation in the first dimension of the 2D-PAGE? Explain your choice.
3. Why do we need to (a) rehydrate the gel strip (b) at room temperature and (c) overnight?
4. Why is the gel strip soaked in equilibration buffer before running the second dimension gel?
5. What are PI and MW of the protein samples in IEF standards and 2D SDS-PAGE standards that you will be loading in order to interpret the experimental result?

Materials and equipment

First dimension

- Immobiline™ DryStrip (ready-cut IPG strips; 7 cm, pH 3–10 or pH 3–10 NL): pH 3–10 (GE Healthcare Life Sciences, Cat. No. 17-6001-11); pH 3–10 NL (GE Healthcare Life Sciences, Cat. No. 17-6001-12)
- Rehydration buffer 1 (2D RB1): 8 M urea, 2% CHAPS, 2% IPG buffer, 15 mM DTT
- IPG buffer pH 3–10 (GE Healthcare Life Sciences, Cat. No. 17-6000-87); IPG buffer pH 3–10 NL (GE Healthcare Life Sciences, Cat. No. 17-6000-88)
- Immobiline™ DryStrip re-swelling tray
- Immobiline™ DryStrip cover fluid (mineral oil: GE Healthcare Life Sciences)
- Immobiline™ DryStrip kit (GE Healthcare Life Sciences)
- Multiphor™ II electrophoresis system (GE Healthcare Life Sciences), EPS 3501 XL power Unit
- MultiTemp III thermostatic circulator (GE Healthcare Life Sciences)
- IEF standards (Bio-Rad, Cat. No. 161-0310)
- 2D SDS-PAGE standards (Bio-Rad, Cat. No. 161-0320)

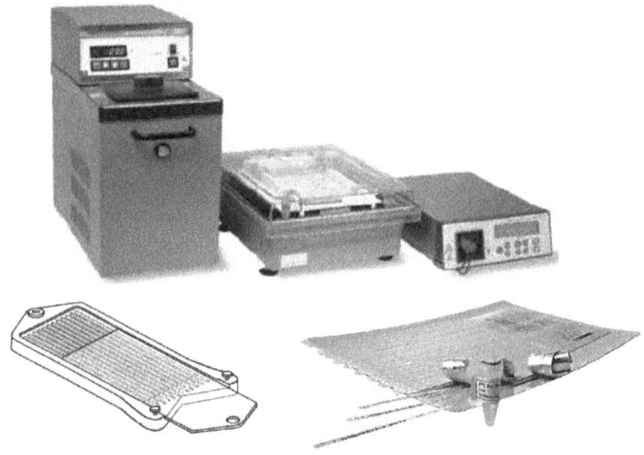

Second dimension

- 4–15% precast SDS-PAGE gel (Bio-Rad, Cat. No. 456-1081)
- SDS-PAGE running buffer
- Mini-Protean® 3 gel electrophoresis system (Bio-Rad)
- GelCode® blue staining solution
- SDS-PAGE molecular weight marker (Bio-Rad, Cat. No. 161-0375)

Procedure (Day 1)

Part I. Rehydration of IPG DryStrip (instructor will do this part)

1. Pipette 125 µL of rehydration buffer (8 M urea, 2% CHAPS, 2% IPG buffer, 15 mM DTT, 0.002% bromophenol blue) into the Immobiline DryStrip (IPG) reswelling tray on a flat level bench.

**Deliver the solution slowly at one point to minimize the introduction of air bubbles.*

**The volume should not exceed the maximum volume determined for each Immobiline DryStrip size (125 μL for 7 cm IPG strip).*

**IPG strip gels must be rehydrated with a matching IPG buffer (pH 3–10 NL, pH 3–10).*

2. Peel off the foil cover slowly from DryStrip, starting from the anode (arrow pointed end) using fine pointed forceps.

3. Place the strip slowly in the tray channel with the *gel side down*. Avoid trapping air bubbles under the strip.
4. Overlay the strip immediately with 200 to 400 μL of mineral oil to prevent evaporation.
5. Rehydrate at room temperature overnight (10 to 24 h).

**The protein sample can be included in the rehydration buffer to load on to the IPG strip (rehydration loading).*

**Ensure that the IPG strip gels are rehydrated along its entire length.*

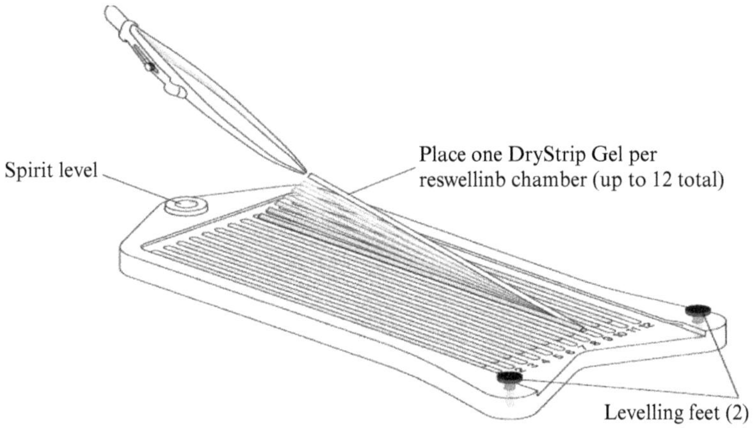

Procedure (day 2)

Part II. Rehydration of proteins

1. Calculate the volume of rehydration buffer to solubilize the protein pellet to yield a final concentration of 3 μg/μL on the basis of the protein assay (Part D of Experiment 50). Assume that the recovery yield of the precipitated protein is 60%. *Calculation is needed in the pre-lab notebook.*
2. Add the calculated amount of rehydration buffer. Pipette up and down repeatedly, vortex the samples, and place on a tube rocker for 30 min at room temperature.

**Do not heat the solution. It will take time for the pellet to be solubilized. Be sure that the sample is completely and stably solubilized.*

3. Spin the tubes at maximum speed at room temperature for 10 min.
4. Transfer the supernatant to a new labeled tube.

**About 100 μL (~300 μg) of total protein will be added to the IPG strip at step 6 of Part III.*

Part III. First dimension (IEF)

1. Prepare the first dimension run; check for proper connections and set up (instructor will perform this step).

(a) Make sure that the red bridging cable in Multiphor II is connected.
(b) Set the temperature on the MultiTemp III Thermostatic Circulator cooling unit to 20 °C at least 30 min before use.
(c) Place the cooling plate in the Multiphor II Electrophoresis unit and make sure the unit is leveled.
(d) Pipette ~ 3 to 4 mL of DryStrip cover fluid (paraffin oil: Amersham Pharmacia Biotech) on to the cooling plate (Fig. 1).

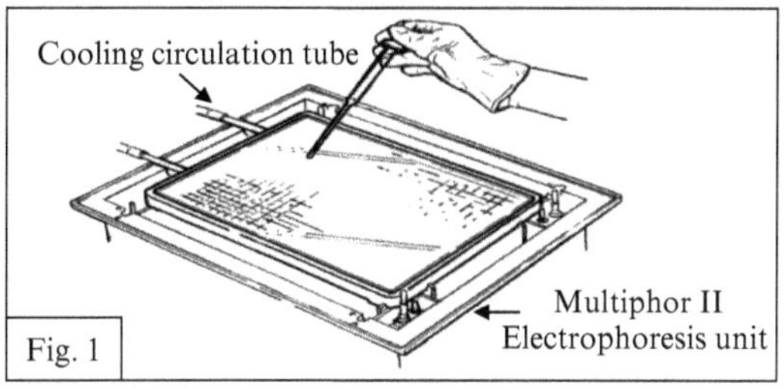

Fig. 1

(e) Position the Immobiline DryStrip tray and electrode holder on the cooling plate (Fig. 2).
(f) Make sure that the red (anodic) electrode connection of the tray is positioned at the top of the plate near the cooling tubes.
(g) Make sure that there are no large bubbles between the tray and the cooling plate; small bubbles are all right.

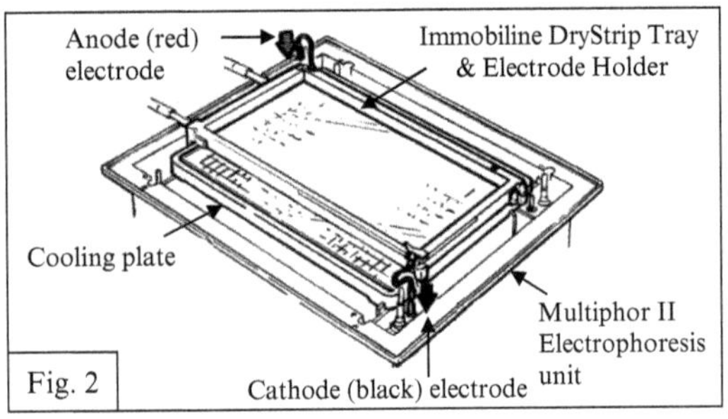

Fig. 2

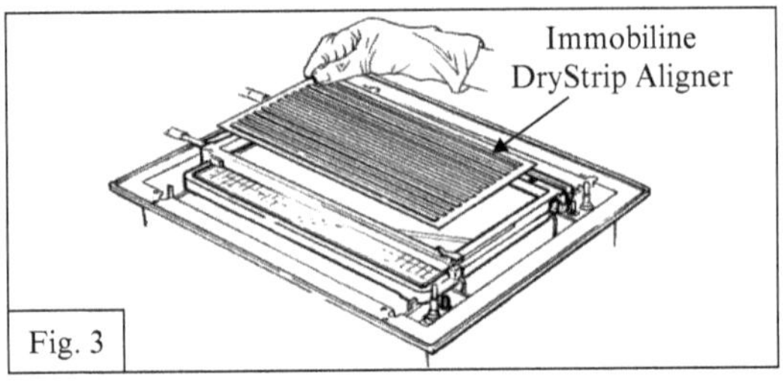

Fig. 3

**The DryStrip Cover fluid at this point serves as an insulating fluid to ensure good contact between the cooling plate and the tray.*

(h) Connect the red (anode) and black (cathode) electrode leads on the tray to the Multiphor II Electrophoresis unit (Fig. 2).

(i) Pour about 10 mL of DryStrip cover fluid into the tray.

**The DryStrip cover fluid protects the strips from evaporation and the samples from oxidation.*

(j) Place the Immobiline DryStrip aligner with 12 grooves side up into the tray on top of the oil (Fig. 3).

**Avoid getting oil on top of the aligner at this point.*

After setting up, the system looks like a picture below.

2. Cut two IEF electrode strips (paper strips prepared from 2 mm thick filter paper) to a length of ~11 cm each, soak each one into dH_2O briefly, and remove excess water by blotting on to paper tissue on a clean flat surface (Fig. 4).

**It is important to remove any excess water; it should just be damp and evenly soaked.*

**Make sure that the electrodes (anode and cathode), sample cup bar, and sample cups are ready.*

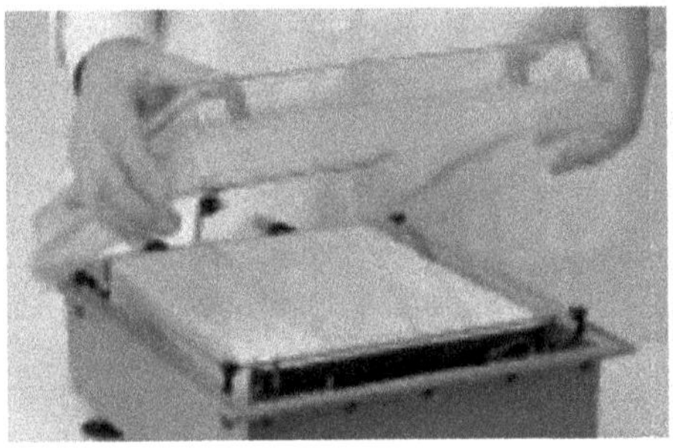

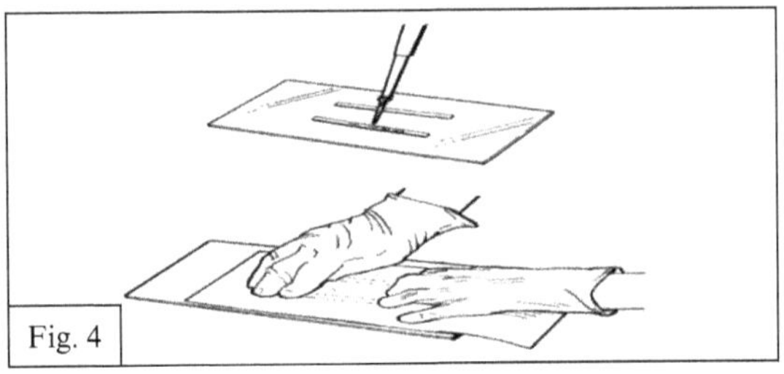

Fig. 4

3. Transfer the rehydrated IPG strips to DryStrip aligner, place the wet IEF electrode strip on to each end of the IPG strips, and put the electrodes on the IEF electrode strips, as detailed below.

(a) Using clean forceps rinse the rehydrated IPG gel strips with deionized water briefly and place them on a sheet of water-saturated filter paper with the gel side up.

(b) Wet a second sheet of filter paper with deionized water, blot it to remove excess water, and put it on to the surface of the IPG gel strips. Blot them gently for a few seconds to remove excess rehydration solution and to prevent urea crystallization on the surface of the gel during IEF focusing.

(c) Immediately transfer the rehydrated IPG strips (*gel side up* and *acidic end (+) facing towards the anode*) to adjacent grooves of the aligner in the tray (Fig. 5).

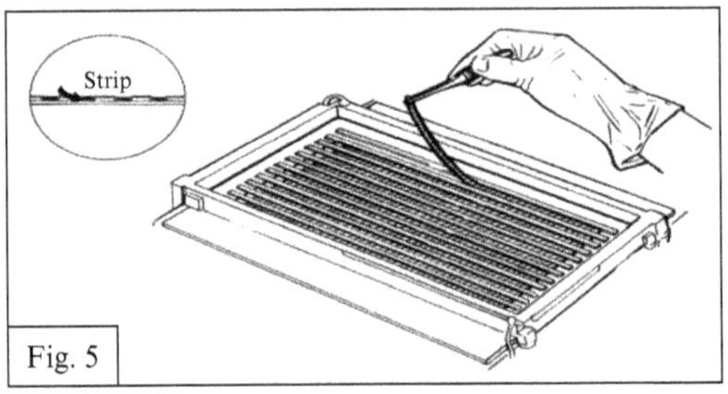

Fig. 5

(d) Place the strips with acidic end (+) at the top of the tray toward the red electrode (anode +). The basic end (–) of the strip should be at the bottom of the tray toward the black electrode (cathode –). Align both ends of the IPG strips to be lined up.

(e) Place each of the two moistened IEF electrode strips on to the top of the acidic and basic ends of the aligned IPG strips (Fig. 6).

**The IEF electrode strips should be at least partially on top of the gel surface with no air bubbles trapped beneath them. The distance between the two electrode strips depends on the length of the IPG gel strip.*

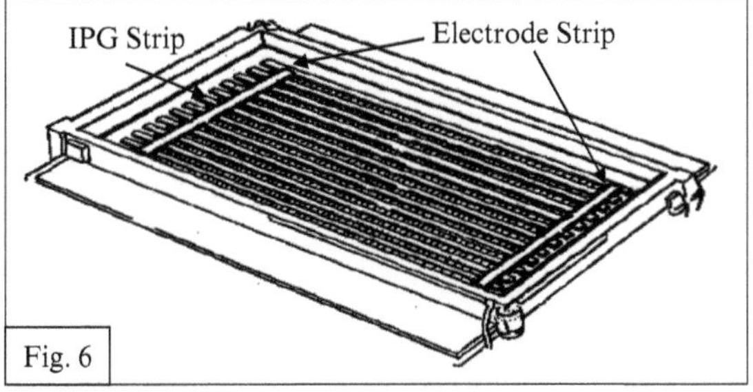

Fig. 6

(f) Position the electrode over the center of the IEF electrode strip and press down on top of the electrode strips (Fig. 7). **Each electrode has a side marked red (anode) or black (cathode). Make sure the marked side (metal band within the electrode) contacts the metal band along the side of the tray that makes the electrical contact.*

**Once the electrodes are in place, make sure the IPG strips have remained aligned in their grooves.*

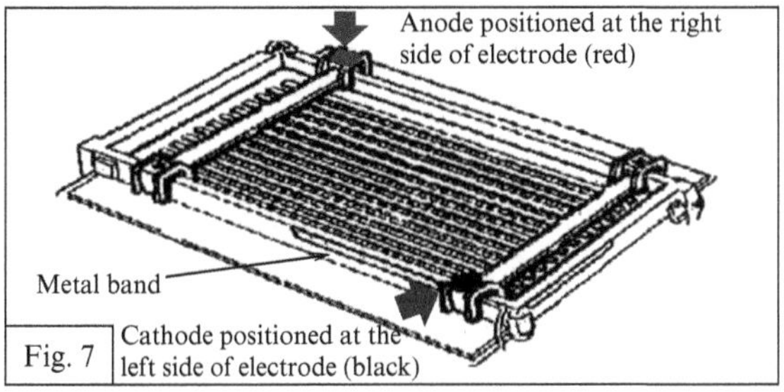

Fig. 7

4. Place the sample cup and sample cup bar over the rehydrated IPG strip as detailed below.

(a) Put the sample cups high enough on the sample cup bar and place in a desired position (middle or near the anode or cathode) where sample cups in the sample cup bar must face toward the electrode (Fig. 8).

**The sample application point affects resolution. When the proteins of interest have acidic pIs, the sample application near the cathode is recommended, and vice versa.*

**Anodic sample application is necessary with a pH 6–11 gradient and cathodic sample application is necessary for a pH 3–7 gradient.*

**The sample application points can be changed based on the result of spot distribution patterns.*

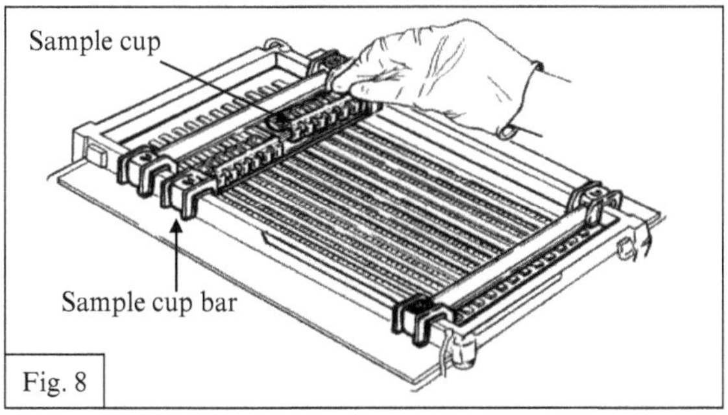

Fig. 8

(b) Align the sample cup to the position right above each IPG strip and gently press the sample cup down to ensure good contact with each strip (Fig. 9).

**This is the most critical part of the setup.*

**Check that the IPG strips are in their correct and straight position in the strip aligner.*

**Make sure that the sample cups are correctly aligned with the IPG strips.*

**Do not press the sample cups so hard that they push through the IPG strip gel.*

**Make sure that the bottom of the sample cups are flat against the gel surface.*

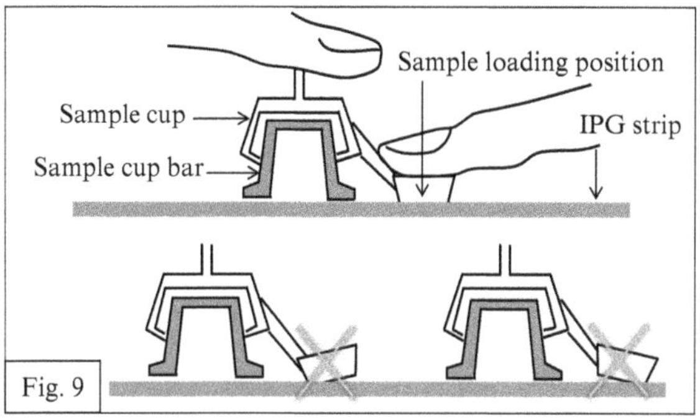

Fig. 9

5. Once the sample cups are properly positioned, pour 70 to 80 mL of mineral oil (Immobiline DryStrip cover fluid) so that the IPG gel strips can be completely covered (Fig. 10). If the oil leaks into the sample cups, reposition the sample cups, pipette the oil out, and check for leakage again.

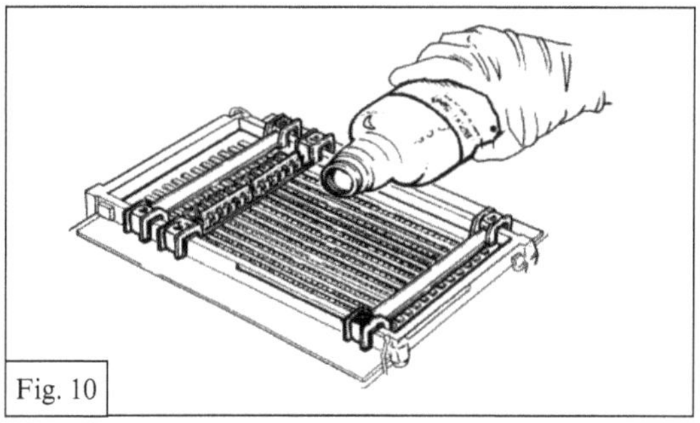

Fig. 10

**You can also check for leakage by loading 20 µL of 0.01 % BPB in rehydration buffer. In this case, the BPB solution should be removed again before loading the protein sample.*

6. Pipette the protein samples (up to 100 µL; maximum concentration 10 mg/mL) into the cups to lie down under the surface of the DryStrip cover fluid (Fig. 11). The sample should sink to the bottom of the cup. Alternatively, load the protein sample first and overlay with DrySrip cover fluid. Do not touch the sample cup during sample loading! Check for leakage!

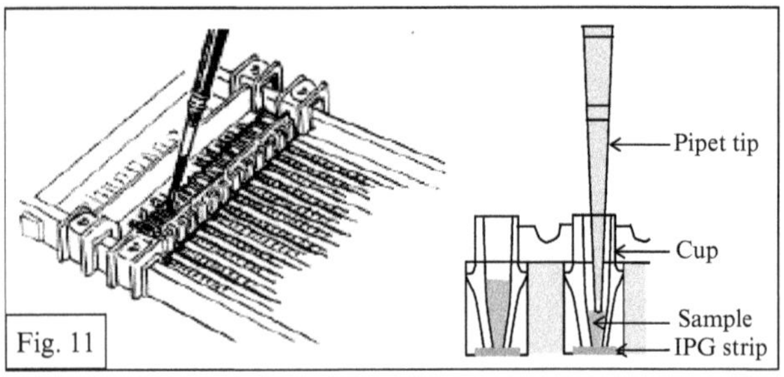

Fig. 11

**Protein precipitation and aggregation at the application point can sometimes be avoided by applying the sample in dilute solutions (60 to 100 µg of protein per 100 µL).*

7. Fill up each sample cup with a few drops of DryStrip cover fluid to completely cover the sample cup.

**The Immobiline DryStrip gels are submerged under a layer of Immobiline DryStrip cover fluid to prevent drying of the gel, precipitation of the components of rehydration solution, and diffusion of gas into the gel.*

8. Close the lid of the Multiphor II electrophoresis chamber and check for correct full assembly.
9. Connect the leads on the lid to the power supply. Set the power supply as instructed below and start the power.

Programmable Parameters (EPS 3501 XL power unit)

Phase	Voltage (V)	Current (mA)	Power (W)	Duration (h:min)	Timer, run end (kVh)
1	200	2	5	0:01	0.001
2	3500	2	5	1:30	2.8
3	3500	2	5	0:35–1:05	2.2–3.7
Total				2:05–2:35	5–6.5

**Note: the above voltage set is for 7 cm long, pH ranges 3–10 IPG (Immobbiline® DryStrip) strip. For all immobiline DryStrip gels, 20 °C, 0.05 mA/strip, Power 5 W total, kilovolt-hour (kVh: amount of kilo voltage in 1 h) values are recommended.*

**Turn on the thermostatic circulator (set at 20 °C) connected to the electrophoresis chamber at least 30 min before running.*

**Disable the start current check using the "MORE" mode of EPS 3501 XL before the run.*

**A good tip to ensure that the IEF is running correctly is to watch the bromophenol blue front. The cathodic end of gel strip begins to lose blue color as it slowly migrates toward the anode (+ red).*

10. After IEF, turn off the power supply and thermostatic circulator. Then remove the electrodes, sample cup bar, and IEF electrode strips from the tray.
11. Lift up to remove the IPG gel strips from the tray using fine pointed forceps. Proceed to do the Part V experiment of the day 3 procedure for a second dimension run.

**Focused gel strips that are not used immediately should be covered well with mineral oil, wrapped in plastic wrap, or placed in a Ziploc bag, and stored at –20 °C in a freezer until use.*

Procedure (Day 3)

Part IV. Preparation of buffers and chemicals

- SDS, Iodoacetamide, DTT, Tris, bromophenol blue, glycerol, urea
- Equilibration base buffer (6 M urea, 30% (w/v) glycerol, 2% (w/v) SDS, and 0.002% bromophenol blue in 50 mM Tris-Cl, pH 8.8)
- 10× SDS running buffer (25 mM Tris, pH 8.3, 192 mM glycine, 0.1% SDS)
- 0.5% agarose solution in 1× SDS running buffer (prepare before use; keep warm at ~55 °C)
- Acrylamide (30%)/bisacrylamide (0.8%)
- Ammonium persulfate, TEMED
- 3 MM paper strip, Kimwipe
- Pre-stained protein MW marker
- GelCode™ Blue stain solution
- Bio-Rad Mini-PROTEAN 3 system components
- Rotary platform shaker

Part V. Equilibration of IPG strips

1. Prepare equilibration buffer I by dissolving 0.15 g of DTT in 15 mL of equilibration base buffer.
2. Take out the IEF-focused IPG gel strips from the freezer and thaw.
3. Remove the excess mineral oil gently by blotting the edge of the strip on to Kimwipe tissue.

**Be careful not to damage the gel.*

4. Carefully place the gel strip in the tray channel with the *gel side up*. Refer to the figure illustrated for rehydration of IPG DryStrip (Part I).
5. Add equilibration buffer I to cover the strips and rock them gently on a platform shaker at room temperature for 10 min.

**Make sure that samples are gently shaken while the IPG strips are facing upwards.*

6. Prepare equilibration buffer II by dissolving 0.6 g of iodoacetamide in 15 mL of equilibration base buffer.
7. Aspirate off equilibration buffer I from the tray using a Pasteur pipette and blot with a piece of filter paper along the edge of the strip to remove excess liquid.
8. Add equilibration buffer II to cover the strips and rock them gently on a platform shaker at room temperature for 10 min.
9. After the second equilibration, rinse the IPG gel strip with 1× SDS-PAGE running buffer to remove DTT and IAA for 10 s, blot the edge of the strip on to Kimwipe tissue, and carefully place the strip on the top of the gel (step 6 of Part VI).

Part VI. Second dimension (SDS-PAGE)

1. Prepare a 10% resolving gel with a two-well (one wide) comb (1.0 mm thickness). Gently overlay dH_2O on to the surface of the gel after pouring.

% Resolving gel*	ddH_2O	30% Acrylamide	1.5M Tris-HCl buffer (pH 8.8)	10% SDS
10%	4.7 mL	2.7 mL	2.5 mL	0.1 mL

*Note: the stacking gel is optional. The percentage of gel can be changed depending on the size of the samples of interest.

2. Set up the Bio-Rad electrophoresis apparatus as in the previous Experiment 42 SDS-PAGE protocol.
3. Wash the top of the gel with a stream of SDS-PAGE running buffer using a Pasteur pipette.

**If a commercial pre-cast gel is used, it is important to remove any excess preservation buffer from the top well by thoroughly flushing out with a stream of SDS-PAGE running buffer before use.*

4. Remove the excess buffer on the flushed top well surface by contacting it with a piece of filter paper at one edge.

**This is important to prevent protein markers from diffusing.*

5. Load a pre-stained protein MW marker (5 μL) on to a 3MM filter paper strip (5 × 5 mm) and insert into the edge of the top of gel.
6. Place the plastic backing of the equilibrated IPG strip on the back plate of the gel cassette using forceps. Push the IPG gel strip between the two plates.

**Make sure you know which IPG strip end near the MW marker is acidic (+) or basic (−).*

**Make sure the strip is in close contact with the top of the gel with no trapped air bubbles.*

7. Pipette the molten (55 to 60 °C) agarose (0.5% in 1× SDS-PAGE running buffer) into the gel cassette to seal the IPG strip to the gel. Hold the cassette at a slight angle as the agarose is being pipetted into it to provide a way for bubbles to escape. There must not be any bubbles trapped in the thin agarose layer between the IPG strip and the top of the gel. Allow the agarose to solidify for at least 5 min.

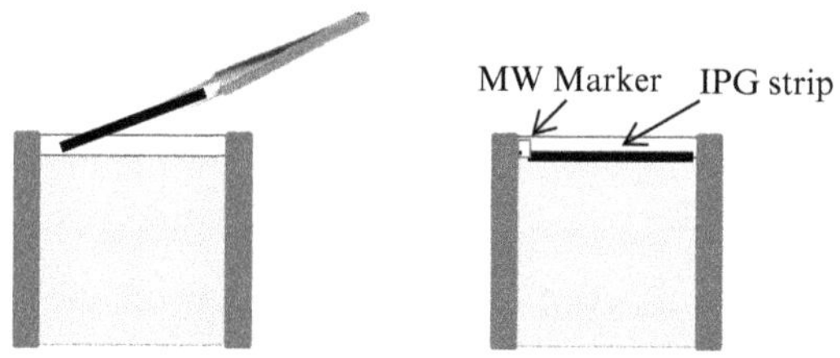

**Gaps between the IPG strip and the gel or damage to the IPG strip during application will result in vertical streaking.*

8. When the agarose has hardened, transfer the cassette to the electrophoresis chamber.

**Agarose sealing solution prevents the IPG gel from moving or floating in the electrophoresis buffer.*

9. Fill the upper and lower buffer chambers with 1× SDS-PAGE buffer.
10. Run the gel at 100 V for 10 min and then at 200 V for about 40 min.

**It is not necessary to remove the IPG gel strips from the surface of the vertical SDS gel after proteins have migrated out of the IPG gel strip. You will see the BPB dye front moving as the gel is running.*

11. Open the cassettes carefully. Use a razor blade to remove the agarose overlay from the gel. Peel the gel off the glass plate carefully by prying the lower edge up with a razor blade.
12. Place in a tray containing dH_2O. Rock on a platform shaker for 10 min. Repeat twice more.

**It is important to wash the gel thoroughly to remove SDS because it interferes with staining.*

13. Dispose of water and add GelCode blue solution. Rock on a platform shaker overnight.
14. Wash the gel in dH_2O to visualize the stained spots.
**If the washed gel has still a high background, especially in the bottom portion of the gel, the gel can be further washed in destaining solution (H_2O, methanol, glacial acetic acid, 50/40/10, v/v/v).*
15. Snap the gel pictures with a digital camera.

**It is important to remember that one spot on a gel may contain more than one protein since multiple proteins can have similar pIs and molecular weights.*

16. Arrange the two treated and untreated samples' gel pictures on the same page with proper labeling of the MW and anode/cathode positions in the lab report.

**Enlarge the pictures if necessary and mark differential spots.*

Post-lab assignment

1. The components of the rehydration buffer are typically urea, CHAPS, IPG buffer, bromophenol blue, and dithiothreitol (DTT). Describe their functions.
2. What is protein carbamylation? How does it occur?
3. There are three types of sample loading on to IPG strips: rehydration loading, cup loading, and paper-bridge loading. Describe the advantage and disadvantage of each type of sample loading.
4. Name some applications of 2D gel electrophoresis.
5. What is proteomics? What other approaches than 2D-GE can be used for proteomics studies? List at least two and briefly explain the principle of each approach.
6. The following diagram is a two-dimensional gel. The − and + signs indicate the polarity of electrodes during each stage of the separation. Explain your answer for each question.

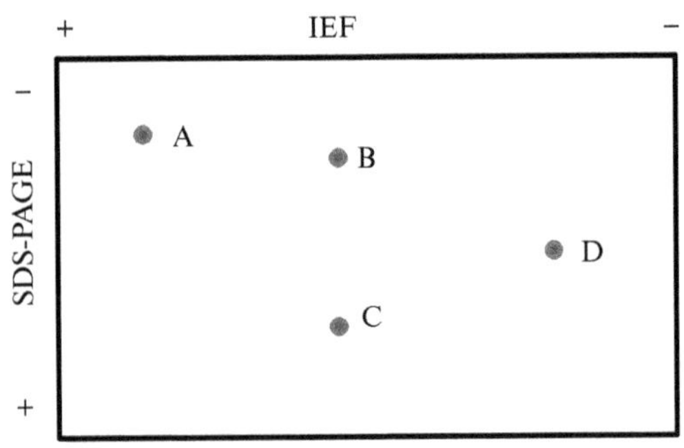

(a) List spots A, B, C, and D in the order of low pI to high pI.

(b) If the native polypeptides of A, B, C, and D are purified by size exclusion chromatography, list them in the order of first to last coming out of the column.

(c) If the native polypeptides of A, B, C, and D are purified by cation exchange chromatography, list them in the order of first to last coming out of the column.

(d) List two independent techniques that can be used to identify the polypeptide of spot A. What is the difference between the two techniques regarding requirement of the protein to be identified?

7. Shown below are the results of a 2D gel separation of protein X.

(a) Indicate the expected approximate positions where (i) the phosphorylated form (■) and (ii) the glycosylated form (♦) of the protein X will migrate. The "−" and "+" signs indicate the polarity of electrodes during the IEF stage of the separation. Explain your choice.

(b) What staining method can be used to detect phosphorylated and glycosylated proteins?

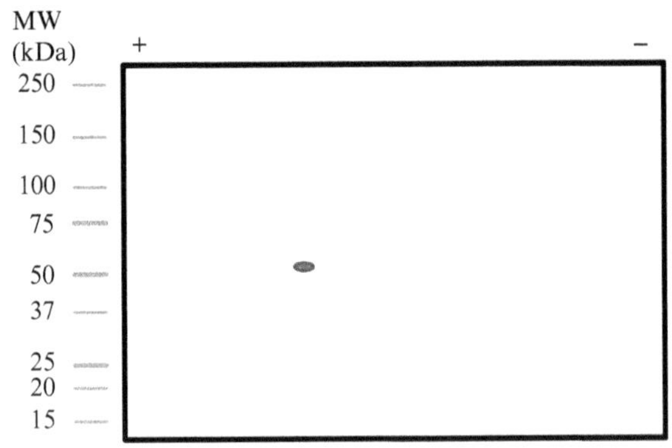

8. What type of IPG strip and sample loading will give the best separation of proteins whose pI ranges from 10 to 11? Explain your choice.

A. IPG 3-11 with sample loaded near to the anode
B. IPG 3-11 with sample loaded near to the cathode
C. IPG 3-11NL with sample loaded near to the anode
D. IPG 3-11NL with sample loaded near to the cathode
E. IPG 7-11 with sample loaded near to the anode
F. IPG 7-11 with sample loaded near to the cathode
G. IPG 7-11NL with sample loaded near to the anode
H. IPG 7-11NL with sample loaded near to the cathode

Further reading

2-D Electrophoresis Principles and Methods, GE Healthcare Handbook 80-6429-60AC.

Görg, A., Drews, O., Lück, C., Weiland, F., and Weiss, F. (2009). 2-DE with IPGs. *Electrophoresis*, **30**: 1–11.

Immobiline® DryStrip Kit for 2-D Electrophoresis with Immobiline® DryStrip and ExcelGel™ SDS Instructions. 18-1038-63 Edition, Amersham Biosciences.

Multiphor II Electrophoresis System User Manual. 18-1103-43 Edition, Amersham Biosciences.

Appendices

Methods in Biotechnology Appendix 1

Instructor notes of MB experiment 1

On each student group workbench

Part I. Pipetting skill

- Spectronic 20 and cuvette
- 70% ethanol wash bottle, paper towel
- P20, P100, and P1000 pipettors
- Parafilm
- Pipette tips, 0.5-mL and 1.5-mL microcentrifuge tubes (non-sterile)
- 34 of borosilicate glass test tubes (13 × 100 mm) (non-sterile) in a test tube rack
- One glass test tube rack; one microcentrifuge tube rack
- 0.6 mL each of 50% glycerol, solutions A, B, C, and D (different colors)
- Waste beaker for disposal of the used micropipette tips
- 0.2 mL of blue dextran solution (10 mg/mL in dH_2O)
- Water carboy filled with dH_2O

Part II. Density versus concentration

- Two analytical balances
- 5 mL of 3 M NaCl
- 4 mL each of unknown NaCl solutions I (1.0 M) and II (0.5 M)
- 12 of 1.5-mL microcentrifuge tubes on the rack

Reminder

- Show demonstration of how to use P20, P200, and P1000 micropipettors.
- Ask students to check if their micropipettors in hand are calibrated using water and balance before they perform the next experiments. At first, this seems to be the easiest way for both students and instructor to recognize their incorrect use when they use calibrated micropipettors.

Instructor notes of MB experiment 2

On each student group workbench

- One Spectronic 20 and one cuvette
- One wash bottle filled with dH_2O
- 19 of borosillicate glass test tubes (13 × 100 mm) on a rack
- 0.8 mL of 3.0×10^{-3} M stock solution of phenolphthalein: Part A: 0.02 mL; Part B: 0.3 + 0.24 + 0.12 + 0.06 + 0.03 mL; Part C: 0.02 mL
- 3 mL each of unknown phenolphthalein solutions #1 (1.0×10^{-4} M) and #2 (0.5×10^{-4} M) diluted in dH_2O: #1: 400 µL of 3.0×10^{-3} M stock + 11.6 mL H_2O; #2: 200 µL of 3.0×10^{-3} M stock + 11.8 mL H_2O
- 2 mL of 1 M NaOH: Part A: 0.06 mL × 2; Part B: 0.03 mL × 6; Part C: 0.15 + 0.03 mL; Part D: 0.3 mL × 3
- pH paper
- 3.5 mL of BSA (high quality: 10 mg/mL in 0.9% NaCl)
- 4 mL of 0.9% NaCl
- ~50 mL of Biuret reagent

■3.0×10^{-3} M stock solution of phenolphthalein (MW = 318.33)

1. Dissolve 0.239 g of phenolphthalein in 100 mL of 100% ethanol (clean beaker!).
2. Bring up to 250 mL with dH_2O.
3. Store in a clean plastic bottle.
4. Label the reagent name, molarity, date of preparation, and person's name who prepared it; 1% stock (0.031 M = 3.1×10^{-2} M).

■Biuret reagent recipe per liter

1. Weigh 1.50 g of copper sulfate pentahydrate ($CuSO_4 \cdot 5\ H_2O$ MW 249.61) or 0.96 g of copper sulfate anhydrous ($CuSO_4$ MW 159.61) with 6.0 g sodium potassium tartrate tetrahydrate ($NaKC_4H_4O_6 \cdot 4\ H_2O$).
2. Dissolve in 500 mL of H_2O.
3. Add 300 mL of 10% NaOH (30 g in 300 mL).
4. Dissolve 1 g of potassium iodide (KI).
5. Make up to the total volume of 1 liter.
6. Store in a plastic dark bottle protected from light.
7. Label the reagent name, date of preparation, and the person's name who prepared it.

Instructor notes of MB experiment 3

On each student group workbench

- One pH meter
- One squeeze wash bottle filled with dH_2O
- Kimwipe
- Water wash bottle
- pH 4, 7, and 10 standard solutions
- 5 mL of 1 M NaOH
- 5 mL of 1 M HCl
- 0.5 mL of concentrated (12 M) HCl
- One 250-mL beaker, two 100-mL and two 50-mL beakers (clean!)

Methods in Biotechnology, First Edition. Seung-Beom Hong, M. Bazlur Rashid and Lory Z Santiago-Vázquez.

Companion Website: www.wiley.com\go\hong\Methodsinbiotechnology

- 100-mL or 200-mL graduated cylinder (clean!)
- Magnetic hot plate stirrer and magnetic bar
- 14 of borosillicate glass test tubes (16 × 150 mm) on a rack

*If your pH electrode does not fit into the glass test tube, transfer to a larger tube and reuse the tube for pH reading without rinsing the tube.

Common materials

- Tris (Trizma base)
- Sodium acetate
- Monobasic NaH_2PO_4
- Dibasic Na_2HPO_4
- Glacial acetic acid (concentrated stock 8 mL)
- Hydrochloric acid (concentrated stock 8 mL)
- Water carboy filled with ddH_2O (MilliQ purified water)
- 3 M KCl (fill-up solution of pH electrode)

Instructor notes of MB experiment 4

On each student group workbench

- One pH meter
- One squeeze wash bottle filled with dH_2O
- Kimwipe
- pH 4, 7, and 10 standard solutions
- Burette (25 mL: clean!), ring stand, burette clamp, funnel
- One 100-mL and one 200-mL beakers (dry and clean!)
- 200-mL graduated cylinder (dry and clean!)
- 10-mL disposable pipette
- 50 mL of HCl solution with unknown concentration (**0.05 M**)
  Concentrated HCl (36% by weight) = 11.6 M
  250 mL of 0.05 M HCl: 1.08 mL of concentrated HCl + 228.4 mL of MilliQ H_2O
- 50 mL of acetic acid (vinegar) solution with unknown concentration (**0.1 M**)
  Glacial acetic acid (99.5% by weight) = 17.4 M
  250 mL of acetic acid: 1.44 mL of acetic acid + 248.56 mL of MilliQ H_2O
- 15 mL of 1.0 M NaOH

Common materials

- Phenolphthalein (0.5% in 50% ethanol) in a dropper bottle:
  50 mL: 0.25 g dissolved in 25 mL of absolute ethanol and make up to 50 mL with MilliQ H_2O
- Phenol red (0.05% in dH_2O) in a dropper bottle
- Water carboy filled with dH_2O

Instructor notes of MB experiment 5

On each student group workbench

- 12 mL of 1% (0.2 g/20 mL) BSA solution in sterile dH_2O
- 8 mL of 1% (0.2 g/20 mL) casein solution (see below)
- 8 mL of 1% (0.2 g/20 mL) gelatin solution (see below)
- 12 mL of 1% (0.2 g/20 mL) γ-globulin solution (see below)
- 5 mL of 0.1 M HCl (concentration 36% = 11.6 M)
- 5 mL of 0.1 M NaOH (MW 40)
- 10 mL of 2% $CuSO_4$ solution
- 5 mL of 2% $AgNO_3$
- 1× PBS (each group needs ~25 mL)
- 6 M guanidine-HCl (each group needs ~6.5 mL)
- 1 mg/mL of hemoglobin in 1× PBS (each group needs ~1.6 mL)
- 5 mL of isopropanol
- 15 mL of acetone (0 °C or −20 °C ON)
- Each 0.472 g of ammonium sulfate crystals in 2 clear 1.5-mL tubes (for 70% saturation)
- Each 0.314 g of ammonium sulfate crystals in 2 clear 1.5-mL tubes (for 50% saturation)
- Each 0.176 g of ammonium sulfate crystals in 2 clear 1.5-mL tubes (for 30% saturation)
- 38 of glass test tubes in tube rack (for Part A Experimentt)
- 28 of clear non-sterile 1.5-mL microcentrifuge tubes (for Part B Experimentt)
- 2 Pasteur pipettes
- Ice box
- pH paper
- Spectronic 20 and cuvette
- Biuret reagent (see Experiment 2 Instructor Notes): 25 mL
- Diluted (1/5) BioRad Bradford Reagent in H_2O: 25 mL
- 1% Casein solution:
  Dilute 5% casein solution (Sigma product #C4765-10 ML) to 1/5 in sterile 1× PBS. Store at 4 °C.
- 1% Gelatin solution (hazy to turbid):
  Dissolve 1 g in 100 mL of *warm* (50–60 °C) sterile 1× PBS. Store at 4 °C.
- 1% γ-Globulin solution (hazy to turbid):
  Dissolve 1 g in sterile 1× PBS and bring up to 100 mL (avoid making excessive foam). Store at 4 °C.
- 1 mg/mL of hemoglobin in 1× PBS (reddish brown):
  Dissolve gently 0.01 g of hemoglobin in sterile 1× PBS, bring up to 10 mL (avoid making excessive foam),and gently suspend by pipetting up and down. Store at 4 °C.

For all groups

TA to work on Part B, step 11 of the lab protocol during the lab exercise.

- 0.132 g of ammonium sulfate crystals in a 1.5-mL tube (23% saturation);
  0.066 g of ammonium sulfate crystals in a 1.5-mL tube (12% saturation);
  0.033 g of ammonium sulfate crystals in a 1.5-mL tube (6% saturation): – control for negative controls of Part B.
- 1× PBS (phosphate buffered saline; 137 mM NaCl, 12 mM phosphate, 2.7 mM KCl, pH 7.4):
  1. Dissolve the following in 800 mL of distilled H_2O.
     - 8 g of NaCl
     - 0.2 g of KCl
     - 1.44 g of Na_2HPO_4 (dibasic anhydrous), OR 1.81 g of $Na_2HPO_4 \cdot 2H_2O$ (dibasic dihydrate), or 2.72 g of $Na_2HPO_4 \cdot 7H_2O$ (dibasic heptahydrate)
     - 0.24 g of KH_2PO_4
  2. Adjust pH to 7.4 with HCl.
  3. Adjust volume to 1 L with additional distilled H_2O.
  4. Sterilize by autoclaving for 20 min on the liquid cycle. Store at room temperature.
- 37 °C, 60 °C, and 95–100 °C (microwave) water bath; water carboy filled with MilliQ H_2O.

Instructor notes of MB experiment 6

On each student group workbench

- 10 mL of *E. coli* HB101 cell (A_{600} of 0.4–0.8):

  For 4 groups, inoculate 0.5 mL of an overnight culture into 50 mL of LB media, grow at 37 °C, shake at 250 rpm for ~ 3 h, and place in a 15-mL sterile conical tube in an ice box.

  Note: other *E. coli* strains (TOP 10, DH10B) able to transport but unable to metabolize L-arabinose can be used instead of HB101.
- Two electrocuvettes in an ice box
- 10 mL of ice-cold sterile H_2O
- 200 µL of ice-cold sterile 10% glycerol
- 2 mL of sterile LB broth medium
- Each 100-mL, 200-mL, and 300-mL Erlenmeyer flasks
- Sterile 1.5-mL microcentrifuge tubes
- 12 Petri plates in a bag
- Ice box
- Chemically-competent, frozen *E. coli* HB101 cells on ice (provide just before use)

Common materials and equipment

- For one negative control: 10 mL of *E. coli* HB101 cells (A_{600} of 0.4 to 0.8) from above
- Water carboy filled with dH_2O
- LB broth powder
- Sterile TE (10 mM Tris, pH 8.0, 1 mM EDTA) buffer
- Bacto-agar
- Turn on 60 °C water bath
- Turn on 37 °C incubator
- Denatured ethanol in beaker, L-shaped glass spreader, alcohol Bunsen burner
- Bio-Rad Gene Pulser Xcell™ with cuvette holder
- Kimwipe
- Microcentrifuge (set up at 4 °C)
- Vortexer

On instruction desk

- Ampicillin stock solution (100 mg/mL in dH_2O: filter-sterilized)
- 20% L-Arabinose (10 g dissolved in dH_2O and bring up to 50 mL: filter-sterilized)
- 20% glucose (10 g dissolved in dH_2O and bring up to 50 mL: filter-sterilized)
- Plasmid DNA pGLO
- Recycling of electro cuvette: after use, wash with 70% ethanol and sterilize by UV irradiation or autoclaving.
- UV transilluminator and sterile toothpicks are needed for the Day 2 procedure.

Instructor notes of MB experiment 7

On each student group workbench

- 10 mL of each overnight LB (+ Amp + L-Ara) culture of transformant HB101(pGLO):

  Amp: 100 µg/mL (final concentration), L-Ara: 1.5×10^{-3} M (final concentration)
- 10 mL of each overnight LB (+ Amp) culture of transformant HB101 (pGLO)
- One hydrophobicity (HIC) column
- 15 mL of equilibration buffer (2 M $(NH_4)_2SO_4$ in TE buffer; autoclaved)
- 1 mL of binding buffer (4 M $(NH_4)_2SO_4$ in TE buffer; autoclaved)
- 5 mL of wash buffer (1.3 MM $(NH_4)_2SO_4$ in TE buffer; autoclaved)
- 2 mL of elution buffer (TE buffer: 10 mM Tris, 1 mM EDTA, pH 8.0; autoclaved)
- 10 mL sterile dH_2O
- 0.5 mL of 1 M HCl and 0.5 mL of 1 M NaOH
- Three sterile test tubes in a tube rack for collection of eluents
- Micropipettor set and sterile microfuge tubes
- One Pasteur pipette
- 590 µL of saturated $(NH_4)_2SO_4$ solution
- 300 µL of 96% ethanol
- 80 µL of 5 M NaCl and 80 µL of **n(1)-**butanol
- pH paper
- Hand-held UV light lamp

Common materials

- Lysozyme (0.05 g in 1 mL of TE buffer; freshly prepared in TE buffer, pH 7.5)
- Protease inhibitor cocktail (Life Technologies, Cat. No. 87786; 100×)

Stock buffers and solutions

- 500 mL of elution buffer (TE buffer): 10 mM Tris-Cl, pH 8.0, 1 mM EDTA, autoclaved

  5 mL of 1 M Tris-Cl (pH 8.0) + 1 mL of 0.5 M EDTA + 496 mL of MilliQ H_2O
- 100 mL of equilibration buffer: 2 M $(NH_4)_2SO_4$ in TE buffer; autoclaved

  13.21 g of $(NH_4)_2SO_4$ dissolved in TE and bring up to 100 mL with TE
- 100 mL of binding buffer: 4 M $(NH_4)_2SO_4$ in TE buffer; autoclaved

  26.42 g of $(NH_4)_2SO_4$ dissolved in TE and bring up to 100 mL with TE
- 100 mL of wash buffer: 1.3 M $(NH_4)_2SO_4$ in TE buffer; autoclaved

  17.18 g of $(NH_4)_2SO_4$ dissolved in TE and bring up to 100 mL with TE
- Saturated $(NH_4)_2SO_4$ (pH 7.8):
  1. Place 50 mL of autoclaved MilliQ water into a 150-mL clean sterile beaker containing a magnetic bar.
  2. Place the beaker on the magnet stirrer hot plate.
  3. Weigh 35 to 40 g of $(NH_4)_2SO_4$ crystals and add to the water slowly while stirring.
  4. Continue to stir until crystals dissolve or a few crystals remain in the beaker.
  5. Cool to room temperature.
  6. Adjust the pH to 7.8 with 10 M NaOH.
  7. Filter crystals out with Whatman No. 1 filter paper; store at room temperature indefinitely.
- **5 M NaCl (sterile)**: dissolve 2.92 g and bring up to 10 mL in MilliQ water; autoclave.
- **96% ethanol**: 0.4 mL of sterile MilliQ water + 9.6 mL of absolute ethanol (Molecular Biology Grade)

■ Regeneration of HIC column

1. Add about 3 column volumes (~4 mL) of wash Bbuffer and drain.

2. Add 5 column volumes (~7 mL) of sterile dH_2O and drain.

3. Add 10 column volumes (~10 mL) of equilibration buffer, drain, retain a small volume of the buffer on the column bed, and close the bottom of the column.

4. Store at 4 °C.

Instructor notes of MB experiment 8

On each student group workbench

Place all protein samples in an ice box.

- Centrifugation-cleared uninduced cell lystae supernatant (Experiment 7, Part B)
- Centrifugation-cleared Ara-induced cell lystae supenartant (Experiment 7, Part B)
- Flowthrough (FT) fraction (Experiment 7, Part D)
- Wash fraction (W) (Experiment 7, Part D)
- Elution fraction 3 (EB I) (Experiment 7, Part D)
- Elution fraction 4 (EB II) (Experiment 7, Part D)
- Organic extraction-purified GFP sample (Experiment 7, Part C)
- Two spin columns
- 2 mL TE buffer (pH 8.0)
- 4–20% Mini-PROTEAN® TGX™ pre-cast gel (Bio-Rad, Cat. No. 456-1093SEDU)
- One power supply
- Safety cap locks for microcentrifuge tubes

Common materials and equipment

- Boiling water bath
- 4× SDS sample loading dye
- GelCode blue (Coomassie blue) staining solution
- 5 mL of Bradford dye (1/50 diluted in water)
- Sephadex G25 swollen in TE buffer (pH 8.0)
- Protein MW marker standard
- 10× SDS running buffer (1 liter):
  30.3 g Tris base
  144.0 g glycine
  10.0 g SDS
  Dissolve and bring the total volume up to 1 liter with dH_2O. Do not adjust the pH. Mix 50 mL of 10× stock with 450 mL of dH_2O for each electrophoresis run.
- 4 × SDS sample buffer:

Composition	per 10 mL
250 mM M Tris-Cl, pH 6.8	2.5 mL of **1.0 M** Tris-Cl (**pH 6.8**)
40 % glycerol	4.0 mL of glycerol
8 % SDS	0.8 g of SDS
0.02% bromophenol blue	0.4 mL of 0.5% bromophenol blue

Add dH_2O to bring up to a final total volume of 9.8 mL.

Combine in a 15-mL conical tube and gently agitate until the components are dissolved. Store at room temperature. Add 20 µL of β-mercaptoethanol to 980 µL of SDS sample buffer just before use.

Instructor notes of MB experiment 9

On each student group workbench

For Part I. Plasmid DNA preparation

- 4 mL overnight culture of *E. coli* HB101 (pGLO) cells grown in LB + 100 µg/mL Amp: each group makes two preps
- 0.5 mL of GTE solution (ice cold)
- 0.7 mL of potassium acetate solution (ice cold)
- 4 mL of 95% ethanol (ice cold)
- 1 mL of 70% ethanol (ice cold)
- 120 µL (30 µL × 4) of TE buffer containing 20 mg/mL of lysozyme
- Sterile 2.0-mL microcentrifuge tubes

For Part II. Chromsomal DNA preparation

- 10 mL overnight culture of *E. coli* HB101 cells grown in LB (each group needs 10 mL of culture)
- Spooling glass rod
- Two sterile glass tube (10 × 75 mm)
- Two sterile large-bore Pasteur pipette (or large-bore P1000 pipette tip)
- TEL buffer (10 mM Tris (pH 8.0), 1 mM EDTA (pH 8.0) + 20 mg/mL lysozyme): each needs 1 mL of group 1

Common materials

- 1 mL of 1 M NaOH
- 0.5 mL of 10% SDS
- DNase-free RNaseA
- TE buffer (10 mM Tris (pH 8.0), 1 mM EDTA (pH 8.0))
- 10% (w/v) sodium dodecyl sulfate (SDS)
- 20 mg/mL of lysozyme in TE (0.02 g in 1 mL TE; prepare just before use)
- 20 mg/mL of proteinase K in TE
- Phenol/chloroform/isoamylalcohol (25:24:1, v/v/v)
- CIA (chloroform/isoamyalcohol, 24:1, v/v)
- 5 M NaCl (autoclaved)
- Isopropanol
- 70% ethanol
- GTE (glucose, Tris, EDTA) cell suspension solution:
  50 mM glucose
  25 mM Tris-HCl (pH 8.0)
  10 mM EDTA (pH 8.0)
  Autoclave and store at 4 °C.
- Potassium acetate neutralizing solution (~pH 4.8):
  11.5 mL of glacial acetic acid
  60.0 mL 5 M potassium acetate (sterile)
  28.5 mL of dH_2O
  Store at 4 °C (′ do not autoclave)
- Denaturing solution (prepare fresh just before use):
  0.6 mL 1 M NaOH + 0.3 mL 10% SDS + 2.1 mL H_2O
- DNAse-free RNAse (2 mg/mL):
  Dissolve 0.005 g RNAse A in 1 mL of 25 mM Tris-Cl, pH 7.5 (final 5 mg/mL).
  Boil for 10 min (to destroy DNAse).
  Add 1 mL of 40–50% sterile glycerol and 0.5 mL of sterile 0.5 M NaCl.
  Store at −20 °C.

Instructor notes of MB experiments 10, 11, and 12

On each student group workbench

- Four 2-mL screw-cap microcentrifuge tubes each containing 1.5 mL of 0.9% NaCl solution
- Ice eox
- Sterile 0.5-mL microcentrifuge tubes
- 15 µL of 5× GoTaq® PCR reaction buffer
- 10 µL of 10 µM F-A25 and R-A25 primers
- 25 µL of 50 mM Tris-HCl (pH 8.0) = EB buffer of plasmid miniprep kit
- 200 µL of nuclease-free sterile dH_2O
- Pipettor set of P10, P20, P100, P1000, and sterile tips
- Sterile 2.0-mL microcentrifuge tubes
- 4 sterile cotton swabs
- 4 thin-walled PCR tubes
- Waste beaker

Common materials and equipment

- GoTaq® DNA polymerase (5 units/µL) in ice box
- 10 mM dNTP mix in ice box
- 10% Chelex® suspension: 10 mL
  Weigh out 1 g of Chelex 100 (100–200 mesh, sodium form). Add 9 mL of 50 mM Tris-HCl (pH 8.0); store at 4 °C.
- *Eco*RI-HF, *Hin*dII-HF, *Pst*I-HF, and *Eco*RV-HF enzymes, and 10× CutSmart™ Buffer (NEB) in ice box
- Quick-load 50-bp DNA ladder (NEB NO473S)
- Agarose (low EEO)
- 10× lithium borate buffer (100 mM, pH 8.2):
  Dissolve 8.4 g of lithium hydroxide monohydrate (or 4.8 g lithium hydroxide anhydrous) in 900 mLH_2O, adjust pH to 8.2 by adding boric acid (~36 g), and bring up to 2 liters. Store at room temperature.
- Thermal cycler (program: see the protocol)
- 37 °C water bath
- Water carboy filled with MilliQ dH_2O
- One 1.5% agarose gel with 14 wells for 13 student samples plus DNA MW marker for Experiment 10.

Instructor notes of MB experiment 13

On each student group workbench

A. ODD assay

- Three agarose plates (20 mL of 1.2% agarose in 1× PBS per Petri dish 100 × 15 mm): NO AIR BUBBLES TRAPPED. SAME THICKNESS: use a sterile 25-mL pipette to place into the Petri plates
- 80 µL of 1% (10 mg/mL) BSA in 1× PBS (0.05 g BSA in 5 mL 1× PBS)
- 60 µL of 100% pig, horse, cow, and rabbit serum in ice box
- 25 µL of anti-bovine serum albumin (BSA) antibody produced in rabbit (Sigma B1520-2ML)
- 25 µL of anti-horse serum antibody produced in rabbit (Sigma H8890-2ML)
- holo-Transferrin bovine (Sigma T1283-50 MG; 50 mg/mL in 1× PBS) in ice box: 50 µL (group 1); 25 µL (group 2); 15 µL (group 3); 10 µL (group 4)
- One glass Pasteur pipette
- One vinyl bag that can hold two Petri plates
- Micropipettor set of P20, P100, P1000, and sterile micropipettor tips

B. ELISA assay

- One microtiter polystyrene plate
- 50 µL each of 0.5% pig, horse, cow, and rabbit serum (diluted in 1× PBS) in ice box
- 50 µL each of 0.1% pig, horse, cow, and rabbit serum (diluted in 1× PBS) in ice box
- 25 mL blocking solution (1× PBS/gelatin)
- 25 mL washing solution (1× PBS/Tween 20)
- 25 mL of 1× PBS
- 5 mL of 0.1 M HCl
- 6 glass test tubes in a rack
- Spectronic 20 and cuvette
- Paper towels

Common materials

- Anti-BSA rabbit whole antiserum (sigma B1520-2ML) in ice box: 100 µL aliquot/group
- 1:1 (v/v) mix of anti-BSA rabbit whole antiserum and anti-horse serum antibody produced in rabbit (Sigma H8890-2ML) in ice box: 50 µL aliquot/group
- Goat anti-rabbit IgG-HRP (1 mg/mL) antibody in ice box: 5×10^4 dilution (20 ng/mL):
  1st, 1 µL Ab + 990 µL blocking buffer, mix well ($1/10^3$ dilution)
  2nd, 100 µL of $1/10^3$ dilution + 900 µL of blocking buffer ($1/10^4$ dilution)
  3rd, 300 µL of $1/10^4$ dilution + 1200 µL of blocking buffer (5×10^4 dilution)
- TMB solution color substrate in ice box (1.3 mL per group: 0.65 mL of 0.4 g/L of TMB + 0.65 mL of 0.02% H_2O_2; prepare fresh just before use)
- Water carboy filled with MilliQ dH_2O
- Platform shaker
- 10× PBS (1 liter):

NaCl	80 g
KCl	2 g
Na_2HPO_4	14.4 g
KH_2PO_4	2.4 g

Add H_2O to 1 liter. The pH of the 10× stock will be about 6.8, but when diluted to 1× PBS (137 mM NaCl, 10 mM phosphate, 2.7 mM KCl) it should change to 7.4. Sterilize by autoclave.

▸ Washing solution:

1× PBS + Tween 20 (0.05%)

Dilute 100 ml of 10× PBS with 895 mL of dH_2O. Add 5 mL of 10% Tween 20 to buffer. This buffer (1× PBS, 0.05% Tween 20) can be stored indefinitely at room temperature.

▶ Blocking solution:
Prepare fresh during the lab exercise: 1× PBS + gelatin (2%); warm to ~55 °C to dissolve gelatin (0.6 g gelatin in 30 mL of 1× PBS in a 50-mL conical plastic tube).

■ Reminder
To reduce the experimental time, conduct Part II, C (ELISA) coating step first and then carry out Part I (ODD assay) during the incubation time of ELISA coating and blocking.

Instructor notes of MB experiments 14 and 15

On each student group workbench (experiment 14)

- 8 Mueller Hinton agar plates (4 mm *uniform thickness*): use a sterile 25-mL pipette to place 25 mL of agar medium per plate into 15× 100-mm Petri plates
- 1 mL of LB cultures of *Escherichia coli, Klebsiella pneumonia, Pseudomonas aeuroginosa,* and *Staphylococcus aureus*: dilute to the same turbidity as the McFarland turbidity standard 0.5 (Fisher Scientific Cat. No. R20410); ~1/10 dilution of overnight culture
- 48 sterile blank paper disks (6 mm diameter, BD Diagnostic Systems) per group
- 25 mL of ice-cold sterile dH_2O in ice box
- 25 mL of methanol in ice box
- 2 × 4 g each of garlic bulb, ginger rhizome, potato tuber, and 2× 1 g each of green tea powder
- Mortar and pestle
- Sterile cheesecloth, funnel, Kimwipe
- Fine-point forceps
- Bunsen burner
- L-shaped glass rod spreader
- Two 15-mL plastic conical centrifuge tubes (each group makes water and methanol extracts from one plant tissue)
- 2-mL microcentrifuge tubes (13)
- Micropipettor set and pipette tips

On each student group workbench (experiment 15)

- Spec 20 and cuvette
- 14 glass tubes (non-sterile) in a rack
- 5-mL pipette aid
- 620 μL of TMB substrate solution (fresh 1:1 mix just before use, 310 μL of 0.4 g/L TMB + 310 μL of 0.02% H_2O_2): for the Part B Experiment
- 520 μL of 1 M HCl
- 320 μL of 0.02% H_2O_2 (undiluted bottle in kit box): for the Part C Experiment
- 360 μL of 0.4 g/L TMB (undiluted bottle in kit box): for the Part C Experiment
- 25 mL of sterile dH_2O at room temperature
- dH_2O wash bottle and 70% wash bottle
- Paper towels

Common materials

- Ampicillin and kanamycin filter disk (instructor)

MB Appendix 2

MB lab math practice problem set

Part I. Metric unit

1. Convert the following:

• 1 gram = ______ mg	1 gram = ______ kg
• 50 mg = ______ g	2 ng = ______ mg
• 1 liter = ______ mL = ____ cm^3 = ______ m^3	100 μL = ______ mL
• 0.1 mg = ____ pg = ______ μg = ______ ng	
• 0.1 L (liter) = ___ mL = ___ μL = ____ nL = ____ pL = ______ fL	
• 0.1 mM = ____ μmole/mL = ______ pmole/μL	
• 1 L × 1 M = ______ mole	1 mL × 1 M = ______ mmole
• 1 L × 1 mM = ______ mmole	1 mL × 1 mM = ______ μmole
• 1 mL of pure water = ______ g	1 L of pure water = ______ g
• 1 ppm (part per million) = ______ %	10% = ______ ppm
• 1 mg/L = ______ ppm in water = ______ %	1 ppm = ________ μg/mL in water
• 1 mg/L = ________ ppb in water	1 ppb = ________ ng/mL in water
• 1 μm (micrometer) = _____ nm (nanometer) = _____ pm (picometer) = _____ angstroms (Å)	

2. Convert the following to scientific notation (exponential form):

- 54678 = __________
- 0.000458 = __________
- 54678 ÷ 0.000458 = __________

3. Convert the following to significant figures:

• 11.0235 + 10.61 = ________	11.0235 + 0.012 = ________
• 12.03 − 8.2145 = ________	12.03 − 0.0125 = ________
• 45.2 ÷ 5.3568 = ________	7.40 ÷ 22 = ________
• 580 × 2.768 = ________	25 × 2.71 = ________

4. Perform the following calculation of error propagation:

- $(3.0 \pm 0.2) + (5.0 \pm 0.6) - (4.54 \pm 0.05)$
- $2 \times (3.25 \pm 0.21)$
- $(2.0 \pm 0.2) + 2 \times (3.0 \pm 0.6)$
- $(2.0 \pm 0.2) \times (3.0 \pm 0.6)$
- $(2.0 \pm 0.2) \div (3.0 \pm 0.6)$

5. Perform the following exponent calculations:

- $\dfrac{(2.2 \times 10^{-7})(1.3 \times 10^{8})}{3.0 \times 10^{-3}}$
- $(6.7 \times 10^{24}) + (3.3 \times 10^{26})$
- $\log_{10}(1.2 \times 10^{-3})$
- Calculate X value for: $10^X = 5$
- Calculate X value for: $2^X = 10^5$
- $(0.96)^{27}$
- Calculate X value for: $X^2 + (4.5 \times 10^{-7})X - (4.5 \times 10^{-8}) = 0$

Part II. Ratio and proportion

1. How many blood cells would be in 1.0 mL of blood if there are about 2×10^4 blood cells in a 10 μL sample?

2. A typical human erythrocyte (red blood cell) has a volume of about 90 fL, and red blood cells make up approximately 40% of blood volume. How many red blood cells are present in 1 mL of blood?

3. What is the mass of 0.75 moles of NaCl if one mole of NaCl has a mass of 58.5 g?

4. A particular laboratory solution contains 5 μL of isopropanol per 10 mL of solution. Express this as a percent solution.

5. How would you prepare 250 mL of 10% (v/v) acetonitrile and 40% (v/v) methanol solution?

6. What is a percent ethanol if a solution contains 15 mL of ethanol per 1 L total volume?

7. How much chemical is present in a 30 μM solution if a 0.5 M solution contains 65 g/L of a chemical compound?

8. How many micrograms of NaCl are present in 25 μL of a solution that is 5 mg/mL NaCl?

9. You have 20 μL of an expensive enzyme with a concentration of 10 000 units/mL. You need to do an experiment that requires a concentration of 1 unit/μL of enzyme in 50 μL total volume. How many tubes can you prepare before you run out of enzyme?

10. How would you prepare a 200 mL solution that has a mixture of phenol-chloroform isoamylalcohol (25:24:1, v/v/v)?

Methods in Biotechnology, First Edition. Seung-Beom Hong, M. Bazlur Rashid and Lory Z Santiago-Vázquez.

Companion Website: www.wiley.com\go\hong\Methodsinbiotechnology

11. A solution with a concentration of 0.14 M is measured to have an absorbance of 0.43. Another solution of the same chemical is measured under the same conditions and has an absorbance of 0.37. What is its concentration?
12. One-tenth mL of a protein solution was diluted with 2.9 mL of water. The A_{280} of the diluted solution was 0.25. How many mL of the original protein solution and water should be mixed to make 1.0 mL of solution of $A_{280} = 0.75$?

Part III. Dilution

1. A stock enzyme solution contains 20 000 units/μL. One μL of the stock is diluted with 99 μL of buffer. What is the final concentration of the enzyme after dilution?
2. You have a very expensive enzyme with 100 units/μL. You need to use the enzyme for 10 individual experiments, each of which requires 1 unit/mL in a 5-mL total reaction volume. Show how you can dilute 0.5 μL of enzyme so that you can use it in your experiment.
3. If you prepare a 1/250 dilution of a 50% solution, what is the final concentration of the solution?
4. How much 1/10 diluted solution can be made if you have 1 mL of original solution?
5. One mL of sample was mixed with 4 mL of buffer, and the 5 mL of the diluted sample was mixed with 25 mL of an assay reagent. The 30 mL contained 10 mg of protein. What was the final concentration of protein in the original sample?
6. If you take a 0.1 mL sample and add 0.9 mL of water and 4.0 mL of reagents, what is the final dilution of the sample?
7. You have a plasmid DNA solution at a concentration of 1 mg/mL. How would the addition of only 0.001 μg of plasmid be accomplished? Assume that it is not possible to accurately measure a volume less than 1 μL.
8. You have a 2.5 M solution of NaCl. You remove 10 μL of the solution and add 490 μL of water. What is the concentration of the diluted salt solution?
9. You have a bacterial culture with 10^8 cells/mL. If you want to get 100 colonies after plating 0.1 mL of the last dilution, how would you dilute the bacterial culture?
10. How would you prepare a final volume of 10 μL of 1× solution using 5× stock solution A?
11. You are making a PCR mixture that will be 1× of all reagents and the final volume will be 200 μL. How would you prepare using the following stocks?

300×	solution A
1000×	solution B
200×	solution C
25×	solution D

12. How would you make a 120-mL agarose gel of 0.7% (w/v) that is 0.5× TBE and 0.5 μg/mL ethidium bromide? TBE (Tris-borate-EDTA) stock is 5× and ethidium bromide stock is 10 mg/mL.
13. How would you make 500 mL of a 10× TE stock? 1× TE is 10 mM Tris and 1 mM EDTA, pH 8.0. Molecular weights of Tris and EDTA are 121.1 and 187, respectively.
14. You want to digest 2 μg of DNA with *Bam*H I enzyme; your DNA stock is 1.2 mg/mL. You must use 1 μL of *Bam*H I and bring your final reaction volume to 20 μL. You also need to have final concentrations of enzyme buffer and BSA (bovine serum albumin) to 1× but their stocks are 10× (enzyme buffer) and 100× (BSA). How would you prepare this mixture?
15. One A_{260} unit of a double-stranded DNA solution equals 50 μg/mL. A spectrophotometer reads 0.057 when you measure a solution that contains 5 μL DNA and 495 μL of dH_2O. What is the concentration of the original DNA stock solution?
16. How would you prepare a 10 mL protein solution of 0.5 mg/mL from the stock that has a concentration of 2 mg/mL?
17. To prepare 100 mL of 1 M acetate buffer (pH 5.2), you need to use 2.182 g of sodium acetate anhydrous ($\rho = 1.528$ g/cm^3) and 4.2 mL of glacial acetic acid. How many mL of H_2O should be added to dissolve the components?

Part IV. Exponential relationship

1. *E. coli* has a doubling time of 25 min at an optimal growth condition. You inoculated a fresh 100 mL of broth culture in a flask with 0.1 mL of *E. coli* cells with 10^6 cells/mL. What is the bacterial cell density (cells/mL) after 6 h?
2. *E. coli* has a doubling time of 25 min at an optimal growth condition. How long should you incubate if you want to get 10^8 cells/mL in a 100-mL culture medium that had been inoculated with 0.1 mL of *E. coli* cells with 10^6 cells/mL?
3. The half-life of ^{32}P isotope is 14 days. You initially received 200 μCi to perform your experiment. How much ^{32}P activity remains after 140 days?

Part V. Concentration

1. Express each of the following for the aqueous 500 mL solution containing 100 g of NaCl.

(a) Molarity	(b) Normality	(c) Percent as weight/volume (w/v)

2. Determine the concentration values for each of the following for the solution containing 10 mg of $Ca(OH)_2$ (molar mass 47) made to 100 mL with distilled water.

(a) mg/dL	(b) Molarity	(c) Normality	(d) Equivalents/L

3. How many grams of solute are required to prepare 500 mL of a 0.1 M solution of sucrose (MW = 342.3)?
4. How many moles are there in 242 mg of Tris (MW = 121.1)?
5. How would you prepare 200 mL of a 0.1 M solution of $CaCl_2$ (MW = 111) using a dihydrate form of $CaCl_2{\bullet}2H_2O$ (MW = 147)?.
6. How would you prepare a 1 liter solution that has 100 ppm of cadmium?
7. How many moles of Na^+ ions are present when 7.102 g of Na_2SO_4 (MW = 142.04) are dissolved to a final volume of 100 mL in dH_2O?
8. How many moles of $CaCl_2$ are there in 14.7 g of $CaCl_2{\bullet}2H_2O$ (MW = 147)?.
9. How would you prepare 3 mL of 0.2 N NaOH from 6 M NaOH?
10. How would you prepare 50 mL of 0.5 N H_2SO_4 from 1 M H_2SO_4?

11. How would you prepare 20 mL of 0.3% NaCl (w/v) from 1% NaCl (w/v)?
12. What molarity is 20% (w/v) NaCl (MW = 58.44) solution?
13. How many mmoles of NH_4^+ ions are present in 264 mg of $(NH_4)_2SO_4$ (MW = 132)?
14. What molarity is 100 ppm of $HgCl_2$ (MW = 271.52) solution?
15. Convert 10 μg/mL of aspirin (MW = 180) into μmole/L.
16. If you dissolve 9 mg of glucose in 5 mL of H_2O, what is the concentration of glucose in the resulting solution? Answer in %, M, mM, and μM. The MW of glucose is 180.
17. The diplod human genome carrying 46 chromosomes has 6.4×10^9 bp. The MW of one bp is 660 g/mole. The volume of human cell nucleus is estimated to be 523×10^{-18} cubic meters (m^3). Calculate the % and molar concentration of the genomic DNA in the cell nucleus.

MB Appendix 3

Answers to MB lab math practice problem set

Part I. Metric unit

1. Convert the following:

• 1 gram = 10^3 mg	1 gram = 10^{-3} kg
• 50 mg = 0.05g	2 ng = 2×10^{-6} mg
• 1 liter = 10^3 mL = 10^3 cm^3 = 10^{-3} m^3	100 µL = 0.1 mL
• 0.1 mg = 10^8 pg = 10^2 µg = 10^5 ng	
• 1 L (liter) = 10^3 mL = 10^6 µL = 10^9 nL = 10^{12} pL = 10^{15} fL	
• 1 mM = 1 µmole/mL = 10^3 pmole/µL	
• 1 L × 1 M = 1 mole	1 mL × 1 M = 1 mmole
• 1 L × 1 mM = 1 mmole	1 mL × 1 mM = 1 µmole
• 1 ml of pure water = 1 g	1 L of pure water = 1 kg
• 1 ppm (part per million) = 10^{-4} %	10% = 10^5 ppm
• 1 mg/L = 1 ppm = 10^{-4}%	1 ppm = 1 µg/mL
• 1 mg/L = 1000 ppb	1 ppb = 1 ng/mL
• 1 µm = 10^3 nm = 10^6 pm = 10^4 angstroms (Å)	

2. Convert the following to scientific notation (exponential form):

- $54678 = 5.4678 \times 10^4$
- $0.000458 = 4.58 \times 10^{-4}$
- $54678 \div 0.000458 = 1.19 \times 10^6$

3. Convert the following to significant figures:

• 11.0235 + 10.61 = 21.6335 = 21.63	11.0235 + 0.012 = 11.0355 = 11.04
• 12.03 − 8.2145 = 3.8155 = 3.82	12.03 − 0.0125 = 12.0175 = 12.02
• 45.2 ÷ 5.3568 = 8.4378733 = 8.44	7.40 ÷ 22 = 0.3363636 = 0.34
• 580 × 2.768 = 209.53757 = 210	25 × 2.71 = 67.75 = 68

4. Perform the following error propagation calculation:

- $(3.0 \pm 0.2) + (5.0 \pm 0.6) - (4.54 \pm 0.05) = 3.5 \pm 0.9$ or 3.5 ± 0.6

Mean = 3.0 + 5.0 − 4.54 = 3.46 = 3.5

$\Delta SD = 0.2 + 0.6 + 0.05 = 0.85 = 0.9$ or $\Delta SD = \sqrt{0.2^2 + 0.6^2 + 0.05^2} = 0.63 = 0.6$

- $2 \times (3.25 \pm 0.21) = 6.50 \pm 0.42$

Mean = 2 × 3.25 = 6.50

$\Delta SD = 2 \times 0.21 = 0.42$

- $(2.0 \pm 0.2) + 2 \times (3.0 \pm 0.6) = 8.0 \pm 1.4$

Mean = 2.0 + 6.0 = 8.0

$\Delta SD = 0.2 + (2 \times 0.6) = 1.4$

- $(2.0 \pm 0.2) \times (3.0 \pm 0.6) = 6.0 \pm 1.2$ or 6.0 ± 1.3

Mean = 2.0 × 3.0 = 6.0

$$\Delta SD = 6.0 \times \sqrt{\left(\frac{0.2}{2.0}\right)^2 + \left(\frac{0.6}{3.0}\right)^2} = 1.3$$

- $(2.0 \pm 0.2) \div (3.0 \pm 0.6) = 0.7 \pm 0.2$

Mean = 2.0 ÷ 3.0 = 0.6667 = 0.7

$$\Delta SD = 0.7 \times \sqrt{\left(\frac{0.2}{2.0}\right)^2 + \left(\frac{0.6}{3.0}\right)^2} = 0.16 = 0.2$$

5. Perform the following exponent calculations:

- $\dfrac{(2.2 \times 10^{-7})(1.3 \times 10^8)}{3.0 \times 10^{-3}} = 9.53 \times 10^3 = 9.5 \times 10^3$
- $(6.7 \times 10^{24}) + (3.3 \times 10^{26}) = 3.367 \times 10^{26} = 3.4 \times 10^{26}$
- $\log (1.2 \times 10^{-3}) = -2.921 = -2.9$
- $10^X = 5$; $X = 0.699 = 0.7$
- $2.0^X = 10^5$; $X = 16.61 = 16.6$
- $(0.96)^{27} = 0.332 = 0.33$
- Using the quadratic equation ($aX^2 + bX + c = 0$) formula,
- $x = \dfrac{-b \pm \sqrt{b^2 - 4ac}}{2a}$,

$a = 1, b = 4.5 \times 10^{-7}, c = -(4.5 \times 10^{-8})$. Thus,

$$X = \frac{-(4.5 \times 10^{-7}) \pm \sqrt{(4.5 \times 10^{-7})^2 + 4(4.5 \times 10^{-8})}}{2}$$

$\approx -2.1 \times 10^{-4}$ or 2.1×10^{-4}

Part II. Ratio and proportion

1. 2×10^6 blood cells
2. 4.4×10^9 RBC/mL
3. $43.875 \approx 44$
4. 0.05%
5. 25 mL of acetonitrile + 100 mL of methanol + 125 mL of dH_2O
6. 1.5%
7. 3.9 mg/L
8. 125 µg
9. 4 tubes
10. 100 mL of phenol + 96 mL of chloroform + 4 mL of isoamylalcohol
11. 0.12 M

Methods in Biotechnology, First Edition. Seung-Beom Hong, M. Bazlur Rashid and Lory Z Santiago-Vázquez.

Companion Website: www.wiley.com\go\hong\Methodsinbiotechnology

Part III. Dilution

1. 200 units/μL

2. Dilute 0.5 μL of enzyme with 9.5 μL of buffer and use 1 μL for each experiment. Each reaction needs 1 units/mL × 5 mL = 5 units, which is 5 units/100 units μL^{-1} = 0.05 μL of enzyme. Therefore, using 1 μL of 1/10 dilution of 0.5 μL of enzyme is equivalent to 0.05 μL of enzyme for each reaction.

3. 0.2%

4. 10 mL

5. 10 mg/mL

6. 1/50

7. Mix 1 μL of DNA with 9 μL of dH_2O, mix 1 μL of the first dilution with 9 μL of dH_2O, mix 1 μL of the second dilution with 9 μL pf dH_2O, and use 1 μL of the third dilution.

8. 0.05 M

9. Transfer 1 mL of culture to 9 mL of broth, mix, and transfer the 1st dilution to 9 mL of broth; repeat this 1/10 dilution 4 more times until the 5th dilution. Plate 0.1 mL of the 5th dilution.

10. Add 2 μL of 5× stock solution to 8 μL of dH_2O.

11. 0.67 μL of A + 0.2 μL of B + 1.0 μL of C + 8.0 μL of D + 190.13 μL of dH_2O

12. **(a)** Combine 0.84 g of agarose + 12 mL 5× TBE + 108 mL of dH_2O.
(b) Heat (microwave) to dissolve the agarose.
(c) Add 6 μL of 10 mg/mL of ethidium bromide and mix.

13. **(a)** Dissolve 6.05 g of Tris + 0.935 g of EDTA in dH_2O up to 400 mL.
(b) Adjust the pH to 8.0 with HCl and add dH_2O up to 500 mL.

14.

dH_2O	13.33 μL (add first)
10× buffer	2.00 μL
10× BSA (1/10 diluted 100× BSA)	2.00 μL
DNA (1.2 mg/mL)	1.67 μL
Bam HI enzyme	1.00 μL (add last)

15. 0.285 mg/mL (concentration = 0.057 × 50 × 100 = 285 μg/mL)

16. 2.5 mL of stock + 7.5 mL of buffer

17. 92.5 mL

Part IV. Exponential relationship

1. 2.2×10^7 cells/mL
$N = N_0 \times 2^n$. This can be rearranged by taking both side of the equation to the base $\log_2$, $\log_2 N = \log_2 N_0 + n \rightarrow n = \log_2 N - \log_2 N_0 = \log_2 (N/N_0)$, where n is the generation number as calculated from incubation time divided by doubling time.
$N_0 = 0.1\ mL \times 10^6$ cells/mL = 10^5 cells. Incubation time = 6 × 60 = 360 min, and so generation number $n = 360/25 = 14.4$. Thus, the final cell number after incubation $N = 10^5 \times 2^{14.4}$. To calculate $2^{14.4} = X$, if we take both sides to the $\log_2$ base, $14.4 = \log_2 X \rightarrow X = 21\ 619 = 2.2 \times 10^4$. Therefore, $N = 10^5 \times (2.2 \times 10^4) = 2.2 \times 10^9$ cells. Because the total cell culture volume is 100.1 ml, cell density after incubation will be (2.2×10^9) cells/100.1 mL = ~2.2×10^7 cells/mL.

2. 6.92 h
$N = 100.1\ mL \times 10^8$ cells/mL = ~10^{10} cells. $N_0 = 0.1\ mL \times 10^6$ cells/mL = 10^5 cells. Thus, $10^{10} = 10^5 \times 2^n \rightarrow 10^5 = 2^n \rightarrow \log_2 (2^n) = \log_2 (10^5) \rightarrow n = \log_2 (100\ 000) = 16.61$. Incubation time/25 min = 16.61 → incubation time = 25 × 16.61 = 415.25 min = 6.92 h.

3. 0.1953 μCi.
$N = N_0 \times (1/2)^n$, where generation number n = time lapsed/half life = 140 days/14 days = 10. Thus, $N = 200 \times (1/2)^{10} = 200 \times (1/1024) = 0.1953$ μCi.

Part V. Concentration

1. **(a)** M = [100 g ÷ (23 + 35.5)] ÷ [0.5 L ÷ 1.0 L] = 3.42 mole/L
(b) N = 3.42 equivalents/L
(c) (100 g/500 mL)(100) = 20%
(d) 100:500 = 1:5

2. **(a)** 10 mg = 10 mg/100 mL/dL = 10 mg/dL (*note:* 1 dL = 100 mL)
(b) Molarity = 0.1/47 = 0.00213 mole/L (100 mg/L = 0.1 g/L)
(c) Normality 0.00213 × 2 = 0.00426 N
(d) Equivalent/L = 0.00426

3. 17.12 g

4. 1.998×10^{-3} moles

5. 2.94 g of $CaCl_2 \cdot 2H_2O$ in 200 mL dH_2O

6. 0.1 g cadmium in 1 liter of dH_2O

7. 1 mole

8. 0.1 mole

9. 0.1 mL of 6 M NaOH + 2.9 mL of dH_2O

10. 12.5 mL of 1 M H_2SO_4 + 37.5 mL of dH_2O

11. 6 mL of 1% NaCl + 14 mL of dH_2O

12. 3.42 M

13. 4 mmole

14. 3.7×10^{-4} M

15. 55.6 μmole/L

16. 0.18%; 0.01 M; 10 mM; 10 000 μM

17. 1.34%; 3.2×10^{-12} M
1 bp = (660 g/mole) ÷ (6.023×10^{23} molecules/mole) = 1.096×10^{-21} g. Thus, mass of human genome in a single-cell nucleus = $(6.4 \times 10^9) \times (1.096 \times 10^{-21}) = 7 \times 10^{-12}$ g. The volume of nucleus = $523 \times 10^{-18}\ m^3 = 523 \times 10^{-15}$ liter = 523×10^{-12} mL. As a result, concentration = $(7 \times 10^{-12}) \div (523 \times 10^{-12})$ = 0.0134 g/mL = 1.34%.
MW of human genome = $(6.4 \times 10^9) \times 660 = 4.224 \times 10^{12}$ g/mole. 1.34% = 13.4 g/liter = (13.4 g/liter) ÷ (4.224×10^{12} g/mole) = 3.2×10^{-12} mole/liter.

MB Appendix 4

Commonly used buffer tables (MB)

0.1 M tris buffer

Dissolve X g of Tris-HCl + Y g of Tris base to a final volume of 1 liter in H_2O.

pH at 25 °C	g/L for 0.1 M solution	
	X g of Tris-HCl (MW 157.6)	Y g of Tris (MW 121.14)
7.0	14.50	0.97
7.1	14.20	1.20
7.2	13.85	1.47
7.3	13.43	1.79
7.4	12.93	2.17
7.5	12.36	2.62
7.6	11.70	3.12
7.7	10.97	3.68
7.8	10.17	4.30
7.9	9.32	4.95
8.0	8.42	5.64
8.1	7.52	6.34
8.2	6.62	7.02
8.3	5.76	7.69
8.4	4.94	8.31
8.5	4.20	8.89
8.6	3.53	9.35
8.7	2.94	9.86
8.8	2.43	10.25
8.9	1.99	10.58
9.0	1.62	10.87

*Tris pK_a at 20 °C = 8.2; pK_a at 25 °C = 8.06; pK_a at 37 °C = 7.72.

*Calculation is based on

$$\mathrm{pH} = 8.06 + \log\frac{[\mathrm{Tris}]}{[\mathrm{Tris \cdot HCl}]}$$

and [Tris] + [Tris-HCl] = 0.1. For example,

$$\log\frac{[\mathrm{Tris}]}{[\mathrm{Tris - HCl}]} = -1.06 \text{ at pH 7.0} \rightarrow \frac{[\mathrm{Tris}]}{[\mathrm{Tris \cdot HCl}]}$$

$$= 0.0871 = \frac{0.1 - [\mathrm{Tris \cdot HCl}]}{[\mathrm{Tris \cdot HCl}]}$$

→ [Tris-HCl] = 0.092 M. Thus, [Tris] = 0.1 – 0.092 = 0.008 M. As a result, Tris-HCl = 157.6 × 0.092 = 14.50 g per liter. Tris = 121.14 × 0.008 = 0.97 g per liter.

*The pH values decrease an average of 0.03 pH units for each °C increase in temperature between 5 °C and 25 °C and an average of 0.025 pH units for each °C increase in temperature between 25 °C and 37 °C.

*Check the pH and adjust it if necessary before bringing to a final volume of 1 liter.

1 M tris buffer

Dissolve 121.14 g of Tris base in 750 mL of H_2O (~910 mL), add X mL of concentrated hydrochloric acid (36%, 11.6 M), and bring to a final volume of 1 liter in H_2O.

pH	X mL of HCl (36%; 11.6 M)
7.0	79.31
7.1	77.59
7.2	75.75
7.3	73.44
7.4	70.73
7.5	67.59
7.6	64.01
7.7	60.01
7.8	55.64
7.9	50.96
8.0	46.08
8.1	41.12
8.2	36.22
8.3	31.49
8.4	27.04
8.5	22.96
8.6	19.30
8.7	16.07
8.8	13.27
8.9	10.89
9.0	8.88

*Tris pK_a at 25 °C = 8.06.

Methods in Biotechnology, First Edition. Seung-Beom Hong, M. Bazlur Rashid and Lory Z Santiago-Vázquez.

Companion Website: www.wiley.com\go\hong\Methodsinbiotechnology

*Calculation is based on

$$pH = 8.06 + \log\frac{[Tris]}{[Tris \cdot HCl]}$$

and [Tris] + [Tris-HCl] = 1. For example,

$$\log\frac{[Tris]}{[Tris - HCl]} = -1.06 \text{ at pH } 7.0 \rightarrow \frac{[Tris]}{[Tris \cdot HCl]} = 0.0871 = \frac{1 - [Tris \cdot HCl]}{[Tris \cdot HCl]}$$

→ [Tris-HCl] = 0.92 M. As a result, 0.92 M HCl is needed to produce 0.92 M Tris-HCl; 0.92 M × 1000 mL = 11.6 M × V → V = 79.31 mL.

*Check the pH and adjust it if necessary before bringing to a final volume of 1 liter.

0.1 M acetate buffer

For 0.2 M acetic acid: dilute 1.16 mL of glacial acetic acid (99.8%, 17.4 M) to a final volume of 100 mL.

For 0.2 M sodium acetate: dissolve 1.64 g of anhydrous (or 2.72 g trihydrate) to a final volume of 100 mL.

For 0.2 M potassium acetate: dissolve 1.96 g of H_3COOK to a final volume of 100 mL.

Mix X mL of 0.2 M acetic acid and Y mL of 0.2 M sodium acetate (X+ Y= 100 mL of 0.2 M) and add 100 mL of H_2O to make a 200 mL of 0.1 M acetate buffer.

pH	*X* mL of 0.2 M glacial acetic acid	*Y* mL of 0.2 M sodium or potassium acetate
3.6	93.5	6.50
3.7	91.8	8.20
3.8	90.0	10.0
3.9	87.6	12.4
4.0	84.9	15.1
4.1	81.7	18.3
4.2	78.0	22.0
4.3	73.8	26.2
4.4	69.1	30.9
4.5	64.0	36.0
4.6	58.6	41.4
4.7	52.9	47.1
4.8	47.0	53.0
4.9	41.5	58.5
5.0	36.0	64
5.1	31.0	69.0
5.2	26.2	73.8
5.3	22.0	78.0
5.4	18.3	81.7
5.5	15.0	85.0
5.6	12.4	87.6

*Acetic acid pK_a at 25 °C is 4.75.

*Calculation is based on

$$pH = 4.75 + \log\frac{[Acetate]}{[Acetic\ acid]}$$

and [Acetate] + [Acetic acid] = 0.1. For example, $\log\frac{[Acetate]}{[Acetic\ acid]} = -1.15\text{atpH}3.6 \rightarrow \frac{[Acetate]}{[Acetic\ acid]} = 0.0692 = \frac{0.1 - [Acetic\ acid]}{[Acetic\ acid]}$ → [Acetic acid] = 0.0934 M. Thus, [sodium acetate] = 0.1 − 0.0934 = 0.0066 M. For acetic acid, 0.0935 M × 200 mL = 0.2 M × V → V = 93.5 mL. For sodium acetate, 0.0065 M × 200 mL = 0.2 M × V → V = 6.5 mL.

*Check the pH and adjust it if necessary before bringing to a final volume of 200 mL.

0.1 M phosphate buffer

Mix X mL of 1 M monobasic + Y mL of 1 M dibasic phosphate and bring to a final volume of 100 mL.

pH	*X* mL of 1 M monobasic phosphate (sodium or potassium)	*Y* mL of 1 M dibasic phosphate (sodium or potassium)
6.0	8.76	1.24
6.1	8.49	1.51
6.2	8.17	1.83
6.3	7.80	2.20
6.4	7.37	2.63
6.5	6.91	3.09
6.6	6.40	3.60
6.7	5.85	4.15
6.8	5.29	4.71
6.9	4.71	5.29
7.0	4.14	5.86
7.1	3.60	6.40
7.2	3.09	6.91
7.3	2.62	7.38
7.4	2.20	7.80
7.5	1.83	8.17
7.6	1.51	8.49
7.7	1.24	8.76
7.8	1.01	8.99

*Monobasic phosphate ($H_2PO_4^-$) pK_{a2} at 20 °C = 6.8–7.2.

*Calculation is based on

$$pH = 6.85 + \log\frac{[Dibasic\ P]}{[Monobasic\ P]}$$

and [Monobasic P] + [Dibasic P] = 0.1. For example,

$$\log\frac{[Dibasic\ P]}{[Monobasic\ P]} = -0.85 \text{ at pH } 6.0 \rightarrow \frac{[Dibasic\ P]}{[Monobasic\ P]} = 0.141 = \frac{0.1 - [Monobasic\ P]}{[Monobasic\ P]}$$

→ [Monobasic P] = 0.0876 M, [Dibasic P] = 0.1 − 0.0876 = 0.0124 M. For monobasic potassium phosphate, 0.0876 M × 100 mL = 1 M × V → V = 8.76 mL. For dibasic potassium phosphate, 0.0124 M × 100 mL = 1 M × V → V = 1.24 mL.

*The pH of phosphate buffer increases to about 0.08 unit when it is diluted with the same volume. For example, dilution of 0.1 M solution (pH 6.8) to 0.05 M by adding the same volume of water would result in pH 6.88.

*The pH of 25 °C buffer will be 0.08 lower than 4 °C and 0.025 higher than 37 °C.

*Check the pH and adjust if necessary.

AMB 1 Appendix 1

Instructor notes of AMB 1 experiments 16 and 17

Each student group workbench

Experiment 16

- Paper towels and 70% ethanol spray bottle
- Two inoculating loops, Bunsen burner
- 1 to 2 mL of overnight broth cultures of *E. coli* MM294 grown at 37 °C in LB medium and *S. cerevisiae* YNN281 grown at 30 °C in YPD medium
- Two LB agar plates
- Two YPDA agar plates (quadrant streak): check pH 5.4 to 5.5 using a pH paper strip
- Two YM agar plates: pH 3.0–4.0
- YM agar (pH 6.2) should be sterilized without pH adjustment, and sterile 1 M HCl added to the sterile molten medium cooled to 45–50 °C (pH 3.0–4.0; check and adjust by pH indicator paper); acidified YM agar should not be heated.
- Two Levine EMB agar plates with lactose: add filter-sterilized sucrose solution after autoclave (add 0.1 mL of 10% lactose to 100-mL autoclave EMB agar medium without lactose; 27.4 g/liter of EMB without lactose).
- Two Levine EMB agar plates with sucrose: add filter-sterilized sucrose solution after autoclave (add 0.1 mL of 10% sucrose to 100-mL autoclave EMB agar medium without lactose; 27.4 g/liter of EMB without lactose).
- One Levine EMB agar plate without carbon source
- Four sterile test tubes containing 2 mL of LB broth

Experiment 17

- Two 500-mL beakers
- One 500-mL graduated cylinder
- Two 200-mL flasks and two 1-L flasks
- 20 Petri plates in a bag, magnetic stir plate and stir bar

Common materials and equipment

- Streaked plates of *E. coli* MM294 and *S. cerevisiae* YNN281
- For instructor's demonstration, 1 LB agar plate and 1 LB broth
- Bacto-Agar (Difco)
- YPD powder (Clontech, Cat. No. 630409):10 g/L of yeast extract, 20 g/L of peptone, 20 g/L of dextrose
- Uracil
- Ura dropout supplement (Clontech Cat. No. 630416)
- YNB base with ammonium sulfate (Thermo Fisher Scientific, Cat. No. Q300-09)
- Water carboy filled with dH_2O
- Analytical balance, magnetic stir plate and stir bar
- Spatula, weigh boats, labeling tape, aluminum foil
- 37 °C and 30 °C culture incubators and shakers
- Water baths (55 °C)
- Autoclave (reserve notice)
- Laminar flow hood cabinet

Instructor notes of AMB 1 experiments 18 and 19

- Prepare the time point cultures (Experiment 18) as follows:

1. Inoculate *S. cerevisiae* YNN281 into 2 mL each of YPDA and YNB/glucose + uracil (= SD + uracil) liquid media.
2. Shake-incubate overnight (16 to 18 h) at 30 °C, 300 rpm (~10^8 to 5×10^7 cells/mL).

**Doubling time is ~90 to 100 min in YPDA medium and ~200 min in YNB/glucose medium. $A_{600} = 1.0$ for ~3×10^7 cells/mL.*

**YNN281 growth at 30 °C, 250 rpm for 16 h (6:00 pm to 10:00 am) gave ~2.35 of A_{600} (=7×10^7 cells/mL) when the single colony cells are inoculated into 50 mL of YPD medium in a 250-mL flask).*

3. Inoculate 0.2 mL of each overnight culture into each 100 mL of YPDA and SD + uracil liquid media (from Experiment 17) in a 500-mL sterile Erlenmeyer culture flask (→1.4×10^5 cells/mL). A 90 mL culture is needed for 4 student groups.

**YPDA and SD + uracil overnight cultures should be used to inoculate into YPDA and SD+ uracil media, respectively.*

4. Swirl the cultures to mix, take aliquots of the YPDA and SD + uracil culture into sterile test tubes (label the time point and medium type), and store at 4 °C (→0 h time point).
5. Incubate the culture flasks at 30 °C with shaking at 250 rpm.

*Incubation time: 0 h, 4 h, 8 h, 12 h, 24 h, 28 h. If incubation begins at 10:00 or 09:30 am, the corresponding times to take aliquots are:

Methods in Biotechnology, First Edition. Seung-Beom Hong, M. Bazlur Rashid and Lory Z Santiago-Vázquez.

Companion Website: www.wiley.com\go\hong\Methodsinbiotechnology

(a) YPDA culture:
0 h: 10:00 am (9:30 am) → take 4 mL per group (16 mL/4 groups)
4 h: 02:00 pm (1:30 pm) → take 4 mL per group (16 mL/4 groups)
6 h: 04:00 pm (3:30 pm) → take 4 mL per group (16 mL/4 groups)
8 h: 06:00 pm (5:30 pm) → take 4 mL per group (16 mL/4 groups)
12 h: 10:00 pm (9:30 pm) → take 4 mL per group (16 mL/4 groups)

(b) SD + uracil culture:
0 h: 10:00 am (9:30 am) → take 4 mL per group (16 mL/4 groups)
6 h: 04:00 pm (3:30 pm) → take 4 mL per group (16 mL/4 groups)
8 h: 06:00 pm (5:30 pm) → take 4 mL per group (16 mL/4 groups)
12 h: 10:00 pm (9:30 pm) → take 4 mL per group (16 mL/4 groups)
24 h: 10:00 am (9:30 am) → take 2 mL per group (8 mL/4 groups)

All aliquots of cultures are stored in the tubes labeled with incubation time and type of medium at 4 °C until use.

- *E. coli* DH5α (pRY121) and DH5α (pUC19) streaked plate on LB agar media.
- Prepare 40 mL of LB (or TB) of *E. coli* DH5α (pRY121) and 40 mL of LB (or TB) of DH5α (pUC19) cultures (Experiment 19) as follows:

1. Inoculate a colony of DH5α (pRY121) and DH5α (pUC19) into 2 mL of LB broth (+2 μL of 100 mg/mL ampicillin) and shake-incubate at 37 °C, at 200 rpm overnight.
2. Inoculate 0.2 mL of fresh overnight culture into 40 mL each of TB and LB broth (+40 μL of 100 mg/mL ampicillin) and shake-incubate at 37 °C, at 200 rpm overnight.
3. Store the culture at 4 °C.

Each student group workbench

Experiment 18 (growth curve)

- Paper towels and 70% ethanol spray bottle
- Each *S. cerevisiae* YNN281 culture grown in YPDA and YNB/glucose liquid media taken at time points after incubation
- Sterile water (~10 mL) in a conical tube for dilution of yeast culture cells
- Autoclaved 1.5 mL microfuge tubes for dilution
- 10 YPDA agar plates (prepared from Experiment 17, Part A) for a viable cell count
- 10 YNB/glucose agar plates (prepared from Experiment 17, Part B) for a viable cell count
- 10 mL of each YPD and YNB/glucose medium for blank and diluent
- Spectronic 20 and cuvette
- Denatured ethanol in beaker, glass spreader, Bunsen burner
- Container filled with sterile 1.5-mL microcentrifuge tubes

Experiment 19 (mini plasmid prep)

Each group will prepare four minipreps using about 2 mL each of LB (or TB) culture for pUC19 isolation and 4 mL each of LB (or TB) culture for pRY121 isolation.

- 4 spin columns of QIAprep Spin Miniprep kit (Catalog No. 27104)
- 4 of 2-mL sterile microcentrifuge tubes for cell harvest
- 8 mL LB or TB broth (+8 μL of 100 mg/mL ampicillin) of *E. coli* DH5α (pRY121) cultures
- 4 mL LB or TB broth (+8 μL of 100 mg/mL ampicillin) of E. coli DH5α (pUC19) cultures
- One 0.5% agarose gel in 1× lithium borate buffer containing 0.5 μg/mL of ethidium bromide
- 10× lithium borate buffer (100 mM, pH 8.2):
  Dissolve 8.4 g of lithium hydroxide monohydrate (4.8 g lithium hydroxide anhydrous) in 900 mL of H_2O, adjust the pH to 8.2 by adding boric acid (~36 g), and bring up to 2 liter. Store at room temperature.
- Horizontal gel electrophoresis and power supply

For all groups

- Hot (70 °C) 10 mL of $T_{10}E_{0.1}$ buffer (10 mL of EB buffer, pH 8.5 + 2 μL of 0.5 M EDTA)
- Buffers: P1, P2, N3, PB, PE, and EB (QIAprep Spin Miniprep Kit)

*Before the Experiment 18 exercise, the instructor explains how to dilute cell culture based on A_{600} values using 1.5-mL microcentrifuge tubes in order to get ~50 colony-forming units (cfu) when 0.1 mL of dilution is plated.

Instructor notes of AMB 1 experiments 20 and 21

Each student group workbench

Experiment 20 (Restriction Digestion, Purification, Concentration, and Quantification of DNA)

- Paper towels and 70% ethanol spray bottle
- pUC19 plasmid mini-prep DNA (from Experiment 19)

Experiment 21 (PCR)

- Four 0.2 mL thin-walled PCR tubes in ice bath
- Nuclease-free sterile dH_2O in a sterile microcentrifuge tube
- 5× colorless GoTaq® reaction buffer (Promega)
- 5× Phusion HF buffer (NEB)
- dNTP mix (10 mM)
- 10 μM LacZ primer 1 + 2 mix (equal volumes of 1/5 dilution of 100 μM of each stock); see Materials and equipment of Experiment 21
- 10 μM GFP SLIC primer 1 + 2 mix; see Materials and equipment of Experiment 21
- pRY121 plasmid (~1 ng/μL: diluted mini-prep DNA)
- *E. coli* genome DNA (MB Experiment 9, Part II)
- P10, P20, and P200 micropipettors and autoclaved micropipettor tips
- Horizontal agarose gel electrophoresis apparatus and power supply

Common materials and equipment

Experiment 20

- Phenol:chloroform:isoamylalcohol, 25:24:1 (v/v/v)
- CIA:chloroform-isoamylalcohol, 24:1 (v/v)
- 40 mL isopropanol in a conical tube
- 40 mL of 70 or 75% ethanol in a conical tube
- **3 M sodium acetate** (adjust to pH 5.2 with glacial acetic acid):

1. Dissolve 40.81 g of sodium acetate•$3H_2O$ or 24.61 g of sodium acetate anhydrous in 80 mL of H_2O.
2. Adjust pH to 5.2 with glacial acetic acid.
3. Bring up to 100 mL.
4. Sterilize by autoclaving.

- **TE buffer** (pH 8.0, 10 mM Tris, 1 mM EDTA):
  1 mL of 1 M Tris (pH 8.0) plus 0.2 mL of 0.5 M EDTA (pH 8.0) plus 98 mL of dH_2O. Sterilize by autoclaving.
- **PE buffer** (10 mM Tris-HCl, pH 7.5, 80% ethanol):
  10 mL of 10 mM Tris-HCl (pH 7.5) + 40 mL of absolute ethanol (molecular biology grade)
- **Elution buffer**: 10 mM Tris-HCl (pH 8.5) or $T_{10}E_{0.1}$ buffer (pH 8.5)
- **PB buffer** (5 M guanidine-HCl, 20 mM Tris-HCl pH 6.6, 30% ethanol) 50 mL:
  1. Prepare 50 mL of 20 mM Tris-HCl (pH 6.6) by adding 10 mL of sterile 0.1 M Tris-Cl (pH 6.6) to 40 mL of sterile dH_2O.
  2. Dissolve 23.9 g of guanidine hydrochloride (sigma G3272; MW 95.53) and bring up to a final 35 mL with the above Tris solution in a sterile conical tube.
  3. Add 15 mL of absolute ethanol (molecular biology grade). No autoclave. Label and store at room temperature.
- **Silica suspension** (100 mg/mL in PB buffer):
  1. Add 1 g of silica (Sigma S-5631) each into two sterile conical tubes, each of which contains 7.5 mL of sterile dH_2O, vortex for 30 s, spin at 2000× g for 2 min (use tabletop centrifuge), and remove the supernatant and any fine particles at the surface (first wash).
  2. Repeat the wash twice more.
  3. Add 10 mL of PB buffer to each tube. No autoclave. Label and store at 4 °C.
- *Sma*I, *Eco*RI-HF and *Hin*dIII-HF enzymes in ice bath
- 10× NEB buffer 4
- 6× or 10× DNA loading dye
- 1 kb DNA ladder (NEB)
- Agarose (low EEO type)
- Ethidium bromide (10 mg/mL in sterile H_2O; store at 4 °C in dark bottle)
- 25 °C and 37 °C water bath
- **10× Lithium borate buffer** (2 L; 100 mM, pH 8.2):
  1. To 900 mL of dH_2O, add 8.392 g of lithium hydroxide monohydrate (MW 43.96) or 4.79 g of lithium hydroxide anhydrous (MW 23.95).
  2. Adjust pH to 8.2 by adding boric acid (~36 g). Add water to bring up to 2 L. Store at room temperature.

Experiment 21

- GoTaq® DNA polymerase (Promega, Cat. No. M3001; 5 U/μL)
- Phusion® Hot Start II DNA polymerase (NEB, Cat. No. F-549; 2U/μL)
- Three thermal cyclers or one thermal cycler with three independently controlled blocks programed before class
  - Taq_LacZ and Taq_Genome program: takes about 1 h 13 min to complete.
  - Phusion-LacZ program: takes about 40 min to complete.
  - Phusion-GFP program: takes about 50 min to complete.
- **Reminder**

To reduce the experimental time, set up four PCR reactions (Experiment 21) first and then carry out restriction enzyme digestion (15 min) followed by agarose gel preparation and DNA purification (Experiment 20) during PCR cycling. DNA purification by phenol-CIA/ethanol precipitation or the silica method each takes about 35 min. Agarose gel solution is poured into a gel casting tray and kept at 4 °C to expedite gel solidification. Each group makes one gel with 10 wells for 7 DNA samples plus DNA marker.

Lane 1: 1-kb DNA ladder (NEB 0.5 μg/5 μL)
Lane 2: pUC19/SmaI (phenol-CIA/ethanol precipitation purified)
Lane 3: pUC19/SmaI (silica-purified)
Lane 4: pUC19/EcoRI-HindIII (silica-purified)
Lane 5: Taq_LacZ/pRY121 PCR DNA
Lane 6: Taq_LacZ/genome PCR DNA
Lane 7: Phusion_LacZ/pRY121 PCR DNA
Lane 8: Phusion_GFP/pGLO PCR DNA

Electrophoresis takes 20–25 min at 300 V in 1× lithium borate buffer.

Instructor notes of AMB 1 experiment 22

Each student group workbench

- pUC19/*Sma*I (Experiment 20 student sample) in ice box
- pUC19/*Eco*RI + *Hin*dIII (Experiment 20 student sample) in ice box
- Four PCR samples (from Experiment 21) in ice box: Taq_LacZ/pRY121, Tac_LacZ/Genome, Phusion_LacZ/pRY121, Phusion_GFP/pGLO
- Nuclease-free sterile H_2O in ice box
- Two thin-walled 0.2-mL PCR tubes (SLIC and CPEC reactions) in ice box
- 4 Silica Econo Mini Spin columns (Epoch Life Science)
- Containers filled with sterile 1.5-mL and 0.5-mL microcentrifuge tubes
- 10 LB/Amp (100 μg/mL)/X-β-Gal (40 μg/mL)/IPTG (0.1 mM) agar plates
- Denatured ethanol in a beaker, glass spreader, Bunsen burner

Common materials and equipment

- PB buffer (see Experiments 20 and 21 Instructor Notes)
- PE buffer (eee Experiments 20 and 21 Instructor Notes)
- **QG buffer** (6 M guanidine thiocyanate, 50 mM Tris-HCl, pH 6.6, 20 mM EDTA, pH 8.0, 0.006% cresol red):
  1. Prepare 50 mL of 50 mM Tris-HCl (pH 6.6) by adding 25 mL of 0.1 M Tris-Cl (pH 6.6) to 25 mL of sterile dH_2O.
  2. Dissolve 35.45 g of guanidine thiocyanate (Sigma G9277-100G, MW 118.16) in the above solution and bring up to a final 50 mL in a sterile conical tube.
  3. Add 0.003 g of cresol red (use a clean spatula to measure) and dissolve to mix.

  No autoclave. Label and store at room temperature.
- Elution buffer (10 mM Tris-HCl, pH 8.5)
- 3 M sodium acetate (pH 5.2) (see Experiments 20 and 21 Instructor Notes)
- 10× ithium Borate buffer (see Experiments 20 and 21 Instructor Notes)
- 0.5 M EDTA (pH 8.0)
- Agarose (low EEO), ethidium bromide (10 mg/mL)
- 6× or 10× DNA loading dye

- DNA marker standards (NEB 1 kb DNA ladder and pUC19/*Sma*I)
- 0.5% agarose gel in 1× lithium borate buffer (see Experiments 20 and 21 Instructor Notes) and EtBr (0.5 µg/mL) for two gels with 14 wells per 4 groups: one gel for gel purification of the PCR product and the second gel for quantification of the T-vector and PCR product.
- Razor blades
- 25–30 °C, 37 °C, 42 °C, and 65 °C water baths
- Thermo cycler programed before class (see Experiment 22, step 35 in Part I, CPEC)
- The following materials are placed in an ice box on the instructor's desk:
- Terminal transferase (NEB, Cat. No. M0315S, 20 units/µL)
- 10× TdT buffer
- 2′,3′-Dideoxythymidine-5′-triphosphate
- $CoCl_2$ (2.5 mM)
- T4 DNA polymerase (NEB, Cat. No. M0203S, 3 units/µL)
- Antarctic phosphatase (NEB, Cat. No. M0289S, 5 units/µL) and 10× phosphatase reaction buffer
- Quick DNA ligase (NEB, Cat. No. M2200S, 2000 units/µL); 2× ligation buffer
- 10× BSA (4 aliquots)
- 10× NEB buffer 2
- Phusion® DNA polymerase (2 U/µL) and 5× Phusion HF buffer
- Competent DH5α or TOPO10 cell (see Part J of Experiment 22): 4 aliquots of 100 µL for each student group + 2 aliquots for positive and negative controls.
- **$CaCl_2$ solution I** (autoclave, store at 4 °C): 100 mL

60 mM $CaCl_2$	6 mL of 1 M $CaCl_2$
Water	94 mL of dH_2O

- **$CaCl_2$ solution II** (autoclave, store at 4 °C): 100 mL

60 mM $CaCl_2$	6 mL of 1 M $CaCl_2$
15% glycerol	15 mL glycerol
Water	78 mL of dH_2O
→ Autoclave, cool, and aseptically add	
10 mM HEPES (pH 7.5)	1 mL of 0.1 M HEPES (pH 7.5; filter-sterilized)

- **Reminder**

The following order of procedures is recommended.

1. Set up the CPEC reaction and run the PCR cycle (Part I).
2. Set up the SLIC reaction (Part H) and keep the sample in ice after completion.
3. Set up the T-vector reaction using a pre-determined volume of quantified pUC19/*Sma*I (Part B).
4. Set up the dephosphorylation reaction using quantified pUC19/*Sma*I (Part C).
5. The PCR reaction in Experiment 21 typically yields a single band so it is not necessary to gel-purify. Skip Part D.
6. Perform spin column purification of the CPEC PCR sample and T-vector sample (Part E).
7. Estimate the concentrations of the purified PCR and T-vector DNA using NanoDrop 2000; TA runs student samples in agarose gel for verification (Part F).
8. Set up TA and blunt end cloning reactions (Part G).
9. Perform *E. coli* transformation (Part K).

Instructor notes of AMB 1 experiment 23

Each student group workbench

- 5 LB/Km (30 µg/mL)/X-β-Gal (40 µg/mL)/L-arabinose (2×10^{-3} M) agar plates (4 experiments and one negative control)
- Two LB/Amp (100 µg/mL)/X-β-Gal (40 µg/mL)/IPTG (0.1 mM) agar plates (positive controls shared by all groups)
- 5 thin-walled PCR tubes in ice box
- Nuclease-free sterile H_2O in ice box
- 5× Phusion HF buffer
- Plasmid DNAs: pUC19 (1 ng/µL), pGLO (1 ng/µL), pGBKT7 (1 ng/µL)
- Primer 1 + 2 mix (25 µM): 5 µL primer 1 (100 µM) + 5 µL primer 2 (100 µM) + 10 µL H_2O; see Materials and equipment of Experiment 23
- Primer 3 + 4 mix (25 µM): 5 µL primer 3 (100 µM) + 5 µL primer 4 (100 µM) + 10 µL H_2O; see Materials and equipment of Experiment 23
- Primer 5 + 6 mix (25 µM): 5 µL primer 1 (100 µM) + 5 µL primer 2 (100 µM) + 10 µL H_2O; see Materials and equipment of Experiment 23
- P10, P20, P200, P1000 micropipettors
- Sterile 1.5-mL microcentrifuges
- 4 silica membrane mini-spin columns
- Mini agarose gel electrophoresis apparatus and power supply

Common materials and equipment

- Phusion® high-fidelity DNA polymerase (2 U/µL; NEB, Cat. No. M0530S) in ice box
- *Nhe*I-HF™ (NEB, Cat. No. R3131S; 20 units/µL) in ice box
- *Cla*I (NEB, Cat. No. R0197S; 5 units/µL) in ice box
- Gibson assembly master mix (NEB Cat. No. E2611S) in ice box
- DNA standard markers (pUC19/SmaI and 50-bp DNA ladder, NEB N0473S)
- *E. coli* TOPO10 competent cells (–80 °C freezer): each group needs 3 aliquots (100 µL each) and 2 aliquots (100 µL each) for two positive controls shared by all groups.
- PB buffer (see Experiments 20 and 21 Instructor Notes)
- PE buffer (see Experiments 20 and 21 Instructor Notes)
- EB buffer (see Experiments 20 and 21 Instructor Notes)
- QG buffer (see Experiment 22 Instructor Notes)
- Isopropanol
- 10× lithium borate buffer (see Experiments 20 and 21 Instructor Notes)
- Agarose (low EEO type)
- 10× DNA loading dye, ethidium bromide (10 mg/mL)
- Standard DNA reference (pUC19/*Sma*I and 50-bp DNA ladder, NEB N0473S)
- LB or SOC medium
- Razor blade
- Four agarose gels (0.5%) *in 1× lithium borate buffer* and EtBr (0.5 µg/mL):
  - Two gels with large 10-well combs (to hold 50 µL sample) (4 groups) for gel extraction (Parts D and E)
  - Two gels with small 14-well combs (4 groups) for DNA quantification (Part F).
  - Each group has 4 samples, each with a 50-µL reaction volume to be loaded on gel for gel purification and for quantification (Part F): LacZα_PCR, Km^R_PCR, pUC Ori_PCR, and pGLO/*Cla*I + *Nhe*I.
- Two water baths set to 37 °C and 50 °C before class

- Two thermal cyclers programed before class:
  - LacZ alpha program (Part A): takes about 32 min.
  - Km/Or program (Part B): takes 36 min.
  - CPEC assembly program (Part H): takes about 46 min.
  - Gibson assembly (Part G): takes 60 min.
- NanoDrop 2000 spectrophotometer
- **Reminder**

The following order of procedures is recommended.

1. Instructor turns on two thermal cyclers in a ready-to-use mode before the lab starts.

2. Set up the three PCR reactions (LacZα, Km^R, and pUCOri) and run PCR cycles (~50 min).

3. Set up restriction enzyme digestion (pGLO/NheI + ClaI) and continue the incubation at 37 °C until the PCR cycle is completed.

4. Load all PCR and restriction DNA samples onto pre-made agarose gels and run at 300 V for 15 min.

5. Perform the spin column purification of gel-extracted PCR and restriction enzyme fragments (~20 min).

6. Estimate the concentrations of the purified PCR and restriction DNA fragments using NanoDrop 2000 (~10 min); teaching assistant runs student samples in agarose gel for verification.

7. Set up Gibson and CPEC assembly reactions and run the PCR cycle (1 h).

8. Use the assembly reaction samples for transformation or store at −20 °C until use.

The *E. coli* transformation protocol (~1 h 40 min) can be followed on the next week lab session once competent cells and transforming DNA samples are ready.

Instructor's notes of AMB 1 experiments 23.I, 24, 25, and 26

Each student group workbench

- Competent *E. coli* cells (3 of 100 µL aliquots) in ice box (Experiment 23, Part I)
- 5 LB/Km/X-β-Gal/L-arabinose agar plates (Experiment 23, Part I)
- One LB/Amp/X-β-Gal/IPTG agar plate (Experiment 23, Part I)
- pRY121 plasmid (80 µL miniprep DNA needed) for restriction digestions and probe labeling
- *E. coli* genomic DNA
- Nuclease-free sterile H_2O
- 10× CutSmart™ buffer (NEB)
- 10× DNA loading dye
- Sterile 1.5-ml microcentrifuge tubes
- P10, P20, P200, P1000 micropipettor and sterile tips
- 0.5% agarose gel in 1× lithium borate buffer; pour gel solution into tray at 4 °C 30 min before the class begins.
- Agarose gel electrophoresis apparatus and power supply
- Empty plastic box or container able to hold one mini-gel (for denaturation and neutralization)
- Pyrex glass dish and support box (to set up the Southern blot transfer)

Common materials and equipment

- Enzymes *Bam*HI-HF, *Eco*RI-HF, *Hin*dIII-HF, and *Pst*I-HF in ice box
- Bionick™ labeling system kit (Life Technologies) component in ice box: 10× dNTP mix, 10× enzyme mix, control pBR322 DNA, stop buffer, and nuclease-free dH_2O
- DNA marker (NEB 1 kb DNA ladder)
- One Protran® (Whatman) NC membrane
- Whatman 3MM paper
- A stack of paper towels
- Scissors or a paper cutter, ruler, forceps, overhead projector transparency sheets, tape
- Swollen Sephadex G-50 resin (see "Stock Solutions" below)
- SDS column buffer (see "Stock Solutions" below)
- Sterile glass wool
- 15-mL conical tubes
- Disposable 1-mL syringes
- 37 °C water bath
- Two thermal cyclers programmed before class
  - Program (16 °C for 1 h) (nick translation reaction)
  - Program (Experiment 21, Phusion-LacZ) (PCR labeling)

Stock solutions

- **1× Denaturing solution** (0.5 M NaOH, 1.5 M NaCl): 10 g of NaOH, 43.83 g of NaCl in 500 mL of MilliQ water; autoclave.
- **1× Neutralizing solution** (1 M Tris-Cl, pH 7.5, 1.5 M NaCl): 121.14 g of Tris base, 87.66 g of NaCl in 800 mL of ultrapure water; adjust the pH to 7.5 with concentrated HCl; bring up to 1 liter with MilliQ H_2O; autoclave.
- **20× SSC** (3 M NaCl, 0.3 M sodium acetate, pH 7.0): 175.3 g of NaCl, 88.2 g of trisodium citrate; adjust pH to 7.0 with a few drops of 10 M NaOH; add ultrapure water to 1 liter total; autoclave.
- **SDS column buffer** (500 mL):

	Final concentration
25 mL of 1 M Tris-HCl (pH 8.0)	50 mM
20 mL of 0.5 M EDTA (pH 8.0)	2 mM
125 mL of 1 M NaCl	250 mM
5 mL of 10% SDS	0.1%
325 mL of dH_2O	

- **Fully swollen Sephadex G-50 resin**:
  - Add 5 g of Sephadex G-50 beads to 125 mL of sterile H_2O in a beaker to swell.
  - Swelling time depends on temperature (60 to 65 °C for 2 h).
  - Use SDS column buffer to wash and equilibrate (3 × 125 mL).
  - Store at 4 °C until use.
- **Reminder**

The following order of procedures is recommended.

1. Begin with *E. coli* transformation starting incubation in ice 30 min if the Part I step of Experiment 23 has not been carried out.

2. Set up the restriction digestion reaction and incubate for 15 min.

3. During the restriction incubation time, set up nick translation and PCR labeling reactions and run PCR (Parts I and II of Experiment 26).

4. Run the restriction samples in the pre-made agarose gel and take a photograph.

5. Cut out any unused gel portion and measure the gel dimension using a ruler.

6. Perform the denaturation, neutralization, and equilibration step (Experiment 25).

7. During the gel treatment time, cut the Whatman 3MM paper, nitrocellulose membrane, and stack of paper towels that match the dimensions of the treated gel.

8. Prepare the Sephadex G50 spin column and purify the labeled probes (step 4 of Experiment 26, Part I).
9. Set up the Southern blot sandwich (Experiment 25).
10. Spread transforming mixtures of Gibson and CPEC assembly reactions on to the selection plates.

Instructor notes of AMB 1 experiments 27 and 28.A and B

Each student group workbench

- Biotin-labeled pRY121 DNA sample and biotin-labeled control pBR322 DNA in ice box
- A strip of nylon or Protran® NC membrane (8 × 3 cm)
- Microcentrifuge tubes
- Safety cap lock (use for denaturing biotinylated probed DNAs in a boiling water bath)
- 0.5 mL of 6× SSC (diluted in sterile water from 20× SSC stock)
- A clean Petri dish (standard size)

Common materials and equipment

- Streptavidin-AP conjugate (2 mg/mL; Thermo Fisher Scientific): add 1.0 µL to 15 mL of wash buffer 1 just before use and mix well per group.
- 1-Step™ NBT/BCIP (250 mL; Thermo Fisher Scientific): use directly without dilution.
- UV cross-linker
- Hybridization oven, glass roller bottle

Stock solutions

Experiment 27

- **Wash buffer 1:** 0.1 M Tris-Cl (pH 7.5), 0.1 M NaCl, 2 mM $MgCl_2$ (1 liter: 100 mL of Tris-HCl (pH 7.5), 100 mL of 1 M NaCl, 2 mL of 1 M $MgCl_2$, 798 mL H_2O)
- **Blocking buffer 2:** 3 % BSA (or gelatin) in wash buffer 1 (add BSA just before use)
- **Detection buffer 3:** 0.1 M Tris-Cl (pH 9.5), 0.1 M NaCl, 50 mM $MgCl_2$ (1 liter:100 mL of 1 M Tris-HCl (pH 9.5), 100 mL of 1 M NaCl, 50 mL of 1 M $MgCl_2$, 750 mL of H_2O)
- **TE buffer** (pH 8.0)**:** 10 mM Tris-HCl (pH 8.0), 1 mM EDTA

Experiment 28, Parts A and B

- **20× SSC:** 175.3 g of NaCl, 88.2 g of trisodium citrate; adjust pH to 7.0 with a few drops of 10 M NaOH; add ultrapure water to 1 liter total; autoclave.
- **50× Denhardt's solution:** 5 g of Ficoll, 5 g of PVP (polyvinylpyrrolidone), 5 g of BSA, dH_2O to 500 mL

*Do not add all at once. Do not add water to the powder, but slowly add powder to 400 mL of ultrapure MilliQ water in a stirring beaker; add one by one after each small aliquot is dissolved.

*DO NOT HEAT to dissolve. After completely dissolved at room temperature, filter-sterilize, dispense 50-mL aliquots, and store at −20 °C.

- **1.5 × Prehybridization buffer:** 29.25 g of NaCl, 3.0 g of Tris, 0.175 g of EDTA, 0.5 g of SDS, 0.5 g of sodium pyrophosphate in 250 mL of MilliQ H_2O; adjust pH to 7.9 with HCl and add ultrapure water to 375 mL total; autoclave.
- **10% SDS:** 20 g of SDS in 200 mL of ultrapure water
- **Salmon sperm DNA:**
  1. Weigh out 250 mg of dried salmon sperm DNA (Sigma Product # D1626-250MG). Wear gloves and cut the white threads with a clean scissor to very small pieces.
  2. Put them in a 50-mL conical tube; add 26 mL of TE buffer and vortex vigorously.
  3. Place the tube in a 65 °C water bath overnight to rehydrate.
  4. Sonicate at the highest power until the trapped air bubbles can easily move up to the top.
  5. Dispense 1 mL (10 mg/mL) of aliquots in screw cap tubes and store at −20 °C.
  6. Before use, boil for 5–10 min again and cool on ice.
- **Reminder**

Students are asked to perform the washing step (Experiment 28) on the next day.

Instructor notes of AMB 1 experiments 28.C and 30.A and B

Each student group workbench

- Ice box
- pRY121 plasmid DNA (40 µL of mini-prep)
- 10× NEB buffer 4
- One silica membrane spin column

Common materials and equipment

Experiment 28, Part C

- Washing buffer 1 (see the Instructor Notes of Experiments 27 and 28)
- Blocking Buffer 2 (see the Instructor Notes of Experiments 27 and 28)
- Detection Buffer 3 (see the Instructor Notes of Experiments 27 and 28)
- Streptavidin-AP conjugate (store at −20 °C: 2 mg/mL)
- 1-Step™ NBT/BCIP solution (store at 4 °C)

Experiment 30, Part I, A and B

- One 0.5% agarose gel (10-well comb) in 1× lithium borate buffer
- QG buffer (see the Instructor Notes of Experiment 22)
- EB buffer (see the Instructor Notes of Experiment 22)
- *Sma*I, *Sac*I, and *Pst*I enzymes in ice box
- 37 °C and 60 °C water baths
- Speed vacuum concentrator
- NanoDrop 2000 spectrophotometer
- **Reminder**

During the Southern band detection step (Experiment 28), carry out Part I of Experiment 30. The isolated and concentrated mixtures of overlapping linear DNA fragments are used to transform into yeast cells for recombination cloning two weeks later.

Instructor notes of AMB 1 experiment 29

Each student group workbench

- Paper towels and 70% ethanol spray bottle

• UBC6 primer mix (10 μM) and PDC5 primer mix (10 μM) in ice box
• RNase-free sterile 1.5-mL and 0.5-mL microcentrifuge tubes; microcentrifuge rack
• RNase-free sterile pipette tips
• RNase-free H_2O
• Thin-walled PCR tubes (one group with 5 tubes; each of three other groups with 4 tubes)
• P10, P20, P200, P1000 micropipettors and pipette tips
• *S. cerevisiae* YNN281 cultures (3 mL) grown in YNB + 2 % ethanol
• *S. cerevisiae* YNN281 cultures (3 mL) grown in YNB + 2 % ethanol + 4% glucose

Common materials and equipment

• **Uninduced and glucose-induced yeast cells** that were pre-grown in YNB minimal medium containing ethanol as the sole carbon source (see Part A of Experiment 29 protocol).
• Absolute ethanol (molecular biology grade)
• **Yeast cell suspension buffer** (1 M sorbitol, 0.1 M EDTA, pH 8.0; 0.1% β-mercaptoethanol and 0.5 units/μL of lyticase):
  – 1.82 g of sorbitol dissolved in 6 mL of H_2O + 2 mL of 0.5 M EDTA (pH 8.0); bring up to 10 mL; filter-sterilize 10 mL of 1 M sorbitol/0.1 M EDTA (pH 8.0).
  – Just before class starts, dissolve 0.01 g (=2600 units) in 5 mL of yeast cell suspension buffer and add 10.3 μL of β-mercaptoethanol (48.7%; provided in an SV total RNA isolation kit).
• **SV total RNA Isolation Kit** (10 preps; Promega):
  – RNA lysis buffer (10 mL: just before class starts, add 200 μL of β-mercaptoethanol (48.7%) to 10 mL → final concentration of 0.97%)
  – RNA dilution buffer (blue)
  – RNA wash solution: absolute ethanol (molecular biology grade) is added to a final 59.8%; 1.5 mL is needed per sample.
  – DNase incubation mix: 40 μL of yellow core buffer + 5 μL of 0.09 M $MnCl_2$ + 5 μL of DNase I enzyme per sample; 50 μL is needed per sample.
  – DNase stop solution: absolute ethanol is added to a final 57.1%; 200 μL is needed per sample.
  – Nuclease-free water.
• **Verso 1-Step RT-PCR kit** (Thermo Fisher Scientific Cat. No. AB1454LDA)
• Thermal cycler (program: Part D): takes ~2 hour to complete PCR cycling.
• Mini-agarose gel electrophoresis apparatus and power supply
• DNA size marker (NEB quick-load 50 bp DNA ladder)
• Agarose, 10× lithium borate buffer
• 6× DNA loading dye
• GelRed (10,000×): diluted to 2.5× in dH_2O
• One black 96-well microplate: fill ABCD rows of 12 column wells with 99 μL of 1× TE buffer (pH 8.0)
• Quant-iT PicoGreen® dsDNA assay kit with λ DNA standard (100 μg/mL; Thermo Fisher Scientific, Cat. No. P7589)
• 1× TE buffer (pH 8.0)
• Tecan Infinite M200 Spectrafluor microplate reader
• **Reminder**
The following order of procedures is recommended.
1. Read A_{600} of induced and uninduced yeast cultures and determine the volume for 2×10^7 cells before the lab exercise begins.
2. Perform the Part B (RNA isolation) exercise. For 4 student groups, groups 1 and 3 use YNB + 2% ethanol culture cells; groups 2 and 4 use YNB + 2% ethanol + 4% glucose culture cells to prepare RNA. Determine the concentrations by NanoDrop 2000.
3. Set up RT-PCR reactions and run PCR (Parts C and D).
4. During PCR, prepare 1.8% agarose gel in 1× lithium borate buffer without GelRed dye. Each group has six RT-PCR samples to run (see step 29 of Part E).
5. Take out 6 μL of each RT-PCR sample at the 25th and 30th cycles and transfer to labeled 0.5-mL tubes in ice bath. Transfer 1 μL of each cycle sample to the assigned well of black microplate (step 20 of Part D). Combine the remaining 5 μL of each cycle PDC5 and UBC6 samples and add 2 μL of DNA loading dye.
6. During PCR, prepare the low and high range standard DNA solutions in the assigned plate wells (steps 2 and 3 of Part G).
7. Once the last PCR cycle is finished, prepare gel loading and carry out agarose gel electrophoresis (steps 28 and 29 of Part E). Also prepare PicoGreen assay samples and conduct Tecan microplate reader analysis (steps 4 to 8 of Part G).

Instructor notes of AMB 1 experiments 30. C and D, 31, and 32

Each student group workbench

Experiment 30, Part II, C and D (Yeast Transformation)
• Spin column-purified pRY121 DNA fragments (Part I, B)
• 10 mL of YNN281 cells as prepared in steps 1 to 4 of Part II, C
• Sterile microcentrifuge tubes
• 5 mL of sterile dH_2O
• 10 μL salmon sperm DNA (10 mg/mL) in a microcentrifuge tube
• 200 μL TE (pH 8.0) buffer
• Four uracil dropout YNB + glucose agar plates
• One uracil dropout YNB + galactose + raffinose agar plate
• One uracil dropout YNB + galactose + glucose agar plate
• **YNB –Ura/+glucose agar plate:**

- Uracil dropout supplement:	0.77 g/liter (0.077%)
- YNB:	70 g/liter (0.67 %)
- Glucose (dextrose):	20.0 g/liter (2 %)
- Agar:	20.0 g/liter (2 %)

• **YNB –Ura/+galactose/+raffinose agar plate:**

- Uracil dropout supplement:	0.77 g/liter (0.077%)
- YNB:	6.70 g/liter (0.67%)
- Galactose:	10.0 g/liter (1 %)
- Raffinose:	10.0 g/liter (1 %)
- Agar:	20.0 g/liter (2 %)

• **YNB –Ura/+ galactose/+glucose agar plate:**

- Uracil dropout supplement:	0.77 g/liter (0.077%)
- YNB:	6.70 g/liter (0.67%)
- Galactose:	10.0 g/liter (1 %)
- Glucose:	10.0 g/liter (1 %)
- Agar:	20.0 g/liter (2 %)

Stock Solutions

- 50% Polyethylene glycol 4000 or 3500 (warm to dissolve; filter sterilize)
- 1 M Tris-HCl pH 8.0 (autoclave)
- 1 M lithium acetate pH 7.5 (dissolve 10.2 g in 80 mL H_2O, adjust to pH 7.5 with dilute acetic acid, bring up to 100 mL; autoclave)
- 0.5 M EDTA pH 8.0 (autoclave)
- Salmon sperm DNA (sonicated and heat-denatured, 10 mg/mL)
- pRY121 plasmid DNA
- Glycerol (autoclave)
- DMSO
- **1× TEL** (10 mM Tris-HCl, pH 8.0, 1 mM EDTA, 0.1 M lithium acetate).

Aseptically prepare fresh from stock solutions just before class. No autoclave.

0.1 mL of 1 M Tris-Cl (pH8.0) + 0.02 mL of 0.5 M EDTA (pH 8.0) + 1.0 mL of 1 M LiAc + 9.88 mL of sterile MilliQ H_2O (total 10 mL).

- **PEG-TEL** (40% PEG-4000 or 3350, 10 mM Tris-HCl pH 8.0, 1 mM EDTA, 0.1 M lithium acetate). Aseptically prepare fresh from stock solutions just before class. No autoclave.

8 mL of 50% PEG-4000 or 3350 (filter-sterilized) + 0.1 mL of 1 M Tris-Cl (pH8.0) + 0.02 mL of 0.5 M EDTA (pH 8.0) + 1.0 mL of 1 M LiAc + 0.88 mL of sterile MilliQ H_2O (total 10 mL).

Experiment 31 (Plasmid Isolation from Yeast)

- 30 mL of overnight culture of YNN281 (pRY121) grown in YPDA medium
- 180 µL volume of acid-washed glass beads (0.45–0.53 mm)
- Plasmid isolation kit (Epoch Life Science, Cat. No. 2160250)

Common materials and equipment

- **Lyticase solution** (5 mg/mL in TE buffer, 0.05 g in 10 mL TE; store at 4 °C for up to one month or at −20 °C for up to 6 months; if colloidal precipitates appear, mix by inversion before using)
- **Breaking buffer** (2% Triton X-100; 1% SDS; 100 mM NaCl; 10 mM Tris-Cl pH 8; 1 mM EDTA): 1 mL of Triton X-100 + 0.05 mL of 10% SDS + 5 mL of 1 M NaCl + 0.5 mL of 1 M Tris-HCl (pH 8.0) + 0.1 mL of 0.5 M EDTA (pH 8.0) + 43.35 mL of MilliQ H_2O (total 50 mL)
- Phenol/chloroform/isoamylalcohol (25:24:1, v/v/v) mixture
- Vortexer
- One 0.5% agarose gel in 1× lithium borate buffer (+0.5 µg/mL of ethidium bromide)
- 30 °C incubator

Experiment 32 (*E. coli* Transformation)

- Ice box
- Three 100 µL aliquots of competent *E. coli* HB101 cells as prepared in Part A of Experiment 32.
- Sterile 1.5-mL microcentrifuge tubes
- Six LB agar plates containing 100 µg/mL of ampicillin
- Denatured ethanol in beaker, L-shaped glass spreader, alcohol Bunsen burner

Common materials and equipment

- pRY121 plasmid DNA
- LB or SOC broth
- Two 100 µL aliquots of competent *E. coli* HB101 cells for positive and negative control
- Two LB/Amp (100 µg/mL) agar plates for positive and negative control
- 42 °C water bath; 37 °C incubator
- **Reminder**

1. Calculate the suspension volume of 1× TEL to get 10^9 cells/mL based on both an A_{600} reading and a 10 mL of cell culture harvest (step 7 of Part II, C, in Experiment 30), and perform yeast transformation immediately.

2. During the incubation period of yeast transformation, perform yeast plasmid isolation using three different protocols from each 3 mL of the yeast culture harvest.

3. As soon as yeast plasmid isolation is finished, carry out *E. coli* transformation using pre-made competent *E. coli* cells.

Instructor notes of AMB 1 experiments 33 and 34

Each student group workbench

Experiment 33

- Six Whatman 3MM paper circles cut to fit inside Petri plates
- Three Petri plates (100 × 15 mm)
- Student streaked YNB −Ura/+ glucose plate (Experiment 30, Part II, D)
- Student streaked YNB −Ura/+ galactose/+ raffinose plate (Experiment 30, Part II, D)
- Student streaked YNB −Ura/+ galactose/+glucose plate (Experiment 30, Part II, D)

Stock solutions

- **Z buffer** (without β-mercaptoethanol): *1 liter (do not autoclave)*

- 60 mM $Na_2HPO_4 \cdot 7H_2O$	16 g (or 8.53 g Na_2HPO_4)
- 40 mM $NaH_2PO_4 \cdot H_2O$	5.52 g (or 4.8 g NaH_2PO_4)
- 10 mM KCl	0.75 g
- 1 mM $MgSO_4 \cdot 7H_2O$	0.25 g (or 0.12 g $MgSO_4$)
- Adjust pH to 7.0; store at room temperature	
- 50 mM β-mercaptoethanol	2.7 mL (add just before use)

- 1 M Na_2CO_3, 100 mL
- 2% X-β-Gal (0.02 g dissolved in 1 ml of DMF (*N,N*-dimethyformamide); store at −20 °C)
- Liquid nitrogen

Experiment 34

- 60 mL of 1/5 diluted Bradford reagent (for 4 groups, mix 50 mL of concentrated Bradford reagent with 200 mL of dH_2O and filter through Whatman paper)
- 6.5 mL of Lowery reagent A and 55 mL of reagent B
- 60 ml of BCA reagent A and B mix (60 mL of reagent A + 1.2 mL of reagent B)
- 1.2 mL BSA (1 mg/mL in sterile dH_2O)
- Test samples: - BSA (60 µL of 1.5 mg/mL in 1× PBS)
  - Globulin (60 µL of 1.5 mg/mL in 1× PBS)

- Gelatin (60 µL of 1.5 mg/mL in 1× PBS)
- Lysozyme (60 µL of 1.5 mg/mL in 1× PBS)
- *E. coli* cell-free whole lysate (50 µL)

• Spectronic 20 and a cuvette, Kimwipe
• Test tube rack for 13 × 100 mm test tubes (54)
• 70% ethanol and water wash bottles to rinse out cuvette

Stock solutions

Make 1-mL aliquots of all proteins, place in a storage box, and store at −20 °C.

• BSA standard (10 mL of 1 mg/mL): 0.01 g in 10 mL of sterile dH_2O
• BSA unknown test solution (10 mL of 1.5 mg/mL): 0.015 g in 10 mL of 1× PBS
• γ-Globulin test solution (10 mL of 1.5 mg/mL): 0.015 g in 10 mL of 1× PBS
• Gelatin test solution (10 mL of 1.5 mg/mL): 0.015 g in 10 mL of 1× PBS
• Lysozyme test solution (10 mL of 1.5 mg/mL): 0.015 g in 10 mL of 1× PBS
• Bradford reagent (Bio-Rad)
• Lowery DC reagents A and B (Bio-Rad)
• BCA reagents A and B (Thermo Fisher Scientific)
• **10× PBS**: 80 g of NaCl + 2 g of KCl + 14.4 g of Na_2HPO_4 + 2.4 g of KH_2PO_4 in 1 liter. The pH of the 10× stock will be about 6.8. When diluted to 1× PBS, it should change to 7.4. Sterilize by autoclaving.
• *E. coli* cell-free lysate: prepare as follows.

1. Harvest 2 × 5 mL of overnight culture in 2 × 2.0-mL microcentrifuge tubes by spinning at 13 000 rpm for 1 min.
2. Discard the supernatant and resuspend each tube cell pellet completely in 0.5 mL of 1× PBS buffer by pipetting up and down with the pipette.
3. Centrifuge at top speed for 1 min, discard the supernatant, and resuspend each tube cell pellet completely in 0.5 mL of 1× PBS buffer by pipetting up and down with the pipette.
4. Add 200 µL of lysozyme (50 mg/mL). Gently mix the contents by inverting the tube several times and incubate at room temperature for 5 min.
5. Place the tube in liquid nitrogen for 1 min and thaw at 37 °C in a water bath; repeat once more.
6. As soon as the cells are thawed, add 7 µL of protease inhibitor cocktail (100 ×).
7. Centrifuge at top speed for 10 minutes at 4 °C to pellet-insoluble cellular debris.
8. Collect the clarified supernatant and make 50 µL aliquots; store at −80 °C.

• Reminder

If students failed to obtain yeast transformants (Experiment 30, Part II, C and D):

• Prepare about 1 cm diameter patches of yeast YNN281 (pRY121) and YNN281 at least 2 cm apart from each other near the center of each of three plates.
• Incubate the plates at 30 °C for 1 to 2 days. Dense patches of cells should be present by the end of this time.

1. Start with X-gal filter lift assay.
2. As soon as X-gal filter is incubated, perform the protein assays. Each of Bradford, Lowy, and BCA assays requires 18 test tubes for BSA calibration standards and test samples of unknown concentrations.

Instructor notes of AMB 1 experiment 35

Each student group workbench

Part I

• Each 15 mL of induced and uninduced cultures of YNN281 (pRY121)
• 30 mL of complete Z buffer (see below)
• 20 µL of toluene and 20 µL of 5% SDS
• 10 mL of 1 M Na_2CO_3
• 2.0-mL microcentrifuge tubes
• 5.0 mL of ONPG (4 mg/mL in the Z buffer) in ice box
• Spectronic 20 and cuvette, Kimwipe

Part II

• 1.5-mL breaking buffer in ice box
• 2 mL of sterile H_2O
• 0.5 mL of BSA (1 mg/mL in the breaking buffer)
• Two 1.5-mL microcentrifuge tubes, each containing 265 µL volume of acid-washed glass beads (0.45 to 0.53 mm)
• 15 glass test tubes in a rack for Bradford assay
• 50 mL of Bradford reagent diluted to 1/5 in MilliQ H_2O

Common materials

• **Yeast cultures**

1. Prepare the following two media.
   • **Induction YNB media** (100 mL)**:** 0.077 g of uracil dropout (Ura DO) supplement + 0.67 g of YNB + 1 g of galactose + 1 g of raffinose
   • **Non-induction YNB media**: 0.077 g of uracil dropout (Ura DO) supplement + 0.67 g YNB + 2 g glucose
2. Inoculate 0.5 mL overnight culture into two 250-mL flasks containing 50 mL each of induced and uninduced YNB medium and incubate at 250 rpm, at 30 °C overnight.

• **Breaking buffer** (100 mM Tris-HCl, pH 8.0, 1 mM dithiothreitol (DTT), 20% glycerol, 1 mM EDTA, 10 mM $MgCl_2$):
10 mL of 0.1 M Tris-HCl (pH 8.0) + 0.1 mL of 1 M DTT + 20 mL of glycerol + 0.2 mL of 0.5 M EDTA + 1 mL of 1 M $MgCl_2$ + 68.7 mL of dH_2O (total 100 mL).
• **Z buffer** (60 mM Na_2HPO_4, 40 mM NaH_2PO_4, 10 mM KCl, and 1 mM $MgSO_4$): 4.26 g of Na_2HPO_4 (8.04 g of $Na_2HPO_4 \cdot 7H_2O$) + 2.76 g of $NaH_2PO_4 \cdot H_2O$ (2.4 g NaH_2PO_4) + 0.375 g KCl + 0.123 g $MgSO_4 \cdot 7H_2O$ (0.06 g of $MgSO_4$); adjust to pH 7.0; bring to 500 mL with dH_2O. Do not autoclave. Store at 4 °C. For complete Z buffer, add 0.14 mL of ß-mercaptoethanol to a filter-sterilized 50 mL Z-buffer just before use.
• **PMSF** (phenylmethylsulfonyl fluoride; 100 mM): dissolve 0.174 g in 10 mL in isopropanol in a 37 °C water bath and store at −20 °C. Redissolve the crystals by swirling in a warm water bath before use.
• 28 °C water bath

• Reminder

Cells for both Parts I and II experiments are harvested at the same time and kept in ice once the cell pellet was suspended in the assay buffers. The following order of procedures is recommended.

1. Harvest 5 mL each of induced and uninduced cultures (Part I experiment) and 10 mL each of induced and uninduced cultures (Part II experiment) in four separate labeled conical tubes.
2. Suspend each Part I cell pellet in 1 mL of Z buffer and each Part II cell pellet in 1 mL of H_2O, transfer them to four labeled 2.0-mL microcentrifuge tubes, spin, and resuspend the Part I cell pellet in 400 µL of Z buffer and the Part II cell pellet in 250 µL of breaking buffer.
3. Perform the Part I experiment first until processed samples are incubated at 37 °C (step 7).
4. During the 30-min incubation time of the Part I sample, process the Part II samples to obtain the clarified cell-free extracts (steps 6 to 9 in Part II).
5. Set up the reaction samples for both Parts I and II, add ONPG to initiate the β-gal assay, and incubate at 28 °C. A strong yellow color development typically occurs within 30 min.
6. During the ONPG reaction time, read A_{600} of the cell suspension (step 5 of Part I) and perform the Bradford protein assay (Part II, C).
7. Read A_{420} of the ONPG reaction samples of Parts I and II.

Instructor notes of AMB 1 experiments 36 and 37

Each student group workbench

Experiment 36

- A glass column (1.0 × 30 cm) packed with 24-mL column volume of Sephadex G100 beads that were equilibrated with 1× PBS I at room temperature in a ring stand
- 50 mL of 1× PBS (degassed by filtering under vacuum)
- 125 µL each of loading samples I and II (see below)
- 1 mL of 10 % hydrogen peroxide H_2O_2 (333 µL of 30% H_2O_2 + 667 µL of H_2O)
- 60 of 1.5-mL microcentrifuge tubes (non-sterile)
- Two plastic Pasteur pipettes
- Microcentrifuge rack
- 50-mL graduated cylinder
- One 10-mL glass test tube
- Waste beaker
- Spectronic 20 and cuvette, Kimwipe

Experiment 37

- 20 mL of buffer A (see below)
- 2 mL of buffer B (see below)
- 125 µL of loading sample (a mixture of catalase, hemoglobin, and cytochrome c)
- One Poly-Prep chromatography column (Bio-Rad, Cat. No. 731-1550)
- 5 mL of DEAE-cellulose resin fully hydrated in buffer A
- Eight 2.0-mL and 32 1.5-mL microcentrifuge tubes (non-sterile)

Stock materials

- 10× PBS:
  80 g of NaCl + 2 g of KCl + 14.4 g of Na_2HPO_4 + 2.4 g of KH_2PO_4 in 1 liter. Sterilize by autoclaving.
- Loading sample I of Experiment 36 (20 mg/mL of blue dextran):
  Dissolve 0.02 g in 1.0 mL of 1× PBS and pass through a 0.45-µm filter.
- Loading sample II of Experiment 36 (protein mixtures):
  Dissolve 0.01 g of catalase + 0.005 g of cytochrome c + 0.005 g of hemoglobin + 0.001 g of vitamin B12 in 1× PBS and bring up to 1 mL (prepare fresh before use).
- Buffer A (20 mM Tris-HCl, pH 8.0; autoclaved)
- Buffer B (20 mM Tris-HCl, pH 8.0, 2.0 M NaCl; autoclaved)
- Loading sample of Experiment 37:
  Dissolve 0.005 g of catalase + 0.002 g of hemoglobin + 0.001 g of cytochrome c in buffer A and bring up to 1 mL (prepare fresh before use).
- Sephadex G-100 fully hydrated in 1× PBS:
  1. Add about 14 g to 500 mL of buffer; carefully stir the suspension with a glass rod; let the gel fully swell for 5 h in a 90 °C water bath, stirring as needed to keep the gel in suspension. Do not use a magnetic stirrer.
  2. Allow the gel particles to settle, and then remove fines and broken beads by decanting the hazy solution. Repeat decantation (4 to 5 times if necessary) until the gel bed settles to a sharp zone.
  3. Dilute so that the final slurry is 75% settled gel and 25% buffer by volume.
  4. Add sodium azide to a concentration of 0.02–0.05%.
  5. Degas for 10 min before packing.
- DEAE-Sephadex (pre-swollen):
  Wash with excess buffer A until the pH reaches 8.0, resuspend the settled resin in buffer A, and degas 10 min before packing.
- **Reminder**

The following order of procedures is recommended.
1. Before the lab starts, set up the 24-mL Sephadex column that is fully equilibrated with 1× PBS buffer.
2. Students begin with loading blue dextran on to the column bed (Parts B and C) and determine the void volume.
3. As soon as blue dextran reaches the bottom portion of the column but has still not eluted yet, close the outlet, remove the buffer from the top, and load the second sample of protein mixture.
4. During fractionation, perform the catalase activity test of 0.5-mL elution fractions (Part E) and A_{410} reading (Part F) at the same time to finish Experiment 36.
5. One student packs a 1-mL ion-exchange column and a second student prepares the step elution buffer.
6. Load the protein sample and carry out the step elution.
7. Conduct a catalase activity test to finish Experiment 37.

AMB 1 Appendix 2

AMB 1 Lab math practice problem set

Part I. Metric unit conversion

1. Convert the following:

- 1 gram = ______ mg 1 gram = ______ kg
- 1 liter = ______ mL 2 ng = ______ mg
- 50 mg = ______ g 100 µL = ______ mL
- 1 mg = ____ pg = ______ µg = _____ ng
- 1 L (liter) = ___ mL = ___ µL = ____ nL = ____ pL = _____ fL
- 1 mM = _____ µmole/mL = _____ pmole/µL
- 1 L × 1 M = ______ mole 1 mL × 1 M = ______ mmole
- 1 L × 1 mM = ______ mmole 1 mL × 1 mM = ______ µmole
- 1 mL of pure water = ______ g 1 L of pure water = ______ g
- 1 ppm (part per million) = ______ % 10% = ______ ppm
- 1 mg/L in water = _____ % = _____ ppm 100 ppm = ________ µg/mL in water
- Calculate the slope of 100% efficiency when % Efficiency = $(10^{-1/\text{slope}} - 1) \times 100$

Part II. Dilution

1. You have a series of 10 tubes, each of which contain 4 µL. You add 1 µL of labeled DNA to the first tube, mix and take 1 µL of this mixture and add to the second tube, mix and take 1 µL of this mixture and add to the next. If you continue this process with each tube, what is the dilution of the serum in the last tube?

2. Eight test tubes are placed in a rack. To the first tube add 3 mL of saline solution. To each of the remaining seven tubes add 2 ml of saline solution. To the first tube add 1 mL of serum and mix well. Transfer 2 mL of tube #1 to tube #2 and mix well, 2 mL of the contents of tube #2 is then transferred to tube #3, and the procedure is repeated for the remaining tubes. What is the dilution of serum in tube #8?

3. You had 30 mg/mL of glucose solution. You took 1 mL of that and diluted to 2 mL. Then 0.5 mL of it was added to 4.5 mL of saline solution, and 2 mL of this solution was diluted to 10 mL. What is the final concentration of glucose?

4. If you use 5 mL of stock glucose containing 10 mg/mL and dilute to 100 mL, what is the concentration of glucose per mL in the dilution?

5. Overnight yeast culture was found to contain about 5×10^7 cells/mL. If you want to inoculate the yeast cells into a fresh 100 mL medium to a final density of about 2.5×10^3 cells/mL, how much do you have to inoculate?

6. The average number of colonies appearing on three independent agar plates is 250 after 0.1 mL each of 1:1000 diluted yeast culture cells were plated on to the three plates and incubated. What is the cell density of the original yeast cell culture?

7. How would you prepare 25 mL of a 1/10 dilution of blood using PBS buffer?

8. How would you prepare 400 µL of a 1/100 dilution of antibody using PBS buffer?

9. How would you prepare 150 mL of a 1/250 dilution of dye solution using water?

10. You have a stock of a buffer solution that is used in experiments. The stock is 10 times (10×) more concentrated than required. How would you prepare 100 mL of the buffer solution at the right concentration?

11. You take 100 mL of suspended bacteria and grind them into a cell "paste." Then you remove 10 µL of these cells and perform a test that tells you how much protein is present. Your test shows that 6 µg of protein are present in the 10 µL of cell paste. What was the concentration of protein in the cell paste? Express your answer in mg/mL.

12. If you take a 0.1-mL sample and add 0.9 mL of water and 3.0 mL of reagents, what is the final dilution of the sample?

13. You want to set up a PCR reaction mixture by adding 0.001 µg of plasmid DNA to a tube. You have 1 mL of plasmid at a concentration of 1 mg/mL. How would you add 0.001 µg of plasmid? Assume that it is not possible to accurately measure a volume less than 1 µL.

14. You have 100 µL of a protein sample. The sample was diluted by taking 5 µL of the sample and adding 20 µL of buffer. A protein assay was performed on the diluted sample. The concentration of protein in the diluted sample was 90 mg/mL. What was the concentration of protein in the original, undiluted sample? How much total protein was present altogether in the original, undiluted sample?

Methods in Biotechnology, First Edition. Seung-Beom Hong, M. Bazlur Rashid and Lory Z Santiago-Vázquez.

Companion Website: www.wiley.com\go\hong\Methodsinbiotechnology

15. You have a DNA standard of 100 ng/μL. You want to prepare a series of diluted standard with concentrations of 50, 12.5, 3.13, 0.78, 0.20, 0.048, and 0.012 ng/μL by transferring 10 μL from one tube to the next for each sample. How would you prepare the entire dilution series using TE buffer?

Part III. Percent solution

1. If you have 10 g of $MgCl_2$ in a solution of 500 mL, what is the concentration % (w/v)?
2. How would you make 100 mL of a 70% (v/v) ethanol solution?
3. You weighed out 4.5 g of $CaCl_2$ and diluted it to 250 mL with distilled water. What is the percent solution (w/v)?
4. You need to make 100 mL of 50% $CaCl_2$ (MW = 111) solution, but all we have is $CaCl_2 \cdot 5H_2O$ (MW = 201) on the shelf. How much do you have to weigh out the hydrated form to make a 50% $CaCl_2$ solution?
5. How do you prepare a sterile 10% SDS solution? Where should you store the stock solution? What happens if the stock solution is kept in the cold?
6. What is the concentration of SDS if you add 10 mL of 10% SDS to 90 mL of water?
7. How would you make 100 mL of a solution that is 5% (v/v) acetic acid and 5% methanol?

Part IV. Ratio and proportions

1. Double-stranded DNA consists of two long strands. An adenine (A) on one strand is always paired with a thymine (T) on the opposite strand and a guanine (G) on one strand is always paired with a cytosine (C) on the opposite strand. If a purified sample of DNA contains 21% thymine, what are the percentages of the other A, G, and C bases?
2. What is the percentage of a solution that contains 200 mL of acetone per liter?
3. How much ethanol is present in 200 mL of a 25% solution of ethanol?
4. The density of glycerol at 20 °C is 1.26 g/mL. What is the volume of 87.0 g of glycerol?
5. The density of β-mercaptoethanol is 1.1 g/mL. If you need to add 13.5 g of this chemical to a solution, how much volume should you measure out?
6. A recipe to make 6× gel loading buffer is 0.25% bromophenol blue, 0.25% xylene cyanol, and 40% sucrose in water. Express the amounts of each component needed in mg to prepare 10 mL of 6× gel loading dye.
7. You prepared Tris-HCl buffer at pH 7.5 at room temperature (25 °C). Tris buffer increases 0.3 pH units for every 10 °C decrease in temperature. What is the expected pH if the Tris-HCl solution is kept in a cold room (4 °C)?
8. How much 95% alcohol is needed to make 500 mL of 70% alcohol?
9. One-tenth mL of a protein solution was diluted with 2.9 mL of water. The A_{280} of the diluted solution was 0.25. How many mL of the original protein solution and water should be mixed to make 1.0 mL of solution of A_{280} = 0.75?

Part V. Mole and molarity

1. What is the MW in grams if one molecule weighs 6.664×10^{-20} g?
2. What is the weight in grams of 1×10^{23} molecules of NaCl (MW = 58)?
3. What molarity (M) gives a 0.9% glucose (MW = 180) solution?
4. What molarity (M) gives a 90 mg/mL of glucose (MW = 180) solution?
5. If 150 g/liter of a particular chemical is required to obtain a 0.5 M solution, how much of this chemical is required to obtain a 30 μM solution?
6. You want to prepare 500 mL of a 1.5 M solution of NaCl. You have powder, a beaker, a graduated cylinder, and a bottle. How you would prepare the solution (MW of NaCl = 58)?
7. You wish to prepare a 500 mL of 10 mM solution of KCl. Starting from KCl powder, how would you prepare the solution (MW of KCl = 74)?
8. You have a solution of 1 M Tris-HCl. How would you prepare 20 mL each of 0.5 M, 0.15 M, and 1 mM Tris?
9. You are given a 10% SDS stock solution. The molecular weight of NaOH is 40. How do you make a 100-mL solution that consists of 1 M NaOH and 1% SDS?
10. Let's say you are TA for a biotechnology class. Your professor asked you to prepare detection buffer 3 (0.1 M Tris-Cl (pH 9.5), 0.1 M NaCl, 50 mM $MgCl_2$) for 6 section classes. Each section needs 500 mL of buffer 3. How will you prepare the buffer from stock solutions of 1 M Tris-HCl (pH 9.5), 1 M NaCl, and 1 M $MgCl_2$?
11. Let's say you are TA for a biotechnology class. Your professor asked you to prepare Sephadex G-50 column buffer (50 mM Tris-HCl (pH 7.9), 2 mM EDTA (pH 8.0), 250 mM NaCl) for 6 section classes. Each section needs 100 mL of the buffer. How will you prepare the buffer from stock solutions of 1 M Tris-HCl (pH 9.5), 1 M NaCl, and 0.5 M EDTA (pH 8.0)?
12. What is the osmolarity of a 1.0 M solution of ethyl alcohol (C_2H_6O) in water?
13. What is the osmolarity of a 0.75 M solution of $MgCl_2$ in water?
14. How many grams of copper sulfate, $CuSO_4$, anhydrous (99 % purity, MW 159.6), do you need to make 1 liter of a 5% solution? What is the molarity of this solution?
15. Anhydrous copper sulfate is not available in question 14. How many grams of copper sulfate, pentahydrate (purity 99 %, FW 249.7), do you need to make an equal molar solution?
16. What is the molarity of solution in which 100 mL of HCl (MW 36.5, ρ =1.19 g/mL, % assay = 37) is added to water to give a final volume of 0.8 liter?

Part VI. Applications

1. How do you prepare a sterile 10 mL of 100 mg/mL ampicillin and where do you have to store the stock solution?
2. Let's say you are TA for a biotechnology class. Your professor asked you to prepare 100 LB agar (1% tryptone, 0.5% yeast extract, 1% NaCl, 1.5% Bacto-agar) plates containing 100 μg/mL

of ampicillin. You assume each Petri plate needs 25 mL and you have 100 mg/mL of ampicillin stock solution. Describe in detail where and how to prepare the plates including instruments to be used.

3. You are doing a large-scale plasmid DNA extraction from an *E. coli* culture. Let's say the protocol called for your cell suspension to have a concentration of lysozyme at 2 mg/mL. Your suspension has a total volume of 10 mL. The lysozyme stock tube has a concentration of 20 mg/mL. How much lysozyme should you add?

4. You have already determined the doubling time of 24 min for a certain *E. coli* strain. You also determined the cell density of 10^7 cells/mL by plate counting. You diluted the cell to 1:100, inoculated 0.1 mL of the diluted cells into 9.9 mL of the same medium, and incubated under the same growth condition that was used to determine the doubling time and cell density. How long should you incubate to reach a cell density of 32 000 cells/mL?

5. Let's say the A_{600} of the 5 mL overnight culture at 10:00 a.m. is 1.0 and you want to harvest 10 mL of a culture of *E. coli* cells at A_{600} = 0.8 at 2:00 p.m. on the same day. You know *E. coli* doubles every 30 min under the growth conditions used. How much volume of the 5 mL culture would you need to inoculate into a fresh 10 mL medium? Assume that only 50% of cells in the 5 mL overnight culture are alive.

6. Calculate the doubling time (dt) and growth rate constant (*K*) when a bacterial cell population in the exponential phase increases from 5×10^6 cells/mL to 7×10^8 cells/mL in a 6-h incubation.

7. Calculate the theoretical total DNA yield (μg) and DNA concentration (ng/μL) assuming that you recover 100% of the pure *E. coli* genomic DNA (4.6×10^6 bp) in 0.1 mL of a final DNA sample from a 2-mL overnight culture of 2×10^9 cells/mL. The MW of the DNA base pair is 660 g/mole.

8. The iron complex of *o*-phenanthroline (molecular weight 236) has a molar extinct coefficient of 10 000 at 525 nm. If the absorbance of 0.01 is the lowest detectable signal, what concentration in ppm can be detected in a 1-cm cell?

9. Absorbance A_{570} of an unknown BPB solution was found to be 0.445. Calculate the concentration in mg/mL of the unknown BPB (MW = 670) solution using the standard calibration curve, as shown below.

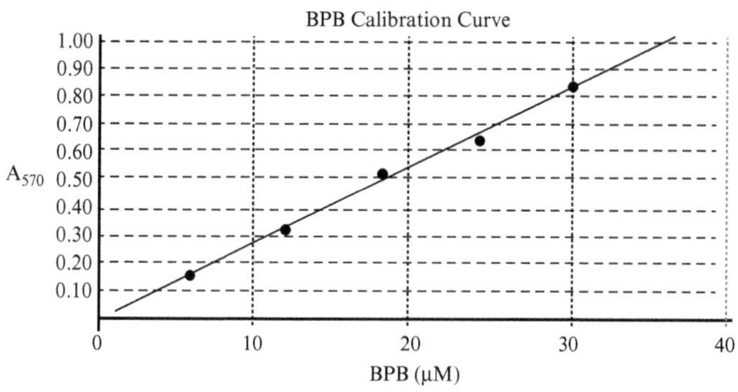

10. A β-galactosidase assay was performed in a 10-mL volume containing 8 mL of Z buffer (pH 7.0), 1 mL of 7.5 mM ONPG solution, and 1 mL of $1/10^4$ diluted crude extract. At each time point, 1 mL of the reaction sample was read at 420 nm in a 1-cm pathlength cuvette. A_{420} readings are as follows: 15 s = 0.059, 30 s = 0.084, 45 s = 0.109, 60 s = 0.134, 75 s = 0.159. The assay setup is illustrated as shown below.

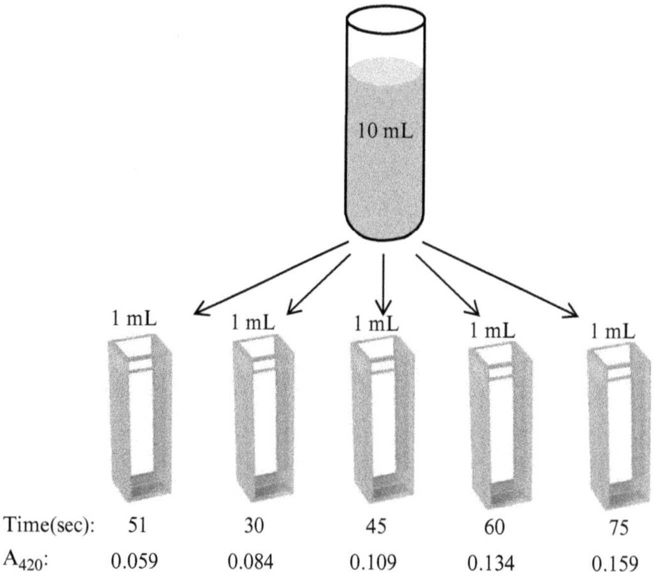

A standard 0.1 mM solution of *o*-nitrophenol (ONP) in the Z buffer has an absorbance 0.45 at 420 nm in a 1.0-cm pathlength cuvette used for this assay.

(a) What increase in A_{420} value/min should correspond to one unit of β-galactosidase under this assay condition?

(b) What is the specific activity of the extract when the protein concentration of the undiluted crude extract was 13.3 mg/mL?

11. The human genome is 3×10^9 bp. The MW of 1-bp dsDNA = 660 g/mole. If you have 1 microgram of human genomic DNA, how many copies of the genome do you have?

12. Your DNA concentration is 10 mg/mL. How many pg/μL is this?

13. The spectrometric readings at A_{260} of 1:100 diluted DNA sample is 0.15. What is the DNA concentration before dilution?

14. The spectrometric readings at A_{260} of 1:100 diluted RNA sample is 0.25. What is the RNA concentration before dilution?

15. A standard curve for real-time qPCR is shown below.

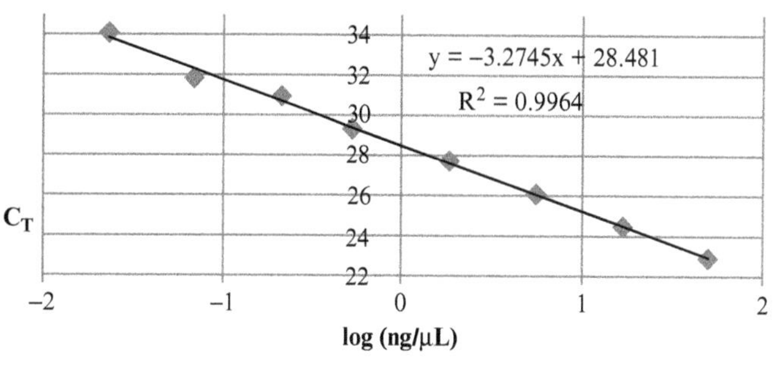

(a) What is the concentration of your target sample when the C_T value is 25.4284?

(b) Calculate the amplification efficiency using the following formula:

$$\% \text{ Efficiency} = \left(10^{-1/\text{slope}} - 1\right) \times 100$$

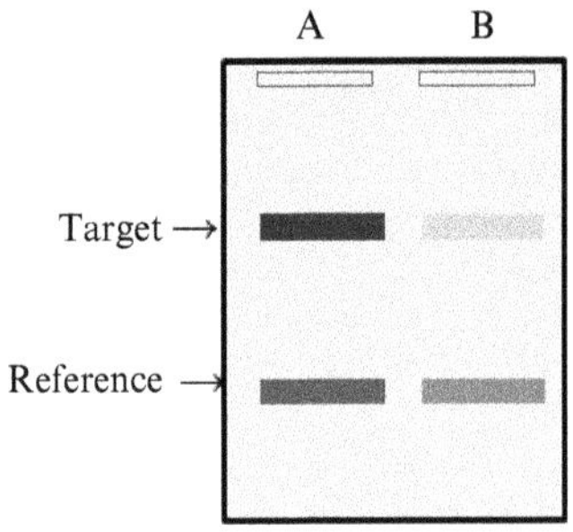

	A. Test (Treated sample)	B. Calibrator (Untreated sample)
Target	50.69	3.53
Reference	30.17	25.62

16. Densitometric analysis data of RT-PCR samples of your target gene and internal reference gene on the agarose gel are shown above.

Based on the table data, calculate the fold change in target gene expression after treatment.

17. You prepared the total RNA from hydrogen peroxide (H_2O_2) treated cells and untreated cells. Then, you set up a dilution series of a known amount of total RNA with RT-qPCR master mix and primer sets specific for target and reference genes in a microtiter plate to establish the standard curves of target genes and reference genes. At the same time, you set up triplicate reactions using the same amount of the treated (test) and untreated (calibrator control) RNA and the unique primer sets for target and reference genes in the same microtiter plate. After PCR, the following data were obtained.

(a) Based on the table and standard curve data, calculate the fold change in the target gene expression after treatment.

(b) According to the Pfaffl method, the fold change can be calculated using the following equation:

$$\text{Fold change} = \frac{(E_{\text{target}})^{\Delta C_{T\ \text{target}}}}{(E_{\text{ref}})^{\Delta C_{T\ \text{reference}}}}$$

where $E = 10^{-1/\text{slope}}$ and $\Delta C_T = C_{T\ \text{calibrator}} - C_{T\ \text{test}}$.

Based on the Pfaffl equation formula, calculate the fold change in target gene expression after treatment.

	C_T values of test (treated sample)	C_T values of calibrator (untreated sample)
Target gene	25.68 25.79 25.85	27.45 27.05 27.32
Reference gene	30.75 30.68 31.05	25.41 25.25 25.35

(c) According to the $\Delta\Delta C_T$ method, fold change can be calculated using the following equation:

Fold change $= 2^{-\Delta\Delta C_T}$, where $\Delta\Delta C_T = \Delta C_{T\ \text{test}} - \Delta C_{T\ \text{calibrator}}$

$\Delta C_{T\ \text{test}} = (C_{T\ \text{target}} - C_{T\ \text{reference}})$ in treated test sample

$\Delta C_{T\ \text{calibrator}} = (C_{T\ \text{target}} - C_{T\ \text{reference}})$ in untreated calibrator sample

Based on the $\Delta\Delta C_T$ formula, calculate the fold change in target gene expression after treatment.

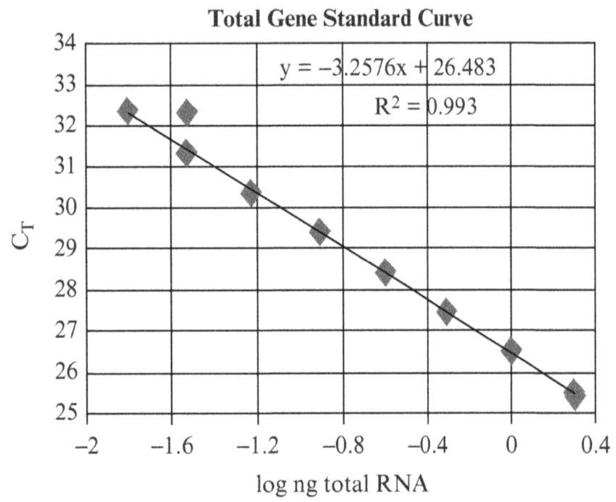

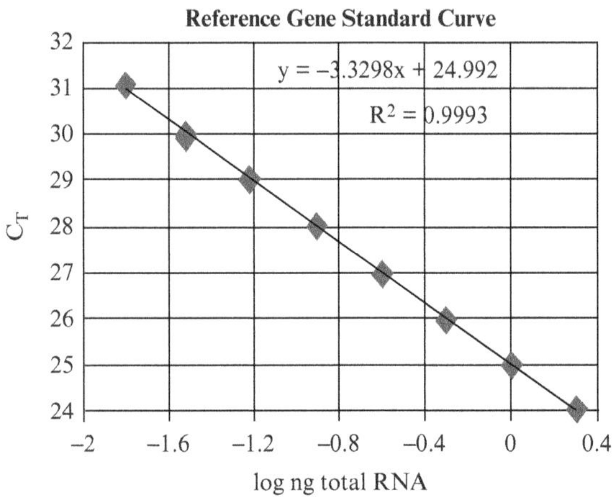

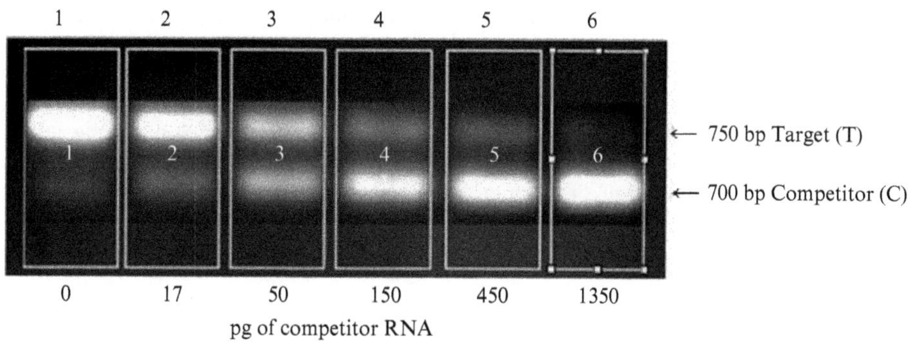

18. To quantitate a target RNA present in a cell, competitive RT-PCR was performed by placing a constant amount (10 ng) of total RNA and increasing amounts of competitor mRNA (0, 17, 50, 150, 450, and 1350 pg) into six PCR tubes containing RT-PCR master mix and the same set of primers. RT-PCR of target and competitor RNA produces 750 and 700 bp DNA fragments, respectively. The synthetic competitor RNA with virtually the same GC content and sequences as those of target RNA was obtained from *in vitro* transcription of cloned plasmid DNA. After RT-PCR, the reaction products were run on to agarose gel and the gel was stained with GelRed dye. Fluorescent bands were visualized under UV light, and the intensity of each band in a gel photograph was quantitated by an ImageJ scan analysis. The obtained results are shown below.

Band	T1	T2	T3	T4	T5	T6	C1	C2	C3	C4	C5	C6
Density	22.52	17.26	5.31	2.62	1.75	0.68	-	2.18	4.08	9.88	14.63	19.09

Based on the scanning density data, calculate the quantity of target mRNA present in 10 ng of total RNA isolated from cells.

19. How would you prepare 10^4 copies of a 272-bp ds DNA molecule for a copy number standard of PCR using the stock DNA solution of 1 mg/mL? The MW of 1-bp dsDNA is 660 g/mole.

20. The genome size (the total number of DNA base pairs in one copy of a haploid genome) of a human (*Homo sapiens*) is about 3.0 billion base pairs. Human beta-globin cDNA is 626 bp. The MW of 1-bp dsDNA = 660 g/mole. You load 10 μg of human genomic DNA (isolated from blood cells) cut with restriction enzyme and separated the DNA fragments on agarose gel to detect the beta-globin gene by Southern blot analysis. In addition, you want to load a positive control sample of a cloned beta-globin gene on the gel.

- **(a)** If you want to load the amount corresponding to a single copy of the beta-globin gene in the gel, how much of the cDNA should you use?
- **(b)** If you are required to, how would you prepare to load a single copy amount of the beta-globin gene using a purified beta-globin cDNA with a concentration of 2 μg/μL?
- **(c)** If you need to, how would you prepare to load 10 copy amounts of the beta-globin gene?

21. The human haploid genome is 3.3×10^9 base pairs. Assuming a random distribution of bases, what size of probe would be expected to have a single target in the haploid genome?

22. You want to digest DNA with *Hin*dIII and *Eco*RI enzymes in a total of 20-μL reaction volume. However, when you look at the enzyme catalog book, you found that *Hin*d III and *Eco*RI enzymes require 50 mM and 100 mM of NaCl in a reaction buffer, respectively. Therefore, you decided to set up the reaction with the *Hin*dIII enzyme first and then with the *Eco*RI enzyme. After incubation of the first *Hin*dIII digestion in a 20-μL reaction, how much of 2 M NaCl solution should you add to initiate the *Eco*RI digestion reaction with 1 μL of the *Eco*RI enzyme?

23. Bacteriophage λ DNA is 48.5 kb in total size. You cut 1 μg of λ DNA with the *Hin*dIII enzyme and use it for a DNA size maker in agarose gel electrophoresis. After staining the gel, you found the band intensity of 1 μL of your PCR product to be about the same as that of 4.4 kb of λ DNA cut with *Hin*dIII. Your total PCR reaction volume is 50 μL. What is the total amount (μg) of the PCR product after amplification?

24. How many mL of dialysis buffer would be needed to remove about 90% of the NaCl from a 10-mL solution initially containing 5 mM NaCl?

25. You must make 1 liter of 0.2 M acetic acid. All you have available is concentrated glacial acetic acid (assay value 98%, specific gravity is 1.05). How many milliliters are needed to make this solution?

26. Recombinant plasmid (vector + insert) library was transformed into yeast for amplification. The yeast genome contains 12 495 682 base pairs. A plate can hold approximately 200 transformants. If the average insert size was 2 kb, how many plates do you have to make in order to have a 99% probability of getting the correct fragment carrying a target gene of interest?

27. The *E. coli* genome size is 4.64×10^6 base pairs. A genomic library is constructed by digesting the DNA with *Hin*dIII (a 6-bp cutter). How many clones should we screen for a 99% probability of finding a particular gene of interest?

28. PCR was performed with 1 microgram of a 2.0 kb duplex DNA template. How many moles of 580-bp target PCR DNA duplex would be produced after 20 cycles, assuming that the amplification efficiency is 80%? The molar mass of 1-bp dsDNA is 660 g/mole.

29. When a target DNA fragment reaches 10^{12} molecules, its amplification efficiency drops dramatically and stops accumulating PCR fragments, reaching the plateau phase. You are to PCR-amplify a 580-bp target DNA from a 14-kb plasmid DNA template.

- **(a)** What amount of template DNA should you use to produce 10^{12} copies of the target DNA molecules after 20 cycles, assuming that the overall amplification efficiency is 80%? The molar mass of 1-bp dsDNA is 660 g/mole.

(b) How would you dilute the template DNA in order to add 2 μL of the diluted DNA to the PCR reaction mix of (a) if the concentration of the DNA template is 0.61 μg/μL?

30. You have a solution of 265 ng/uL of a linearized 3686-bp plasmid DNA. How many copies of the plasmid are there per microliter so that you can use for a Q-PCR standard curve after a serial dilution of this plasmid. The MW of the 1-bp plasmid is 660 g/mole.

31. You have received 3 FAM (5,6-carboxyfluorescein)-labeled peptides from a company. The vial is labeled as Fam-Tat 1 mg, MW = 1979.22 g/mole. You dissolve it in 1 mL of PBS solution to make a stock solution. You need 100 μL of 50 μM Fam-Tat solution for your experiment. How would you prepare this from the stock solution?

AMB 1 Appendix 3

Answers to AMB 1 Lab math practice problem set

Part I. Metric unit conversion

1. Convert the following:

• 1 gram = 10^3 mg	1 gram = 10^{-3} kg
• 1 liter = 10^3 mL = 10^3 cm^3	2 ng = 2×10^{-6} mg
• 50 mg = 0.005 g	100 μL = 0.1 mL
• 1 mg = 10^9 pg = 10^3 μg = 10^6 ng	
• 1 L (liter) = 10^3 mL = 10^6 μL = 10^9 nL = 10^{12} pL = 10^{15} fL	
• 1 mM = 1 μmole/mL = 10^3 pmole/μL	
• 1 L × 1 M = 1 mole	1 mL × 1 M = 1 mmole
• 1 L × 1 mM = 1 mmole	1 mL × 1 mM = 1 μmole
• 1 mL of pure water = 1 g	1 L of pure water = 1 kg
• 1 ppm (part per million) = 10^{-4} %	10% = 10^5 ppm
• 1 mg/L = 10^{-4}% = 1 ppm	100 ppm = 100 μg/mL
• -3.32	

Part II. Dilution

1. $1/5^{10}$
2. $1/2^9$
3. 0.3 mg/mL
4. 0.5 mg/mL
5. 5 μL
6. 2.5×10^6 cells/mL
7. 2.5 mL of blood + 22.5 mL of PBS
8. 4 μL of antibody + 396 μL of PBS
9. 0.6 mL of dye + 149.4 mL of water
10. 10 mL of 10× buffer + 90 mL of water
11. 0.6 mg/mL
12. 1/40
13. After a series of three sequential dilutions of 1:10 (i.e., 1 μL of plasmid + 9 μL of water), 1 μL of solution in the third tube contains 0.001 μg plasmid.
14. Concentration = 450 mg/mL. Total protein = 45 mg.
15. Transfer 10 μL from each tube to the next tube containing 30 μL of TE buffer. (DF = 50/12.5 ≈ 12.5/3.13 ≈ 3.13/0.78 ≈ 0.78/0.20 ≈ 0.20/0.049 ≈ 0.049/0.012 ≈ 4. Thus dilution should be ¼ in each tube.)

Part III. Percent solution

1. 2%
2. 70 mL of absolute ethanol + 30 mL of dH_2O
3. 1.8%
4. 90.5 g
50% = 50 g/100 ml, molar ratio = 201 pentahydrate/111 anhydrous. Thus, 201/111 = X/50 g → X = 90.5.
5. 10 g of SDS is heat-dissolved in 80 mL of dH_2O; the mixture is brought up to 100 mL, autoclaved, and stored at room temperature. If kept cold, SDS will precipitate.
6. 1%
7. 5 mL of acetic acid + 5 mL of methanol + 90 mL of dH_2O

Part IV. Ratio and proportions

1. 21% A, 29% G, 29% C
2. 20%
3. 50%
4. 69 mL
5. 12.27 mL
6. 0.025 g of BPB, 0.025 g of XC, and 4 g of sucrose
7. pH 8.13
8. 368.4 mL of 95% ethanol + 131.6 mL of dH_2O
9. 0.1 mL of protein solution + 0.9 mL of dH_2O

Part V. Mole and molarity

1. 4×10^4 g
2. 9.63 g
3. 0.05 M
4. 0.5 M
5. 0.009 g/liter
6. 43.5 g + ~400 mL of dH_2O: dissolve in a beaker; bring up to 500 mL in a graduated cylinder.
7. 0.37 g + ~400 mL dH_2O: dissolve in a beaker; bring up to 500 mL in a graduated cylinder.
8. 20 mL of 0.5 M Tris-Cl: 10 mL of 1 M Tris-Cl + 10 mL of dH_2O
20 mL of 0.15 M Tris-Cl: 3 mL of 1 M Tris-Cl + 17 mL of dH_2O
20 mL of 1 mM Tris-Cl: 0.02 mL of 1 M Tris-Cl + 19.98 mL of dH_2O

Methods in Biotechnology, First Edition. Seung-Beom Hong, M. Bazlur Rashid and Lory Z Santiago-Vázquez.

Companion Website: www.wiley.com\go\hong\Methodsinbiotechnology

9. 4 g of NaOH is dissolved in 80 mL of dH_2O; bring up to 90 mL of dH_2O and add 10 mL of 10% SDS.
10. 300 mL of 1 M Tris-Cl (pH 9), 5300 mL of 1 M NaCl, 150 mL of 1 M $MgCl_2$, and 2250 mL dH_2O.
11. 30 mL of 1 M Tris-Cl (pH 7.9), 2.4 mL of 0.5 M EDTA, 150 mL of 1 M NaCl, and 417.6 mL of dH_2O.
12. Ethyl alcohol is a covalent compound that does not ionize in water. Each mole of C_2H_6O that dissolves produces only one mole of particles. Therefore, osmolarity = 1.0 M × 1 mole of particles/molar compound = 1.0 osmole C_2H_6O.
13. When 1 mole of $MgCl_2$ dissolves in water, 3 moles of ions are produced. $MgCl_2 \rightarrow Mg^{+2} + 2\ Cl^-$. Therefore, osmolarity = 0.75 M × 3 moles ions/molar compound = 2.25 osmole/L of $MgCl_2$.
14. 50.5 g/L, 0.32 M
For 100% pure $CuSO_4$, 5% = 50 g/L. There is an inverse relationship between the gram weight needed and the purity ratio; for 99% purity, 50 × (1/0.99) = 50.5 g is needed in 1 L. Molarity = 50.5 g/L ÷ 159.6 g/mole = 0.32 mole/L.
15. 79.8 g/L
For 100 %, $g = V \times MW \times M = 1 \times 249.7 \times 0.32 = 79$ g. Because the pentahydrate form is 99% pure, g = 79 × (1/0.99) = 79.8 g is needed per liter.
16. 1.5 M
For 100% pure HCl, 100 mL of HCl = 100 × 1.19 = 119 g; 37% 100 mL of HCl = 119 × 0.37 = 44.03 g HCl is present in 0.8 liter. $g = V \times MW \times M \rightarrow 44.03 = 0.8 \times 36.5 \times M = 29.2\ M \rightarrow M = 44.03/29.2 = 1.5$.

Part VI. Applications

1. Dissolve 1 g of ampicillin (sodium salt) in 8 mL of dH_2O, bring up to 10 mL, filter-sterilize (0.2–0.4 μm pore), and store in a refrigerator.
2. **(a)** 25 g of tryptone + 12.5 g of yeast extract + 25 g of NaCl + add dH_2O, dissolve, and bring up to 2.5 liters.
(b) Dispense 500 mL into each of 5 flasks (1 liter capacity).
(c) Add 37.5 g of Bacto-agar to each flask.
(d) Autoclave at 121 °C for 15 to 20 min.
(e) Cool to 50–55 °C in a water bath.
(f) Add 0.5 mL of 100 mg/mL ampicillin to each flask and swirl gently to mix.
(g) Pour immediately into Petri plates in a sterile hood.
(h) Let the agar solidify and dry for about 5 to 10 min until tiny water droplets disappear.
(i) Cover the lid and put into a plastic bag.
(j) Store at 4 °C.
3. 1 mL of 20 mg/mL of lysozyme
4. 120 min
5. $\Delta T = 40$ h and dt = 0.5 h. Therefore, $n = 4/0.5 = 8$ in $N_f = 2^n \times N_i$. The A_{600} value is equivalent to cell density. Therefore, the final cell number in 10 mL $N_f = 10 \times 0.8 = 8 \rightarrow N_i = 8/2^8 = 8/256 = 0.03$ (initial cell number in 10 mL). As a result, V mL × 1.0 = 0.03 → V = 0.03 mL inoculation volume for 100% live cells. Because only 50% cell are live, $V = 0.03 \times 2 = 0.06$ mL inoculation volume is needed to inoculate.
6. $K = (\log N - \log N_0)\ (t - t_0) = [\log (7 \times 10^8) - \log (5 \times 10^6)]/8 = 2.146/8 = 0.268$/h and $dt = 0.301/K = 0.301/0.268\ h^{-1} = 1.12$ h.
7. The MW of *E. coli* chromosomal DNA = $(4.6 \times 10^6) \times 660 = 3.0 \times 10^9$ g/mole. The mass of a single copy of an *E. coli* chromosome = $(3.0 \times 10^9)/(6 \times 10^{23}) = 5 \times 10^{-15}$ g. The total number of cells = $(2 \times 10^9) \times 2 = 4 \times 10^9$ cells. The total mass of an *E. coli* chromosome from 4×10^9 cells = $(5 \times 10^{-15}) \times (4 \times 10^9) = 20 \times 10^{-6}$ g. As a result, a theoretical total DNA yield at 100% recovery is 20 μg and concentration = 20 μg/0.1 mL = 20 000 ng/100 μl = 200 ng/μL.
8. $A = \varepsilon cl$, $0.01 = 10\ 000 \times c \times 1$, $c = 10^{-6}\ M^{-1}\ cm^{-1}$. Because 1 M = 236 g/L, $c = 10^{-6} \times 236$ g/L $= 236 \times 10^{-3}$ mg/L = 0.236 ppm.
9. $A = \varepsilon cl$, where εl = slope of line. Slope = $\Delta Y/\Delta X = 0.54/20 = 0.027$, $A = 0.445$, so $c = 0.445/0.027 = 16.48$ μM $= 16.48 \times 10^{-6}$M $= (16.48 \times 10^{-6} \times 670)$ g/L = 0.011 g/L = 0.011 mg/mL.
10. **(a)** Because A_{420} of 0.1 mM ONP is 0.45, $A = \varepsilon cl \rightarrow \varepsilon = 0.45/(0.0001\ M \times 1\ cm) = 4500\ M^{-1}\ cm^{-1}$. Because the total reaction volume = 10 mL and 1 unit = 1 μmole/min, 1 μmole of ONP in 10 mL = 10^{-6} mole/10^{-2} L = 10^{-4} M. As a result, $A = \varepsilon cl = 4500\ M^{-1}\ cm^{-1} \times 10^{-4}\ M \times 1\ cm = 0.45$ increases in A_{420}/min correspond to 1 unit.
(b) ΔA_{420}/time between 15 and 75 s = 0.1/min; ΔA_{420}/time between 30 and 60 s = 0.05/0.5 min = 0.1/min; ΔA_{420}/time between 45 and 75 s = 0.05/0.5 min = 0.1/min. This corresponds to 0.1/0.45 = 0.222 units. Because $1/10^4$ diluted enzyme is 13.3×10^{-4} mg/mL and 1 mL of this diluted enzyme is used, specific activity = 0.222 units/(13.3×10^{-4}) mg = 167 units/mg. Alternatively, the following formula can be used: *Specific activity* = ΔA_{sample} × *Assay volume (in L)* × *(10^6 μmole/mole)/ε* × $\Delta_{time\ (min)}$ × *protein (mg) used in the assay* = 0.1 × (10^{-2} L) × (10^6 μmole/mole)/(4500 L/mole) × 1 (min) × (13.3×10^{-4} mg) = 167 μmole/min/mg
11. Human genome = $3 \times 10^9 \times 660 = 1980 \times 10^9$ g/mole; 1 copy of genome = $1980 \times 10^9/6 \times 10^{23} = 330 \times 10^{-14}$ g; 1 μg = 1×10^{-6} g = $10^{-6}/330 \times 10^{-14} = 3 \times 10^5$ copies.
12. 10^7 pg/μL
13. 0.75 mg/mL (A_{260} = 1 is equivalent to 50 μg/mL DNA; 0.15 × 50 × 100 = 750 μg/mL).
14. 1 mg/mL (A_{260} = 1 is equivalent to 40 μg/mL RNA; 0.25 × 40 × 100 = 1000 μg/mL).
15. **(a)** 8.56 ng/μL
(b) 102%
16. Relative fold change in target = target quantity in test/target quantity in calibrator = 50.69/3.53 = 14.36.
Relative fold change in reference = reference quantity in test/reference quantity in calibrator = 30.17/15.62 = 1.93.
Corrected fold change in target = relative fold change in target/relative fold change in reference = 14.36/1.93 = 7.44.
17. **(a)** Standard curve method Based on the regression line formulas, log ngRNA of target = $(C_T - 26.483)/-3.2576$ and log ngRNA of reference = $(C_T - 24.992)/-3.3298$.

Sample	C_T	log ng RNA	ng RNA	Mean	SD	CV (SD/mean)
Target-test	25.68	0.247	1.77		0.107	0.06
Target-test	25.79	0.213	1.63	1.65		
Target-test	25.85	0.194	1.56			
Target-calibr.	27.45	−0.297	0.51		0.083	0.14
Target-calibr.	27.05	−0.174	0.67	0.58		
Target-calibr.	27.32	−0.257	0.55			
Ref.-test	30.75	−1.729	0.019		0.003	0.17
Ref.-test	30.68	−1.708	0.02	0.018		
Ref.-test	31.05	−1.819	0.015			
Ref.-calibr.	25.41	−0.126	0.75		0.046	0.06
Ref.-calibr.	25.25	−0.077	0.84	0.79		
Ref.-calibr.	25.35	−0.108	0.78			

Relative fold change in target = (1.65 ± 0.107)/(0.58 ± 0.083):
Mean = 1.65/0.58 = 2.84.

$$\Delta SD \text{ (standard deviation for the new mean)} = \text{Mean} \times \sqrt{CV1^2 + CV2^2} = 2.84 \times (\sqrt{0.06^2 + 0.14^2}) = 0.67$$

Thus, the fold change = 2.84 ± 0.67.
Relative fold change in reference = (0.018 ± 0.003)/(0.79 ± 0.046):
Mean = 0.018/0.79 = 0.023.

$$\Delta SD = \text{Mean} \times CV = 0.023 \times (\sqrt{0.17^2 + 0.06^2}) = 0.0041$$

Thus, the fold change = 0.023 ± 0.004.
As a result, the corrected fold change in the target gene expression = (2.84 ± 0.67)/(0.023 ± 0.004):
Mean = 2.84/0.023 = 123.

$$\Delta\Delta SD = 123 \times \sqrt{\frac{0.67}{2.84}^2 + \frac{0.004}{0.023}^2} = 36$$

The target gene is 123 ± 36 times more expressed in response to the H_2O_2 treatment.

(b) Pfaffl method

$$\text{Efficiency of target} = 10^{-1/-3.2576} = 10^{0.31} = 2.042$$

$$\text{Efficiency of reference} = 10^{-1/-3.3298} = 10^{0.3} = 1.99$$

	E	$C_{T\,test}$	$C_{T\,cal}$	ΔC_T ($C_{T\,cal} - C_{T\,test}$)	$E^{\Delta CT}$	$E^{\Delta CT\,target}/E^{\Delta CT\,reference}$
Target 1		25.68	27.45	1.77	3.538	141.52
Target 2	2.042	25.79	27.05	1.26	2.459	102.46
Target 3		25.85	27.32	1.47	2.856	142.8
Ref. 1		30.75	25.41	−5.34	0.025	
Ref. 2	1.995	30.68	25.25	−5.43	0.024	
Ref. 3		31.05	25.35	−5.7	0.020	

The mean ± SD of 141.52, 102.46, and 142.8 is 129 ± 23. The target gene is 129 ± 23 times more expressed in response to H_2O_2 treatment.

(c) $\Delta\Delta C_T$ method

	$C_{T\,test}$	$\Delta C_{T\,test}$	$C_{T\,cal}$	$\Delta C_{T\,cal}$	$\Delta\Delta C_T$
Target	25.77 ± 0.086	−5.06 ± 0.22	27.27 ± 0.20	1.93 ± 0.22	−6.99 ± 0.31
Reference	30.83 ± 0.20		25.34 ± 0.081		

$$\Delta C_{T\,test} = C_{T\,target} - C_{T\,reference};\ \Delta C_{T\,cal} = C_{T\,target} - C_{T\,reference}$$

$$\Delta\Delta C_T = \Delta C_{T\,test} - \Delta C_{T\,cal}$$

At mean of 6.99, $2^{-\Delta\Delta C_T} = -2^{6.99} = 127$. At SD of − 0.31, $2^{-\Delta\Delta C_T} = 2^{7.3} = 158$. At SD of + 0.31, $2^{-\Delta\Delta C_T} = 2^{6.68} = 103$.
Thus, the target gene is expressed 127-fold more with a range from 158- to 103-fold difference in response to treatment.

18.

pg RNA	Competitor (C)	Target (T)	Corrected C*	C/T ratio	log pgRNA	log C/T
0	0	22.52	0			
17	2.18	17.26	2.34	0.14	1.23	−0.854
50	4.08	5.31	4.37	0.82	1.67	−0.086
150	9.88	2.62	10.59	4.04	2.18	0.606
450	14.63	1.75	15.67	8.95	2.65	0.952
1350	19.09	0.68	20.45	30.07	3.13	1.478

*Corrected $C = C$ density × (750/700) because a smaller fragment will stain with less intensity than a larger fragment.

*Log C/T (y axis) and log pgRNA (x axis) data are used to prepare a scatter plot using Microsoft Excel as shown below.

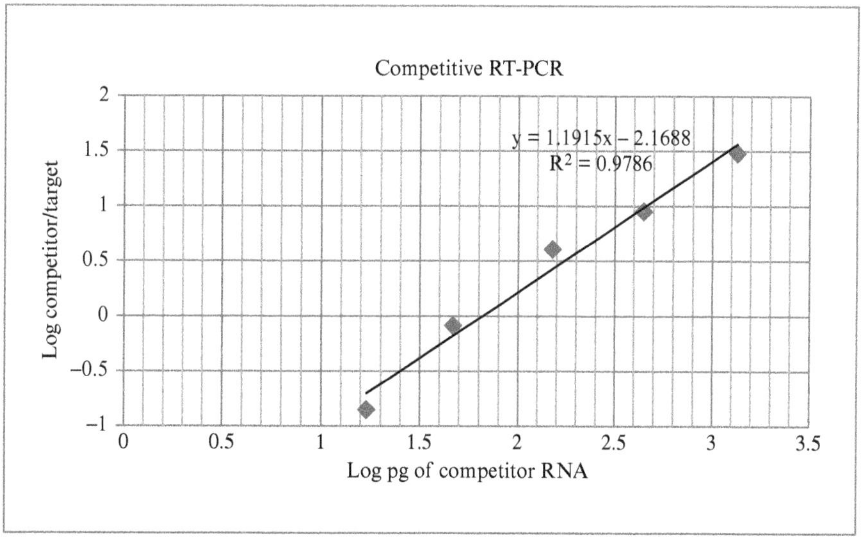

When the concentrations of the target and competitor are equal, then the C/T ratio is 1 and thus y (log 1) = 0 = 1.1915x − 2.1688 (linear regression equation) → x = 2.1688/1.1915 = 1.82 → $\log_{10}$ pgRNA = 1.82. As a result, antilog$_{10}$1.82 = 66 pg of target mRNA is present in 10 ng of total RNA.

19. MW of a 272-bp DNA = 272 × 660 = 1.8×10^5 g/mole. Therefore, the mass of a single molecule = 1.8×10^5 g/(6×10^{23}) = 3×10^{-19} g. The mass of 10^4 copies = (3×10^{-19}) × 10^4 = 3×10^{-15} g. Stock DNA concentration = 1 mg/mL = 1 μg/μL = 10^{-6} g/μL. DF = 10^{-6}/(3×10^{-15}) = 1/3 × 10^9. This means a 1/3 dilution (1 μL of DNA + 2 μL of H_2O) followed by 9 times a consecutive serial 1/10 dilution.

20. **(a)** Since the beta-globin gene is present in a diploid status in genome of human blood cells calculation should be based on diploid size 2 × (3×10^9) = 6×10^9 bp. Now 10 μg human DNA = 10×10^{-6} g/(6×10^9 × 660) g/mole = 2.5×10^{-18} mole; this mole corresponds to the amount of a single-copy gene in 10 μg human DNA. A 626-bp globin DNA = 626 × 660 = 4.1×10^5 g/mole. Thus, 4.1×10^5 g/mole × 2.5×10^{-18} mole = 10.25 × 10^{-13} g = ~1.0 pg should be used for one copy number standard.

Alternatively, (6×10^9)bp/(10×10^{-6})g = 626 bp/mass of beta-globin → mass of beta-globin = (626 × 10×10^{-6})/(6×10^9) = 1043 × 10^{-15}g = 1.043 × 10^{-12}g = ~1.0 pg should be used for one copy number standard.

(b) Make a ½ dilution (1 μL of cDNA + 1 μL of TE buffer) followed by three sequential series of 1/100 dilution (i.e., 1 μL of cDNA + 99 μL of TE buffer) and use 1.0 μL of the final diluted sample (2 pg/μL).

(c) Make 1/20 dilution (1 μL cDNA + 19 μL TE buffer) followed by two consecutive 1/100 dilutions and use 1.0 μL of the final diluted sample (20 pg/μL).

21. $33 \times 10^9/(4)^x = 1 \rightarrow X \log_{10}{}^4 = \log_{10}(3.3 \times 10^9) \rightarrow 0.6X = 9.519 \rightarrow X = 16$ bp.

22. The unknown volume X of 2 M NaCl should be added to a final concentration of 100 mM when the M1V1 = M2V2 formula is used. Because the final concentration of volume X should be 100 mM, the formula is written as (50 mM × 20 μL) + (100 mM × X μL) = 2000 mM × X. This will give 0.53 μL. However, even this is not an accurate mathematical answer because we did not take the 1 μL of enzyme addition into consideration. If we consider this, then (50 mM × 20 μL) + 100 mM × (X + 1 μL) = 2000 mM × X. This will give you the X value of 0.58 μL. However, in reality, the 0.08-μL volume is far less than a pipetting error. Accordingly, 0.5-μL pipetting may actually lead to a more accurate result. A slight mathematical mistake offsets a slight pipetting error to be made perfect.

23. 4.54 μg: (4.4/48.5) × 1 = 0.09 μg of DNA is present in a 4.4-kb DNA band. Because band intensity is proportional to the amount of DNA, 0.09 μg of DNA is present in 1 μL of PCR product. As a result, the total amount = 0.09 μg/μL × 50 μL = 4.54 μg.

24. To remove 90% NaCl, 5 mM × 0.1 = 0.5 mM NaCl should remain after dialysis. Thus, 5 mM × 10 mL = 0.5 mM × (V + 10) mL → V = 90 mL is required.

25. The molar mass of acetic acid CH_3COOH = 60 g/mole. Therefore, (0.2 mole/L)(1 L)(60 g/mole) = 12 g of acetic acid is needed and 12 g/1.05 g/mL = 11.43 mL of pure acetic acid must be taken. Because the acid is 98% pure, one must take (11.43) × (100/98) = 11.663 mL. Dilute this to 1000 mL (use distilled water) to make the solution.

26. The number of clones to be screened to obtain the recombinant DNA at a certain probability is given by the equation: N = In (1 − P)/In (1 − f), where In = $\log_e$, N = number of independent clones needed to isolate a specific DNA segment with probability P, and f = the size (bp) of the average of cloned insert divided by the size (bp) of the target genome.

N = In (1 − 0.99)/In [1 − (2000/12 495 682)] = In (0.01)/In (1 − 0.0001601) = In (0.01)/In (0.9998) = −4.60517/−0.0002 = 23025.85. Thus, the number of plates = 23026/200 = 115.

27. A 6-bp cutter will cut at a frequency of $(1/4)^6$ = 1/4096. That is, on average, 4096 bp *Hin*dIII fragments will be present along the genomic DNA of a random sequence.

N = In (1 − 0.99)/In (1 − 4096/4.64 × 10^6) = In (0.01)/In (1 − 0.000883) = In (0.01)/In (0.999117) = −4.60517/−0.00088 = 5233 clones.

28. The MW of 2 kb DNA = 2000 × 660 = 1.32×10^6 g/mole; 1 μg of 2-kb DNA = 1×10^{-6} g/1.32×10^6 g/mole = 7.6×10^{-13} mole; $N = N_0 (1 + E)^n$, where N is the final amount of PCR DNA after n cycles, N_0 is the initial amount of template DNA, and E is the amplification efficiency. Thus, N = (7.6×10^{-13}) × $(1.8)^{20}$ = 9.7×10^{-12} mole.

29. **(a)** The MW of a 14-kb DNA template = 14000 × 660 = 9.24 × 10^6 g/mole; 1 mole = 6×10^{23} molecules. Therefore, 1 molecule = (9.24×10^6)/(6×10^{23}) g = 1.54×10^{-17} g; $N = N_0 (1 + E)^n$, where N is 10^{12}, E is 0.8, and n is 20. Thus, N_0 = $10^{12}/1.8^{20}$ = 10^{12}/125 892.54 = 7 943 282 molecules = 7 943 282 × (1.54×10^{-17}) g = 12 232 654 × 10^{-17} g = 122 × 10^{-12} g = 122 pg.

(b) After dilution of template DNA, concentration must be 122 pg/2 μL = 61 pg/μL. Before dilution, stock DNA contains 0.61 μg/μL = 0.61 × 10^6 = 61 × 10^4 pg/μL. Thus, stock DNA is diluted 1/10^4 (e.g., 4 consecutive serial dilutions of 1 μL of DNA + 9 μL of H_2O).

30. The MW of plasmid DNA = 3686 × 660 = 2 432 760 g/mole.

265 ng = (265 × 10^{-9})/(2.43276 × 10^6) = 1.09 × 10^{-17} mole = (1.09 × 10^{-17}) × (6 x 10^{23}) = 6.54 × 10^6 copies per μL.

31. 1 mg/mL of stock solution = 1 g/L = 1/1979.22 = 5×10^{-4} M = 500 μM.

500 μM × V = 50 μM × 100 μL → V = 10 μL stock is needed. Thus, 10 μL of stock is mixed with 90 μL of 1× PBS.

AMB 1 Appendix 4

Band Quantification by ImageJ Program

1. Double click on the ImageJ icon and the ImageJ window will appear on the desktop; do not enlarge this window. Note that this window has a Menu Bar, a Tool Bar, and a Status Bar.

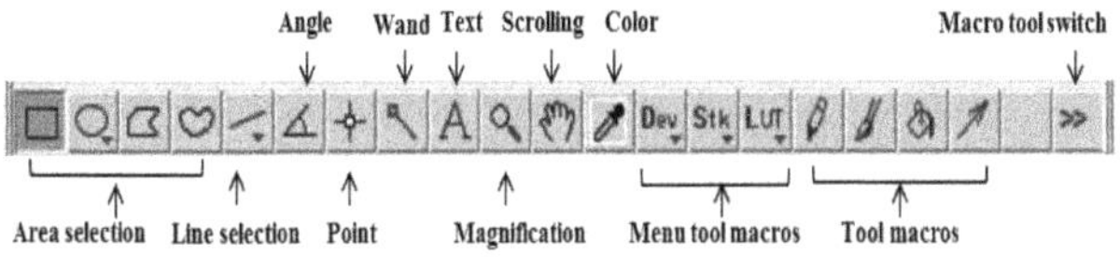

2. From the Menu Bar, click "Analyze" → click submenu "Gels ▸" → click "Gel Analyzer Options" → check (√) "Uncalibrated OD" and "label with percentages." Click "OK."

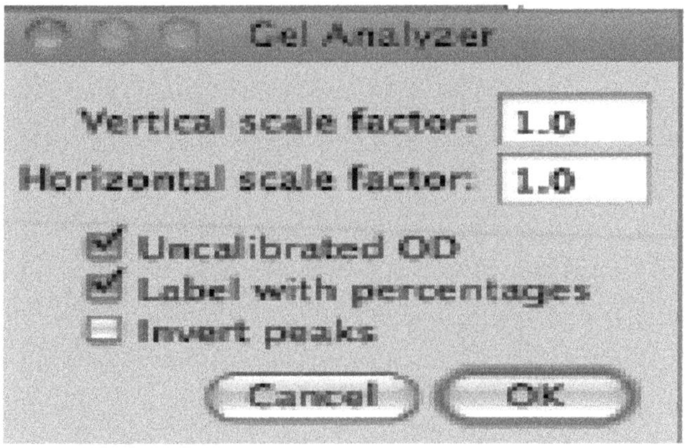

3. From the Menu Bar, click "File" → "Open" → locate and select the stored image file (TIFF, GIF, and JPEG) of interest from the computer. Click "Open."

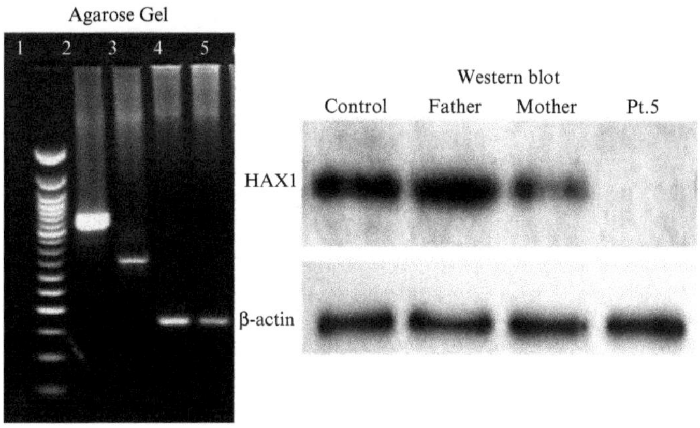

4. The gel analysis routinely requires the image to be a gray-scale image. To do this, from the Menu Bar, click "Image" → "Type ▸" → select "8-bit." This must be selected every time you begin.

Optional step: if the image, especially for the Western blot, is too dark or too light to analyze properly, go to "Image" → "Adjust" → "Brightness/Contrast" and adjust the slider bars until the bands on the image are visible. The pixel values will not be changed, just the view. Below is an example of image adjusted pictures.

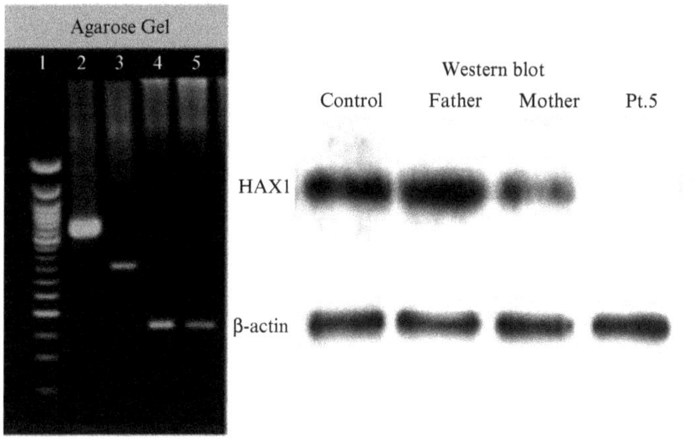

5. From the Menu Bar, click the "Analyze" → click "Calibrate" → select "Uncalibrated OD" option from the Function dropdown box (uncheck if "Inverted Peaks" is checked; if this

Methods in Biotechnology, First Edition. Seung-Beom Hong, M. Bazlur Rashid and Lory Z Santiago-Vázquez.

Companion Website: www.wiley.com\go\hong\Methodsinbiotechnology

is checked, upside-down peaks will be displayed) → click OK. A logarithmic plot should appear; move it to the side. This must be selected every time you begin.

6. From the Tool Bar, select "Rectangular Selection Tool" ( ). If necessary, you may select "Magnifying Glass Tool" ( ) to expand the image file.

7. Use the Rectangular Selection Tool to draw the first lane. Your rectangle should be tall and narrow to enclose a single lane because ImageJ assumes that lanes are vertical and individual bands are horizontal. Note that location (X, Y), width (w), and height (h) values are displayed in the Status Bar. Remember the w and h values for the next lane selection for the same dimension.

8. From the Menu Bar, click "Analyze" → select the submenu "Gels ▸" → click "Select First Lane." The selected first lane will be highlighted with number 1 in the middle.

9. Move the rectangular selection tool and outline the next lane to the same w and h as the first one. Note that the rectangular dimension must be the same as the previous ones in all the lanes. Instead of drawing a new rectangle, you can use your mouse cursor to click the center and hold within the highlighted box and drag it over to the next lane.

10. From the Menu Bar, click "Analyze" → select the submenu "Gels ▸" → click "Select Next Lane." The selected 2nd lane will be highlighted with number 2 in the middle.

11. Repeat the previous steps 8 and 9 until all the lanes are highlighted. Note that if you accidently click the mouse cursor outside the highlighted box, the click–hold–drag function no longer works. Also you cannot do the next step 12 of plotting lanes. If this happens, you must draw a new rectangular box of the same dimension over the next lane.

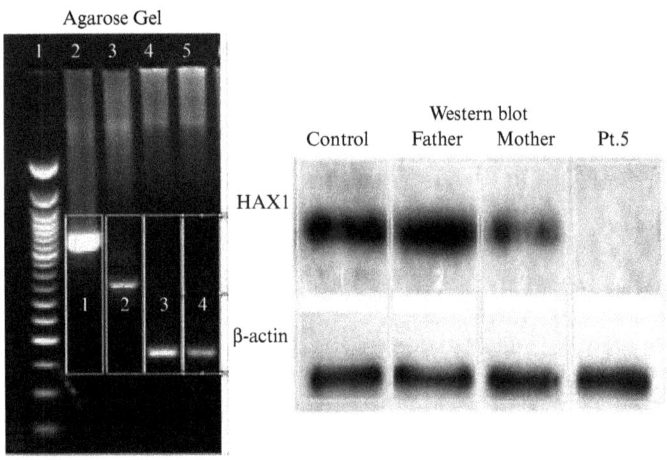

12. From the Menu Bar, click "Analyze" → select submenu "Gels ▸" → click "Plot Lanes" to generate the profile plots of lanes. The Plots window showing the peaks of all lanes will appear on the desktop; the first and last lane peaks will show up in the top and bottom panel, respectively. Higher and wider peaks represent darker and wider bands on the original gel, respectively. If you want to plot again, then click "Re-plot Lanes". Images of actual gels or blots have some background signal noise, so the peaks usually do not reach down to the baseline of the profile plot. Below is the example for agarose gel and Western blot.

Depending on the position of the rectangle, peaks are sometimes shown in the inverted position. In this case, click "Analyze" → click submenu "Gels ▸" → click "Gel Analyzer Options" → check (√) "Invert Peaks."

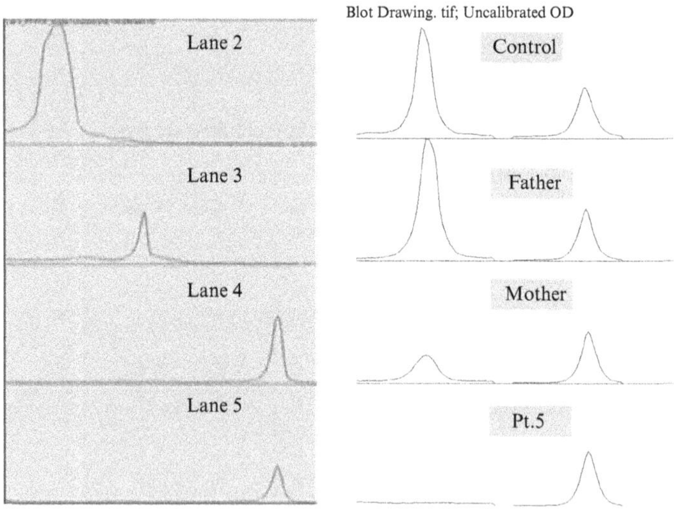

13. From the Tool Bar, select "Line Selection Tool" ( ) and draw a line across the base of the peaks of interest to enclose the peaks. *It is important to completely enclose each peak. An unclosed smaller peak may have a larger peak area than a closed larger peak.*

To access to all the lanes, scroll down the image vertically. Make sure that each peak must be in a completely closed area. You may use the Magnifying Glass Tool ( ) to see more expanded peaks. If the peak is completely closed off, you may not need this step. This rarely happens in agarose gels but often happens in blots.

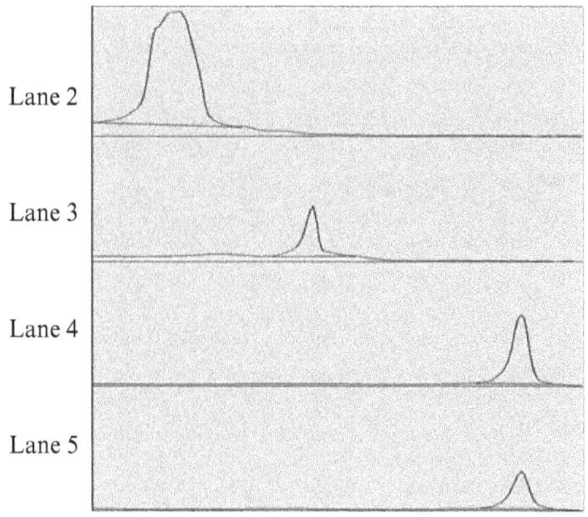

14. From the Tool Bar, select "Wand Tool" ( ) and click the inside of each peak in the first lane to measure the size. The peak will be highlighted and the Results window showing the

selected numbers and areas will appear on the desktop. Note that if you click the same peak twice, its result will show twice. To avoid confusing data in the Results window, select each peak in the order of the top (first lane) panel to the bottom (last lane) panel. If you click a wrong place or make a mistake, you may clear the results by clicking the submenu "Clear Results" from the "Analyze" Menu Bar or "Clear" from the Edit menu of the Results window.

15. Once the area of all the peaks are displayed on the Results window, click "Analyze" from Menu → select submenu "Gels ▸" → click "Label Peaks" to label each measured peak with its size as a percent of the total size of all the measured peaks. Now percent values appear on the Results window, in which you can save the data from the File menu (File → Save as → Excel data).

Below is the image scan example of agarose gel.

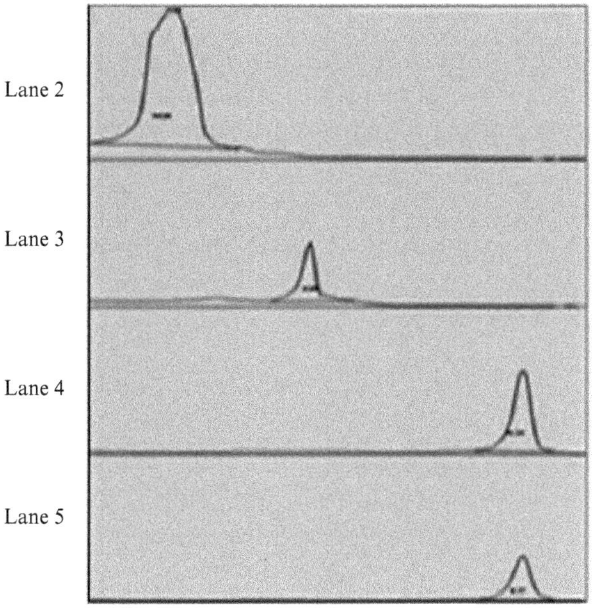

Peak	Area	Percent
1	19722.36	64.653
2	3032.205	9.94
3	4953.861	16.239
4	2796.74	9.168
Total	30505.17	100

Below is the image scan example of Western blot.

Remember that the "Area" and "Percent" values are relative values, based only on the highlighted peaks and are displayed in the order of your selection with the Wand tool.

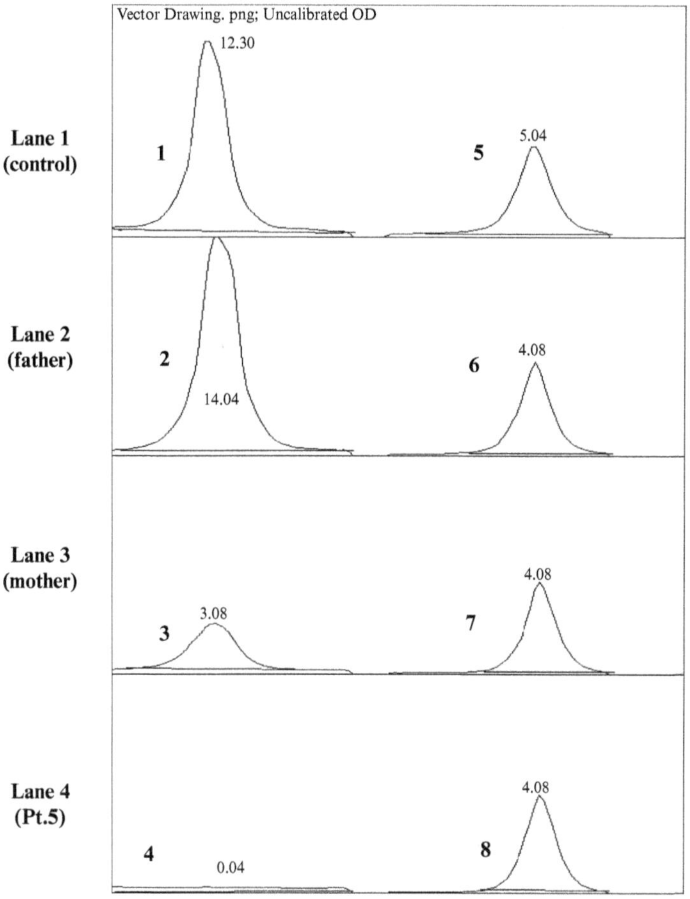

Peak	Area	Precent
1	1.33	11.705
2	1.649	14.517
3	0.371	3.266
4	0.112	0.986
5	0.606	5.336
6	0.54	4.756
7	0.525	4.624
8	0.546	4.81
Total	5.679	50

16. In the example of agarose gel, there is no standard DNA. Therefore, just calculate the relative fold changes from percent values. For example, lane 1 DNA is 6.5 times (64.653 ÷ 9.94) as many as lane 2 DNA; lane 3 DNA is 1.8 times (16.24 ÷ 9.17) as many as lane 4 DNA.

If you include the molecular standard marker in your analysis, you will have the following gel picture obtained from steps 1 to 9.

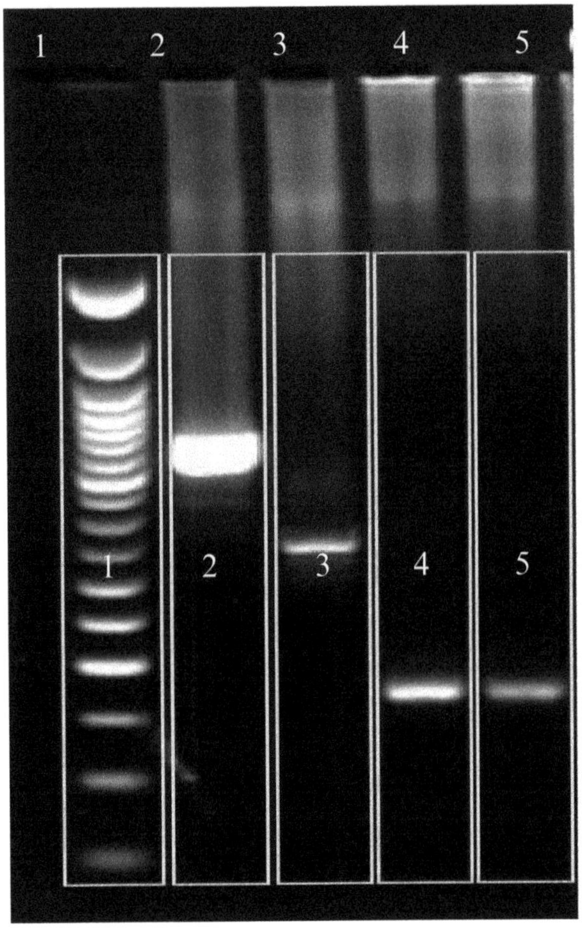

Following steps 10 to 13, the results would be like those shown below.

Data analysis for gel without internal reference control

If you calculate the relative fold changes from the percent values, the 13th peak DNA is 6.6 times (29.49 ÷ 4.468) as many as the 14th peak DNA; the 15th peak DNA is 1.8 times (6.952 ÷ 3.864) as many as the 16th peak DNA. Note that peaks 13, 14, 15, and 16 DNAs are the same as lanes 2, 3, 4, and 5 DNAs in the previous analysis, and that the fold changes are the same as those of the previous analysis (64.653 ÷ 9.94; 16.238 ÷ 9.168).

Because the DNA amount of each band of MW marker DNA is known, you can estimate the amounts of experimental products. For example, if the 9th peak contains 107 ng DNA, the 13th peak has $107 \times (29.49/4.152) = 760$ ng; the 14th peak has $107 \times (4.468/4.152) = 115$ ng; the 15th peak has $107 \times (6.952/4.152) = 179$ ng; and the 16th peak has $107 \times (3.864/4.152) = 99.6$ ng.

You may use any peak in the first lane of standard DNA. If you use the first peak as a standard, you can make a new column for the relative density and amount in the Excel spreadsheet.

- The relative density is calculated by dividing the percent value for each sample by the percent value for the standard (the 9th peak in this case, 4.152).
- The amount is calculated by multiplying the standard (107 ng of the 9th peak in this case) by relative density.

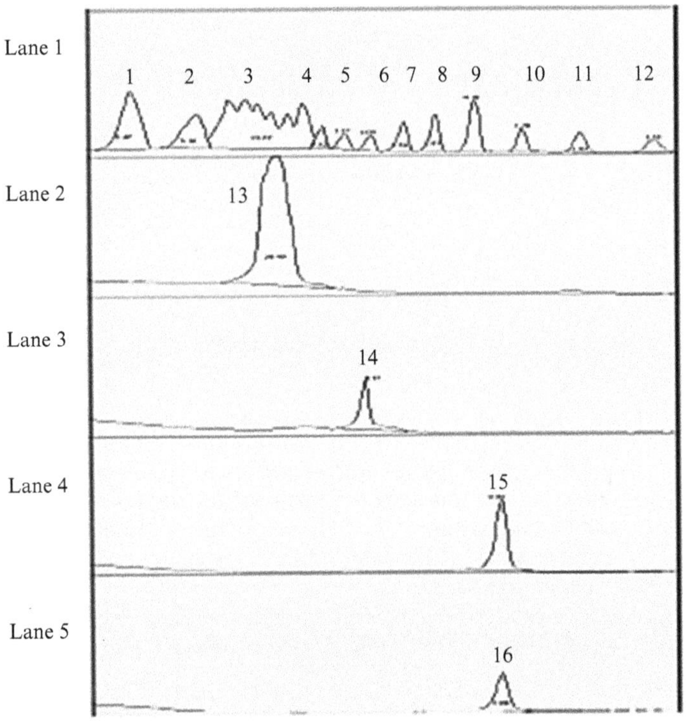

Peak	Area	Percent
1	3220.305	8.465
2	2017.062	5.302
3	9003.631	23.668
4	576.971	1.517
5	444.799	1.169
6	404.385	1.063
7	832.213	2.188
8	977.213	2.569
9	1579.335	4.152
10	647.213	1.701
11	710.749	1.868
12	594.87	1.564
13	11218.32	29.49
14	1699.719	4.468
15	2644.477	6.952
16	1470.062	3.864
Total	38041.32	100

If you want to compare the density of samples on multiple gels or blots, you have to use the same standard sample on every gel to provide a common reference when you calculate relative density values.

	Area	Percent	Relative density	ng
1	3220.305	8.465	2.04	
2	2017.062	5.302	1.28	
3	9003.631	23.668	5.70	
4	576.971	1.157	0.37	
5	444.799	1.169	0.28	
6	404.385	1.063	0.26	
7	832.213	2.188	0.53	
8	977.213	2.569	0.62	
9	1579.334	4.152	1.00	107
10	647.213	1.701	0.41	43.87
11	710.749	1.868	0.45	48.15
12	594.87	1.564	0.38	40.66
13	11218.32	29.49	7.10	759.7
14	1699.719	4.468	1.08	115.56
15	2644.477	6.952	1.67	178.69
16	1470.062	3.864	0.93	99.51
Total	38041.32	100		

Data analysis for gel or blot with internal reference control

For doing a relative comparison, you have to divide the peak values of your target protein (or DNA) by the peak values of your control (calibrator) reference. In the above example of a Western blot plot, peaks 1, 2, 3, and 4 are the target protein and peaks 5, 6, 7, and 8 are the reference protein in lanes 1, 2, 3, and 4, respectively.

First, calculate relative density values for each of the target bands by dividing each percent or area value of the target band by the percent or area value of the control (calibrator) target band (peak 1, area 1.330, percent 11.705).

1. The relative fold change in the target = target quantity in the test/target quantity in the calibrator.

Second, calculate relative density values for each of the reference bands by dividing each percent or area value of reference bands by the percent or area value of the control (calibrator) reference (peak 5, area 0.606, percent 5.336).

2. The relative fold change in the reference = reference quantity in the test/reference quantity in the calibrator.

Lastly, calculate adjusted density values by dividing the relative density of each target protein by the relative density of each loading reference control.

3. The corrected fold change in the target = relative fold change in the target/relative fold change in the reference.

The adjusted density indicates a 39% increase in the father's target protein, a 68% decrease in the mother's target protein, and a 91% decrease in the Part 5 target protein, as compared to the normal control.

Sample	Peak	Area	Percent	Relative density (relative fold change)	Adjusted density (corrected fold change)
Control Target	1	1.330	11.705	1	1
Father Target	2	1.649	14.517	1.24	1.39
Mother Target	3	0.371	3.266	0.28	0.32
Part 5 Target	4	0.112	0.986	0.08	0.09
Control Reference	5	0.606	5.336	1	
Father Reference	6	0.540	4.756	0.89	
Mother Reference	7	0.525	4.624	0.87	
Part 5 Reference	8	0.546	4.810	0.90	

Useful downloads

ImageJ is a digital image processing program that was developed at the National Institutes of Health, USA. It is a public freeware and you may download this program. The ImageJ web site has instructions for use of the program and links to useful resources.

- ImageJ Program Download:
  http://rsb.info.nih.gov/ij/download.html
  Mac OS X:
  Download ImageJ 1.46 (5.7 MB) as a double-clickable Mac OS X application.
  Windows:
  Download ImageJ 1.46 bundled with 64-bit Java (24 MB; requires 64-bit Windows) or with 32-bit Java (28 MB). Before downloading, check your computer's system type to determine whether it has a 64-bit or 32-bit Operating System.
- ImageJ User Guide Book:
  http://rsb.info.nih.gov/ij/docs/guide/user-guide-USbooklet.pdf
- Tutoring videos are available from the following web sites:
  https://www.youtube.com/watch?v=JlR5v-DsTds (quantify protein bands)
  https://www.youtube.com/watch?v=MINYgJeSKGY (Western Blot analysis Part 1)
  https://www.youtube.com/watch?v=Nu-b2DOhosI (Western Blot analysis Part 2)

AMB 1 Appendix 5

Distributor addresses

Microbiological strains

• *E. coli* **MM294** (F^- *glnX44(AS), λ⁻, rfbC1, spoT1, thiE1, hsdR17, creC510*): *E. coli* Genetic Stock Center, **CGSC No. 6315** (http://cgsc.biology.yale.edu/Strain.php?ID=5439)

• *S. cerevisiae* **YNN 281** (*MATa, trp1Δ, his3Δ200, ura3-52, lys2-801(amber mutation), ade2*

• *1(ochre mutation), gal, mal, CUP(r))*: American Type Culture Collection ATCC® Number 204661 (http://www.atcc.org/Products/All/204661.aspx)

• *S. cerevisiae* **BY4741** (*MATa, his3Δ1, leu2Δ0, met15Δ0, ura3Δ0*): American Type Culture Collection ATCC® Number 201388 (http://www.atcc.org/Products/All/201388.aspx)

• *E. coli* **TOP10** (*Δ(ara-leu) 7697 araD139 fhuA ΔlacX74 galK16 galE15 e14- ϕ80dlacZΔM15 recA1 relA1 endA1 nupG rpsL (StrR) rph spoT1 Δ(mrr-hsdRMS-mcrBC)*): New England Biolab NEB, Cat. No. C30191 (NEB 10-beta competent *E. coli* high efficiency) (https://www.neb.com/products/c3019-neb-10-beta-competent-e-coli-high-efficiency)

Plasmid DNAs

• **pRY121:** ATCC® Number 37658 (http://www.atcc.org/Products/All/37658.aspx)

• **pUC19:** NEB, Cat. No. N3041S (https://www.neb.com/products/n3041-puc19-vector)

• **pGLO:** Bio-Rad, Cat. No. 16604505 (http://www.bio-rad.com/en-us/sku/1660405-pglo-plasmid?parentCategoryGUID=2)

• **pGBKT7:** Clontech, Cat. No. 630443 (http://www.clontech.com/US/Products/Protein_Interactions_and_Profiling/Yeast_Two- Hybrid/Vectors)

Methods in Biotechnology, First Edition. Seung-Beom Hong, M. Bazlur Rashid and Lory Z Santiago-Vázquez.

Companion Website: www.wiley.com\go\hong\Methodsinbiotechnology

AMB 2 Appendix 1

Instructor notes of AMB 2 experiment 38

Each student group workbench

- Paper towels and 70% ethanol spray bottle
- One 200-mL graduated cylinder
- One 500-mL flask (or bottle), two 250-mL flasks, one 100-mL flask
- 16 Petri plates in a plastic bag

Common materials and equipment

- Bacto-agar
- TB powder (if absent, Tryptone, yeast extract, KH_2PO_4 monobasic, K_2HPO_4 dibasic)
- Glycerol
- LB powder (if absent, Tryptone, yeast extract, NaCl)
- Two 500-mL flasks (or bottles) for TB and LB broth
- Ampicillin (100 mg/mL; 1.0 g in 10 mL of dH_2O; filter sterilize)
- Chloramphenicol (30 mg/mL in 100% ethanol; no filter sterilization, store at –20 °C)
- X-β-Gal (40 mg/mL in dimethylformamide; dissolve 0.04 g in 1 mL of DMF; no filter sterilization; wrapped in aluminum foil to prevent damage by light and should be stored at –20 °C
- IPTG (isopropyl-β-D thiogalactopyranoside, 0.1 M: dissolve 0.0238 g in 1 mL of water, filter-sterilized, and store at –20 °C)
- L-arabinose (1.5 M): dissolve 7.5 g arabinose in dH_2O, bring up to 50 mL, filter-sterilize, and store at 4 °C
- Water carboy filled with dH_2O
- 60 °C water bath
- Autoclave (reserved notice in login sheet)
- Heat-resistant gloves
- Laminar flow hood

• Reminder

In the first class, the instructor should talk about aseptic techniques and autoclave usage according to the operating manual and safety policy.

Instructor notes of AMB 2 experiment 39

Each student group workbench

- P10, P20, and P200 micropipettors and pipette tips
- Four LB/amp/X-β-Gal plates (prepared by student in Experiment 38)
- Four LB/amp/X-β-Gal/IPTG plates (prepared by student in Experiment 38)
- Thin-walled PCR tubes, ice box
- Phosphorylated primer mix (25 µM)
- 10 mM dNTP
- Nuclease-free H_2O
- 5× Phusion enzyme buffer

Common materials and equipment

- Thermal cycler (program: Part A, step 2): sign up login sheet
- Sterile 1.5-mL microcentrifuge tubes
- Competent *E. coli* DH5α or TOPO10 cell (take out just before use from the –80 °C freezer)
- pUC19WHITE and pUC 19 plasmid
- 10× lithium borate buffer (100 mM, pH 8.2):
  Dissolve 8.4 g of lithium hydroxide monohydrate (or 4.8 g of lithium hydroxide anhydrous) in 900 mL of H_2O, adjust the pH to 8.2 by adding boric acid (~36 g) and bring up to 2 liter. Store at room temperature.
- 25 µM primer mix: 10 µL of each of 100 µM stock + 20 µL of H_2O

*5′-Phosphorylated primer 1 stock (100 µM): GTCGACTCTAGAGGATCCCCGGGT

*5′-Phosphorylated primer 2 stock (100 µM): CTGCAGGCATGCTAGCTTGGCGTA

- Agarose, ethidium bromide (10 mg/mL), DNA loading dye (10× or 6×)
- DNA size marker (1 kb DNA ladder)
- One horizontal agarose gel electrophoresis apparatus with a 10-well comb
- LB or SOC medium, water carboy filled with deionized water
- 42 °C water bath, 37 °C shake incubator
- Glass spreader; Bunsen burner filled with 100% denatured ethanol
- Phusion® high-fidelity DNA polymerase (NEB, Cat. No. M0530S)
- Quick Ligation™ kit (NEB, Cat. No. M2200S)

Methods in Biotechnology, First Edition. Seung-Beom Hong, M. Bazlur Rashid and Lory Z Santiago-Vázquez.

Companion Website: www.wiley.com\go\hong\Methodsinbiotechnology

• Reminder

The approximate time required for each step is as follows: 35 min PCR cycle, 20 min agarose gel electrophoresis, 5 min ligation reaction, and 1 h 20 min of transformation.

Instructor notes of AMB 2 experiment 40

Each student group workbench

- Two thin-walled PCR tubes on ice box
- Nuclease-free sterile H_2O
- 10× NEB buffer 4
- pGLO plasmid (1 ng/μL)
- pCAL-kc plasmid (1 μg/10 μL)
- Four LB + Amp (100 μg/mL) + Cm (30 μg/mL) + IPTG (0.1 mM) agar plates
- Two LB + Amp (100 μg/mL) + Cm (30 μg/mL) + arabinose (2×10^{-3} M) agar plates
- Glass spreader, Bunsen burner

Common materials and equipment

- Thermo cycler I (program: Part I, A): sign up login sheet
- Thermo cycler II (program: Part II, A): sign up login sheet
- NanoDrop 2000 spectrophotometer
- PCR master mix (Part I, A) for 25 μM primer mix (FNheI + RKpnI)
- PCR master mix (Part II, A) for 25 μM primer mix (SLIC1 + SLIC2)
- 25 μM primer mix (FNheI + RKpnI): 12.5 μL of 100 μM of each stock + 25 μL of sterile H_2O
  FNheI primer: GGAGATATACATATGGCTAGCAAAGGAGAAG
  RKpnI primer: AGTCGAATGGTACCTTTGTAGAGCTCATCCATGCCATG
- 25 μM primer mix (SLIC1 + SLIC2): 12.5 μL of 100 μM of each stock + 25 μL sterile H_2O
  SLIC 1 primer: AAGAAGGAGATATACCATGGCTAGCAAAGGAGAAGAAC
  SLIC 2 primer: CTTTTTCCATCGTCGCTTGTATTTGTAGAGCTCATCCAT
- Phusion® high-fidelity DNA polymerase (NEB, Cat. No. M0530S)
- *Nhe*I-HF™ (NEB Cat. No. R3131S), *Kpn*I-HF™ (NEB, Cat. No. R3142S), *Bam*HI-HF™ (NEB, Cat. No. R3136S), *Dpn*I (NEB, Cat. No. R0176S)
- Standard DNA reference (1 kb DNA ladder)
- Quick T4 DNA ligase and 2× Quick ligation buffer (NEB, Cat. No. M2200S)
- T4 DNA polymerase (NEB, Cat. No. M0203S; 3 U/μL), 10× BSA
- EconoSpin™ mini-spin column (Epoch Life Science, Cat. No. 1920-050): 14 columns/4 groups. Only two groups set up pCALkc/NheI + KpnI: need 4 columns; two other groups need 3 spin columns.
- PX DNA binding buffer (5.5 M guanidine-HCl, 20 mM Tris-HCl, pH 6.6)
- PB buffer (5 M guanidine-HCl, 20 mM Tris-HCl, pH 6.6, 30% isopropanol)
- PE buffer (10 mM Tris-HCl pH 7.5, 80% ethanol):
  For 500 mL, 5 mL of 1 M Tris-Cl (pH 7.5) + 95 mL of MilliQ H_2O + 400 mL of 100% ethanol
- QG buffer (6 M guanidine thiocynate, 50 mM Tris-HCl, pH 7.5, 20 mM EDTA, pH 8.0, 0.006% cresol red)
- $T_{10}E_{0.1}$ buffer (10 mM Tris-HCl pH 8.5, 0.1 mM EDTA): 900 μL of EB buffer + 100 μL of TE buffer
- 10× DNA loading dye (no XC but only BPB dye)
- Agarose, ethidium bromide (10 mg/mL)
- Mini-agarose gel electrophoresis apparatus
- *E. coli* Lemo21 (DE3) competent cells (at −80 °C): 2 × 100 μL of aliquots per group and 2× 100 μL of aliquots for positive and negative controls for the whole group
- SOC medium
- 37 °C and 42 °C and 60 °C water baths
- 10× LB (lithium borate) buffer (see Instructor Notes of Experiment 39)
- 10× TAE (Tris-acetate-EDTA) buffer (48.4 g Tris-base, 11.42 mL glacial acetic acid, 20 mL 0.5 M EDTA (pH 8.0) in 1 liter)

• Reminder

The following order of procedures is recommended.

1. Instructor turns on two thermal cyclers in a ready-to-use mode and prepares PCR master mixes for Parts I and II exercises before the lab starts.

2. Students set up the two PCR reactions (Part I: FNheI + RKpnI/pGLO and Part II: SLIC1 + 2/pGLO) and run PCR cycles (~60 min).

3. During PCR, set up two restriction digestion reactions of vector pCAL-kc (Part I, B) and prepare one 0.5% agarose gel with a 10-well comb in LB buffer (Part I, E) and one 0.7% agarose gel with a large well comb in 1× TAE buffer (Part II, B).

4. When PCR is finished, purify the Part I PCR sample using spin column and set up restriction digestion of the purified PCR DNA (Part I, C and D) (~30 min). Part II PCR DNA is run on to 1× TAE agarose gel (~25 min).

5. Perform spin column purification of both the restriction samples of vector pCALkc and Part I PCR DNA (Part I, C, ~10 min) and gel purification of the Part II PCR DNA (Part II, C, ~15 min).

6. Estimate the concentrations of the purified PCR and restriction DNA fragments using NanoDrop 2000 (~10 min); teaching assistant runs students' samples in 1× LB agarose gel for verification.

7. Set up both ligation (Part I, F) and SLIC (Part II, E).

8. Carry out transformation using chemically competent Lemo 21(DE3) cells (~1 h 20 min).

Instructor notes of AMB 2 experiment 41

Each student group workbench

- Poly-prep chromatography column (Bio-Rad, Cat. No. 731-1550), stand, and clamp
- Two of 2.0-mL microcentrifuge tubes
- 21 of 1.5-mL microcentrifuge tubes
- Three 15-mL conical tubes (lysis, binding, and wash buffer) on ice box
- One glass test tube and one plastic Pasteur pipette
- 15-mL binding/lysis base buffer per group
- 1.5-mL elution buffer I per group
- 1.5-mL elution buffer II per group
- Induced cells of Lemo 21 DE3 (pCAL-GFP) in ice box
- Uninduced cells of Lemo 21 DE3 (pCAL-GFP) in ice box

Common materials and equipment

- Preparation of induced and uniduced cell cultures (see Part I)
- Cell pellet washing buffer (10 mL; autoclave; prepare just before use using sterile stocks): Dissolve 0.6 g of urea in 8 mL of H_2O, add 0.1 mL of TritonX-100 and 0.5 mL of 1 M Tris-Cl (pH 7.5), and bring up to 10 mL.
- Resolubilization buffer (5 mL; no autoclave; prepare just before use):

  Dissolve 2.4 g of urea in 3 mL of H_2O, add 0.25 mL of 1 M Tris-Cl (pH 7.5), and bring up to 5 mL.
- Binding/lysis base buffer (200 mL; autoclave):

  Add 10 mL of 1 M Tris-Cl (pH 7.5), 7.5 mL of 4 M NaCl, 2.0 mL of 0.1 M MgAc, and 4.0 mL of 0.1 M $CaCl_2$ to 176.5 mL of H_2O. For a 10-mL lysis buffer, add 0.002 g of lysozyme and 200 μL of 0.1 M PMSF just before use. For a 10-mL binding buffer, add 10 μL of 1 M imidazole and 7 μL of β-mercaptoethanol before use.
- Washing buffer (100 mL; autoclave):

  Add 10 mL of 1 M Tris-Cl (pH 7.5), 15 mL of 4 M NaCl, 2.0 mL of 0.1 M MgAc, and 0.4 mL of 0.1 M $CaCl_2$ to 72.6 mL H_2O. For a 10-mL washing buffer, add 10 μL of 1 M imidazole and 7 μL of β-mercaptoethanol before use.
- Elution buffer I (50 mL; autoclave):

  Add 5 mL of 1 M Tris-Cl (pH 7.5), 7.5 mL of 4 M NaCl, and 0.5 mL of 0.2 M EGTA to 37 mL of H_2O. For a 1.5-mL elution buffer I, add 30 μL of 0.1 M PMSF and 1 μL of β-mercaptoethanol just before use.
- Elution buffer II (50 mL; autoclave):

  Add 5 mL of 1 M Tris-Cl (pH 7.5), 12.5 mL of 4 M NaCl, and 0.5 mL of 0.2 M EGTA to 32 mL of H_2O. For a 1.5-mL elution buffer I, add 30 μL of 0.1 M PMSF and 1 μL of β-mercaptoethanol just before use.
- Regeneration buffer 1 (50 mL; autoclave):

  Add 5 mL of 1 M $(NH_4)_2CO_3$ (pH 8.6) and 0.5 mL of 0.2 M EGTA to 44.5 mL of H_2O.
- Regeneration buffer 2 (50 mL; autoclave):

  Add 12.5 mL of 4 M NaCl and 1 mL of 0.1 M $CaCl_2$ to 36.5 mLof H_2O.
- Regeneration buffer 3 (50 mL; autoclave): the same as elution buffer 2
- 20% ethanol (40 mL MilliQ H_2O + 10 mL absolute ethanol)
- Calmodulin Sepharose™ 4B resin (GE Healthcare 17-0529-01) equilibration:
  1. Decant the storage ethanol from the settled resin. Resuspend the resin in 5 bed volumes of the binding buffer. Allow the slurry to settle.
  2. Decant the supernatant again from the resin. Resuspend the resin in 5 bed volumes of the binding buffer.
  3. To complete the equilibration, again allow the resin to settle, decant the supernatant, and add an equal volume of the binding buffer.
  4. Degas the resin slurry and store at 4 °C.
- Lysozyme
- 0.1 M PMSF
- β-Mercaptoethanol (14.3 M)
- Sonicator
- A storage freezer box: each group has 4 aliquots of lysates (IS, IP, US, UP) and 8 column fractions (FT, W, EBI-1, 2, 3; EBII-1, 2, 3)

• Reminder

1. Students are asked to label all the tubes to be needed before the lab begins:
   - Three 15-mL conical tubes (to prepare lysis, binding, and wash buffers)
   - Two 2.0-mL microcentrifuge tubes (to centrifuge the sonicated lysates of induced and uninduced cells)
   - Four 1.5-mL microcentrifuge tubes (Part II, B: IS and US)
   - Four 1.5-mL microcentrifuge tubes (Part II, C: IP and UP)
   - Two 1.5-mL microcentrifuge tubes (EBI and EBII buffers)
   - Eight 1.5-mL microcentrifuge tubes (FT, W, EBI-1, 2, 3; EBII-1, 2, 3 fractions)
2. Calmodulin resin is expensive, but the resin regenerated more than three times must not be used.

Instructor notes of AMB 2 experiments 42 and 43A

Each student group workbench

- Protein samples (from Experiment 41) in ice bucket
- Two Amicon Centricon 10 units (Millipore, Cat. No. 4205)
- 9 Safety cap locks and microcentrifuge tubes (IS, US, IP, UP, FT, W, EBI, EBII, GFP/β-Gal)
- Two pipette box lids (for staining and blot transfer equilibration)
- Two sets of clean spacer and short glass plates, two 10-well combs
- Bio-Rad mini-protean 3-cell components (one casting stand, two casting frames, two clamping frames, two electrode assemblies, one mini tank)
- One 25-mL Erlenmeyer side-arm vacuum flask with a stopper
- P20, P200, and P1000 micropipettors and pipette tips
- 5-mL disposable pipettes with a pipet pump aid

Common materials and equipment

- A sharp razor blade
- Protran nitrocellulose membrane and Whatman 3MM filter papers for electroblotting
- Pre-stained protein molecular weight marker (Thermo Fisher Scientific, Cat. No. 26619)
- Acrylamide (30%)/bisacrylamide (0.8%)
- TEMED
- 10% APS (prepare just before use): 0.1 g of ammonium persulfate in 1.0 mL of H_2O
- β-Mercaptoethanol
- Coomassie blue staining solution: GelCode blue (Thermo Fisher Scientific, Cat. No. 24590)
- Water-saturated butanol
- 1.5 M Tris-HCl (pH 8.8): dissolve 18.15 g of Tris base in 60 mL of H_2O, adjust to pH 8.8 with 6 M HCl, and bring to 100 mL
- 0.5 M Tris-HCl (pH 6.8): dissolve 6 g of Tris base in 60 mL of H_2O, adjust to pH 8.8 with 6 M HCl, and bring to 100 mL
- 10% (w/v) SDS: heat to dissolve 10 g of SDS (molecular biology grade high purity) in 80 mL of H_2O and bring to 100 mL of H_2O
- 10× SDS electrophoresis buffer: dissolve 30.3 g of Tris base, 144 g of glycine, and 10 g of SDS in 800 mL of H_2O and bring to 1 liter (no pH adjustment)
- 4× SDS sample buffer: dissolve 0.8 g of SDS, 4 mL of glycerol, and 0.4 mL of 0.5% bromophenol blue in 5 mL of 0.5 M Tris-Cl

(pH 6.8). Add 20 µL of β-mercaptoethanol to 980 µL of 4× SDS sample buffer just before use
- Electroblot transfer buffer (25 mM Tris base, 192 mM glycine, 20% methanol): dissolve 0.3 g of Tris base, 7.2 g of glycine and 100 mL of methanol (high-quality reagent grade) in 300 mL of H_2O and bring to 500 mL (no pH adjustment)
- Round floating 1.5-mL microcentrifuge tube racks
- Boiling water bath

• Reminder

1. Three fractions of EBI and EBII are pooled together and concentrated using Amicon centricon (45 to 50 min).
2. During centrifugation time, each group prepares two gel cassette sandwiches (one for staining and the other for electroblotting). A Bio-Rad spacer plate is a glass plate with permanently bonded gel spacer of either 1.00 or 0.75 mm thickness. Therefore, combs with the same thickness as the spacer must be chosen. We routinely added double amounts of 10% APS and TEMED to resolving and stacking gels to reduce the polymerization time. Gel was polymerized within 5 min after APS and TEMED were added. Thus, it is important to quickly pour the pre-determined gel volume into a glass plate sandwich and overlay the top of resolving gel with water-saturated butanol.
3. After polymerization, measure the width and length of resolving gel for electroblotting. If a 3.3-mL resolving gel solution is used in a 0.75-mm spacer thickness plate, the gel dimension would be 9.5 × 7.0 cm.
4. During polymerization, prepare loading samples. Electrophoresis takes ~40 min at 200 V.
5. During electrophoresis, cut one nitrocellulose membrane and 3MM Whatman papers.
6. After electrophoresis, one gel is washed three times in deionized water before Coomassie blue staining, and a second gel is soaked in blot transfer buffer before electroblotting. Staining is performed overnight and electroblotting takes 10 to 12 min at 10 V.

Instructor notes of AMB 2 experiment 43.B and 44

Experiment 43, Part I, B
Each student group workbench

- 50 mL blocking solution
- 100 mL washing solution
- A clean plastic box (12 × 9 cm)
- 15 mL of primary antibody (add 2 µL of antibody to 15 mL of blocking solution just before use)
- 15 mL of secondary antibody (add 1.5 µL of antibody to 15 mL of blocking solution just before use)

Common materials and equipment

- Mouse anti-β-galactosidase antibody (primary antibody: Promega, Cat. No. Z3781)
- Mouse anti-GFP antibody (primary antibody: Living Colors® A.v. Monoclonal Antibody (JL-8); mouse IgG2a isotype; 1 mg/mL; Clontech, Cat. No. 632380 or 632381)
- Anti-mouse IgG (H + L)-AP conjugate (secondary antibody: Promega, Cat. No. Z3721)
- NBT/BCIP for the conjugated alkaline phosphatase substrate solution (Fisher Scientific, Cat. No. PI-34042)
- TBS (Tris-buffered saline) buffer: dissolve 18 g of NaCl and 2.42 g of Tris base (or 20 mL of 1 M Tris-Cl, pH 7.5) in 800 mL of H_2O, adjust the pH to 7.5 with HCl, and bring to 1 liter
- Washing solution (TTBS): 100 mL of TBS + 0.05 mL of Tween 20 (prepare fresh before use)
- Blocking solution (TTBS/milk): dissolve 5 g of non-fat dry milk in 95 mL of TBS and add 50 µL of Tween 20 (prepare fresh before use)
- Rotatory platform shaker

Experiment 44
Each student group workbench

- Inoculating loops
- Alcohol lamp

Common materials and equipment

- Yeast strain Y187 streaked on a YPDA plate
- Yeast strain Y2H Gold streaked on a YPDA plate
- YPD medium (Clontech)
- Minimal SD base medium (Clontech, Cat. No. 630411) containing 2% glucose
- Dropout supplements (Clontech): –Trp, –Leu, –Leu/–Trp, –Ura, –Trp/–Ura
- Arginine DO supplement (CSM-ARG, MP Biomedicals, Cat. No. 114510-112)
- Uracil
- L-canavanine (60 mg/mL in H_2O; filter-sterilized)
- 0.2% adenine hemisulfate (dissolve 0.1 g in 50 mL of H_2O, filter-sterilize, and store at 4 °C)
- Bacto-Agar (Difco)
- Water carboy filled with deionized H_2O
- Four 1-liter flasks, two 500-mL flasks, seven 250-mL flasks
- 109 Petri plates (100 × 15 mm)
- Weight balances, weighing boats, spatula, aluminum foil, labeling tape
- Autoclave (reservation notice in login sheet)
- 30 °C incubator
- Laminar flow hood

• Reminder

The antibody detection (Experiment 43, Part I, B) procedure takes about 3 hours. During membrane processing time, each student group prepares the assigned media. During autoclaving time, students label the Petri plates.

All the plates are stored at room temperature rather than at 4 °C so that any contaminating microbes can be detected over a week period.

Instructor notes of AMB 2 experiment 45

Each student group workbench

- Three SD –Leu dropout plates (Part I, C)
- Three SD –Trp dropout plates (Part I, C)
- Two SD –Leu/–Trp dropout plates (Part I, C)
- Five SD –Trp/–Ura dropout plates (Part II, C)

- 25-mL YPDA cultures (A_{600} of 1.0 to 2.0) of yeast Y187 and Y2HGold strains
- 25-mL YPDA cultures of yeast BY4733 and BY4733 (p414-TEF1p Cas9-CYC1.t) strains
- Four 50-mL sterile conical tubes for cell harvest
- 20 mL of sterile H_2O
- Sterile 1.5-mL microcentrifuge tubes (8)
- Glass road spreader in denatured ethanol beaker
- Bunsen burner

Common materials and equipment

- pGBKT7-53 DNA (0.1 µg/µL)
- pGADT7-T DNA (0.1 µg/µL)
- pCL1 DNA (0.1 µg/µL)
- p414-TEF1p-Cas9-CYC1.t
- p426-SNR5p-gRNA.CAN1.y-Sup4t
- Salmon sperm carrier DNA (10 mg/mL)
- Four SD-Leu and two SD-Trp dropout plates (positive and negative control plates)
- DPB 276 strain streaked on to aYPDA plate (Part II, C)
- 42 °C water bath
- Table top centrifuge, microcentrifuge
- 50% PEG 3350: 10 g in 20 mL of dH_2O; autoclave
- DMSO (dimethyl sulfoxide)
- Sterile H2O
- 10× TE buffer: 20 mL of 1 M Tris-HCl (pH 7.5) + 4 mL of 0.5 M EDTA (pH 8.0) + 176 mL of H_2O; autoclave
- 10× LiAc (1 M lithium acetate): dissolve 10.2 g in 80 mL of H_2O, adjust to pH 7.5 with dilute acetic acid, bring up to 100 mL, and autoclave
- 1.1× TE/1× LiAc (prepare aseptically just before use): 1.1 mL of 10× TE + 1.1 mL of 10× LiAc + 7.8 mL of sterile H_2O
- PEG/LiAc solution (prepare aseptically just before use): 8 mL of 50% PEG + 1 mL of 10× TE + 1 mL of 10× LiAc
- YPDA cultures of Y187, Y2H Gold, BY4733, and BY4733 (p414-TEF1p Cas9-CYC1.t)

• Reminder

1. Before lab starts, read A_{600} of yeast YPDA cultures of Y187, Y2H Gold, BY4733, and BY4733 (p414-TEF1p Cas9-CYC1.t) and calculate the volumes of 1.1 × TE/LiAc buffer to resuspend the cell pellets to a final cell density of 10^9 cells/mL.

2. Yeast transformation for Y2H assay and genome editing of *CAN1* locus is performed at the same time.

3. Students should organize and label the plates according to the type of nutritional selection marker and transforming strains.

Instructor notes of AMB 2 experiment 46

Each student group workbench (Part I. Yeast mating)

- Two sterile cotton swabs
- 50 µL of 50% PEG6000
- One YPDA plate
- One SD –Leu/–Trp plate
- Sterile 1.5-mL microcentrifuge tubes
- P20, P200, and P1000 micropipettors and pipette tips

Common materials and equipment (Part I)

- Y2HGold (pGBKT7-53) and Y187 (pGADT7-T) cultures grown overnight in 10 mL of YPDA medium. Read A_{600} of 1/10 dilution (0.3 mL of culture + 2.7 mL of YPDA)
- SD –Leu dropout plates (two per group for Day 2)
- SD –Trp dropout plates (two per group for Day 2)
- SD –Leu–Trp dropout plates (two per group for Day 2)
- Sterile H_2O (10 mL per group for Day 2)
- Kanamycin sulfate (50 mg/mL)
- 0.5 McFarland standard (Day 2)
- QDO/X-α-Gal plates (two per group for Days 3 and 4)
- YPD plate (one per group for Days 3 and 4)
- Sterile 1.5-mL microcentrifuge tubes (for making dilutions)
- Inoculation loops

Each student group workbench (Part II. Genomic PCR)

- YPDA culture or colonies of BY4733, canavanine-resistant (*can1*) Trp^+Ura^+ (Experiment 45 student plates) and canavanine sensitive (CAN1) Trp^+Ura^+ BY4733 (control), and canavanine-resistant DPB276 (control)
- 400 µL of 200 mM LiAc + 1% SDS solution
- Four PCR tubes on a rack in ice box
- P10, P20, P200, and P1000 micropipettor and pipettor tips
- PCR master mix (100 µL)

Common materials and equipment (Part II)

- 200 mM LiAc + 1% SDS solution (2 mL of 1 M LiAc + 1 mL of 10% SDS + 7 mL H_2O)
- 100% (molecular biology grade) and 70% etanol, TE (pH 8.0) buffer
- 10× LB buffer (lithium borate 100 mM, pH 8.2; see Instructor Notes of Experiment 39)
- Agarose, ethidium bromide (10 mg/mL)
- PB buffer (5 M guanidine-HCl, 20 mM Tris-HCl, pH 6.6, 30% isopropanol)
- PE buffer (10 mM Tris-HCl pH 7.5, 80% ethanol)
- QG buffer (6 M guanidine thiocynate, 50 mM Tris-HCl, pH 6.6, 20 mM EDTA, pH 8.0, 0.006% cresol red)
- EB buffer (10 mM Tris-HCl, pH 8.5)
- TE buffer (pH 8.0)
- EconoSpin™ mini spin column (Epoch Life Science, Cat. No. 1920-050)
- Two mini-gels with large tooth comb (each 60 mL of 0.5% agarose containing 0.5 µg/mL of EtBr in 1× lithium borate buffer); prepare during the class
- 70 °C water bath
- NanoDrop 2000 spectrophotometer
- Thermal cycler (program: see the protocol): sign up login sheet
- Genomic PCR master mix (see the protocol for Part II, B)
- DNA size marker (1 kb DNA ladder: 0.5 µg/5 µL)

• Reminder

1. Yeast genomic DNA isolation followed by PCR (Part II) is performed first. During the PCR cycle (about 50 min), yeast mating is conducted and two agarose gels are prepared. After PCR, run PCR samples in agarose gel, gel-purify the PCR fragments, and determine the concentrations.

2. Remind students to come to the lab tomorrow in order to conduct dilutions of yeast mating cells and plating on selection plates (Part I, A).

Instructor notes of AMB 2 experiment 47

Each student group workbench (Part I. colony PCR)

- Colonies from QDO plates (from Experiment 46)
- Mini-agarose gel electrophoresis equipment and 12-well comb
- Eight thin-walled PCR tubes in ice box (Groups 1 and 2)
- Six thin-walled PCR tubes in ice box (Groups 3 and 4)
- P10, P20, and P200 micropipettors and pipette tips
- Three 0.5-mL microcentrifuge tubes, each coniaining 10 μL volume of acid-washed glass beads (0.45–0.53 mm)
- 100 μL TE (10 mM Tris-Cl, pH 8.0, 1 mM EDTA) buffer containing 2% Triton X-100
- 100 μL 5× GoTaq® green buffer
- Nuclease-free H_2O
- 10 μL of dNTP mix (10 mM)
- 10 μL of GAD primer mix (10 μM): see Materials and equipment
- 10 μL GBK primer mix (10 μM): see Materials and equipment

Common materials and equipment (Part I. colony PCR)

- GoTaq® DNA polymerase (Promega, Cat. No. M3001)
- GADT7-T (10 ng/μL) and GBKT7-53 (10 ng/μL) plasmid DNA
- Thermal cycler (program: see the Part I protocol): sign up login sheet

Each student group workbench (Part II. cycle sequencing)

- Gel-purified PCR DNA of *can1* and *CAN1* (Experiment 46)

Common materials and equipment (Part II. Cycle sequencing)

- M13F/CAN1 F + R primer mix (0.8 μM each)
- Control CEPH primer (0.8 μM each); if this is not available, pUC19 DNA can be used with regular M13 F/R primer set
- BigDye® direct cycle sequencing kit (Thermo Fisher Scientific, Cat. No. 4458689):
- BigDye direct PCR master mix
- BigDye direct sequencing master mix
- BigDye direct M13 Fwd primer
- BigDye direct M13 Rev primer
- Control DNA CEPH (50 ng/μL)

  Note: this kit does not provide the primers for control CEPH DNA. Contact the manufacturer for the primer. Alternatively, pUC19 DNA can be used with a regular M13F/R primer set.
- Nuclease-free sterile H_2O
- One flat 96-well PCR plate for the whole group
- One thermal sealing film for the whole group
- 96-Well plate tabletop centrifuge
- Thermal cycler (program: see the Part II protocol)

• Reminder

We routinely followed the following order of procedures.

1. The instructor prepares the PCR master mix and students set up PCR reactions in a 96-well plate for cycle DNA sequencing (Part II). It takes about 60 min to complete the PCR cycle.
2. During the PCR cycling time, prepare yeast colony DNA, set up yeast colony PCR reactions and run PCR (Part I). The instructor may prepare yeast colony PCR master mix to save time. A yeast colony PCR cycle takes about 2 h 30 min.
3. During the PCR cycle, each student group prepares a 0.5% agarose gel in 1× LB buffer with a 14-well comb; the gel solution is poured into a gel tray mold at 4 °C to shorten the hardening time.
4. After PCR is complete, the 96-well plate is briefly centrifuged and stored at −20 °C. Colony PCR samples are run on to agarose gel either at 300 V for 30 min or at 25 V overnight (~14 h).

Instructor notes of AMB 2 experiments 48 and 49

Each student group workbench (experiment 48)

- Cycle DNA sequencing reaction samples (from Experiment 47, Part II)

Common materials and equipment (experiment 48)

- BigDye XTerminator purification kit (Thermo Fisher Scientific, Cat. No. 4376486)
- MicroAmp® clear adhesive film
- MicroAmp® optical 96-well reaction plate)
- 96-Well plate centrifuge
- Vortexer
- POP-7™ polymer separation matrix (Thermo Fisher Scientific, Cat. No. 4393708)
- Anode buffer (Thermo Fisher Scientific, Cat. No. 4393925)
- Cathode buffer (Thermo Fisher Scientific, Cat. No. 4408256)
- 3500xL genetic analyzer capillary array, 50 cm (Thermo Fischer Scientific, Cat. No. 4404689)
- 96-Well (standard) eetainer and base set
- Septa for 3500/3500xL genetic analyzers (Thermo Fisher Scientific, Cat. No. 4410701)
- ABI 3500xL genetic analyzer: sign up login sheet

Each student group workbench (experiment 49)

Part I

- Four YNB/–Ura/+glucose (= SD/–Ura) agar plates
- Two YNB/–Ura/+galactose/raffinose agar plates
- 25 mL of YPDA culture (A_{600} of 1.0 to 2.0) of BY4741 in a 50-mL conical tube
- 25 mL of YPDA culture (A_{600} of 1.0 to 2.0) of DPB260 in a 50-mL conical tube
- Four sterile 1.5-mL microcentrifuge tubes
- Glass spreader in an ethanol beaker; alcohol lamp

Part II

- One YNB/–Ura/+glucose (= SD/–Ura)
- One YNB/–Ura/+5-FOA/+glucose (= SD/–Ura/+5-FOA)
- One YNB/–Ura/+galactose/raffinose
- One YNB/+ Ura/+galactose/raffinose

- One YNB/–Ura/+5-FOA/+galactose/+raffinose
- 1 mL culture of DPB276 strain grown in YNB/+Ura/+galactose/ +raffinose

Common materials and equipment (experiment 49)

- Two YNB/–Ura/+glucose agar plates (for the two negative controls)
- Plasmid pYES2.1-GFP_{uv}
- Salmon sperm carrier DNA (10 mg/mL)
- DMSO
- 10× TE buffer (see Experiment 45 Instructor Notes)
- 10× LiAc (see Experiment 45 Instructor Notes)
- 1.1× TE/1× LiAc (see Experiment 45 Instructor Notes)
- PEG/LiAc solution (see Experiment 45 Instructor Notes)
- 42 °C water bath
- Tabletop centrifuge; microcentrifuge

Agar plates

- YNB/–Ura/+glucose agar:

- Uracil dropout supplement	0.77 g/liter (0.077%)
- YNB + $(NH_4)_2SO_4$	6.70 g/liter (0.67 %)
- Glucose (dextrose)	20.0 g/liter (2 %)
- Agar	20.0 g/liter (2%)

**YNB and glucose components can be replaced with 26.5 g/liter of SD base.*

- YNB/–Ura/+galactose/raffinose agar:

- Uracil dropout supplement	0.77 g/liter (0.077%)
- YNB + $(NH_4)_2SO_4$	6.70 g/liter (0.67%)
- Galactose	10.0 g/liter (1 %)
- Raffinose	10.0 g/liter (1 %)
- Agar	20.0 g/liter (2 %)

- YNB/+Ura/+galactose/raffinose agar:

- Uracil dropout supplement	0.77 g/liter (0.077%)
- Uracil	0.02 g/liter (0.002%)
- YNB + $(NH_4)_2SO_4$	6.70 g/liter (0.67%)
- Galactose	10.0 g/liter (1 %)
- Raffinose	10.0 g/liter (1 %)
- Agar	20.0 g/liter (2 %)

- YNB/–Ura/+5-FOA/+glucose (= SD/–Ura/+5-FOA) agar:
  1. Autoclave 2 grams of agar in 50 mL of ddH_2O.
  2. Add the following components into a glass beaker, heat to dissolve, bring to 50 mL with H_2O, and filter-sterilize.

- Uracil dropout supplement	0.077 g
- YNB + $(NH_4)_2SO_4$	0.67 g
- Glucose	2.00 g
- 5-FOA	0.10 g (0.1%)

Notes:

**YNB and glucose components can be replaced with 2.65 g SD base.*

**Warm up the above 5-FOA mixture in step 2 at 55 to 65 °C on a hot plate, stir until it dissolves (~30 to 60 min).*

**After the 5-FOA containing mixture is completely dissolved, sterilize it by filter.*

**Place both autoclaved agar and the 5-FOA containing solution in a water bath (50 to 60 °C) for about 10 min.*

**Do not adjust the pH of the 5-FOA containing medium. Only the protonated form of 5-FOA is permeable.*

3. Add sterilized solution to the autoclaved agar solution, mix thoroughly, and pour into four plates. Store the plates at 4 °C (for long-term storage) or at room temperature (48 h).

- YNB/–Ura/+5-FOA/+galactose + raffinose agar 100 mL:
  1. Autoclave 2 grams of agar in 50 mL of ddH_2O.
  2. Add the following components into a glass beaker, dissolve in a hot (55 to 65 °C) plate, bring to 50 mL with H_2O, and filter-sterilize.

- Uracil dropout supplement	0.077 g
- YNB + $(NH_4)_2SO_4$	0.67 g
- Galactose:	1.00 g
- 5-FOA	0.10 g (0.1%)

3. Add the filter-sterilized solution to the autoclaved agar solution, mix thoroughly, and pour into four plates.

• Reminder

The following order of procedures is recommended.

1. Before the lab starts, the instructor launches an application of ABI 3500xL, checks the system status, and prepares the instrument. The instructor also prepares a pre-mix for DNA sequencing sample purification (Experiment 48, Part I) and reads A_{600} of cell cultures (Experiment 49, Part I, A) before the lab starts.

2. The DNA sequencing sample clean-up reaction is performed first. During the 25-min clean-up time, yeast transformation is carried out.

3. During the incubation time of the yeast cell–DNA mixture, the cleaned-up DNA sequencing sample plate is loaded into the instrument and the plate is linked for running.

4. During DNA sequencing electrophoresis, finish the remaining steps of yeast transformation (Part I) and the growth test (Part II).

Instructor notes of AMB 2 experiment 50

Each student group workbench

- 5 mL each of "H_2O_2-treated" and "untreated control" *E. coli* culture (prepare according to the protocol)
- 3 mL of acetone (HPLC grade, 100% cold –20 °C)
- 1.5-mL microcentrifuge tubes
- 9 glass test tubes (13 × 100 mm)
- Vortex mixer
- 1020 (510 × 2) μL of *RC* Reagent I
- 1020 (510 × 2) μL of *RC* Reagent II

- 1040 (520 × 2) μL of Reagent A′
- 8.2 mL of *DC* Reagent B
- Spectronic 20 and cuvette

Common materials and equipment

- Liquid nitrogen
- TCA (100% trichloroacetic acid: 1 g of TCA + 454.5 μL of sterile MilliQ H_2O; prepare fresh just before use): each group needs ~0.2 mL
- BugBuster™ protein extraction reagent (Novagen, Cat. No. 70584-3, 100 mL)
- Protease inhibitor cocktail (Thermo Fisher Scientific, Cat. No. 87786; 100×)
- BSA (1 mg/mL): 0.002 g in 2 mL of BugBuster™ protein extraction reagent
- RC DC protein assay reagent (Bio-Rad, Cat. No. 500-0121):
- *RC* Reagent I
- *RC* Reagent II
- *DC* Reagent S
- *DC* Reagent A
- *DC* Reagent B
- Reagent A′: add 10.2 μL of *DC* Reagent S to each 510 μL of *DC* Reagent A
- Microcentrifuge at 4 °C

Instructor notes of AMB 2 experiment 51

Each student group workbench (IFF)

- TCA/acetone-precipitated protein samples (Experiment 50)

Common materials and equipment (IEF)

- Rehydration buffer (8 M urea, 2% CHAPS, 0.5% IPG buffer, 0.002% bromophenol blue, 0.015 M DTT): vortex to dissolve 0.96 g of ultrapure urea and 0.04 g of CHAPS in 1.4 mL of H_2O, add 40 μL of IPG buffer (pH 3–10), 4 μL of 1% BPB, and 30 μL of 1 M DTT, and bring up to 2 mL
- Rehydrateded Immobiline™ DryStrip (ready-cut IPG strips: 7 cm, pH 3–10): 125 μL of rehydration buffer is needed for each IPG strip
- Reswelling tray
- Immobiline™ DryStrip cover fluid (mineral oil)
- Immobiline™ Drystrip cover fluid (mineral oil)
- Immobiline™ DryStrip kit (anode electrode; cathode electrode; sample cup bar; sample cups; tray and electrode holder; DryStrip aligners; IEF electrode strips)
- Forceps
- Multiphor™ II electrophoresis system (Amersham Pharmacia Biotech)
- Blotting filter paper
- IEF standard (Bio-Rad, Cat. No. 161-0310)
- Bio-Rad 2-D SDS-PAGE standards (Bio-Rad, Cat. No. 161-0320)

Each student group workbench (SDS-PAGE)

- Two IEF-IPG strips (H_2O_2 stressed and unstressed)

Common materials and equipment (SDS-PAGE)

- Reswelling tray
- Equilibration base buffer: dissolve 36 g of urea in 3.35 mL of 1.5 M Tris-Cl (pH 8.8), 30 mL of glycerol, 20 mL of 10% SDS, and 0.1 mL of 1% bromphenol blue and bring up to 100 mL
- Dithiothreitol (DTT)
- Idoacetamide (IAA)
- Ammonium persulfate (APS)
- TEMED
- Acrylamide (30%)/bisacrylamide (0.8%)
- Agarose sealing solution: 0.25 g in 50 mL of 1× SDS electrophoresis buffer
- Bio-Rad Mini-PROTEAN 3 system components
- Forceps
- Pre-stained protein molecular weight marker
- 10% APS (prepare just before use): 0.1 g of ammonium persulfate in 1.0 mL of H_2O
- 1.5 M Tris-HCl (pH 8.8): dissolve 18.15 g of Tris base in 60 mL of H_2O, adjust to pH 8.8 with 6 M HCl, and bring to 100 mL
- 0.5 M Tris-HCl (pH 6.8): dissolve 6 g of Tris base in 60 mL of H_2O, adjust to pH 8.8 with 6 M HCl, and bring to 100 mL
- 10% (w/v) SDS: heat to dissolve 10 g of SDS (molecular biology grade high purity) in 80 mL of H_2O and bring to 100 mL H_2O
- 10× SDS electrophoresis buffer: dissolve 30.3 g of Tris base, 144 g of glycine, and 10 g of SDS in 800 mL of H_2O and bring to 1 liter (no pH adjustment)
- Coomassie blue staining solution: GelCode Blue (Life Technologies)

• Reminder

1. The instructor must rehydrate the IPG strip on the day before the IEF experiment. A thermostatic circulator cooling unit (set to 20 °C) should be turned on at least 30 min before use. The IEF time for the IPG strip (7 cm) takes about 3 hours to finish and is programmed to automatically power off with a continuous cooling mode. Next day, IPG strips are recovered and stored at –20 °C until use.

2. We routinely used a commercial pre-cast gel (Bio-Rad, Cat. No. 456-1081) for the second dimension. If the SDS-PAGE gel is made, it is recommended to use a 1-mm thickness spacer plate for making the gel.

AMB 2 Appendix 2

AMB 2 Lab math practice problem set

Part I. Metric unit conversion

1. Convert the following:

• 1 gram = _______ mg	1 gram = _______ kg
• 1 liter = _______ mL = _________ cm^3	2 ng = _______ mg
• 50 mg = _______ g	100 µL = _______ mL
• 1 mg = _____ pg = _______ µg = _______ ng	
• 1 L (liter) = ___ mL = ____ µL = _____ nL = _____ pL = ______ fL	
• 1 mM = _____ µmole/mL = ______ pmole/µL	
• 1 L × 1 M = ______ mole	1 mL × 1 M = ______ mmole
• 1 L × 1 mM = ______ mmole	1 mL × 1 mM = ______ µmole
• 1 mL of pure water = ______ g	1 L of pure water = _______ g
• 1 ppm (part per million) = ________ %	10% = ________ ppm
• 1 mg/L in water = _________ ppm	100 ppm = _________ µg/mL in water
• 1 mg/kg = __________ ppm	

Part II. Dilution

1. You have a series of 10 tubes, each of which contains 4 µL. You add 1 µL of labeled DNA to the first tube, mix and take 1 µL of this mixture and add to the second tube, mix and take 1 µL of this mixture and add to the next. If you continue this process with each tube, what is the dilution of the serum in the last tube?

2. Eight test tubes are placed in a rack. To the first tube add 3 mL of saline solution. To each of the remaining seven tubes add 2 mL of saline solution. To the first tube add 1 mL of serum and mix well. Transfer 2 mL of tube #1 to tube #2 and mix well, 2 mL of the contents of tube #2 are then transferred to tube #3, and the procedure is repeated for the remaining tubes. What is the dilution of serum in tube #8?

3. You had 30 mg/mL of glucose solution. You took 1 mL of that and diluted to 2 mL. Then 0.5 mL of it was added to 4.5 mL of saline solution, and 2 mL of this solution was diluted to 10 mL. What is the final concentration of glucose?

4. If you use 5 mL of stock glucose containing 10 mg/mL and dilute to 100 mL, what is the concentration of glucose per mL in the dilution?

5. Overnight yeast culture was found to contain about 5×10^7 cells/mL. If you want to inoculate the yeast cells into a fresh 100-mL medium to a final density of about 2.5×10^3 cells/mL, how much do you have to inoculate?

6. The average number of colonies appeared on three independent agar plates is 250 after 0.1 mL each of 1:1000 diluted yeast culture cells were plated on to the three plates and incubated. What is the cell density of the original yeast cell culture?

7. How would you prepare 25 mL of a 1/10 dilution of blood using PBS buffer?

8. How would you prepare 400 µL of a 1/100 dilution of antibody using PBS buffer?

9. How would you prepare 100 mL of a 1/100 dye solution from 1/40 dye solution?

10. You have a stock of a buffer solution that is used in experiments. The stock is 10 times (10×) more concentrated than it is used. How would you prepare 100 mL of the buffer solution at the right working concentration?

11. You take 100 mL of suspended bacteria and grind them into a cell "paste." Then you remove 10 µL of these cells and perform a test that tells you how much protein is present. Your test shows that 6 µg of protein are present in the 10 µL of cell paste. What was the concentration of protein in the cell paste? Express your answer in mg/mL.

12. If you take a 0.1 mL sample and add 0.9 mL of water and 3.0 mL of reagents, what is the final dilution of the sample?

13. You want to set up a PCR reaction mixture by adding 0.001 µg of plasmid DNA to a tube. You have 1 mL of plasmid at a concentration of 1 mg/mL. How would you add 0.001 µg of plasmid? Assume that it is not possible to accurately measure a volume less than 1 µL.

14. You have 100 µL of a protein sample. The sample was diluted by taking 5 µL of the sample and adding 20 µL of buffer. A protein assay was performed on the diluted sample. The concentration of protein in the diluted sample was 9 mg/mL. What was the concentration of protein in the original, undiluted sample? How much total protein was present altogether in the original, undiluted sample?

15. You have 100 µL of a protein sample. The sample was diluted by taking 5 µL of the sample and adding 20 µL of buffer. A protein assay was performed on the diluted sample. The total amount of protein in the diluted sample was 9 mg. What was the concentration of protein in the original, undiluted sample? How much total protein was present altogether in the original, undiluted sample?

Methods in Biotechnology, First Edition. Seung-Beom Hong, M. Bazlur Rashid and Lory Z Santiago-Vázquez.

Companion Website: www.wiley.com\go\hong\Methodsinbiotechnology

16. An oligonucleotide stock has a concentration of 60 pmole/μL. You want to use 3.2 picomoles of the oligonucleotide in 2 μL volume as a primer in a DNA sequencing reaction. How would you prepare to get the desired 3.2 picomoles in 2 μL volume?
17. How much 8.0 M urea will you add to 50 mL of a solution such that the final concentration is 2.0 M?

Part III. Percent solution

1. If you have 10 g of $MgCl_2$ in a solution of 500 mL, what is the concentration as a percent (w/v)?
2. How would you make 100 mL of 70% (v/v) ethanol solution?
3. You weighed out 4.5 g of $CaCl_2$ and diluted it to 250 mL with distilled water. What is the percent solution (w/v)?
4. You need to make 100 mL of 50 % $CaCl_2$ (MW = 111) solution, but all we have is $CaCl_2 \cdot 5H_2O$ (MW = 201; % assay = 99%) on the shelf. How much do you have to weigh out the hydrated form to make a 50% $CaCl_2$ solution?
5. How do you prepare a sterile 10% SDS solution? Where should you store the stock solution? What happens if the stock solution is kept in the cold?
6. What is the concentration of SDS if you add 10 mL of 10% SDS to 90 mL of water?
7. How would you make 100 mL of a solution that is 5% (v/v) acetic acid and 5% methanol?
8. The formula weight of $CuSO_4 \cdot 5H_2O$ is 249.7. What is the molecular weight of the anhydrous form of copper sulfate and what is the % assay of the hydrate form?
9. How would you make 100 mL of 20% (w/v) solution of HCl (MW = 36.5, % assay = 38, specific gravity = 1.19)?
10. How would you make 100 mL of 10 mM H_2O_2 from 30% (v/v) H_2O_2 (specific gravity = 1.46) solution?
11. The ELISA result shows that 0.001% serum solution is the lowest detection limit. Calculate the detection sensitivity of ELISA in terms of μg/mL assuming that 75% of serum is IgG.

Part IV. Ratio and proportions

1. Double-stranded DNA consists of two long strands. An adenine (A) on one strand is always paired with a thymine (T) on the opposite strand and a guanine (G) on one strand is always paired with a cytosine (C) on the opposite strand. If a purified sample of DNA contains 21% thymine, what are the percentages of the other A, G, and C bases?
2. What is the percentage of a solution that contains 200 mL of acetone per liter?
3. How much ethanol is present in 200 mL of a 25 % solution of ethanol?
4. The density of glycerol at 20 °C is 1.26 g/mL. What is the volume of 87.0 g of glycerol?
5. The density of β-mercaptoethanol is 1.1 g/mL. If you need to add 13.5 g of this chemical to a solution, how much volume should you measure out?
6. A recipe to make 6× gel loading buffer is 0.25% bromophenol blue, 0.25% xylene cyanol, and 40% sucrose in water. Express the amounts of each component needed in mg to prepare 10 mL of 6× gel loading dye.
7. You prepared Tris-HCl buffer at pH 7.5 at room temperature (25 °C). Tris buffer increases 0.3 pH units for every 10 °C decrease in temperature. What is the expected pH if the Tris-HCl solution is kept in a cold room (4 °C)?
8. How would you make 500 mL of 70% alcohol using 95% ethanol?
9. How would you make 1 liter of 1000 ppm standard of Na using the salt NaCl (atomic mass of Na = 23; atomic mass of Cl is 35.45) and water?
10. What is the ppm of Cd for 1 M $CdCl_2$ (atomic mass of Cd = 112.4; atomic mass of Cl = 35.45) solution in water?
11. Coliform bacteria are randomly distributed in a certain river at an average concentration of 1 per 20 mL of water. If we draw from the river a test tube containing 10 mL of water, what is the chance that the sample contains exactly 2 coliform bacteria? (*Hint: this is related to the Poisson distribution, and so compute Poisson probability* $p(k) = r^k/(k!)(e^r)$, where *e* is the constant 2.71828, *k* is the number of expected occurrence events, and *r* is the known constant rate (frequency) of a mutually independent, random, rare occurrence event in a large population).
12. If *Taq* polymerase swaps a G for an A in one PCR product at the second PCR cycle, what percentage of DNA strands will have the mutation after 30 cycles of PCR?

Part V. Mole and molarity

1. What is the MW in grams if one molecule weighs 6.664×10^{-20} g?
2. What is the weight in grams of 1×10^{23} molecules of NaCl (MW = 58)?
3. What molarity (M) is 0.9% glucose (MW = 180) solution?
4. What molarity (M) is 90 mg/mL of glucose (MW = 180) solution?
5. 150 g/liter of a particular chemical is required to obtain a 0.5 M solution. How much of this chemical is required to obtain a 30 μM solution?
6. You want to prepare 500 mL of a 1.5 M solution of NaCl. You have powder, a beaker, a graduated cylinder, and a bottle. How would you prepare the solution (MW of NaCl = 58)?
7. You wish to prepare 500 mL of a 10 mM solution of KCl (MW = 74). Starting from KCl powder, how would you prepare the solution?
8. You have a solution of 1 M Tris-HCl. How would you prepare 20 mL of 0.5 M Tris? 0.15 M Tris? 1 mM Tris?
9. You are given a 10% SDS stock solution. The molecular weight of NaOH is 40. How do you make a 100-mL solution that consists of 1 M NaOH and 1% SDS?
10. What is the molar concentration of a 3% (v/v) glycerol (density 1.261 g/cm^3, MW 92.09) solution?
11. Let's say you are TA for a biotechnology class. Your professor asked you to prepare a Sephadex G-50 column buffer (50 mM Tris-HCl (pH 7.9), 2 mM EDTA (pH 8.0), 250 mM NaCl) for 6 section classes. Each section needs 100 mL. How will you prepare the buffer from stock solutions of 1 M Tris-HCl (pH 7.9), 1 M NaCl, and 0.5 M EDTA (pH 8.0)?
12. A solution is prepared by dissolving 0.125 moles of ammonium sulfate, $(NH_4)_2SO_4$, in water to make 100 mL of stock solution. A 25-mL sample of this stock solution is added to 50 mL

of water. Calculate the concentration (M) of ammonium ions (NH_4^+) in the final solution.

13. When you mix 200 mL of dH_2O, 100 mL of 125 mM solution A ($Na_2HPO_4 \cdot 7H_2O$), and 50 mL of 200 mM solution B (NaH_2PO_4), what will be the final molarity for the sodium ion (Na^+) in this mixture?

14. You have only 0.01 g of IAA (indole acetic acid: MW = 175). How would you make 10 mM IAA solution in water with 0.01 g of IAA?

15. Explain how you make a 10 mL of 10 μM solution of IAA (indole acetic acid: MW = 175). Be detailed in the order and amount of additions.

16. One unit of *Taq* DNA polymerase is added to a 50-μL PCR reaction. The enzyme has a specific activity of 292 000 units/mg and a molecular weight of 94 kDa. How many molecules of the enzyme are there in the reaction?

17. How much of a solute (MW = 75, % assay = 85%) is needed to make 500 mL of a 2 M solution?

18. What mL of glacial acetic acid (MW = 60.05, specific gravity = 1.05, % assay = 99) is needed to make 1 L of a 2 M solution?

19. How would you prepare 750 mL of a 2 N solution of H_2SO_4 (MW = 98.08, % assay = 96, specific gravity = 1.84)?

20. Normal human blood osmolarity is between 300 and 310 miliosmole/L. If you choose 308 miliosmole/L for blood osmolarity and prepare a saline (NaCl) solution that is isotonic to blood serum, what percentage of NaCl (MW = 58.44) do you have to make?

21. What is the molar concentration of 50% (v/v) glycerol (MW = 92.09 g/mole; ρ = 1.26 g/mL) solution in water?

22. What is the molar concentration of 50% (w/w) glycerol (MW = 92.09 g/mole; ρ = 1.26 g/mL) solution in water?

23. Assume that a bacterial cell contains equal concentrations of 1000 different enzymes in the cytosol and that each protein has a molecular weight of 100 kDa. Also assume that the cytosol (specific gravity 1.20) is 20% (w/v) soluble protein and that soluble protein consists entirely of enzymes. Calculate the average molar concentration of each enzyme.

Part VI. Applications

1. A rule of thumb is that the PCR reaction should start out with at least 10^4 copies of target DNA. What amount of DNA should you use if you are doing PCR using human diploid genomic DNA (6×10^9 bp)? The average molecular weight of one base pair is 660 g/mole.

2. A primer stock consisting of the two primers needed for a PCR reaction has a concentration of 20 μM. What volume of the primer stock should be added to a 50-μL reaction so that the primer concentration in the reaction is 0.4 μM?

3. You are doing a large-scale plasmid DNA extraction from an *E. coli* culture. Let's say the protocol called for your cell suspension to have a concentration of lysozyme at 2 mg/mL. Your suspension has a total volume of 10 mL. The lysozyme stock tube has a concentration of 20 mg/mL. How much lysozyme should you add?

4. You have already determined the doubling time of 24 min for a certain *E. coli* strain. You also determined the cell density of 10^7 cells/mL by plate counting. You diluted the cell to 1:100, inoculated 0.1 mL of the diluted cells into 9.9 mL of the same medium, and incubated under the same growth condition that was used to determine the doubling time and cell density. How long should you incubate to reach a cell density of 32 000 cells/mL?

5. Let's say the A_{600} of the 5 mL overnight culture at 10:00 a.m. is 1.0 and you want to harvest a 10-mL culture of *E. coli* cells at A_{600} = 0.8 at 2:00 p.m. on the same day. You know *E. coli* doubles every 30 min under the growth conditions used. How much volume of the 5-mL overnight culture would you need to add to a fresh 10 mL of medium?

6. A protein has a MW of 135 000 dalton. What is the concentration of a 2×10^{-4} M solution in mg/mL?

7. Your DNA concentration is 10 mg/mL. How many pg/μL is this?

8. The spectrometric readings at A_{260} of a 1/100 diluted DNA sample is 0.15. What is the DNA concentration before dilution?

9. The spectrometric readings at A_{260} of 1/100 diluted RNA sample is 0.25. What is the RNA concentration before dilution?

10. A stock of human genomic DNA has a concentration of 0.25 mg/mL. You want to cut 10 μg with 20 units of the restriction endonuclease *Eco*RI. The *Eco*RI enzyme has a concentration of 2500 units/mL. (a) What volumes of DNA and enzyme should be used? (b) Set up the enzyme digestion reaction using the minimal total volume and 10× enzyme buffer.

11. The MW of 1-bp dsDNA = 660 g/mole. You have 1 μg of a 3.5-kb dsDNA fragment. How many picomoles (pmole) of ends does this represent?

12. Bacteriophage λ DNA is 48.5 kb in total size. You cut 1 μg of λ DNA with *Hin*dIII enzyme and used it for a DNA size marker in agarose gel electrophoresis. After staining the gel, you found the band intensity of 1 μL of your PCR product to be about the same as that of 4.4 kb of λ DNA cut with *Hin*dIII. Your total PCR reaction volume is 50 μL. What is the total amount (μg) of PCR product after amplification?

13. The error rates of *Taq* DNA polymerase and Phusion high-fidelity DNA polymerase are 2.28×10^{-5} and 4.4×10^{-7} mutation/bp/doubling, respectively. What percent of the 2-kb PCR products synthesized by each of the enzymes contain an error after 30 cycles of PCR assuming that the amplification efficiency is 100%?

14. The dsDNA detection limit of ethidium bromide in agarose gel is 1 to 5 ng/band. (a) What amount of a 6-kb template DNA is needed to detect a 5 ng of 1-kb amplicon after 30 PCR cycles assuming that the amplification efficiency is 50%? (b) How many template molecules are needed? The MW of 1-bp dsDNA is 660 g/mole.

15. A 0.9% NaCl (MW = 58.44) solution is equivalent to the physiological ionic strength. Calculate the ionic strength of each of 0.9% NaCl, 0.1 M sodium acetate (CH_3COONa), and 0.1 M dibasic sodium phosphate (Na_2HPO_4) and compare.

16. Circular polymerase extension cloning (CPEC) calls for the use of equimolar amounts of vector and insert DNAs. (a) If you want to use 200 ng of 2386-bp vector DNA, what amount (ng) of 258-bp insert DNA is needed? The MW of 1-bp dsDNA is 660 g/mole. (b) The DNA concentration of insert DNA is 0.05 μg/μL. How would you dilute the insert DNA so that its volume of 2.2 μL contains the required amount of insert DNA (a) in a total reaction volume of 20 μL?

17. You purified the CBP fusion protein to homogeneity by affinity chromatography and verified the presence a single protein band of 77 kDa in SDS-PAGE. You then sent it out for determination of amino acid composition. The result is shown in the table below. The average of MW of amino acids is 110.

Residue	Content (pmole)
Ala	380
Arg	160
Asp	350
Cys	25
Gly	370
Glu	370
His	120
ILe	180
Leu	300
Lys	100
Met	100
Phe	125
Pro	270
Ser	220
Thr	200
Tyr	60
Val	170

(a) How many Met residues are present in this protein?

(b) What are the maximum possible numbers of disulfide bonds that can be formed in this protein?

18. Silver (Ag^+) binds to thiol groups (–SH) in proteins and forms stable S-Ag and subsequently inactivates enzymes. Therefore, you started to add $AgNO_3$ to 1.0 mL of a solution containing 2.0 mg/mL of a pure enzyme X in order to determine the amount of AgNO3 that completely inhibit the enzyme activity. You found that a total of 0.25 μmole of $AgNO_3$ was required. Assuming that enzyme X contains two cysteines (each with a free –SH group), calculate the molecular weight of X.

19. A neuropeptide of 40 amino acids in length has been purified from a frog. N-terminal and C-terminal sequencing reveals the following information:

NH_3^+ – Met Ala Val Cys Thr Ser ... and ... Leu Ala Leu Phe Thr Asn – COOH

The neuropeptide was digested with two different protease enzymes in the presence and absence of β-mercaptoethanol and was separated by SDS-PAGE. Each peptide fragment was then eluted and sequenced. The SDS-PAGE pattern and amino acid (aa) sequence information are shown below: Lane 1, undigested control 40 aa peptide; Lane 2, protease-cleaved fragment without

SDS-PAGE Results
of Trypsin Digestion

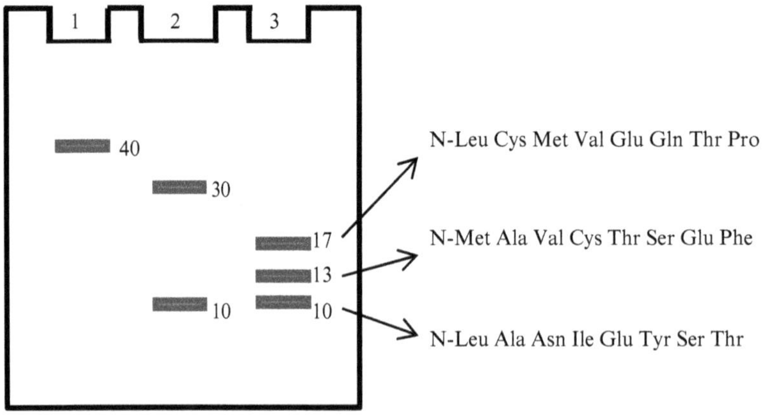

SDS-PAGE Results of
V8 protease Digestion

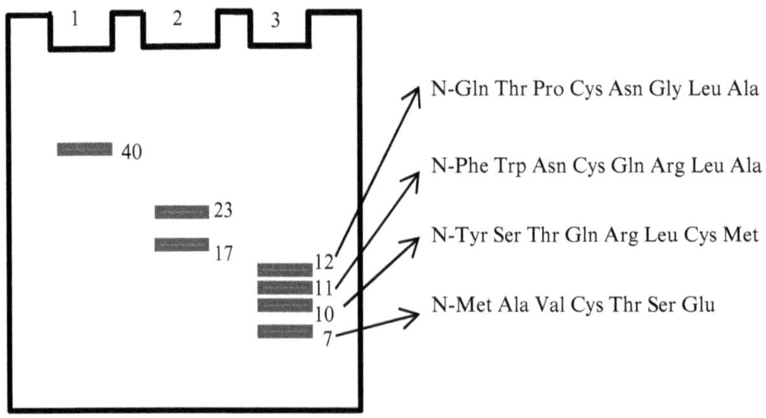

The standard genetic code			
T	C	A	G
T TTT Phe F	TCT Ser S	TAT Tyr Y	TGT Cys C
TTC	TCC	TAC	TGC
TTA Leu L	TCA	TAA STOP	TGG Trp W
TTG	TCG	TAG STOP	TGA STOP
C CTT Leu L	CCT Pro P	CAT His H	CGT Arg R
CTC	CCC	CAC	CGC
CTA	CCA	CAA Gln Q	CGA
CTG	CCG	CAG	CGG
A ATT Ile I	ACT Thr T	AAT Asn N	AGT Ser S
ATC	ACC	AAC	AGC
ATA	ACA	AAA Lys K	AGA Arg R
ATG Met M	ACG	AAG	AGG
G GTT Val V	GCT Ala A	GAT Asp D	GGT Gly G
GTC	GCC	GAC	GGC
GTA	GCA	GAA Glu E	GGA
GTG	GCG	GAG	GGG

treatment of β-mercaptoethanol; Lane 3, protease-cleaved fragment with β-mercaptoethanol treatment.

Determine the sequence of the neuropeptide and assign any disulfide bond mediated by Cys.

20. You have discovered a novel protein that is essential for an unusual cellular process in an exotic species that has not been studied before and has not had its genome sequenced. After isolating a very small amount of this protein, you were able to obtain the following peptide sequence (N-terminal to C-terminal):

Arg–Ser–Asp–Ser–Gly–Pro–Met–Lys–Gln–Tyr–Ile–Cys–Trp–Ser

Given only this sequence, how many different mixtures of a 21-nucleotide-long oligo DNA with the lowest possible degeneracy that could be used as a hybridization probe for screening a cDNA library for this protein? You must show which part of the peptide sequence is used for this probe and how you calculated its degeneracy.

AMB 2 Appendix 3

Answers to AMB 2 Lab math practice problem set

Part I. Metric unit

1. Convert the following:

• 1 gram = 10^3 mg	1 gram = 10^{-3} kg
• 1 liter = 10^3 mL = 10^3 cm^3	2 ng = 2×10^{-6} mg
• 50 mg = 0.005 g	100 μL = 0.1 mL
• 1 mg = 10^9 pg = 10^3μg = 10^6 ng	
• 1 L (liter) = 10^3 mL = 10^6 μL = 10^9 nL = 10^{12} pL = 10^{15} fL	
• 1 mM = 1 μmole/mL = 10^3 pmole/μL	
• 1 L × 1 M = 1 mole	1 mL × 1 M = 1 mmole
• 1 L × 1 mM = 1 mmole	1 mL × 1 mM = 1 μmole
• 1 ml of pure water = 1 g	1 L of pure water = 1 kg
• 1 ppm (part per million) = 10^{-4} %	10% = 10^5 ppm
• 1 mg/L = 1 ppm	100 ppm = 100 μg/mL
• 1 mg/kg = 1 ppm	

Part II. Dilution

1. $1/5^{10}$
2. $1/2^9$
3. 0.3 mg/mL
4. 0.5 mg/mL
5. 5 μL
6. 2.5×10^6 cells/mL
7. 2.5 mL of blood + 22.5 mL of PBS [1/10 = X/25 → X = 2.5 mL of blood]
8. 4 μL of antibody + 396 μL of PBS [1/100 = X/400 → X = 4 μL of antibody]
9. 40 mL of 1/40 dye solution + 60 mL water [1/100 is 2.5 times more dilute than 1/40, so 1 mL of 1/40 + 1.5 mL of water gives 2.5 mL of 1/100; 100/2.5 = 40 times more volume for stock and diluent].
10. 10 mL of 10x buffer + 90 mL of water
11. 0.6 mg/mL
12. 1/40
13. After a series of three sequential dilutions of 1:10 (i.e., 1 μL of plasmid + 9 μL of water), 1 μL of solution in the third tube contains 0.001 μg plasmid.
14. Concentration = 45 mg/mL. Total protein = 4.5 mg
15. Concentration = 1.8 mg/μL. Total protein = 180 mg
16. 1 μL stock is added to 36.5 μL in water (37.5-fold dilution) → 1.6 pmole/μL → 3.2 pmole in 2 μL
17. 16.67 mL [$2 \times (50 + V) = 8 \times V \rightarrow 6V = 100 \rightarrow V = 16.67$ mL]

Part III. Percent solution

1. 2%
2. 70 mL of absolute ethanol + 30 mL dH_2O
3. 1.8%
4. 91.4 g

   50% = 50 g/100 mL, molar ratio = 201 pentahydrate/111 anhydrous. Thus, 201/111 = X/50 g → X = 90.5 g is needed for 100% pure pentahydrate. However, pentahydrate is 99% pure and so 90.5 × (1/0.99) = 91.4 g is needed.
5. 10 g of SDS is heat-dissolved in 80 mL of dH_2O; the mixture is brought up to 100 mL, autoclaved, and stored at room temperature. If kept in the cold, SDS will precipitate.
6. 1%
7. 5 mL of acetic acid + 5 mL of methanol + 90 mL of dH_2O
8. MW of anhydrate form = 159.7; 63.9% assay of hydrate form

   MW of H_2O = 18; $5H_2O$ = 5 × 18 = 90; 249.7 − 90 = 159.7; (159.7/249.7) × 100 = 63.9%
9. 44.2 mL of HCl + 55.8 mL of dH_2O

   20% (w/v) = 20 g/100 mL. For HCl, 1 mL = 1.19 g/mL × 0.38 = 0.4522 g/mL. Therefore, 20 g ÷ 0.4522 g/mL = 44.2 mL
10. H_2O_2 = (1 × 2) + (16 × 2) = 34 g/mole

    30% (v/v) = 30 mL/100 mL = (30 × 1.46) g/100 mL = 43.8 g/100 mL = 438 g/L = 438/34 = 12.88 M = 12 880 mM. Thus, 12 880 × V = 10 mM × 100 mL → V = 0.0776 mL = 77.6 μL of 30% H_2O_2 is added to 100 mL.
11. Serum contains lots of different proteins including IgG. 0.001% serum = 0.001 g in 100 mL = 0.001 mg/mL = 1 μg/mL. Because serum contains 75% IgG that is detected by antibody, detection sensitivity of ELISA is 0.75 μg/mL.

Part IV. Ratio and proportions

1. 21% A, 29% G, 29% C
2. 20%
3. 50%
4. 69 mL
5. 12.27 mL

Methods in Biotechnology, First Edition. Seung-Beom Hong, M. Bazlur Rashid and Lory Z Santiago-Vázquez.

Companion Website: www.wiley.com\go\hong\Methodsinbiotechnology

6. 0.025 g BPB, 0.025 g XC, and 4 g sucrose
7. 8.1
8. 368.4 mL of 95% ethanol + 131.6 mL of dH_2O
9. 1000 ppm = 1 g of Na^+/L of water. The MW of NaCl = 58.45. To get 1 g of Na^+ from NaCl, 58.45/23 = X/1 → X = 58.45/23 = 2.54 g. Hence, weigh out 2.54 g of NaCl and dissolve in 1 liter volume of water.
10. 1 M $CdCl_2$ is equal to 112.4 g of Cd in 1 L = 112.4 g/L = 112 400 ppm
11. Our window of observation is 10 mL. If the concentration is 1 per 20 mL, it is equivalent to 0.5 per 10 mL. Therefore, $r = 0.5$ is the rate of occurrence relevant to our chosen window of 10 mL. As a result, probability is:

$$p(2) = \frac{0.5^2}{2! \times 2.71828^{0.5}} = \frac{0.25}{2} \times \sqrt{2.71828} = 0.0578$$

12. After the 2nd cycle, one of 4 PCR products has a mutation. Therefore, $\frac{1}{4}$ = 25% is mutated irrespective of PCR cycles. Assuming 100% of amplification efficiency, the total number of PCR products = 2^{30}; the number of mutated PCR product = 2^{28} after 30 cycles, so $2^{28}/2^{30}$ = 1/4.

Part V. Mole and molarity

1. 4×10^4 g
2. 9.63 g
3. 0.05 M
4. 0.5 M
5. 0.009 g/liter
6. 43.5 g + ~400 mLof dH_2O: dissolve in beaker; bring up to 500 mL in graduated cylinder
7. 0.37 g + ~400 mL of dH_2O: dissolve in beaker; bring up to 500 mL in graduated cylinder
8. 20 mL of 0.5 M Tris-Cl: 10 mL of 1 M Tris-Cl + 10 mL of dH_2O
20 mL of 0.15 M Tris-Cl: 3 mL of 1 M Tris-Cl + 17 mL of dH_2O
20 mL of 1 mM Tris-Cl: 0.02 mL of 1 M Tris-Cl + 19.98 mL of dH_2O
9. 4 g of NaOH is dissolved in 80 mL dH_2O, bring up to 90 mL of dH_2O, and add 10 mL of 10% SDS
10. 0.41 M [3% = 3 × 1.261 = 3.783 g in 100 mL = 37.83 g/L = 37.83/92.09 = 0.41 M]
11. 30 mL of 1 M Tris-Cl (pH 7.9), 2.4 mL of 0.5M EDTA, 150 ml of 1 M NaCl, and 417.6 mL of dH_2O
12. 0.834 M

0.125 moles/100 mL = 1.25 moles/L = 1.25 M; 25 mL × 1.25 M = 75 mL × M; M = 0.417 molarity (M) of $(NH_4)_2SO_4$. Thus, $[NH_4^+]$ = 0.417 × 2 = 0.834 M. Alternatively, 25 mL contains 0.125/4 = 0.03125 moles of ammonium sulfate. After dilution, the total volume becomes 75 mL. Therefore, molar concentration = 0.03125 moles/75 mL = 4.17×10^{-4} moles/mL = 0.417 mmole/mL = 0.417 M. Two molecules of ammonium ions are associated with one molecule of sulfate ions. Therefore, the amount increases twice but the volume remains unchanged. Thus, $[NH_4^+]$ = 0.417 × 2 = 0.834 M.

13. 0.1 M

Solution A of 125 mM = 125 mmoles/L. Therefore, 100 mL contains 12.5 mmole (125 × 0.1) of A molecules. After dilution to a total volume of 350 mL, this will give a concentration of 12.5 mmoles/0.35 L = 35.7 mmoles/L = 35.7 mM.

Solution B of 200 mM = 200 mmoles/L. Therefore, 50 mL contains 10 mmoles (200 × 0.05) of B molecules. Dilution to a total volume of 350 mL gives 10 mmoles/0.35 L = 28.6 mM. Because $Na_2HPO_4 \cdot 7H_2O$ in solution A contains two Na^+ ions and NaH_2PO_4 in solution B contains one Na^+ ion, the final Na^+ concentration is (35.7 × 2) + 28.6 = 100 mM = 0.1 M.

14. 0.01 g in 5.7 mL of water

g = V[L] × MW [g/mole] × M [mole/L]; 0.01 = V × 175 × 0.01 → V = 5.7 mL. Therefore, dissolve 0.01 g of IAA in water and bring to 5.7 mL.

15. **(a)** Prepare a 10 mM solution by dissolving 0.01 g in water and bringing up to 5.7 mL [0.01 = V × 175 × 0.01 → V = 0.0057 liter = 5.7 mL].

(b) Add 10 μL of 10 mM solution to 9990 μL of water [10 μM × 10 000 μL = 10 000 μM × V → V = 10 μL].

16. 2.2×10^{10} molecules

MW = 94 kDa = 94 000 g/mole; 1 unit = 1 mg ÷ 292 000 units = 34.25×10^{-7} mg = 34.25×10^{-10} g = 34.25×10^{-10} g ÷ 94 000 g/mole = 0.3644×10^{-13} mole. Thus, the number of molecules = (0.3633×10^{-13} mole) × (6×10^{23} molecules/mole) = 2.2×10^{10}.

17. 88.24 g

g = L × MW × M × CF, where L is liter, MW is g/mole, M is mole/liter, and CF is the correction factor =1/purity; g = 0.5 × 75 × 2 × (1/0.85) = 88.24 g.

18. 114.38 mL

g = 1 × 60.05 × 2 × (1/0.99) = 121.3 g, which is equivalent to 121.3 g/1.05 g mL^{-1} = 114.38 mL.

19. 40.76 mL of 2 N H_2SO_4 + 709.24 mL of water

2 N H_2SO_4 = 1 M H_2SO_4

g = 0.75 × 98.08 × 1 × (1/0.9808) = 75 g, which is equivalent to 75 g/1.84 g/mL = 40.76 mL.

20. NaCl dissociates into Na^+ and Cl^- in solution, and so 308 mOsm/2 = 154 mmole/L = 0.154 mole/L of NaCl. This concentration requires 0.154 mole/L × 58.44 g/mole = ~9.0 g/L = 0.9 g/100 mL = 0.9%.

21. One liter of 50% (v/v) glycerol contains 500 mL of glycerol whose mass is 500 × 1.26 = 630 g. Thus, 630 g/L/93.09 g/mole = 6.77 mole/L = 6.77 M.

22. One kg of 50% (w/w) glycerol contains 500 g water and 500 g glycerol. The volume of 500 g water is 500 mL, whereas the volume of glycerol is 500/1.26 = 396.83 mL. Therefore, the total volume of 1 kg of 50% (w/w) glycerol is 896.83 mL, and glycerol concentration is 500/0.8963 = 557.85 g/L. Thus, 557.85 g/L/92.09 g/mole = 6.06 mole/L = 6.06 M.

23. The total concentration of protein in cytosol is 1.2 g/mL × 0.2 = 0.24 g/mL = 240 g/L = 240 g/L ÷ 100 000 g/mole = 2.4×10^{-3} mole/L = 2.4×10^{-3} M.

Part VI. Applications

1. MW = ($6 \times 10^9 \times 660$) g/mole → one copy = ($6 \times 10^9 \times 660$) ÷ (6×10^{23}) = 6.6×10^{-12} g = 6.6 pg. Thus, 10^4 copies = 6.6×10^4 pg = 66 ng.
2. 1 μL
3. 1 mL of 20 mg/mL of lysozyme
4. 120 min
5. 0.03 mL

ΔT = 4 h; DT = 0.5 h → n = 4/0.5 = 8. A_{600} is equivalent to cell density and so $N_f = 0.8 \times 10 = 8 = N_i \times 2^8 \rightarrow N_i = 8/256 = 0.03125$. Because overnight culture cell density A_{600} = 1.0 → 0.03125/1.0 = ~0.03 mL volume. Therefore, you need to add 30 μL.

6. 27 mg/mL [dalton = g/mole; 1 M = 135 000 g/mole = 135 000 g/L; 135 000 × 2 × 10^{-4} g/L = 27 g/L]

7. 10^7 pg

8. 0.75 mg/mL

9. 1 mg/mL

10. **(a)** 40 μL of DNA and 8 μL of *Eco*RI enzyme

(b) The minimal total volume is 10 times of enzyme volume, i.e., 80 μL: 24 μL of dH_2O (add first), 8 μL of 10× buffer, 40 μL of DNA, and 8 μL of *Eco*RI enzyme (add last)

11. 0.86

3.5 kb = 3500 × 660 = 2.31 × 10^6 g/mole; 1 μg = 1 × 10^{-6}/2.31 × 10^6 = 0.43 × 10^{-12} mole = 0.43 pmole of 3.5-kb DNA present. Therefore, DNA ends = 2 × 0.43 = 0.86 pmole ends.

12. 4.54 μg

13. Mutation rate (frequency) = ER (error rate) × bp (target DNA length) × *n* (number of doubling cycles) × AE (amplification efficiency). For Taq DNA polymerase, (2.28 × 10^{-5}) × 2000 × 30 × 1 = 1.368; i.e., 136.8% of PCR products contain one nucleotide error. This means that every PCR product molecule contains an average of 1.4 errors. For Phusion high-fidelity DNA polymerase, (4.4 × 10^{-7}) × 2000 × 30 × 1 = 0.0264; i.e. 2.64% of PCR products contain one nucleotide error. This means that 97.36% of PCR products contain no errors.

14. **(a)** $N_f = N_i \times (1 + E)^n \rightarrow 5\text{ ng} = N_i \times 1.5^{30} \rightarrow N_i = 5/1.5^{30}$ = 5/190 546 = 26 × 10^{-6} ng = 26 fg.

(b) MW of a 6-kb DNA template = 6000 × 660 = 4 × 10^6 g/mole. Thus, 26 fg of template = 26 × 10^{-15} g = (26 × 10^{-15}) ÷ (4 × 10^6) = 6.5 × 10^{-9} mole = (6.5 × 10^{-9}) × 6 × 10^{23} = 39 × 10^{14} molecules.

15. Ionic strength $I = \frac{1}{2}\Sigma c_i Z_i^2$, where c_i is the molar concentration (M) of ionic species and Z_i is the net charge of the species.

- For 0.9% NaCl (=9/58.44 = 0.15 M), $I = \frac{1}{2}([Na^+] \times 1^2 + [Cl^-] \times 1^2) = \frac{1}{2}(0.15 \times 1 + 0.15 \times 1) = 0.15$ M.
- For 0.1 M CH_3COONa, $I = \frac{1}{2}([Na^+] \times 1^2 + [CH_3COO^-] \times 1^2) = \frac{1}{2}(0.1 \times 1 + 0.1 \times 1) = 0.1$ M (smaller than physiological ionic strength).
- For 0.1 M Na_2HPO_4, $I = \frac{1}{2}([Na^+] \times 1^2 + [HPO_4^{2-}] \times 2^2) = \frac{1}{2}(0.2 \times 1 + 0.1 \times 4) = 0.3$ M (larger than physiological ionic strength).

16. **(a)** 22 ng

MW of 2386-bp vector DNA = 2386 × 660 = 1 574 760 g/mole ≈ 1.57 × 10^6 g/mole.

200 ng of vector DNA = 200 × 10^{-9} g = (2 × 10^{-7})/(1.57 × 10^6) = 1.27 × 10^{-13} mole.

MW of 258-bp insert DNA = 258 × 660 = 170 280 g/mole ≈ 1.7 × 10^5 g/mole.

Thus, 1.27 × 10^{-13} mole of insert DNA = 1.27 × 10^{-13} × 1.7 × 10^5 ≈ 2.2 × 10^{-8} g = 22 ng.

(b) Insert DNA = 50 ng/μL. Therefore, 1/50 dilution (i.e., 1 μL of insert DNA + 49 μL water) yields 1 ng/μL; 2.2 μL of this 1/50 dilution results in 2.2 ng DNA.

17. **(a)** 20 Met

The total amino acid content is 3500 pmole and the total number of amino acids is 77 000/110 = 700 residues. Met residue is 100 pmole. Thus, Met = 700 × (100/3500) ≈ 20.

(b) 10 disulfide bonds

Cys reside is 25 pmole = 700 × (25/3500) ≈ 5. Because only 2 combinations are possible for disulfide bonds, 5 × 2 = 10.

18. Total mass of protein X used in the assay = 1.0 mL × 2.0 mg/mL = 2 mg = 0.002 g. Total moles of protein X bound to Ag^+ = 0.25/2 = 0.125 μmole, because the ratio of Ag^+ to protein is 2:1. Thus, 0.125 μmole is equivalent to 0.002 g of protein X. As a result, MW of protein X = 2 × 10^{-3} g/0.125 × 10^{-6} mole = 16 000 g/mole = 16 kDa.

19. The S–S bond occurs between trypsin 17 and 13, between V8 protease 12 and 11 fragments, and between V8 protease 10 and 7 fragments according to the SDS-PAGE analysis. However, the

NH_3^+-Met Ala Val Cys Thr Ser... and ... Leu Ala Leu Phe Thr Asn-COOH

Trypsin 17: N- Leu Cys Met Val Glu Gln Thr Pro

Trypsin 13: N-Met Ala Val CysThr Ser Glu Phe

Trypsin 10: N- Leu Ala Asn Ile Glu Tyr Ser Thr

V8 protease 12: N-Gln Thr Pro Cys Asn Gly Leu Ala

V8 protease11: N-Phe Trp Asn Cys Gln Arg Leu Ala

V8 protease10: N-Tyr Ser Thr Gln Arg Leu Cys Met

V8 protease 7: N-Met Ala Val Cys Thr Ser Glu

NH_3^+- Met Ala Val Cys Thr Ser Glu Phe Trp Asn Cys Gln Arg Leu Ala Asn Ile Glu Tyr Ser Thr

Gln Arg Leu Cys Met Val Glu Gln Thr Pro Cys Asn Gly Leu Ala Leu Phe Thr Asn-COOH

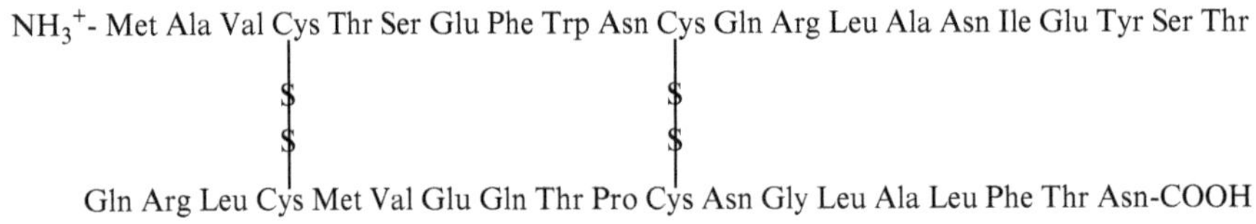

S–S bond between V8 protease 10 and 7 fragments is the same as the one between trypsin 17 and 13 fragments, based on the sequence. Thus, the disulfide bond formation can be assigned as shown above.

20. The degeneracy is related to the number of possible sequences that encode the same sequence. This is determined from the product of the degeneracy of each codon:

Arg	Ser	Asp	Ser	Gly	Pro	<u>Met</u>	<u>Lys</u>	<u>Gln</u>	<u>Tyr</u>	<u>Ile</u>	<u>Cys</u>	<u>Trp</u>	Ser
6	6	2	6	4	4	1	2	2	2	3	2	1	6

The lowest 7 aa codon degeneracies in the underlined row is the best (lowest degeneracy) overall since the overall degeneracy is the product of those codon degeneracies = $1 \times 2 \times 2 \times 2 \times 3 \times 2 \times 1 = 48$.

AMB 2 Appendix 4

Colony counting with ImageJ program

1. Open the picture from the File menu in ImageJ.

Note: it is important to use a good image picture to minimize faulty results. When you take a picture, open the Petri plate cover.

2. From the Menu Bar, click "Image" → "Type" →"8 bit." This will create a gray scale (0 to 255 shades) image.
3. Click "Image" → "Adjust" → "Threshold" to highlight the Petri plate you want to count. The threshold window will pop up.
4. Using the lower slide bar in the Threshold window, carefully adjust the background red color so that colonies stand out and everything present in the original image appears in the threshold image.
5. Click "Apply" in the Threshold window. This will convert an 8-bit gray image to a binary image of the black = 0 and white = 255.
6. From the Menu Bar, click "Process" → "Binary" → "Watershed." This will cut the merged colonies apart. This step may be omitted if colonies are well separated.

7. From Tool Bar, select Oval ( ) or Freehand ( ) and outline the area in which you want to count colonies.
8. From the Menu Bar, select "Analyze" → "Analyze Particles." Check the Display Results, Clear Results, and Summarize. Click "OK" →. A window will appear showing how many particles (colonies) are in the chosen area.

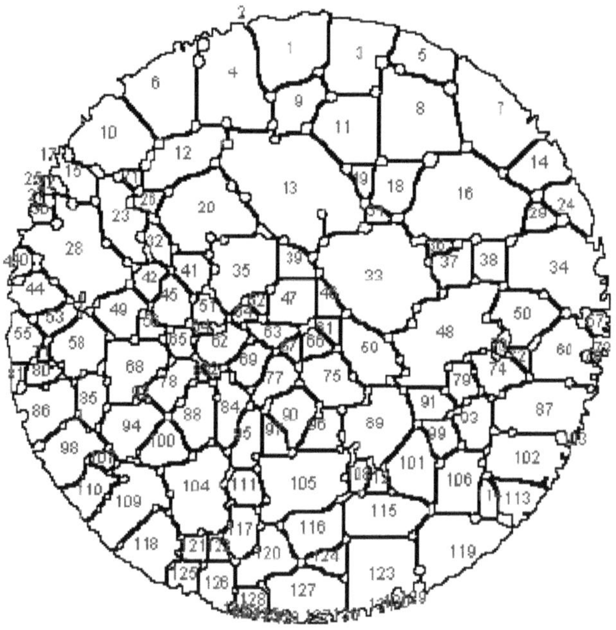

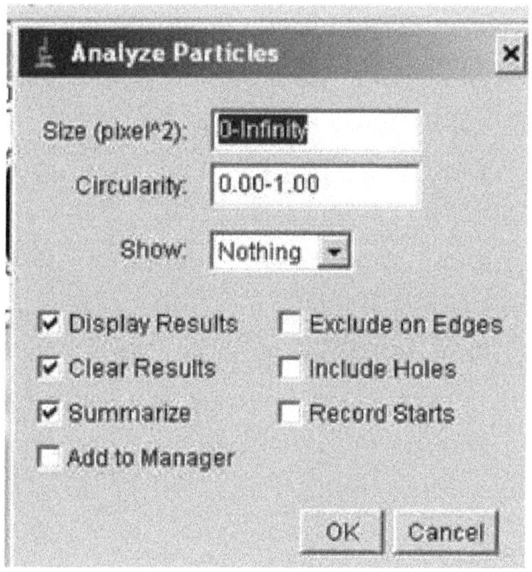

Count	Total Area	Average Size	%Area	Mean
175	83070	474.686	84.174	255

Methods in Biotechnology, First Edition. Seung-Beom Hong, M. Bazlur Rashid and Lory Z Santiago-Vázquez.

Companion Website: www.wiley.com\go\hong\Methodsinbiotechnology

AMB 2 Appendix 5

Bacteria and yeast genetic nomenclature

Bacteria genetic nomenclature

(a) Bacterial gene symbols consist of three lower-case italic letters, with a suffixed different capital letter to distinguish different genes affecting the same metabolic pathway function. For example, *trpA* and *trpB* are required for tryptophan synthesis but are different wild-type genes involving the same tryptophan biosynthetic pathway in *E. coli*.
(b) A mutation or allele is denoted by a suffixed italic number. For example, *trpA1*and *trpA2* are specific mutations in the *trpA* gene where 1 and 2 denote unique allele numbers.
(c) If the exact mutated gene locus is unknown, the capital letter is replaced with a hyphen. For example, *trp-1* is a mutated *trp* locus where the exact position affected is unknown.
(d) Deletion of a gene is indicated by Δ followed by the gene symbol. For example, Δ(*lac-proAB*) denotes that genes from *lac* to *proAB* are deleted.
(e) Insertion is denoted by a gene symbol where insertion took place, followed by the :: symbol and the inserted gene symbol. For example, *trypA::Tn10* indicates that *Tn10* is inserted into the *trypA* gene.
(f) Protein products are indicated by the corresponding gene symbol but with an initial capital letter and in non-italic type. For example, TrpA and TrpB are proteins encoded by *trpA* and *trpB*, respectively.
(g) Phenotypes are designated by a three-letter symbol with an initial capital and in non-italic type; a wild-type phenotype is denoted by a superscript plus sign ($^+$), whereas a mutant phenotype is indicated by a superscript minus sign ($^-$). For example, Trp$^+$ indicates a strain with a wild-type normal tryptophan metabolism, whereas Trp$^-$ indicates a strain with a mutant defective tryptophan metabolism.
(h) Genotypes are designated by a three lower-case italic letters; a wild-type genotype is denoted by a superscript plus sign ($^+$), whereas a mutant genotype is indicated by a superscript minus sign ($^-$). For example, *trp*$^+$ indicates a strain with wild-type *trp* genes, whereas *trp*$^-$ indicates a strain with mutant *trp* genes with a defective tryptophan metabolism.

Yeast genetic nomenclature

The rules for genetic nomenclature for diploid yeast differ from the rules for haploid bacteria.
(a) Yeast gene or locus symbols comprise three italic letters and a suffixed number; the letter is upper case and lower case for dominant alleles and recessive alleles, respectively. Wild-type genes are designated with a superscript plus ($^+$). For example, *ADE1* and *ADE2* are required for adenine synthesis but are different wild-type genes involving the same adenine biosynthetic pathway in yeast. *ADE2* denotes a dominant allele, whereas *ade2* indicates a recessive mutant allele. *ADE2*$^+$ denotes a wild-type allele.
(b) Mutant alleles are designated by the gene symbol followed by a hyphenated italic Arabic number. For example, *ade2-101* is a specific recessive mutation in the *ADE2* gene where number 101 denotes a unique allele number.
(c) Insertion is denoted by the gene symbol where insertion took place, followed by the *::* symbol and the inserted gene symbol. For example, *arg2::LEU2* indicates that a functional *LEU2* gene is inserted into the *ARG2* locus, and *arg2* becomes non-functional and is recessive. *ARG2::LEU2* indicates that a functional *LEU2* gene is inserted into the *ARG2* locus, and *ARG2* remains functional and dominant.
(d) Deletion of a gene is indicated by Δ preceded by the gene symbol. For example, *trp1*Δ denotes that the *TRP1* gene is completely or partially deleted and becomes non-functional and recessive.
(e) Protein products are designated by the corresponding gene symbol in non-italic type with an initial capital letter, suffixed by a lower-case "p" for protein. For example, Ade2p is the protein encoded by the *ADE2* gene.
(f) Phenotypes are denoted by three letter gene symbols in non-italic type with an initial capital letter; a superscript plus or minus denotes the wild-type or mutant, respectively. For example, Arg$^+$ denotes a wild-type strain, whereas Arg$^-$ denotes an auxotrophic mutant strain requiring arginine.
(g) To describe a genome location of a gene or ORF (open reading frame), each ORF is assigned a systematic name in the form of a seven-character alphanumeric formula. The first three letters define the host origin (Y for yeast), the specific chromosome (A through P to denote chromosomes I through XVI), and whether the ORF lies to the left (L) or right (R) of the centromere. This

Methods in Biotechnology, First Edition. Seung-Beom Hong, M. Bazlur Rashid and Lory Z Santiago-Vázquez.

Companion Website: www.wiley.com\go\hong\Methodsinbiotechnology

is followed by three digits listing the relative ORF position from the centromere. Finally, the letter W or C is inserted to indicate expression from the Watson (5′ to 3′) or Crick (3′ to 5′) strand of DNA.

For example, YKL074C defines the 74th ORF (or gene) to the left of the centromere of chromosome XI and denotes expression (coding strand) from the Crick strand. YNL323W defines the 323rd ORF (or gene) to the left of the centromere of chromosome XIV and denotes expression (coding strand) from the Watson strand.

Further reading

Casali, N. (2003). *Escherichia coli* host strains. In *Methods in Molecular Biology* (Edited by N. Casali and A. Preston), Vol **235**, pp. 27–48. Humana Press Inc., Totowa, NJ.

http://www.sci.sdsu.edu/~smaloy/MicrobialGenetics/topics/mutations/nomenclature-v3.pdf

http://www.swan.ac.uk/genetics/big322/Yeast%20Book/6.html

http://dbb.urmc.rochester.edu/labs/Sherman_f/yeast/6.html

AMB 2 Appendix 6

Plasmid maps

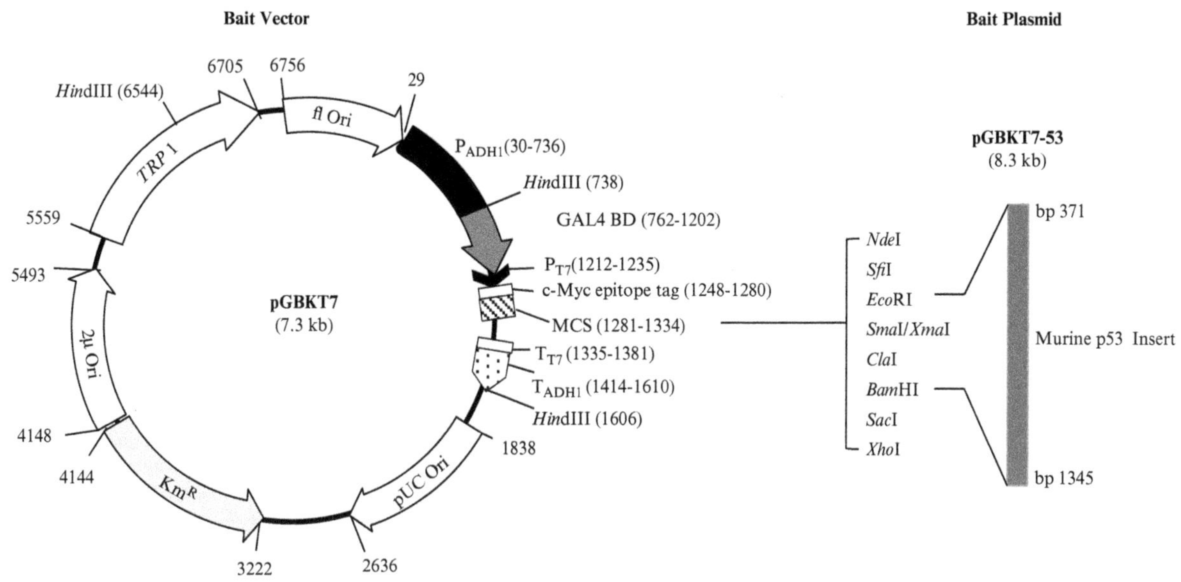

1155 TCATCGGAAGAGAGTAGTAACAAAGGTCAAAGACAGTTGACTGTATCGCCGGAATTTGTA

Part of GAL4 BD (DNA binding domain)

F-Primer

1215 ATACGACTCACTATAGGGCGAGCCGCCATCATGGAGGAGCAGAAGCTGATCTCAGAGGAG

p_{T7} (T7 promoter) c-Myc Epitope Tag

1275 GACCTGCATATGGCCATGGAGGCCGAATTCCCGGGGATCCGTCGACCTGCAGCGGCCGCA

*Nde*I *Sfi*I *Eco*RI *Sma*I/*Xma*I *Bam*HI *Sal*I *Pst*I

1326 TAACTAGCATAACCCCTTGGGGCCTCTAAACGGGTCTTGAGGGGTTTTTTGCGCGCTTGC

P_{T7} (T7 terminator)

R-Primer

1410 AGCCAAGCTAATTCCGGGCGAATTTCTTATGATTTATGATTTTTATTATTAAATAAGTTA

Methods in Biotechnology, First Edition. Seung-Beom Hong, M. Bazlur Rashid and Lory Z Santiago-Vázquez.

Companion Website: www.wiley.com\go\hong\Methodsinbiotechnology

The murine p53 cDNA (GenBank Accession No., K01700) was cloned into the *Eco*RI and *Bam*HI sites of pGBKT7 (7.3 kb), resulting in the in-frame fusion of murine P53 (a.a 72-390) to the GAL4 DNA-BD (1-147). The bait insert plasmid is now called pGBKT7-53.

The SV40 large T antigen (GenBank Locus SV4CG) was cloned into the *Eco*RI and *Xho*I sites of the pGADT7 (8.0 kb) vector, resulting in the in-frame fusion of the SV T antigen (a.a. 87-708) to the GAL4 AD (a.a. 768-881). The prey insert plasmid is now called pGADT7-T.

For more information about pGBKT7-53 and pGADT7-T, refer to Matchmaker™ BioSensor Kit User Manual (Protocol No. PT3584-1 (PR14058), Cat. No. 630446, and Protocol No. 3249-5 (PR8Z151); Cat. No. K1612-1), Clontech Laboratories, Inc. and Matchmaker GAL4 Two-Hybrid Vectors Handbook (Protocol No. PT3062-1, Version No. PR6X890), Clontech Laboratories, Inc.

Plasmid pCL1 contains a gene encoding the full-length, wild-type GAL4 protein under the control of the constitutive promoter of the yeast alcohol dehydrogenase gene (P_{ADH1}). The

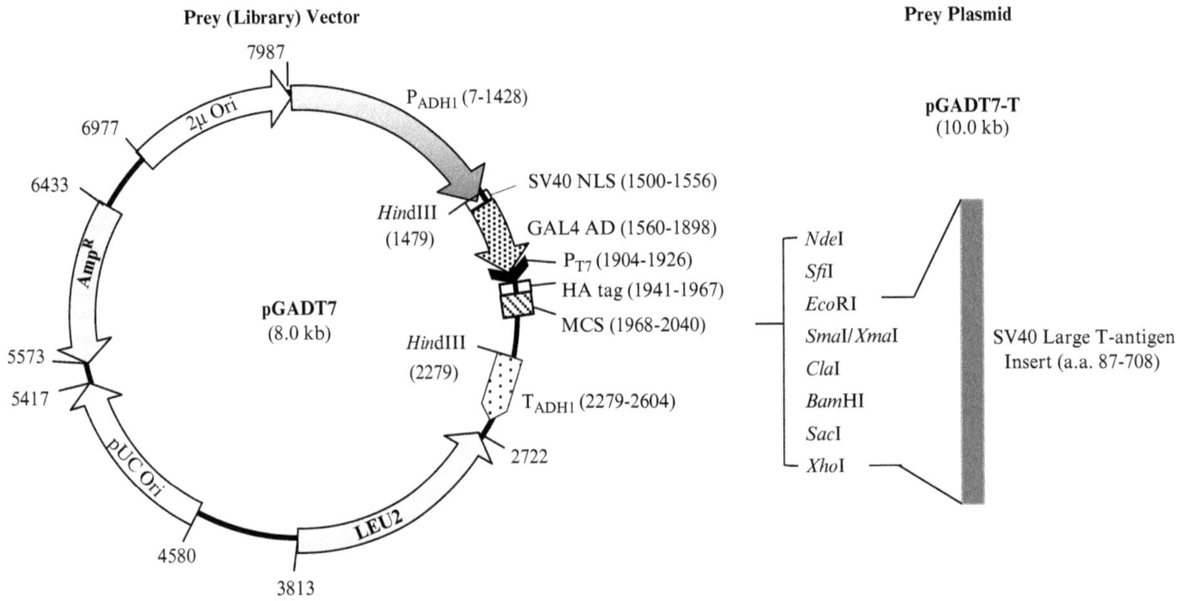

5'AD Insert Screening F primer

1857 CTATTCGATGATGAAGATACCCCACCAAACCCAAAAAAAGAGATCTTTAATACGACT

C-terminal portion of GAL4 AD (activation domain) P_{T7}(T7

1914 CACTATAGGGCGAGCGCCGCCATGGAGTACCCATACGACGTACCAGATTACGCT CAT

promoter) Start for P_{T7} HA Epitope Tag *Nde*I

1971 ATGGCCATGGAGGCCAGTGAATTCCACCCGGGTGGGCATCGATACGGGATCCATCGA

*Sfi*I *Eco*RI *Sma*I/*Xma*I *Cla*I *Bam*HI *Sac*I

2028 GCTCGAGCTGCAGATGAATCGTAGATACTGAAAAACCCCGCAAGTTCACTTCAACTG

*Xho*I 3'AD Insert Screening R primer

2085 TGCATCGTGCACCATCT

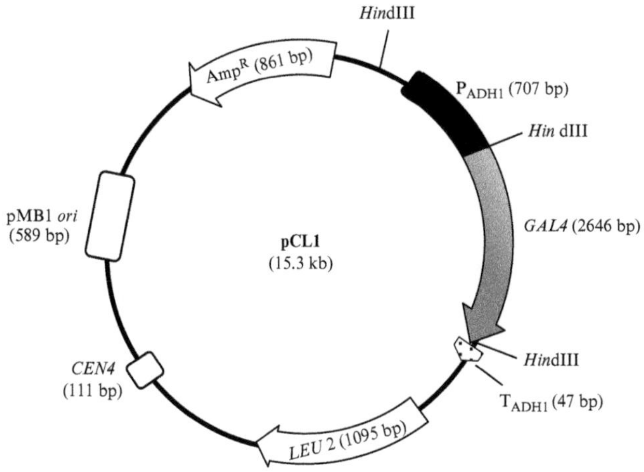

GAL4 protein activates any reporter genes under the control of a *GAL4*-responsive upstream activating DNA sequence element (UAS). Reproduced with Permission of Takara-Clontech from MATCHAKER GAL4 Two-Hybrid Vectors Handbook (Protocol No. PT3062-1, Version No. PR6X890), Clontech Laboratories, Inc. and Fields, S. and Song, O. (1989). A novel genetic system to detect protein–protein interactions. *Nature*, **340**: 245–246.

AMB 2 Appendix 7

Distributor addresses

Microbiological strains

- *E. coli* **DH5α** (*F-, endA1, glnV44, thi-1, recA1, relA1, gyrA96, deoR, nupG, Φ80dlacZΔM15, Δ(lacZYA-argF)U169, hsdR17(rK- mK+), λ–*): NEB, Cat. No. C29871
- *E. coli* **TOP10** (*Δ(ara-leu) 7697 araD139 fhuA ΔlacX74 galK16 galE15 e14- ϕ80dlacZΔM15 recA1 relA1 endA1 nupG rpsL (StrR) rph spoT1 Δ(mrr-hsdRMS-mcrBC)*): NEB, Cat. No. C30191 (NEB, 10-beta competent *E. coli* high efficiency) (https://www.neb.com/products/c3019-neb-10-beta -competent-e-coli-high-efficiency)
- *E. coli* **HB101 K-12** (*F⁻ mcrB mrr hsdS20(r_B^- m_B^-) recA13 leuB6 ara-14 proA2 lacY1 galK2 xyl-5 mtl-1 rpsL20(Sm^R) glnV44 λ⁻*): Bio-Rad, Cat. No. 166-0408EDU
- *E. coli* **Lemo21 (DE3)** (*fhuA2 [lon] ompT gal (λ DE3) [dcm] ΔhsdS/ pLemo*(Cam^R) *λ DE3 = λ sBamHIo ΔEcoRI-B int::(lacI::PlacUV5::T7 gene1) i21 Δnin5*, pLemo = pACYC184-*PrhaBAD-lysY*): NEB, Cat. No. C2528H (https://www.neb.com/products/c2528-lemo21de3-competent-e-coli)
- *S. cerevisiae* **Y187** (*MATα, ura3-52, his3-200, ade2-101, trp1-901, leu2- 3,112, met–, gal4Δ, gal80Δ,URA3::GAL1$_{UAS}$ -GAL1$_{TATA}$-lacZ, MEL1*): Clontech, Cat. No. 639457
- *S. cerevisiae* **Y2HGold** (*MATa, trp1-901, leu2-3, 112, ura3-52, his3-200, gal4Δ, gal80Δ, LYS2::GAL1$_{UAS}$-GAL1$_{TATA}$-HIS3, GAL2$_{UAS}$-GAL2$_{TATA}$-ADE2, URA3::MEL1$_{UAS}$-MEL1$_{TATA}$, AUR1-C, MEL1*): Clontech, Cat. No. 630498
- *S. cerevisae* **BY4733** (MATa *ade2Δ::hisG, his3Δ200, leu2Δ0, lys2delta0, met15Δ0, trp1Δ63, ura3Δ0*): ATCC 200895 (http://www .atcc.org/Products/All/200895.aspx)
- *S. cerevisae* **DPB276** (*LEU2::pTEF-Dcr1, TRP1::pTEF-Ago1, can1-100, ade2-1, HIS3::pGAL1-hpSC_URA3*): Dr. David Bartel, Whitehead Institute/MIT/HHMI (dbartel@wi.mit.edu)
- *S. cerevisae* **DPB260** (*MATα, LEU2::pTEF-Dcr1, TRP1::pTEF-Ago1, can1-100, ura3::EGFP(S65T)-KanMX6, ade2-1, HIS3::pGAL1-strong SC_GFP*): Dr. David Bartel, Whitehead Institute/MIT/HHMI (dbartel@wi.mit.edu)
- *S. cerevisae* **BY4741** (*MATa, his3Δ1, leuΔ0, met15Δ0, ura3Δ0*): ATCC 201388 (http://www.atcc.org/Products/All/201388.aspx)

Plasmid DNAs

- **pUC19:** NEB, Cat. No. N3041S (https://www.neb.com/ products/n3041-puc19-vector)
- **pUCWHITE:** Department of Biotechnology, University of Houston Clear Lake, Houston, TX 77058
- **pGLO**: Bio-Rad, Cat. No. 16604505 (http://www.bio-rad.com /en-us/sku/1660405-pglo-plasmid?parentCategoryGUID=2)
- **pCAL-kc**: Agilent Technologies, Cat. No. 214301
- **pCAL-GFP$_{uv}$:** Department of Biotechnology, University of Houston Clear Lake, Houston, TX 77058
- **pGBKT7:** Clontech Matchmaker GAL4 two-hybrid system 3 and libraries kit (PT3247-1)
- **pGADT7:** Clontech Matchmaker GAL4 two-hybrid system 3 and libraries kit (PT3247-1)
- **pGBKT7-53:** Clontech Matchmaker GAL4 two-hybrid system 3 and libraries kit (PT3247-1)
- **pGADT7-T:** Clontech Matchmaker GAL4 two-hybrid system 3 and libraries kit (PT3247-1)
- **pCL1:** Clontech Matchmaker GAL4 two-hybrid system 3 and libraries kit (PT3247-1)
- **p414-TEF1p-Cas9-CYC1t:** http://www.addgene.org/43802/
- **p426-SNR52p-gRNA.CAN1.Y-SUP4t:** http://www.addgene .org/43803/
- **pYES2.1-GFP$_{uv}$:** Department of Biotechnology, University of Houston Clear Lake, Houston, TX 77058

Methods in Biotechnology, First Edition. Seung-Beom Hong, M. Bazlur Rashid and Lory Z Santiago-Vázquez.

Companion Website: www.wiley.com\go\hong\Methodsinbiotechnology

Glossary

Terms and Definitions

Absorbance: A quantitative measure of the amount of a given wavelength light absorbed by a substance; also called optical density.

Absorption spectrum: A pattern of absorbance of a given sample when exposed to specific wavelengths or ranges of wavelengths.

Absorptivity coefficient: A light absorbance value of a particular substance measured at a specific wavelength under specific conditions of pH, temperature, and solvent.

Accession number: Unique identification letter and number code assigned to every DNA sequence deposited in GenBank database.

Acid dissociation constant: The equilibrium constant (K_a) for the reaction in which a weak acid is in equilibrium with its conjugate base in aqueous solution.

Agar: A purified extract of seaweed algae used for a solidifying agent of culture media.

Agarose: A chemically inert, electrically neutral polysaccharide polymer purified from agar that is suitable for electrophoresis.

Allele: One of the different forms of a gene located at a single locus on a chromosome.

***Alu* element:** A short stretch of repeated DNA characterized by the presence of the *Alu* (*Arthrobacter luteus*) restriction enzyme site.

Ampholyte: The mixture of small oligo-molecules (300–1000 Da) that contain both acidic and basic buffering groups with closely spaced isolectric points (pIs).

Anabolic: The biochemical process involved in the synthesis of metabolites.

Analyte: A particular compound that will be measured and identified.

Annotation: Adding relevant information such as gene name, deduced amino acid sequence, promoter region, or other comments to the database entry of the DNA sequence.

Anode: A positive electrode.

Antibiotic: A compound that kills or inhibits the growth of bacteria.

Antibody: A protein molecule produced by plasma cells in the immune system that recognizes and binds to an antigen with high specificity and affinity; also known as immunoglobulin.

Antigen: A molecule that is recognized by and bound to an antibody.

Amino acid: The basic building block of protein that is composed of an α-amino group, a carboxyl group, and a side chain group.

Amplification: The production of many identical copies of a specific DNA fragment from a template DNA; can be *in vivo* or *in vitro*.

Apoptosis: Programmed cell death in which a cell triggers its own death in response to an external or internal signal when a cell is no longer needed due to its unhealthy or old status.

Argonaute (AGO): A family of RNase H (ssRNase) protein that cleaves the target mRNA strand complementary to their bound siRNA or their bound miRNA.

Aseptic technique: A set of specific procedures to reduce or eliminate the chance of contamination of biological samples and instruments from microorganisms.

Aspirate: Remove a liquid layer of supernatant using a pipette by hand or suction vacuum.

ATP (adenosine triphosphate): A ribonucleotide 5′-triphosphate that captures a cellular energy in the bonds between the phosphates during catabolic processes and releases its energy during anabolic processes.

Autoradiography: An imaging technique to detect the intensity and position of a radioactivity signal on an X-ray film.

Auxotroph: A mutant organism defective in either synthesis or breakdown of a particular biomolecule due to the loss or absence of an anabolic enzyme and catabolic enzyme, respectively. Humans are an anabolic auxotroph for vitamin and essential amino acids. Yeast is a catabolic auxotroph for lactose.

Base dissociation constant: The equilibrium constant (K_b) for the reaction in which a weak base is in equilibrium with its conjugate acid in aqueous solution.

Basic local alignment search tool (BLAST): Internet-based method to search against GenBank database to find similar DNA or protein sequences by rapid sequence comparison.

Beer's Law: A rule stating that light absorbance (A) is directly proportional to the concentration (c) of the absorbing substance and the path length (l) of the light through the solution substance having an extinction coefficient (ε); also known as the Beer–Lambert Law: $A = \varepsilon cl$.

BioBrick: Restriction enzyme fragments carrying a functional unit such as a promoter, terminator, coding sequence, ribosome binding site, replication origin, plasmid backbone, etc. BioBrick parts are matched and mixed to assemble new genetic circuits through compatible cohesive ends. Standard BioBrick parts are available from iGEM registry organization (http://parts.igem.org/Main_Page).

Biuret: A mixture of chemical compounds that is used to detect the presence of peptide bonds based on appearance of violet color and to assess the concentration of proteins based on color intensity.

Blank: A reference that contains no analyte but contains the solvent and any reagents besides a test substance.

Blunt end: The end of linear double-stranded DNA that has no single-stranded sequence in both 3′ and 5′ end.

Methods in Biotechnology, First Edition. Seung-Beom Hong, M. Bazlur Rashid and Lory Z Santiago-Vázquez.

Companion Website: www.wiley.com\go\hong\Methodsinbiotechnology

Borosillicate: A salt of boric and silicic acids making glasses resistant to thermal shock.

Bradford dye: Coomassie blue dye that is used to determine the concentration of proteins based on a shift in the absorption spectrum of the dye–protein complex.

Buffer: Aqueous solution consisting of a mixture of a weak acid and its conjugate base, or vice versa, to maintain a nearly constant pH.

Burette: A graduated glass tube, commonly having a stopcock at the bottom, used for dispensing measured amounts of a chemical solution.

Calibration: (1) Adjustment of a measuring device for accurate volumes (micropipettor), weight (balance), or pH (meter) with respect to numerical readings. (2) Use of standards with known amounts of analyte to establish the linear relationship between the spectrophotometric absorbance and the analyte concentration.

Canavanine: A toxic analog of arginine amino acid; often used for a negative selection of wild-type cell producing arginine permease (*CAN*1) and for a positive selection of arginine permease-deficient mutant (*can1*).

Cas9 protein: *Streptococcus pyogenes* endonuclease that produces double-stranded breaks at specific sites targeted by the crRNA-guide sequence. The *Cas* gene is associated with the CRISPR locus as a functional part of the adaptive immune system to resist invading foreign genetic elements.

Catabolic: The biochemical process involved in the degradation of metabolites.

Catabolite repression: Repression of catabolism of other carbon compounds when a preferred (most easily metabolizable) carbon source is utilized.

Catalysis: Increase in the rate of a biochemical reaction.

Cathode: A negative electrode.

Chaotropic: The ability of a substance to interfere with hydrogen bonds and van der Waals forces.

Chelator: A compound able to bind to and remove metal ions from solution.

Chelex: A chelating ion exchange resin able to bind metal ions.

Chi (χ^2)-square test: A statistical test used to compare observed data with calculated data we would expect to obtain according to the null hypothesis stating that there is no significant difference between the expected and observed results.

Chromatogram: The visual representation of individual peaks separated by chromatography or four nucleotide peaks in a DNA sequencing sample produced by a sequencing machine.

Chromatography: A process in which complex mixtures of molecules are separated on a column containing a mobile phase (solvent) and a stationary phase (resin).

Chromogenic: Producing a color.

Circular polymerase extension cloning (CPEC): A directional cloning technique in which both the linearized vector and PCR fragments that have a short 15- to 25-bp flanking overlapping vector DNA sequence at their ends are directly subject to PCR amplification without adding primers using a high-fidelity Phusion polymerase.

***Cis*-acting (regulatory) element:** Non-coding DNA sequence that is bound by one or more *trans*-acting factors such as a transcriptional activator or repressor and a DNA nicking or replication enzyme. It is located near a gene they regulate on the same DNA.

Clarification: Removal of unwanted matter by centrifugation to produce supernatant or by filtration to produce filtrate.

Clone: An exact copy of biological materials such as a DNA segment, a whole cell, or a complete organism.

Cloning: Producing many copies of identical DNA molecules, cells, or organisms from a single DNA molecule, cell, or organism, respectively.

Codon: A consecutive 3-nuclotide sequence in mRNA that codes for a specific amino acid.

Cofactor: An inorganic ion required for enzyme activity.

Colony: A visible mass of microorganisms that originated from multiplication of a single cell on an agar plate.

Colorimeter: A type of spectrophotometer that measures the amount of a specific visible wavelength of light transmitted through a color sample. See also "Spectronic 20."

Competent: Able to take up exogenous DNA and is thus transformable.

Complementary: The status in which nucleotide sequences can base-pair with each other.

Complementary DNA (cDNA): A synthetic DNA made from a specific mRNA by the action of reverse transcriptase enzyme.

Conjugate acid: A base that gained a hydrogen ion (H^+).

Conjugate base: An acid that lost a hydrogen ion (H^-).

Conjugated antibody: An antibody linked covalently to a color-producing enzyme or fluorescent dye label for detection of an antigen.

Consensus sequence: The best-fit common sequence in which each position represents the base/amino acid that most frequently occurs when many aligned sequences are compared.

Constitutive: Always expressed regardless of conditions.

Cosmid: A hybrid vector DNA that contains a lambda bacteriophage *cos* sequence essential for phage packaging in a plasmid.

CRISPR (clustered regularly interspaced short palindromic repeat): Bacterial or archaea DNA locus that contains an array of short direct repeats separated by short spacers of similar length that was derived from degradation of foreign DNAs such as phages and plasmids by a Cas–crRNA complex.

CRISPR RNA (crRNA): The mature short RNA fragments processed from a long pre-crRNA that resulted from transcription of a CRISPR repeat-spacer array DNA sequence. crRNA is the transcript of spacer sequence and requires tracrRNA to bind to Cas protein. The crRNA-tacrRNA complex directs Cas9 to target a specific invading foreign DNA and makes site-specific dsDNA breaks.

Cuvette: A sample tube that is designed to fit a spectrophotometer or electroporator.

Cycle threshold (C_T): The number of PCR cycles at which signal intensity is above the background level of intensity with no template control. This value changes depending on the initial target copy number and is thus used to calculate the fold change in real-time quantitative PCR.

Dalton (Da): Unit of atomic or molecular weight; 1 Da is equivalent to 1 g/mole.

Death phase: The end stage of cell growth during which the cell begins to die due to the lack of nutrients and accumulation of wastes.

Degrees of freedom (df): The actual number of values that are free to vary or choose in a statistical calculation.

Denaturation: Structural destruction of a macromolecule.

Denhardt's solution: A mixture of blocking agents used for preventing the non-specific binding of nucleic acids to membrane in Southern blot hybridization experiments.

Dephosphorylation: Removal of a phosphate group from the 5′ end of each strand of dsDNA by treatment with dephosphorylase enzyme.

Dicer: Cytoplasmic RNase III enzyme that cleaves dsRNA and stem-loop pre-miRNA to produce siRNA and miRNA, respectively.

Diploid: A cell having two chromosome sets or an individual having such cells.

Directional cloning: A cloning procedure that allows a target linear DNA fragment to insert into a plasmid vector or join with other linear DNA fragments only in one direction.

Dismutation: A reaction in which oxidized and reduced forms of a chemical species are produced simultaneously.

Dissociation constant: The equilibrium constant when a salt dissociates into its component ions.

Drosha: Nuclear RNase II enzyme that cleaves a hairpin stem-loop structure of primary miRNA (pri-miRNA) to produce precursor miRNA (pre-miRNA).

Dyad symmetry: See palindromic sequence

Electrokinetic injection: A voltage-induced siphoning effect that occurs in the capillary whose one end is placed into the buffer reservoir on the cathode end and the other end into the buffer reservoir containing the sample to be analyzed on the anode end.

Electropherogram: A densitometric intensity plot produced by an automatic DNA sequencing machine that shows sequences.

Electrophoresis: Movement of charged molecules under an electrical field in order to separate mixtures of biomolecules.

Electroporation: Introduction of an exogenous DNA into cells by applying a high-voltage pulse.

End point: The pH point at which a pH indicator changes color in a colorimetric titration.

Enzyme: A protein or RNA molecule that enhances the rate of a specific chemical reaction by lowering activation energy.

Enzyme activity: A measure of the enzyme amount that catalyzes conversion of 1 μmole substrate to product (or the formation of 1 μmole product) per min; often expressed as a unit (U).

Enzyme-linked immunosorbent assay (ELISA): A technique that uses a specific antibody linked to an enzyme and a color-changing substrate to detect an antigen in a liquid sample.

Equilibrium: A balanced state in which forward and reverse chemical reactions proceed at equal rates.

Equilibrium constant: A constant characteristic for each chemical reaction in which all reactants and products are in an equilibrium status.

Equivalence point: The pH point at which chemically equivalent quantities of acid and base have been mixed.

Erlenmeyer flask: A flask that features a flat bottom, a conical body, and a cylindrical neck.

Exon: Any non-intron section of the coding sequence of a gene that will be transcribed into mRNA.

Expressioneering: A cloning technique that utilizes *in vivo* homologous recombination to directly clone a PCR fragment into an expression vector by simply transforming both the linearized vector and PCR fragments that have a short 18 bp flanking overlapping vector DNA sequence into a special *E. coli* strain developed by Lucigen Inc.

Extinction coefficient: See "Absorptivity coefficient."

Facultative: Able to grow in either the presence or absence of oxygen.

FASTA format: The format of sequence presentation for *in silico* analysis in which a greater than (">") symbol beginning in the first column of a single-line description is followed by the lines of single-letter codes of nucleotide or amino acid sequence data.

Fastidious: Able to grow only when specific nutrients are present in a culture medium.

Fluorescence: Emission of light of longer wavelength by excited molecules that absorb light of shorter wavelength.

Fluorophore: A fluorescent molecule that can emit light upon light excitation.

Fractionation: The process of separating molecular components of complex mixtures into fractions based on the different properties such as solubility, net charge, size, affinity, and density.

Gap: The loss of one or more nucleotide sequence in one of the two strands in dsDNA.

Gateway cloning: A directional cloning technique that utilizes *in vitro* recombination enzymes called BP and LR clonasees™ (Invitrogen) and defined *att* sequences that are attached to the ends of both linearized plasmid vector and PCR fragment. The recombination enzymes specifically recognize the *att* sequence and recombine to join the two DNAs.

GelRed™: A non-cytotoxic, non-mutagenic dye that emits fluorescence when bound to dsDNA.

GenBank: Database bank of genetic sequences maintained by a division of the National Institutes of Health.

GenBank format: The format of sequence presentation that begins with the word LOCUS and many annotation lines, followed by single-letter codes of nucleotide or amino acid sequence data. The sequence data begin with a line having "ORIGIN" and ends with a two slashes mark ("//").

Gene conversion: The unidirectional gene transfer process in which a DNA segment containing one allele is lost and replaced by another homologous DNA segment carrying an alternative allele without DNA crossover of strand exchange.

Genetic regulatory circuits: Clusters of genes involved in controlling gene expression through transcription factors (repressor, activator, co-activator) and regulatory DNA binding elements.

Genome: The entire genetic DNA material possessed by a cell.

Genotype: The genetic constitution of an organism dictated by the specific allelic composition of a cell.

Gibson assembly cloning: A directional cloning technique in which two or more liner DNA fragments having at least a 16-bp overlap sequence are joined using T5 exonuclease, Phusion polymerase, and *Taq* ligase. The complementary ssDNA 3′ overhang ends of linear DNA fragments are produced by T5 exonuclease, the gaps occurred after the complementary base pairing is filled by Phusion polymerase, and the nicks made after filling-in is sealed by *Taq* ligase. Multiple DNA fragments having flanking overlaps between them can be joined in a single reaction using this method; also called isothermal assembly cloning.

Golden Gate cloning: A directional assembly cloning technique that utilizes the type IIS restriction enzyme to join multiple DNA fragments in a single step of ligation.

Guide RNA (gRNA): An artificial fusion of crRNA and tracrRNA that confers both targeting specificity and scaffolding/binding ability for Cas9 nuclease; gRNA does not exist in nature; also called "sgRNA (synthetic single-guide RNA)."

Halo: A circular clear zone in the turbid background of a bacterial culture lawn on an agar plate; this results from the death or growth inhibition of bacteria due to the diffusion of antibiotics from a disk placed in the agar plate.

Haploid: A cell having one chromosome set or an organism having such cells.

Hardy–Weinberg equation: The equation used to calculate the genetic variation (i.e., the frequency of alleles) in a population: $p^2 + 2pq + q^2 = 1$, where p and q are the frequencies of the alleles A and a, respectively, as a result of random mating in the absence of mutation, migration, natural selection, or random drift.

Hemizygote: A diploid organism having only a single copy of a gene instead of two copies.

Henderson-Hasselbalch equation: The equation used to calculate, pH, pK_a, and the molar concentrations of acid (HA) and base (A^-) in a solution:

$$\text{pH} = \text{p}K_a + \log \frac{\text{Molarity of base } [A^-]}{\text{Molarity of acid } [HA]}$$

Heterozygote: An individual having a heterozygous gene pair (gene pair of different alleles).

High-dose hook effect: A dosage effect in immunoassay reaction in which a measured level of an excessive amount of antigen exhibits a significantly lower absorbance signal than the actual level present in a sample, and thus detection sensitivity decreases with a high range of concentrations of analyte.

Homologous recombination (HR): A DNA repairing mechanism in which broken double-stranded DNA ends are joined using a homologous DNA template. This repairing is accurate and restores the original nucleotide sequence.

Homozygote: An individual having a homozygous gene pair (a gene pair having identical alleles in the two chromosome sets of the diploid individual).

Host: An organism capable of supporting the replication and growth of smaller organisms or foreign nucleic acid elements as a result of infection or transformation.

Housekeeping genes: Genes that are always expressed (i.e., constitutively expressed) because of their constant demand by cells.

HTML (HyperText Markup Language): A standard language used to create Internet web pages.

Hybridization: The process in which two complementary single strands of nucleic acids are allowed to base-pair each other and form a double-stranded molecule.

Hydrodynamic: The motion (diffusion) of a macromolecule in relation to the aqueous solvent in which it is dissolved or suspended.

Hydrogen bond: The electrostatic attraction between a strongly electronegative atom (oxygen or nitrogen) and a hydrogen atom.

Hydrolysis: Cleavage of a chemical bond in a molecule using water.

Hydrophilic: Polar molecules that dissolve easily in water.

Hydrophobic: Non-polar molecules that are insoluble in water.

HyperText Markup Language: See HTML.

Inducible: Capable of increasing gene expression by adding a small amount of a small inducer molecule or by changing growth conditions.

Induction: An increase in gene expression by the activity of regulatory proteins.

InFusion cloning: A directional cloning technique similar to SLIC that utilizes InFusion™ enzyme having $3' \rightarrow 5'$ exonuclease activity (Clontech).

Inoculum: A biological material introduced into a fresh medium in order to initiate a culture.

***In silico*:** The use of computers for analyses.

Intergenic region: A non-transcribed DNA region that lies between genes.

Intron: An intervening sequence that is initially transcribed from a gene but excised in functional mRNA to be used for translation.

***In vitro*:** In vessels away from an organism.

***In vivo*:** In a living organism.

Ionization constant: See "Dissociation constant."

Isothermal assembly cloning: see "Gibson assembly cloning."

Isoelectric point (pI): The pH at which a solute has no net electric charge and thus does not move in an electric field.

Kimwipe: A disposable wipe paper used for cleaning lab equipment and instruments.

Kinetics: The study of reaction rates.

Kirby–Bauer test: A technique used to determine if a microorganism is susceptible or resistant to antibiotics by culturing the microbe on agar growth media surrounding a test drug and inspecting the clear zone of growth inhibition.

Lag phase: The early stage of growth during which no signs of growth occurs after inoculation of a culture.

LB (Luria–Bertani): A complex rich medium for culturing bacteria.

LB buffer: Lithium borate buffer used for fast agarose gel electrophoresis.

Ligase: An enzyme that catalyzes the joining of two molecules of DNA or RNA at the expense of ATP energy.

Lineweaver–Burk equation: The reciprocal of both sides of the Michaelis–Menten equation so that the values of K_m and V_{max} can be easily determined in a straight line graph plotting the reciprocal of substrate concentration versus the reciprocal of enzyme velocity; also known as the double-reciprocal equation:

$$\frac{1}{V_0} = \frac{K_m}{V_{max}} \frac{1}{[S]} + \frac{1}{V_{max}}$$

Log(arithmic) phase: The stage of explosive growth during which cell division occurs at the maximum rate; also called the exponential phase.

Lysate: A fluid of lysed cells after enzymatic and/or mechanical disruption.

Lysis: Disintegration of the cell wall and membrane to release cellular contents.

Mapping: A graphic drawing of the relative positions of restriction enzyme sites or genes on the DNA segment or entire chromosome.

Maxi-prep: A large-scale procedure for isolation of nucleic acid molecules.

Medium (plural: media): A nutrient with a formulated composition in order to grow organisms in the laboratory.

Metagenome: Genomes of all the individual microorganisms present in an environmental sample.

Michaelis–Menten equation: The mathematical formula of enzyme-catalyzed reactions describing the hyperbolic dependence of the initial reaction velocity V_0 on substrate concentration $[S]$:

$$V_0 = \frac{V_{max}[S]}{K_m + [S]}$$

where V_{max} is the maximum velocity and K_m is the substrate concentration at $\frac{1}{2}V_{max}$.

Micro RNA (miRNA): A small non-coding 20–25 bp long dsRNA molecule with a 3′-end UU overhang produced by Drosha and Dicer enzymes from a large primary mRNA containing a hairpin stem-loop structure.

Microtiter plate: A flat multiwell plastic plate that is used for transferring serially diluted microliter volumes.

Milli-Q water: High-quality pure water produced by successive steps of filtration and deionization process by the Milli-Q® device developed by the Millipore Company.

Minimum inhibitory concentration (MIC): The minimum concentration of a substance necessary for inhibiting the growth of microbes.

Mini-prep: A small-scale procedure for isolation of nucleic acid molecules.

Mutation: An inheritable change in DNA that differs from wild-type DNA.

NanoDrop: A benchtop spectrophotometer that permits a direct measurement of microliter volumes (0.5–2.0 μL) of samples without the use of a cuvette.

Nick: A single-strand break as a result of the lack of a phosphodiester bond between adjacent nucleotides of one strand in a dsDNA.

Non-homologous end joining (NHEJ): A DNA repairing mechanism in which broken double-stranded DNA ends are ligated without a homologous DNA template. This repairing is inaccurate and introduces deletion or insertion mutations.

Normalize: Adjust data for the effects that arise from variation in the assay techniques (non-biological contributions) rather than from biological differences between the samples. For example, using differing numbers of cells or differing amounts of cell extract will affect the final outcomes. Therefore, the final result data must be adjusted based on the quantitative data of cell numbers or extracts.

Nucleotide: A molecule composed of a nitrogen base, a sugar, and a phosphate group as a basic building block of nucleic acids.

Null hypothesis: A statistical hypothesis that investigators want to reject or nullify, typically proposing that there is no statistical significance in a set of given observational data because of chance.

Optical Density: See "Absorbance."

Orthologs: The homologous genes in terms of sequence and function that exist in the genomes of different organisms.

Ouchterlony: A technique that uses a diffusion of both antibody and antigen in agarose gel to detect antibody–antigen interaction by visible precipitation lines; also known as double immunodiffusion.

Overhang: A short single-stranded DNA sequence that extends out from the 3′ or 5′end of a double-stranded DNA. A 5′ overhang end has the single-stranded bases ending with a 5′-phosphate. A 3′ overhang end has the single-stranded bases ending with a 3′-hydroxyl group.

Overlap extension (OE) cloning: A directional cloning technique in which both a circular plasmid vector and a linear DNA insert fragment are mixed and PCR-amplified using the chimeric primers that have a short flanking overlapping sequence between each end of the linear insert fragment and the insertion point of the plasmid vector. After amplification, the parental plasmid vector is removed by *Dpn*I enzyme digestion.

Oxidation: The loss of electrons from a compound.

Oxidizing agent (oxidant): The acceptor of electrons in an oxidation–reduction reaction.

Oxidation–reduction (redox) reaction: A reaction in which electrons are transferred from a donor to an acceptor molecule.

Palindrome: A nucleotide sequence in a dsDNA molecule where the same sequence is found in the opposite direction (an inverted repeat sequence without any intervening sequence inside the repeat sequence) on each strand; also called "dyad symmetry."

PAM (protospacer adjacent motif): A 3-bp NGG nucleotide sequence located at the 3′end of protospacer; this motif is essential for recognition of targeting DNA in CRISPR-Cas 9 gene editing system.

Paralogs: The homologous genes that occur in two or more different positions of the same genome of an organism.

Peptide: Two or more amino acids covalently joined by peptide bonds.

Peptide bond: A linkage between the α-amino group of one amino acid and the α-carboxyl group of another, with the removal of a water molecule.

Permeabilize: To make a cell wall or membrane permeable using surfactant substances.

Peroxidase: An enzyme that catalyzes oxidation of a substrate using hydrogen peroxide as the electron acceptor.

Petri plate: A covered glass or plastic dish used to grow microorganisms on agar-containing culture medium.

pH: The negative logarithm of proton (H^+) concentration of an aqueous solution.

Phenolphthalein: A white crystalline compound that is frequently used as a pH indicator in acid–base titration, displaying a pink color at pH 8.2 to 12.0.

Phenol red: A red crystalline compound that is frequently used as a pH indicator, exhibiting a gradual transition from yellow to red over the pH range 6.8 to 8.2.

Phenotype: The visible characteristics of an organism.

Pheromone: A small diffusible molecule that is secreted from an organism to influence the behavior of the opposite sex.

Phred: A base-calling program that reads DNA sequence chromatogram files and analyzes the peaks to call bases, assigning quality scores to each base call.

Phusion polymerase: A high-fidelity DNA polymerase enzyme that produces a blunt end of PCR fragment.

PicoGreen: A dye compound that has high specificity to dsDNA and emits fluorescence signal upon binding to dsDNA at the detection limit of 25 pg/mL of dsDNA.

pK_a: The negative logarithm of an acid dissociation constant. It is the pH at which a buffer is most resistant to change in pH upon addition of acids or bases.

Plasmid: A small supercoiled circular DNA molecule that replicates independently of chromosome.

Plunger: Part of a micropipettor device that is used to draw or repel the liquid sample in the micropipette tip.

Polymorphism: The occurrence of multiple alleles at a genetic locus.

Polymerase: An enzyme that catalyzes the synthesis of nucleic acids on pre-existing nucleic acid templates.

Polypeptide: A long chain of amino acids linked by peptide bonds.

Primer: A short oligonucleotide to which DNA polymerase enzyme adds new deoxyribonucleotide phosphates.

Priming: Initiation of DNA synthesis from the 3′-OH end of a DNA fragment by DNA polymerase.

Prion: An infectious protein particle that causes degenerative disorders of the brain nerve cells.

Probe: A labeled single-stranded nucleic acid fragment that can be used to identify a specific DNA segment carrying complementary nucleotide sequences by hybridization.

Promoter: A DNA sequence to which RNA polymerase and its associated factors bind to initiate transcription.

Protein: A macromolecule consisting of one or more polypeptide chains.

Protospacer: The genome target DNA of at least 20 bp in length that is complementary to spacer DNA present in a crRNA or gRNA coding sequence.

Pulse-field gel electrophoresis: A technique for separating very large DNA fragments by subjecting agarose gel to electrical fields alternating between different angles.

Quenching: The process that decreases the fluorescence intensity of a molecule.

Radical: A chemical species having an unpaired electron in the outermost shell of the atom; also called a "free radical." It can have a positive, negative, or neutral charge and is highly reactive by losing an electron or gaining or sharing an electron from any neighboring molecule in order to complete its own outer shell.

Reading frame: one of three possible ways of translating a single-stranded nucleotide sequence into codons.

Recombinant DNA: An artificially made hybrid DNA molecule containing two or more sources of genetic materials, typically consisting of vector and insert molecules.

Redox: See "oxidation and reduction."

Reducing agent (reductant): The electron donor on an oxidation–reduction reaction.

Reducing sugar: A sugar that is able to be oxidized and causes the reduction of other substances; monosaccharides or disaccharides have a free aldehyde (–CHO) or ketone (–CO) group.

Reduction: The gain of electrons by a compound or ion.

Rennin: A proteolytic enzyme that is produced by stomach cells in ruminant animals to impede the rapid flowing of liquid milk through the stomach by coagulating milk protein caseins. It is also known as chymosin.

Repression: A decrease in gene expression by the activity of regulatory proteins.

Restriction enzyme: An enzyme that cut both strands of DNA at a specific site; also called restriction endonuclease.

Restriction site: The nucleotide sequence recognized and cut by restriction enzyme.

Retrovirus: An RNA virus carrying reverse transcriptase.

Reverse complementary sequence: The DNA sequence resulting from converting a DNA sequence into its antiparallel reverse strand counterpart.

Reverse transcriptase: An enzyme that catalyzes the synthesis of a DNA from an mRNA template.

RISC (RNA-induced silencing complex): A large ribonucleoprotein silencing complex that consists of argonaute, dicer, and transactivating response RNA-binding protein (TRBP).

RNA interference (RNAi): A post-transcriptional gene silencing mechanism in which specific mRNAs are either degraded or inhibited for translation into protein.

RNA-Seq (RNA sequencing): A transcriptome profiling technique that utilizes next-generation high-throughput DNA sequencing technology; also called "whole transcriptome shotgun sequencing."

Salting in: Increased solubility of a solute as the result of the increase in the ionic strength of a solution.

Salting out: Precipitation of a solute as the result of the increase in the ionic strength of a solution.

Screening: A procedure that allows all organisms to grow but sorts them by differential phenotypes.

Selection: A procedure that favors the growth of a certain organism while inhibiting the growth of others.

Selfing: To fertilize eggs with sperms from the same individual.

Semi-log graph: A graph that has a logarithmic scale on the *y* axis and a linear scale on the *x* axis.

Sequence and ligation-independent cloning (SLIC): A directional cloning technique that utilizes the base-pairing between the complementary ssDNA 5′ overhang ends of both linearized vector and PCR fragments carrying a short 16–25 bp flanking overlapping vector DNA sequence, which are created by a 3′ to 5′ exonuclease activity of T4 DNA polymerase in the absence of dNTPs. Multiple DNA fragments having flanking overlaps between them can be joined in a single reaction using this method.

Serum: The clear liquid separated from clotted blood.

Single guide RNA (sgRNA): See "gRNA."

Singlet oxygen: An excited form of oxygen in which one of the electrons reverses its spin to result in the antiparallel spinning of two electrons in separate orbitals in its outer shell when oxygen absorbs sufficient energy.

Small interfering RNA (siRNA): 20–25 bp long dsRNA molecules with a 3′ end UU overhang produced from long dsRNA molecules by the Dicer enzyme.

SOC (super optimal broth with catabolite repression): A rich medium used in bacterial transformation to increase transformation efficiency.

Solution: A liquid sample of solute(s) dissolved in a solvent.

Solute: A substance dissolved in a given solution.

Solvent: A liquid substance that dissolves a solute.

Somaclonal variation: Spontaneous changes in genetic or phenotypic traits of cloned somatic cells that occurred during tissue culture.

Somatic cell: Any cell in the body except gametes and their precursors.

Sonication: The process in which ultrasound wave energy is used to disrupt the cell wall and membrane or to agitate particles in solution.

Southern blotting: A DNA blot hybridization procedure used in order to detect a specific DNA fragment in complex mixtures of DNA. Typically restriction fragments are separated in agarose gel and transferred to a membrane, which is then hybridized to a complementary labeled, single-stranded nucleic acid probe.

Spacer: Short non-repetitive individual fragments of similar lengths of foreign DNA (resulting from degradation by the Cas–crRNA complex) that are integrated into a CRISPR locus. Each spacer is derived from different sequences of foreign DNAs.

Specific activity: A measure of enzyme activity that catalyzes conversion of 1 μmole substrate to product (or the formation of 1 μmole product) per min per mg of protein (μmole min^{-1} mg^{-1}).

Spectronic 20: A spectrophotometer developed by Bausch & Lomb, Inc. that measures the amount of a visible light wavelength absorbed by a color solution. See also "colorimeter."

Spectrophotometer: An instrument that quantitatively measures the amount of a specific wavelength light passing through a medium. Depending on the wavelength spectrum used, there are ultraviolet, visible, and infrared types.

Spectrum: Electromagnetic radiation separated by different wavelengths of light.

Spheroplast: An osmotically sensitive bacterial or yeast cell that has a cell membrane but lacks a rigid cell wall.

Stationary phase: The stage of stagnant growth during which cells either stop growing or grow very slowly.

Steady state: A flowing state in which all components remain stable at a constant concentration.

Sterilization: The process that removes or kills all forms of life.

Sticky end: Two short single-stranded ends protruding from a linear double-stranded DNA that are complementary to each other, facilitating ligation of the ends; also known as cohesive end.

Stoichiometry: The quantitative molar ratio relationships of reactants and products in a balanced chemical reaction.

Strain: The cells of a single species descended from a common ancestor that retains the original genotype.

Stringency: The conditions of hybridization or washing that influence the strength of base pairing between the probe and target DNA.

Subcloning: Cloning by moving a shorter DNA fragment of interest to a destination vector.

Subculture: Transfer of existing cultured cells to a fresh growth medium.

Substrate: The molecule upon which an enzyme acts to convert into a product.

Successive hybridization assembly (SHA): A directional assembly technique in which multiple DNA fragments that have overlapping sequences are simply treated with denaturation and renaturation in a single reaction tube.

Supernatant: The upper liquid layer separated from a sediment pellet after centrifugation.

Synthetic biology: A branch of biology that focuses on developing new biological systems by mixing and matching different genetic building blocks.

T4 DNA polymerase: Bacteriophage T4 enzyme that has a 5′ → 3′ DNA polymerization activity and a 3′ → 5′ proofreading exonuclease activity but lacks a 5′ → 3′ exonuclease activity; can be used to fill in 5′ overhangs or to remove 3′ overhangs to make blunt ends in the presence of dNTPs. However, the 3′ → 5′ exonuclease activity hydrolyzes both ss and ds DNA in the absence of dNTPs.

T5 exonuclease: An enzyme that has 5′ → 3′ exonuclease activity producing a 3′ overhang end on each strand of dsDNA.

TA cloning: A cloning technique that utilizes the ability of a 3′-adenine (A) overhang tail at each end of PCR fragments to base-pair with a 3′-thymine (T) overhang tail on a linearized vector DNA before joining together with T4 DNA ligase.

Tandem repeats: An array of identical sequences of two or more nucleotides that lie directly adjacent to each other on the chromosome.

***Taq* ligase:** A heat-stable ligase enzyme that seals DNA nicks at 50 °C.

TARE: Reset the zero of the balance scale display to subtract the weight of the empty container from the total weight before the weight of a substance is determined.

TATA box: A short stretch of AT-rich nucleotide sequences that RNA polymerase bind prior to the initiation of transcription.

Template: A molecular mold for the biosynthesis of genetic information molecules such as DNA and RNA.

Titer: The highest dilution (i.e., the lowest concentration) that still shows a positive reading or symptom in a serially diluted solution, such as antibody titer or virus titer. It is expressed as a dilution ratio or dilution factor.

Titrant: An acid or base solution of known concentrations to be used for titration.

Titration: A procedure for carrying out a chemical reaction for neutralization between the acid and base solutions by controlled addition of a known concentration of titrant to an analyte of an unknown concentration.

Titration curve: A plot of pH versus the amounts of titrant added during titration of an analyte.

TOPO TA cloning: A cloning technique similar to TA cloning that utilizes Vaccinia virus topoisomerase for joining two DNA fragments instead of T4 ligase.

Tracking dye: A colored solution to be mixed with a loading sample so that the progress of electrophoresis can be visually monitored.

***Trans*-acting (regulatory) element:** Protein-coding sequences that regulate a distantly located gene. Typically it encodes transcription factors or other DNA binding proteins. See also "*cis*-acting element."

Trans-activating crRNA (tracrRNA): The endogenous bacterial RNA that links the crRNA by base pairing to form a dual RNA molecule (tracrRNA:crRNA), which now directs Cas9 protein to introduce double-stranded DNA breaks at specific sites targeted by the crRNA guide sequence.

Trans-activating response RNA binding Pprotein (TRBP): A cofactor that has three dsRNA binding domains and interacts with dicer.

Transcription: A process in which a particular portion of DNA sequence in one strand is copied into its complementary RNA sequence by RNA polymerase.

Transcriptome: The entire set of RNA molecules present in a given cell or tissue under a certain condition.

Transformation: Introduction of a foreign DNA into a cell, causing the cell to acquire a new phenotype.

Transgenic: Transformed by integration of a new exogenous DNA into the genome of higher eukaryotes.

Transmittance: The ratio of the amount of light transmitted through the sample to that transmitted through the blank solution.

Turnover number (K_{cat}): The number of moles of substrate that react to form product per mole of enzyme per unit time.

T-vector: The cloning vector used for TA cloning in which a thymidine (T) residue is added to the 3′ end of each strand on a linearized vector using TdT enzyme and ddTTP substrate.

Type IIS restriction enzyme: An enzyme that cuts dsDNA outside its recognition site to create 5′ or 3′ overhang cohesive ends.

Ubiquitin: A small highly conserved protein that tags (covalently links to) a cellular target protein for degradation by proteasome.

Ultrafiltration: A technique used to separate macromolecules on the basis of molecular weight cutoff (MWCO) of membrane porosity.

Uracil: A nitrogenous base normally found in RNA but not DNA.

Vector: Autonomously replicating DNA molecule that can carry a foreign DNA fragment for its propagation or expression in a host cell. Plasmids, bacteriophages, plant or animal virus DNAs, cosmids, and bacterial or yeast artificial chromosomes are the major types of vectors.

Western blot: A procedure to detect a specific protein in a complex mixture of proteins. Typically proteins are separated in SDS-polyacrylamide gel and transferred to a membrane, which is then allowed to react with a specific antibody; also called immunoblot.

Whole genome sequencing: Sequencing the entire genome of an organism at a single time using high-throughput sequencing technologies.

Wild type: The genotype or phenotype that is normally found in nature.

Yeast two hybrid: A technique that screens for interacting proteins using bait and prey expression vectors in yeast cells.

Zwitterion: A dipolar ion having both positive and negative charges that are spatially separated.

Abbreviations

A_{260}	Absorbance at 260 nm
Ab	Antibody
Ag	Antigen
Amp	Ampicillin
APS	Ammonium persulfate
Ara	L-arabinose
Arg	Arginine
ATP	Adenosine triphosphate
BCIP	5-Bromo-4-chloro-3-indolylphosphate
BLAST	Basic local alignment search tool
bp	Base pair
β-gal	beta-galactosidae
BPB	Bromophenol blue
BRCA	Breast cancer
BSA	Bovine serum albumin
Cas	CRISPR-associated
CBP	Calmodulin binding peptide
cDNA	Complementary DNA
CDS	Coding DNA sequence
cfu	Colony forming unit
CHAPS	3-[(3-Cholamidopropyl) dimethylammonio]-1-propanesulfonate
CIA	Chloroform-isoamylalcohol
Cm	Chloramphenicol
conc.	Concentration
CPEC	Circular polymerase extension cloning
CrRNA	CRISPR RNA
CRISPR	Clustered regularly interspaced short palindromic repeat
CsCl	Cesium chloride
Da	Dalton
dATP	Deoxyadenosine triphosphate
dCTP	Deoxycytidine triphosphate
dGTP	Deoxyguanosine triphosphate
DEAE	diethylaminoethyl
dH_2O	Deionized water
DMSO	Dimethyl sulfoxide
ddNTP	Dideoxyribonucleotide triphosphates
dNTP	Deoxyribonucleotide triphosphates containing an equimolar mixture of dATP, dCTP, dGTP, and dTTP
ds	Double-stranded
DTT	Dithiothreitol
dTTP	Deoxythymidine triphosphate
DNase	Deoxyribonuclease
DO	Dropout
ϵ	Extinction coefficient
EDTA	Ethylene-diamine-tetraacetic acid
EGTA	Ethylene glycol tetraacetic acid
ELISA	Enzyme-linked immunosorbent assay
EtBr	Ethidium bromide
EtOH	Ethanol
5-FOA	5-Flouro-orotic acid
g	Gravity
GADPH	Glyceraldehyde 3-phosphate dehydrogenase
GFP	Green fluorescent protein
Glc	D-Glucose
gRNA	Guide RNA
GTE	Glucose-Tris-EDTA
h	Hour
His	Histidine
HRP	Horseradish peroxidase
HTML	HyperText Markup Language
HTTP	Hypertext transfer protocol
IAA	Iodoacetamide
IEF	Isoelectric focusing
Ig	Immunoglobulin
IPG	Immobilized pH gradient
IPTG	Isopropyl-1-thio-β-D-galactoside
K_a	Acid discussion constant
K_b	Base discussion constant
Kb	Kilobase
kDa	Kilodalton
Km	Kanamycin
LB buffer	Lithium borate buffer
LB medium	Luria–Bertani medium
LED	Light-emitting diode
Leu	Leucine
LiAc	Lithium acetate
Log	Logarithm
MCS	Multiple cloning site
2(β)-Me	2(β)-Mercaptoethanol
Met	Methionine
MIC	Minimum inhibitory concentration
min	Minute
miRNA	Micro RNA
mRNA	Messenger RNA

Methods in Biotechnology, First Edition. Seung-Beom Hong, M. Bazlur Rashid and Lory Z Santiago-Vázquez.

Companion Website: www.wiley.com\go\hong\Methodsinbiotechnology

MW Molecular weight
NBT Nitro blue tetrazolium chloride
NCBI National Center for Biotechnology Information
NGS Next generation sequencing
NHEJ Non-homologous end joining
nm Nanometer
ntd Nucleotide
OD Optical density
ONPG *Ortho*-nitrophenyl-β-D galactoside
ORF Open reading frame
ori Origin of replication
PAGE Polyacrylamide gel electrophoresis
PBS Phosphate-buffered saline
PCR Polymerase chain reaction
PEG Polyethylene glycol
PMSF Phenymethyl sulfonylfluoride
ppm Parts per million
psi Pound-force per square inch
PVDF Polyvinylidene difluoride
QDO Quadruple dropout
RBS Ribosome binding site
RISC RNA-induced silencing complex
RNase Ribonuclease
rpm Revolutions per minute
RNA Ribonucleic acid
ROS Reactive oxygen species
RT Reverse transcription
s Second
SC Synthetically complete
SD Synthetically defined
SDS sodium dodecyl sulfate
sgRNA Single guide RNA
siRNA Short interfering RNA
SLIC Sequence and ligation independent cloning
SOC Super optimal broth with catabolite repression
SOD Superoxide dismutase
ss Single-stranded
SSC Saline Sodium Citrate
T_a Annealing temperature
T_m Melting temperature
TAE Tris-acetate-EDTA
Taq *Thermus aquaticus* DNA
TBE Tris-borate-EDTA
TBS Tris-buffered saline
TCA Trichloroacetic acid
TE Tris-EDTA buffer
TEL Tris EDTA lithium acetate
TEMED Tetramethylethylenediamine
TMB 3,3′,5,5′-Tetramethylbenzidine
Tris tris(hydromethyl)aminomethane
Tris-Cl Tris-hydrocholroide
Trp Tryptophan
TTBS Tris-Tween 20-buffered saline
Ura Uracil
UV Ultraviolet
V Voltage
V_0 Initial velocity
V_{max} Maximum velocity
vol Volume
wt Weight
X-α-gal 5-Bromo-4-chloro-3-indolyl alpha-D-galactopyranoside
Y2H Yeast two hybrid
YNB Yeast nitrogen base
YPD Yeast extract peptone dextrose
YPDA Yeast extract peptone dextrose adenine hemisulfate

Index

Methods in Biotechnology, First Edition. Seung-Beom Hong, M. Bazlur Rashid and Lory Z Santiago-Vázquez.

Companion Website: www.wiley.com\go\hong\Methodsinbiotechnology

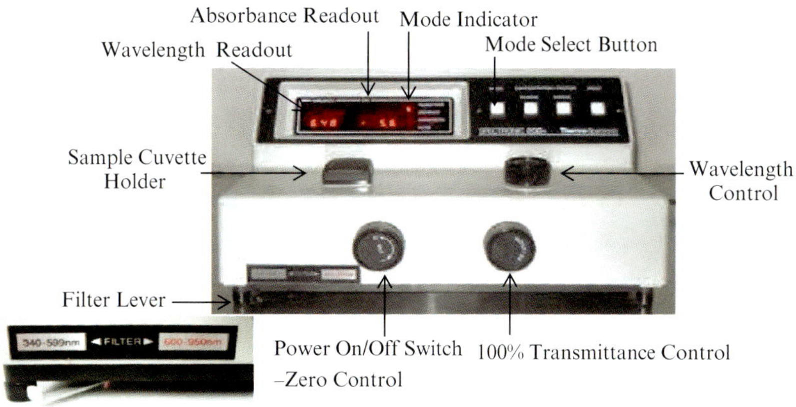

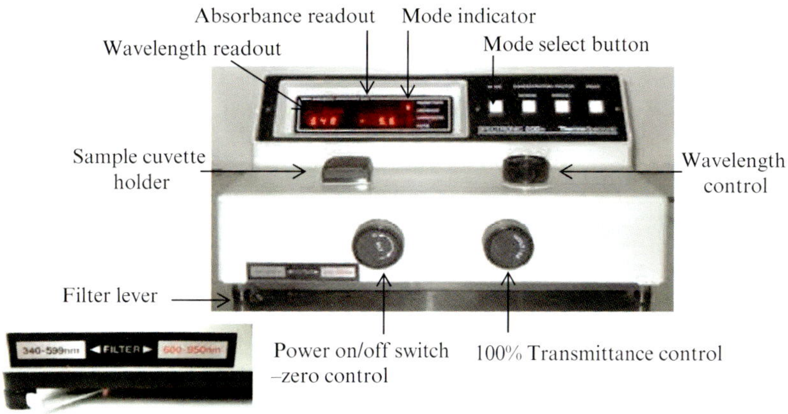

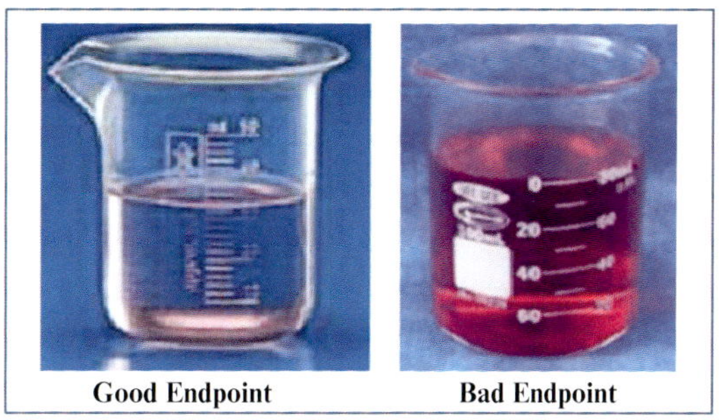

Methods in Biotechnology, First Edition. Seung-Beom Hong, M. Bazlur Rashid and Lory Z Santiago-Vázquez.

Companion Website: www.wiley.com\go\hong\Methodsinbiotechnology

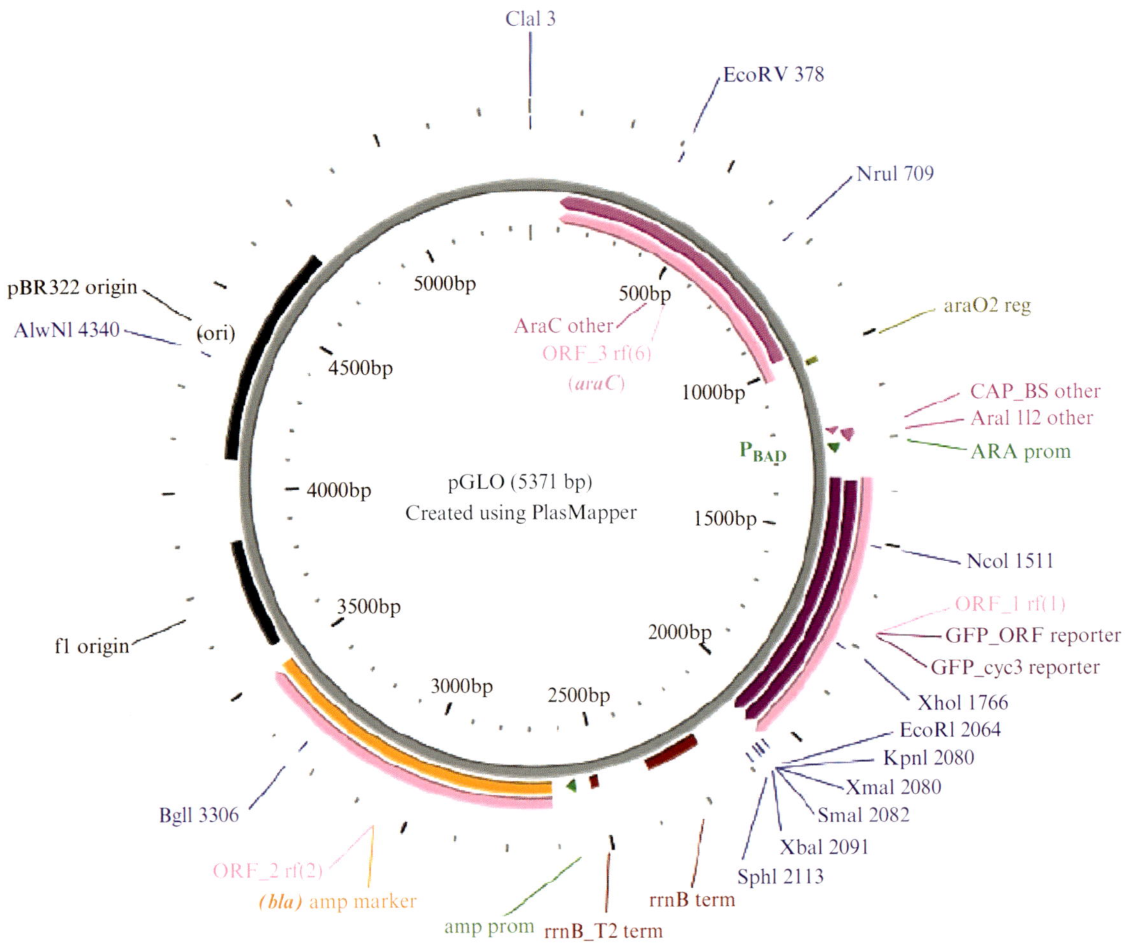

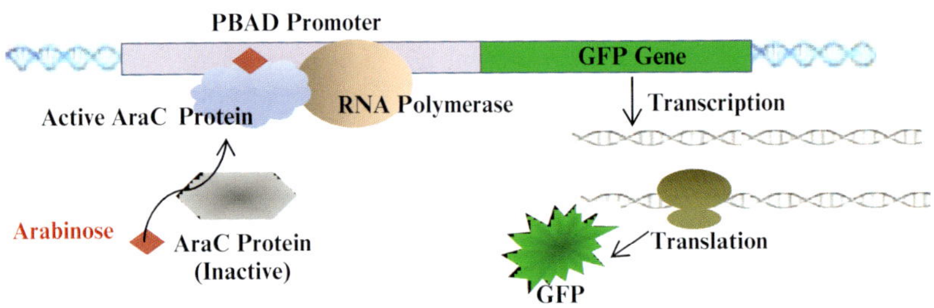

Lewis Structure of Reactive Oxygen Species (•unpaired electron that is unstable)

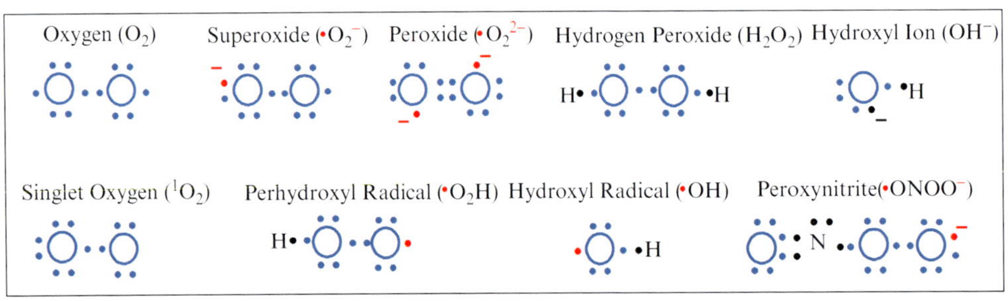

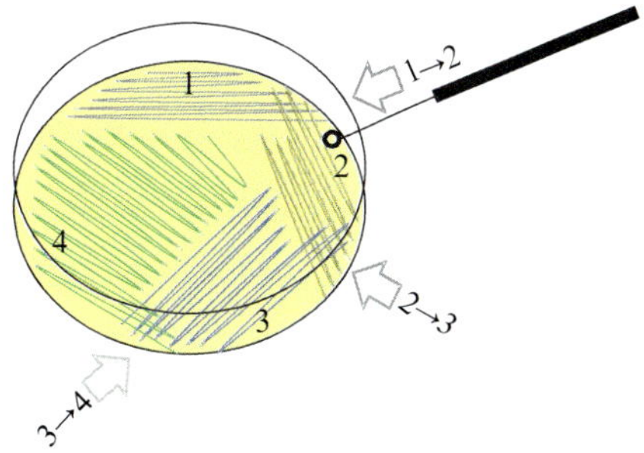

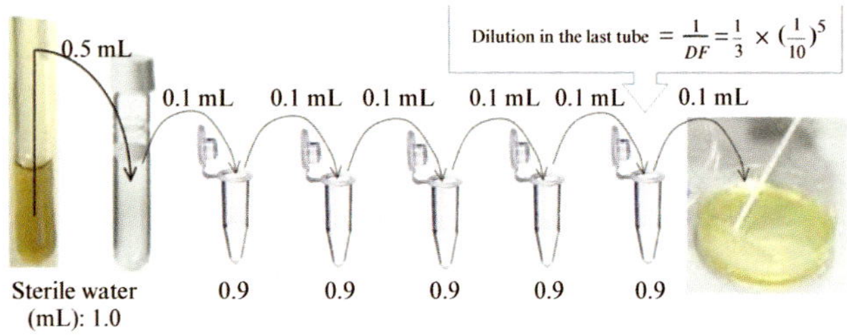

Primer #: NNNNNNNNNNNNNNNNNNNNNNNNNNNNNNNNNNNNNNN

Sbjct 1 (24…5) Minus   Sbjct 2 (670…690) Plus

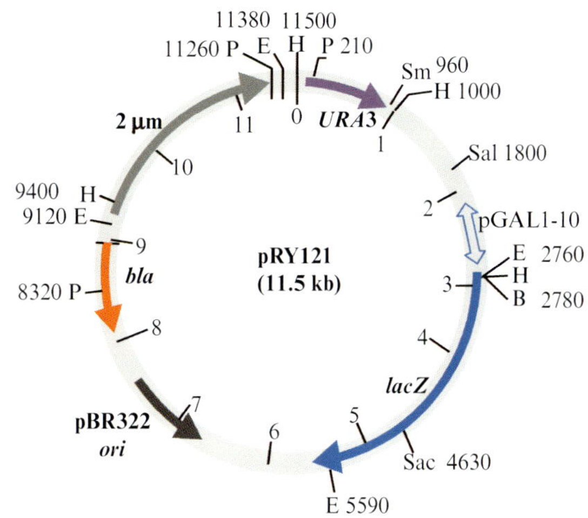

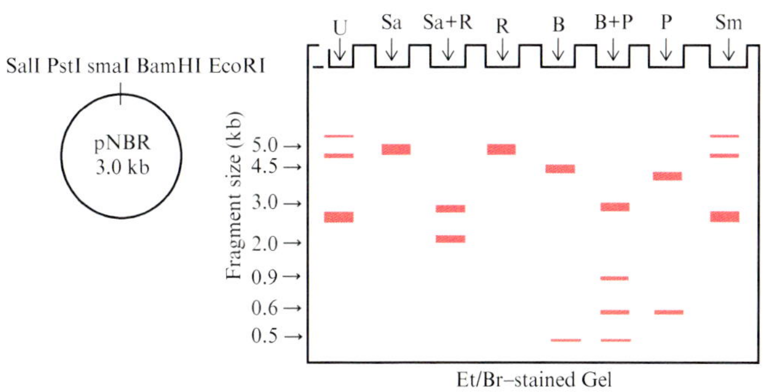

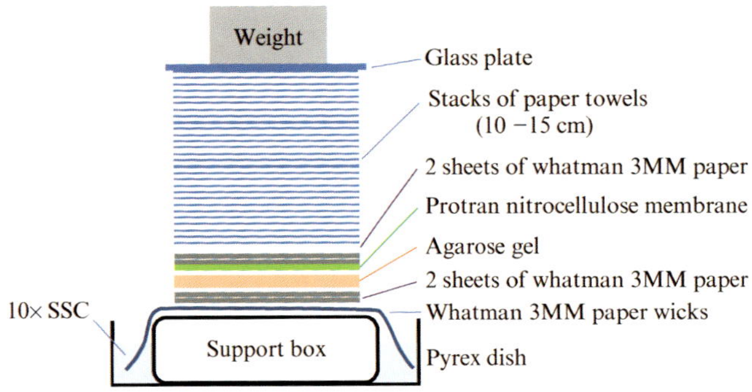
Weight
Glass plate
Stacks of paper towels
(10 −15 cm)
2 sheets of whatman 3MM paper
Protran nitrocellulose membrane
Agarose gel
2 sheets of whatman 3MM paper
Whatman 3MM paper wicks
10× SSC
Support box
Pyrex dish

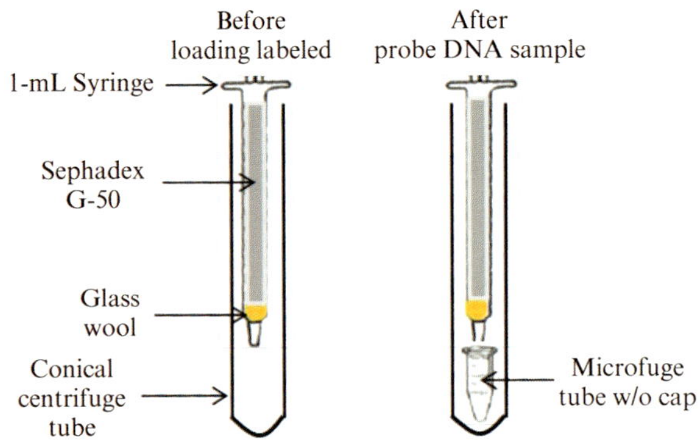
Before
loading labeled
After
probe DNA sample
1-mL Syringe
Sephadex
G-50
Glass
wool
Conical
centrifuge
tube
Microfuge
tube w/o cap

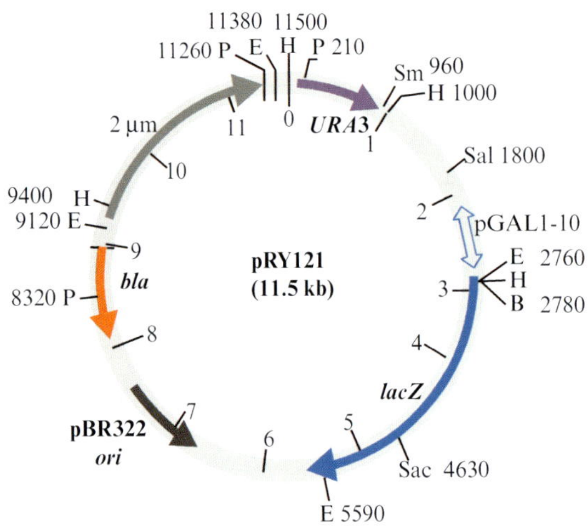
11380 11500
11260 P
E
H
P 210
Sm 960
H 1000
URA3
Sal 1800
pGAL1-10
E 2760
H
B 2780
lacZ
Sac 4630
E 5590
pBR322
ori
8320 P
bla
9120 E
9400 H
2 μm
pRY121
(11.5 kb)
0
1
2
3
4
5
6
7
8
9
10
11

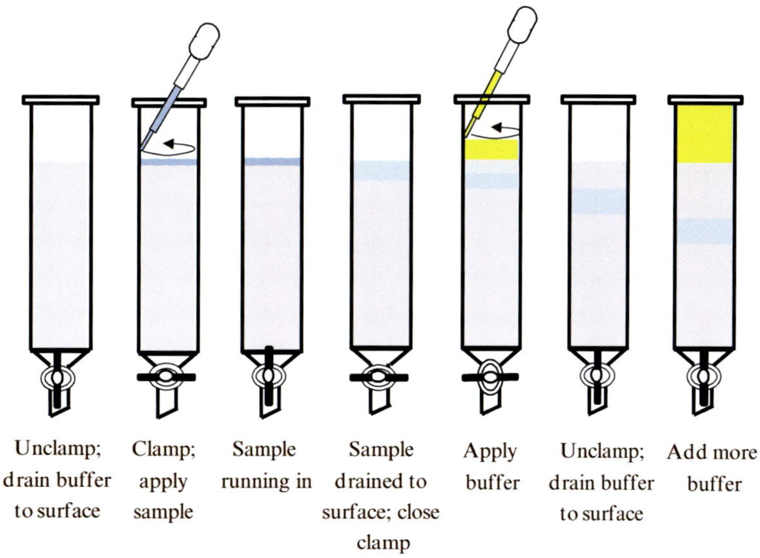

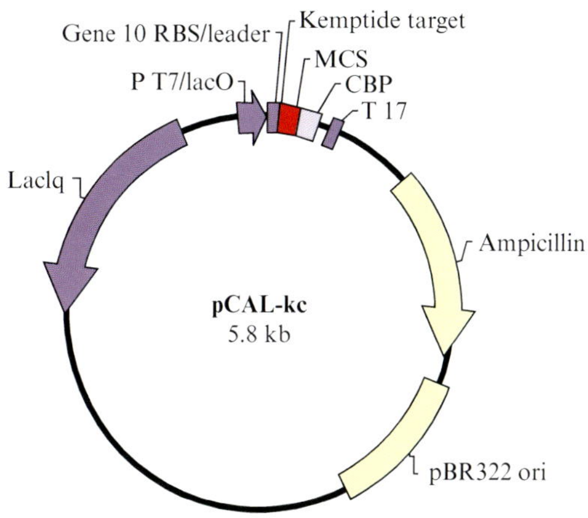

Reverse Complementary Sequence of MCS Region (322-510) in Expression Vector pCALkc (GenBank Acc. No. U36453)

NcoI, NheI, BamHI

TAAGAAGGAGATATACC ATG GCT AGC ATG ACT GGT GGA CAG CAA ATG GGT C GGA TCC
RBS — M A S M T G G Q Q M G * G S
start — T7 gene 10 leader peptide

KpnI

CTT AGA CGC GCA TCA CTT GGT AGA TCC ATG TAT CCA CGT↓GGG AAT GGT ACC AAG
L R R A S L G R S M Y P R G N G T K
Kemptide Target — Thrombin Target

CGA CGA TGG AAA AAG AAT TTC ATA GCC GTC TCA GCA GCC AAC CGC TTT AAG AAA
R R W K K N F I A V S A A N R F K K
Calmodulin Binding Peptide (CBP: 26 aa)

ATC TCA TCC TCC GGG GCA CTT TGA
I S S S G A L stop

* ATG is not in frame with the C-termninal fusion tags
↓ Thrombin Cleavage Site

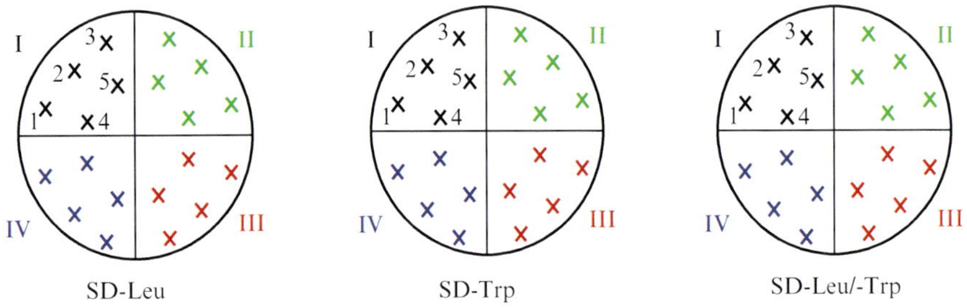

I: Y2HGold (pGBKT-7-53 + pGADT7-T): ✕
II: Y2HGold (pGBKT-7-53): ✕
III: Y187 (pGADT7-T): ✕
IV: Y2HGold (pCL1) or Y187 (pCL1): ✕

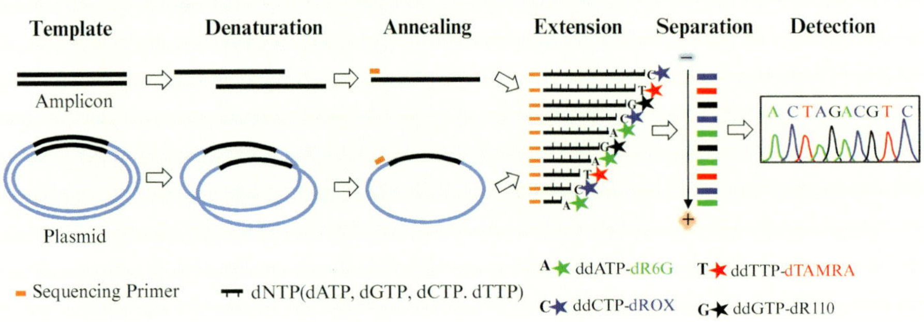

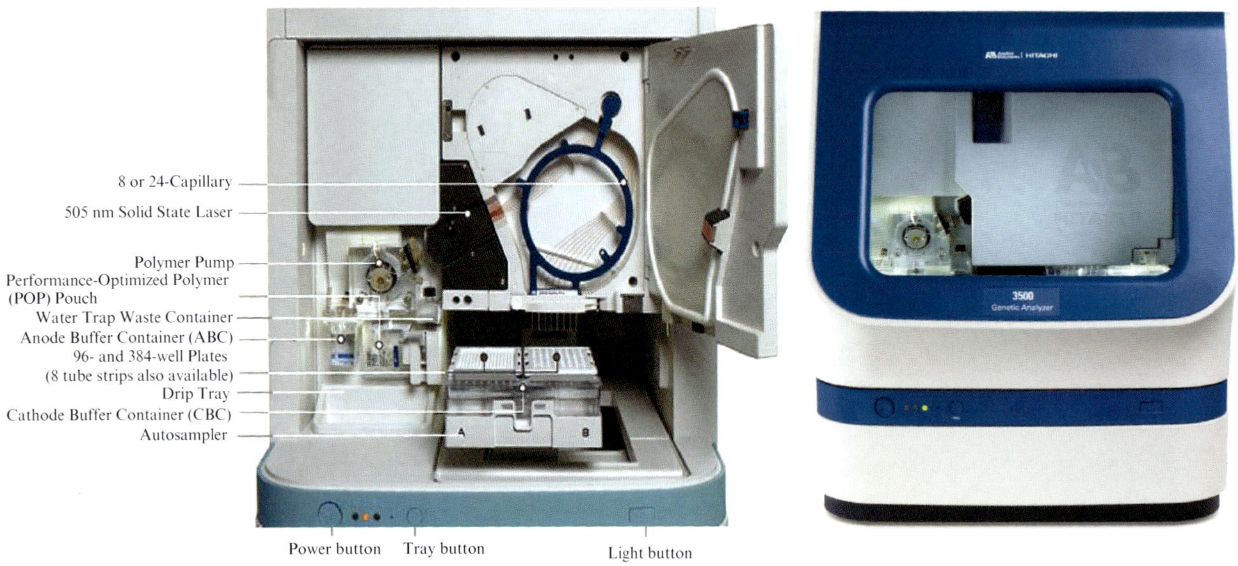

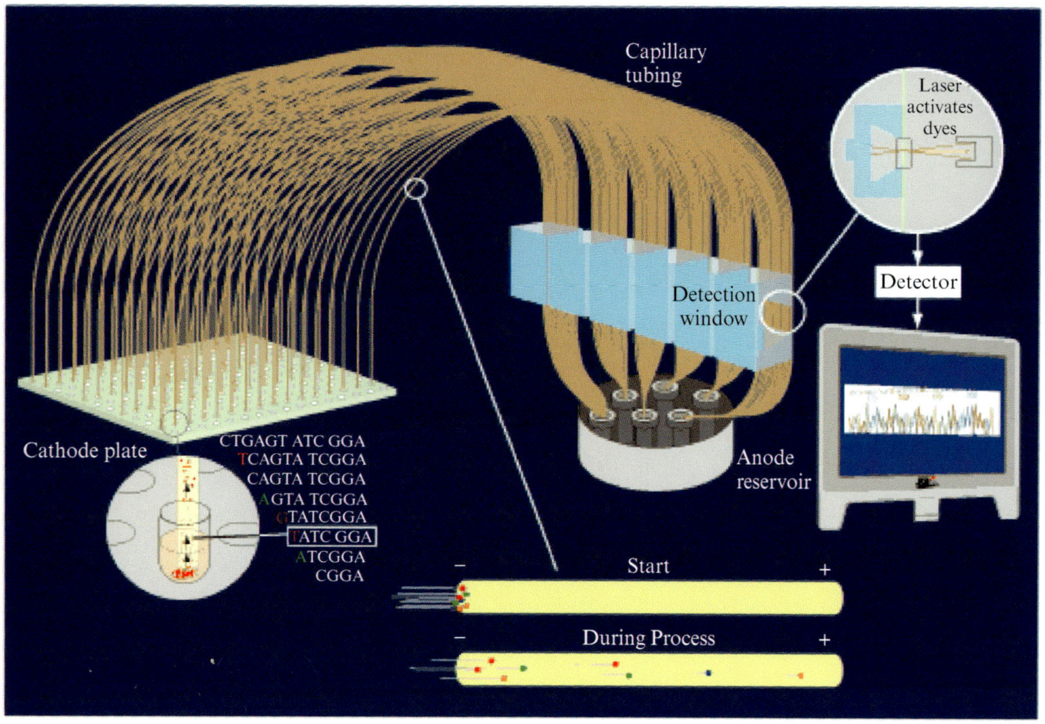
Capillary tubing
Laser activates dyes
Detector
Detection window
Cathode plate
CTGAGT ATC GGA
TCAGTA TCGGA
CAGTA TCGGA
AGTA TCGGA
GTATCGGA
TATC GGA
ATCGGA
CGGA
Anode reservoir
Start
During Process
–
+

Power button
Tray button
Light button

2:45 PM
2:45 PM
2:47 PM

applied biosystems
by life technologies
3500 Series
Data Collection Software 3
The programs included herein are subject to a restricted use license and can only be used in conjunction with this application.
Launching 3500 Series Data Collection Software 3...

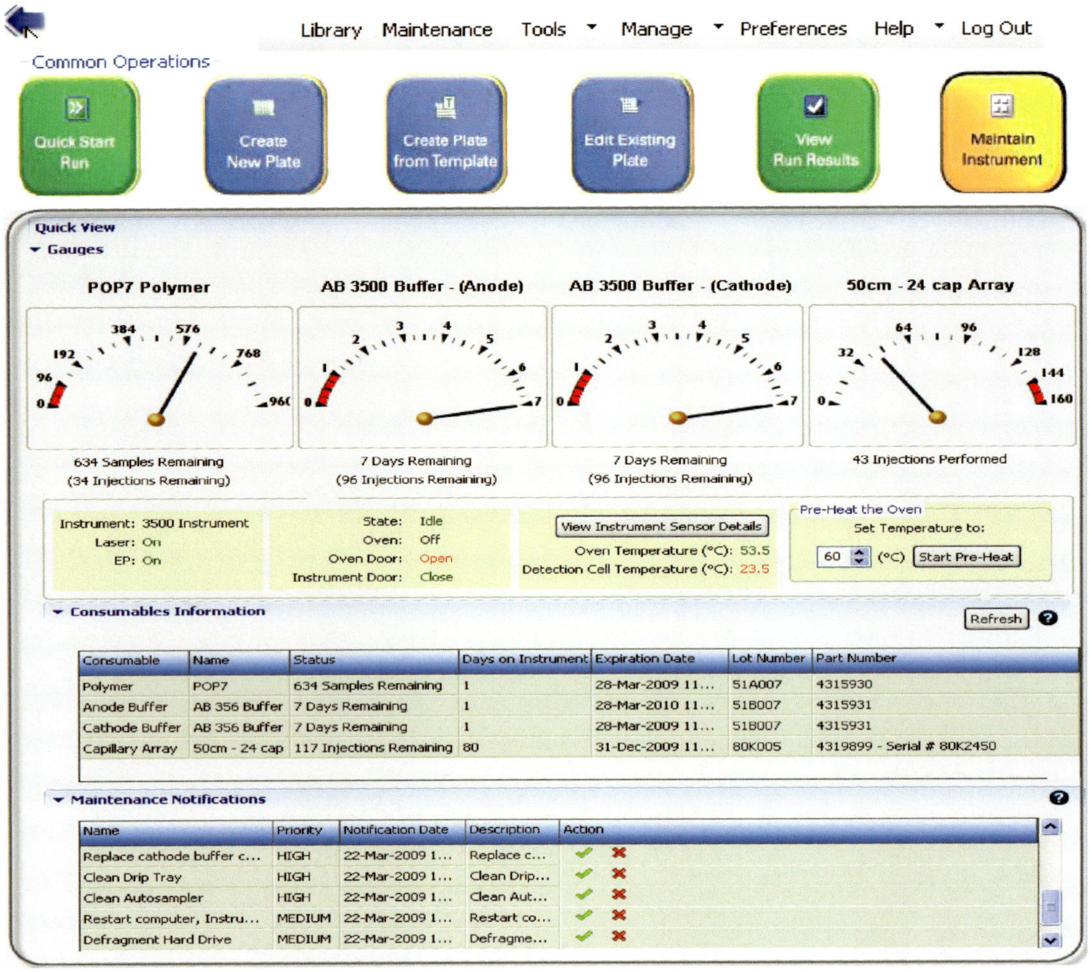
Library
Maintenance
Tools
Manage
Preferences
Help
Log Out
Common Operations
Quick Start Run
Create New Plate
Create Plate from Template
Edit Existing Plate
View Run Results
Maintain Instrument
Quick View
Gauges
POP7 Polymer
AB 3500 Buffer - (Anode)
AB 3500 Buffer - (Cathode)
50cm - 24 cap Array
634 Samples Remaining
(34 Injections Remaining)
7 Days Remaining
(96 Injections Remaining)
7 Days Remaining
(96 Injections Remaining)
43 Injections Performed
Instrument: 3500 Instrument
Laser: On
EP: On
State: Idle
Oven: Off
Oven Door: Open
Instrument Door: Close
View Instrument Sensor Details
Oven Temperature (°C): 53.5
Detection Cell Temperature (°C): 23.5
Pre-Heat the Oven
Set Temperature to:
60 (°C) Start Pre-Heat
Consumables Information
Refresh
Consumable | Name | Status | Days on Instrument | Expiration Date | Lot Number | Part Number
Polymer | POP7 | 634 Samples Remaining | 1 | 28-Mar-2009 11... | 51A007 | 4315930
Anode Buffer | AB 356 Buffer | 7 Days Remaining | 1 | 28-Mar-2010 11... | 51B007 | 4315931
Cathode Buffer | AB 356 Buffer | 7 Days Remaining | 1 | 28-Mar-2009 11... | 51B007 | 4315931
Capillary Array | 50cm - 24 cap | 117 Injections Remaining | 80 | 31-Dec-2009 11... | 80K005 | 4319899 - Serial # 80K2450
Maintenance Notifications
Name | Priority | Notification Date | Description | Action
Replace cathode buffer c... | HIGH | 22-Mar-2009 1... | Replace c...
Clean Drip Tray | HIGH | 22-Mar-2009 1... | Clean Drip...
Clean Autosampler | HIGH | 22-Mar-2009 1... | Clean Aut...
Restart computer, Instru... | MEDIUM | 22-Mar-2009 1... | Restart co...
Defragment Hard Drive | MEDIUM | 22-Mar-2009 1... | Defragme...

Remove Bubbles
REMOVE bubbles from the polymer pump

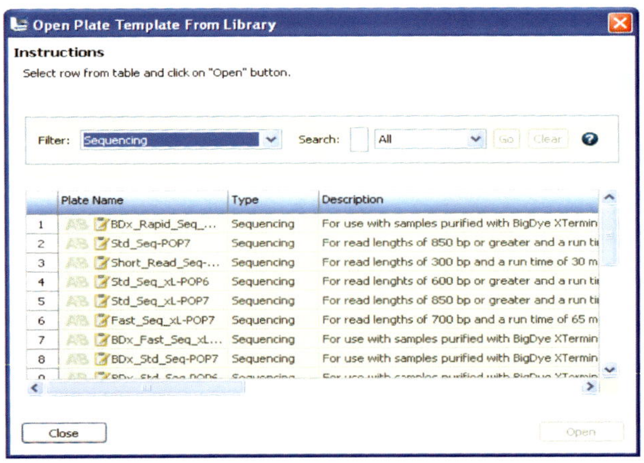
Open Plate Template From Library
Instructions
Select row from table and click on "Open" button.
Filter: Sequencing
Search: All
Go
Clear
Plate Name | Type | Description
1 | BDx_Rapid_Seq_... | Sequencing | For use with samples purified with BigDye XTermin
2 | Std_Seq-POP7 | Sequencing | For read lengths of 850 bp or greater and a run ti
3 | Short_Read_Seq-... | Sequencing | For read lengths of 300 bp and a run time of 30 m
4 | Std_Seq_xL-POP6 | Sequencing | For read lenghts of 600 bp or greater and a run ti
5 | Std_Seq_xL-POP7 | Sequencing | For read lenghts of 850 bp or greater and a run ti
6 | Fast_Seq_xL-POP7 | Sequencing | For read lengths of 700 bp and a run time of 65 m
7 | BDx_Fast_Seq_xL... | Sequencing | For use with samples purified with BigDye XTermin
8 | BDx_Std_Seq-POP7 | Sequencing | For use with samples purified with BigDye XTermin
Close
Open

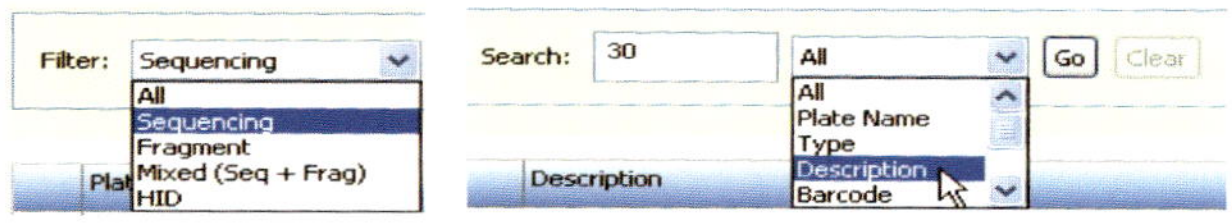
Filter:
Sequencing
All
Sequencing
Fragment
Mixed (Seq + Frag)
HID
Search:
30
All
All
Plate Name
Type
Description
Barcode
Go
Clear
Description

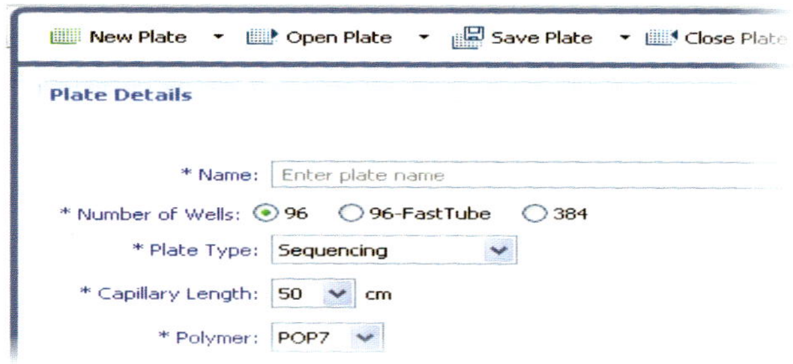
New Plate
Open Plate
Save Plate
Close Plate
Plate Details
* Name: Enter plate name
* Number of Wells: 96 96-FastTube 384
* Plate Type: Sequencing
* Capillary Length: 50 cm
* Polymer: POP7

Setup
Define Plate Properties
Assign Plate Contents

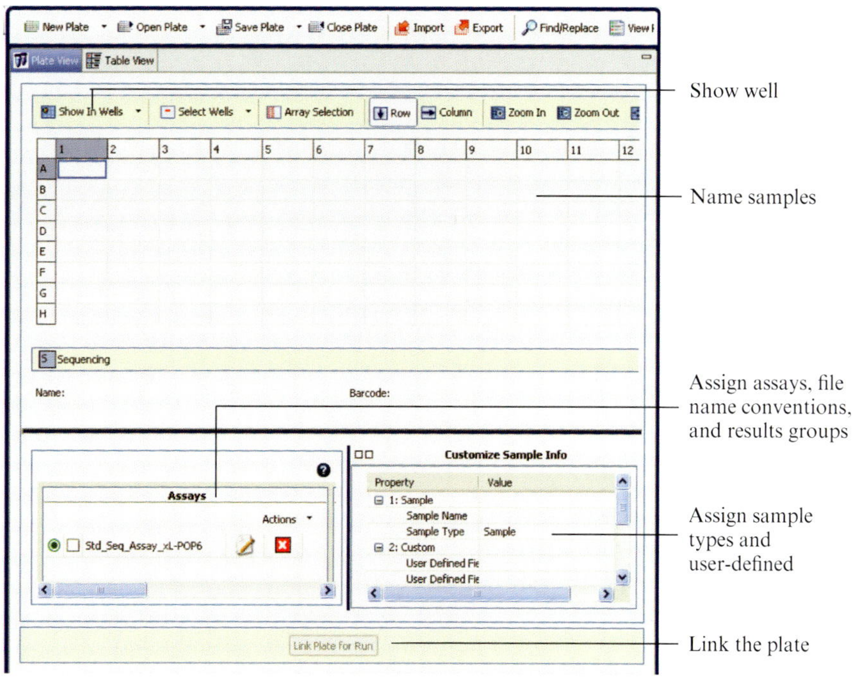
New Plate
Open Plate
Save Plate
Close Plate
Import
Export
Find/Replace
Plate View
Table View
Show In Wells
Select Wells
Array Selection
Row
Column
Zoom In
Zoom Out
Sequencing
Name:
Barcode:
Assays
Actions
Std_Seq_Assay_xL-POP6
Customize Sample Info
Property
Value
1: Sample
Sample Name
Sample Type
Sample
2: Custom
Link Plate for Run
Show well
Name samples
Assign assays, file name conventions, and results groups
Assign sample types and user-defined
Link the plate

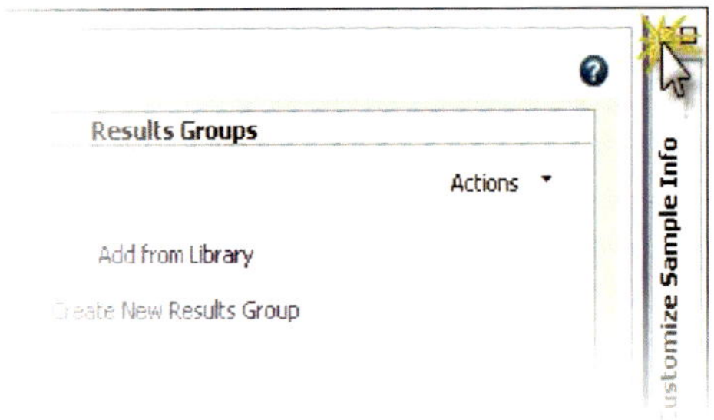
Results Groups
Actions
Add from Library
Create New Results Group
Customize Sample Info

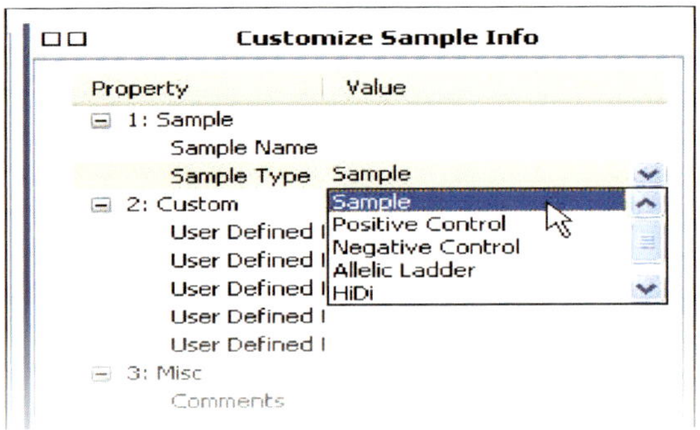
Customize Sample Info
Property
Value
1: Sample
Sample Name
Sample Type
Sample
Sample
Positive Control
Negative Control
Allelic Ladder
HiDi
2: Custom
User Defined
User Defined
User Defined
User Defined
User Defined
3: Misc
Comments

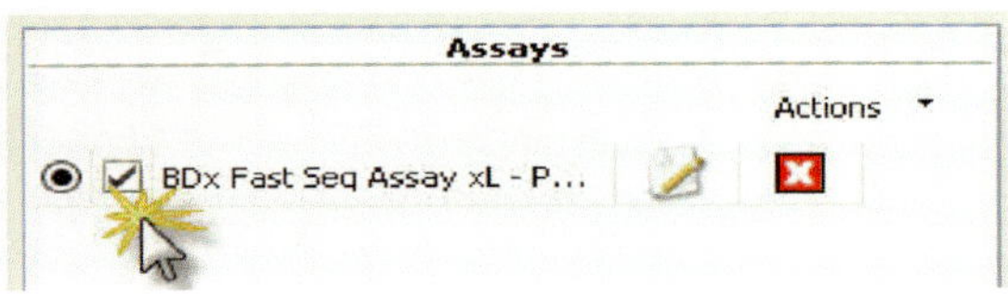
Assays
Actions
BDx Fast Seq Assay xL - P...

File Name Conventions
Actions
My FNC
Results Groups
Actions
My Sequencing Results Gro...

New Plate
Open Plate
Save Plate
Close Plate
Import
Export
Find/Replace
View Plate Grid Report

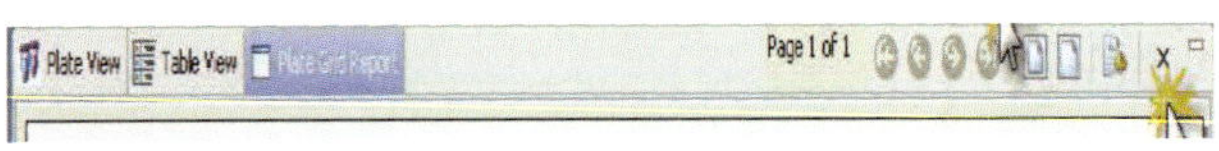
Plate View
Table View
Plate Grid Report
Page 1 of 1